CLIMATE CHANGE AND ITS CAUSES, EFFECTS AND PREDICTION SERIES

COASTAL SENSITIVITY TO SEA LEVEL RISE – FOCUSING ON THE MID-ATLANTIC REGION

Climate Change and its Causes, Effects and Prediction Series

Global Climate Change
Horace M. Karling (Editor)
2001. ISBN: 1-56072-999-6

Global Climate Change Revisited
Harace B. Karling (Editor)
2007. ISBN: 1-59454-039-X

Climate Change Research Progress
Lawrence N. Peretz (Editor)
2008. ISBN: 1-60021-998-5

Climate Change: Financial Risks
US Government Accountability Office
2008. ISBN: 978-1-60456-488-4

Post-Kyoto: Designing the Next International Climate Change Protocol
Matthew Clarke
2008. ISBN: 978-1-60456-840-0

Post-Kyoto: Designing the Next International Climate Change Protocol
Matthew Clarke
2008. ISBN: 978-1-61668-101-2
(Online Book)

The Effects of Climate Change on Agriculture, Land Resources, Biodiversity in the United States
Peter Backlund, Anthony Janeto and David Schimel
2009. ISBN: 978-1-60456-989-6

Economics of Policy Options to Address Climate Change
Gregory N. Bartos
2009. ISBN: 978-1-60692-116-6

Thresholds of Climate Change in Ecosystems
Vicente Orostegui (Editor)
2009. ISBN: 978-1-60741-487-2

Climate Variability, Modeling Tools and Agricultural Decision-Making
Angel Utset (Editor)
2009. ISBN: 978-1-60692-703-8

Climate Variability, Modeling Tools and Agricultural Decision-Making
Angel Utset (Editor)
2009. ISBN: 978-1-60876-791-5
(Online Book)

Emissions Trading: Lessons Learned from the European Union and Kyoto Protocol Climate Change Programs
Ervin Nagy and Gisella Varga (Editors)
2009. ISBN: 978-1-60741-194-9

Constructing Climate Change Legislation: Background and Issues
Gerald P. Overhauser (Editor)
2009. ISBN: 978-1-60692-986-5

Designing Greenhouse Gas Reduction and Regulatory Systems
Sonja Enden (Editor)
2009. ISBM: 978-1-60741-195-6

Global Climate Change: International Perspectives and Responses
Elias D'Angelo (Editor)
2009. ISBN: 978-1-60741-233-5

Disputing Global Warming
Anton Horvath and Boris Molnar (Editors)
2009. ISBN: 978-1-60741-235-9

Disputing Global Warming
Anton Horvath and Boris Molnar (Editors)
2009. ISBN: 978-1-60876-503-4
(Online Book)

Focus on Climate Change and Health
Viroj Wiwanitkit
2009. ISBN: 978-1-60741-247-2

Land Use and Climate Change
Suresh C. Rai
2009. ISBN: 978-1-60741-362-2

Impacts of Climate Change on Transportation and Infrastructure - A Gulf Coast Study
Iason Pavlopoulos (Editor)
2009. ISBN: 978-1-60741-424-7

Coastal Sensitivity to Sea Level Rise - Focusing on the Mid-Atlantic Region
Melvin C. Urajner (Editor)
2009. ISBN: 978-1-60741-440-7

Lightning in the Tropics: From a Source of Fire to a Monitoring System of Climatic Changes
Osmar Pinto, Jr.
2009. ISBN: 978-1-60741-764-4

Carbon Offsets: Examining their Role in Greenhouse Gas Reduction
Karen T. Morningstar (Editor)
2010. ISBN: 978-1-60741-444-5

Carbon Capture and Greenhouse Gases
Imrus Juhász and Gyorgy Halász (Editors)
2010. ISBN: 978-1-60692-089-3

The Science of Climate Change and Policy Implications
Kyle S. Hartzell (Editor)
2010. ISBN: 978-1-60741-448-3

Effects of Climate Change on Energy Production and Use in the U.S.
Alrik M. Solberg (Editor)
2010. ISBN: 978-1-60741-426-1

Policy Option Issues for CO2 Emissions
Nikolaus Vogler (Editor)
2010. ISBN: 978-1-60741-381-3

The Role of Agriculture in Carbon Capture and Climate Change
Arvid Bjurstrom (Editor)
2010. ISBN: 978-1-60741-445-2

Carbon Sequestration: Methods, Modeling and Impacts
Elke Hoch and Siegbert Grunwald (Editors)
2010. ISBN: 978-1-60741-498-8

Climate Change Litigation and Law
Jean-François Masson (Editor)
2010. ISBN: 978-1-60876-089-3

Carbon Tax and Cap-and-Trade Tools: Market-Based Approaches for Controlling Greenhouse Gases
Nelson E. Burney (Editor)
2010. ISBN: 978-1-60876-137-1

Global Change and Forestry: Economic and Policy Impacts and Responses
Jianbang Gan, Stephen Grado and Ian A. Munn (Editors)
2010. ISBN: 978-1-60876-262-0

Climate Change Measures and Trade Considerations
George R. Fried (Editor)
2010. ISBN: 978-1-60876-756-4

Assessing Climate Change Impacts on the United States
Alex B. McNeill (Editor)
2010. ISBN: 978-1-60876-160-9

Assessing Climate Change Impacts on the United States
Alex B. McNeill (Editor)
2010. ISBN: 978-1-60876-160-9
(Online Book)

CLIMATE CHANGE AND ITS CAUSES, EFFECTS AND PREDICTION SERIES

COASTAL SENSITIVITY TO SEA LEVEL RISE – FOCUSING ON THE MID-ATLANTIC REGION

MELVIN C. URAJNER
EDITOR

Nova Science Publishers, Inc.

New York

For permission to use material from this book please contact us:
Telephone 631-231-7269; Fax 631-231-8175
Web Site: http://www.novapublishers.com

Library of Congress Cataloging-in-Publication Data

Coastal sensitivity to sea level rise : focusing on the Mid-Atlantic region / Melvin C. Urajner.
p. cm.
ISBN 978-1-60741-440-7 (hardcover)
1. Sea level--Middle Atlantic States. 2. Coast changes--Middle Atlantic States. I. Urajner, Melvin C.
GC90.M64C63 2009
551.45'80975--dc22

2009034044

Published by Nova Science Publishers, Inc. ✝ New York

CONTENTS

ACKNOWLEDGMENTS

Throughout the process of preparing SAP 4.1, the authors were advised by a Federal Advisory Committee chosen for their diverse perspectives and technical expertise. The Coastal Elevations and Sea-Level Rise Advisory Committee (CESLAC) consisted of: Margaret Davidson (Chairperson), NOAA; Rebecca Beavers, National Park Service; Alan Belensz, New York State Office of the Attorney General; Mark Crowell, Federal Emergency Management Agency; Andrew Garcia, U.S. Army Corps of Engineers; Carl Hershner, Virginia Institute of Marine Science; Julie Hunkins, North Carolina Department of Transportation; Mark Mauriello, New Jersey Department of Environmental Protection; Mark Monmonier, Syracuse University; William S. Nechamen, Association of State Floodplain Managers; Sam Pearsall, Environmental Defense Fund; Anthony Pratt, Coastal States Organization; Greg Rudolph, American Shore and Beach Preservation Association; Harvey Ryland, Institute for Business and Home Safety; Gwynne Schultz, Maryland Department of Natural Resources. Jack Fitzgerald of EPA was the Designated Federal Official for the CESLAC, with support provided by Stratus Consulting.

Technical expert review was provided by: Fred Anders, New York Department of State; Mark Davis, Tulane University; Lesley Ewing, California Coastal Commission; Janet Freedman, Rhode Island Coastal Resources Council; Vivien Gornitz, NASA; Ellen Hartig, New York City Department of Parks & Recreation; Maria Honeycutt, AGI Congressional Fellow; Kurt Kalb, New Jersey Department of Environmental Protection; Stephen Leatherman, Florida International University; Ken Miller, Maryland Department of Natural Resources; Jim O'Connell, University of Hawaii, Sea Grant; Richard Osman, Smithsonian Institution; Marc Perry, U.S. Census Bureau; Chris Spaur, U.S. Army Corps of Engineers; John Teal, Teal Partners; John Thayer, North Carolina Department of Environment and Natural Resources; Dan Trescott, Southwest Florida Regional Planning Council; John Whitehead, Appalachian State University; Rob Young, Western Carolina University. An expert review of an early draft of Chapter 2 was also provided by an interagency geospatial team consisting of Eric Constance, USGS; Todd Davison, NOAA; Dean Gesch, USGS; and Jerry Johnston, EPA.

This report relied heavily on stakeholder involvement that was implemented through a series of three meetings held in the Mid-Atlantic Region (Easton, Maryland; Red Bank, New Jersey; and Plymouth, North Carolina). Many of the comments received and discussion initiated at these meetings helped to define some of the issues addressed in this report. Linda Hamalak of NOAA organized these public meetings and the subsequent author meetings. The author meetings were hosted by the Blackwater National Wildlife Refuge in Maryland; the NOAA National Marine Fisheries Service in Sandy Hook, New Jersey; and the Partnership for the Sounds in Columbia, North Carolina

The authors were also assisted by several of their colleagues at EPA, NOAA, and USGS. The interagency management team of Rona Birnbaum, EPA; Patricia Jellison, USGS; and Michael Szabados, NOAA, were instrumental in advising the authors during the final stages of the report. Rebecca Feldman of NOAA provided key logistical support in addition to her contributions as an author. Karen Scott of EPA managed the expert review process, supported by Perrin Quarles Associates.

The USGS authors thank several colleagues for their reviews, discussions, and contributions to Chapters 1, 3, 4, 13, and 14, as well as other portions of the report. These include: Mark Brinson, East Carolina University; Tom Cronin, USGS; Duncan FitzGerald, Boston University; Virginia Burkett, USGS; Curt Larsen, USGS, (retired); Laura Moore, University of Virginia; Elizabeth Pendleton, USGS; Shea Penland (deceased), University of New Orleans; and Asbury Sallenger, USGS.

Russ Jones of Stratus Consulting coordinated technical and GIS support for several chapters in this report, with support from Jue Wang, Pyramid Systems Inc.; Richard Streeter and Tom Hodgson, Stratus Consulting; and John Herter and Gaurav Sinha, Industrial Economics. Christina Thomas (contractor to Stratus Consulting) edited the expert review draft.

Chapter 3 includes results of a panel assessment. The authors of Chapter 3 thank the panelists for their contributions: Fred Anders, New York State, Dept. of State; K. Eric Anderson, USGS; Mark Byrnes, Applied Coastal Research and Engineering; Donald R. Cahoon, USGS; Stewart Farrell, Richard Stockton College; Duncan FitzGerald, Boston University, Massachusetts; Paul Gayes, Coastal Carolina University; Carl Hobbs, Virginia Institute of Marine Science; Randy McBride, George Mason University; Jesse McNinch, Virginia Institute of Marine Science; Stan Riggs, East Carolina University; Antonio Rodriguez, University North Carolina; Jay Tanski, New York Sea Grant; Art Trembanis, University of Delaware.

Chapter 4 includes results based on a panel assessment. The panel consisted of: Denise Reed, University of New Orleans; Dana Bishara, USGS; Jeffrey Donnelly, Woods Hole Oceanographic Institution; Michael Kearney, University of Maryland; Alexander Kolker, Louisiana Universities Marine Consortium; Lynn Leonard, University of North Carolina-Wilmington; Richard Orson, Orson Environmental Consulting, J. Court Stevenson, University of Maryland. The panel was conducted under contract to EPA, with James G. Titus as the project officer. Jeff DeBlieux of The Nature Conservancy also contributed to portions of Chapter 4.

The review process for SAP 4.1 included a public review of the Second draft. We thank the individuals who commented on this draft. The author team carefully considered all comments submitted, and many resulted in improvements to this Product.

We also thank the team of editors that worked closely with authors to produce this product. This includes Anne Waple, CCSP/STG Inc., and the graphics team at the National Climatic Data Center.

Finally, we are especially grateful to Alan Cohn of EPA for his management of the day-to-day process of developing and producing this report, and providing overall coordination for this effort.

Preface

The U.S. Climate Change Science Program (CCSP) was launched in February 2002 as a collaborative federal interagency program, under a new cabinet-level organization designed to improve the government-wide management and dissemination of climate change science and related technology development. The mission of the CCSP is to "facilitate the creation and application of knowledge of the Earth's global environment through research, observations, decision support, and communication". This Product is one of 21 synthesis and assessment products (SAPs) identified in the 2003 *Strategic Plan for the U.S. Climate Change Science Program*, written to help achieve this mission. The SAPs are intended to support informed discussion and decisions by policymakers, resource managers, stakeholders, the media, and the general public. The products help meet the requirements of the Global Change Research Act of 1990, which directs agencies to "produce information readily usable by policymakers attempting to formulate effective strategies for preventing, mitigating, and adapting to the effects of global change" and to undertake periodic scientific assessments.

One of the major goals within the mission is to understand the sensitivity and adaptability of different natural and managed ecosystems and human systems to climate and related global changes. This SAP (4.1), *Coastal Sensitivity to Sea-Level Rise: A Focus on the Mid-Atlantic Region*, addresses this goal by providing a detailed assessment of the effects of sea-level rise on coastal environments and presenting some of the challenges that need to be addressed in order to adapt to sea-level rise while protecting environmental resources and sustaining economic growth. It is intended to provide the most current knowledge regarding the implications of rising sea level and possible adaptive responses, particularly in the mid-Atlantic region of the United States.

Scope and Approach of This Product

The focus of this Product is to identify and review the potential impacts of future sea-level rise based on present scientific understanding. To do so, this Product evaluates several aspects of sea-level rise impacts to the natural environment and examines the impact to human land development along the coast. In addition, the Product addresses the connection between sea-level rise impacts and current adaptation strategies, and assesses the role of the existing coastal management policies in identifying and responding to potential challenges.

As with other SAPs, the first step in the process of preparing this Product was to publish a draft prospectus listing the questions that the product would seek to answer at the local and mid-Atlantic scale. After public comment, the final prospectus listed ten questions. This Product addresses those ten questions, and answers most of them with specificity. Nevertheless, development of this Product has also highlighted current data and analytical capacity limitations. The analytical presentation in this Product focuses on what characterizations can be provided with sufficient accuracy to be meaningful. For a few questions, the published literature was insufficient to answer the question with great specificity. Nevertheless, the effort to answer the question has identified what information is needed or desirable, and current limitations with regard to available data and tools.

This Product focuses on the U.S. mid-Atlantic coast, which includes the eight states from New York to North Carolina. The Mid-Atlantic is a region where high population density and extensive coastal development is likely to be at increased risk due to sea-level rise. Other coastal regions in the United States, such as the Gulf of Mexico and the Florida coast, are potentially more vulnerable to sea-level rise and have been the focus of other research and assessments, but are outside the scope of this Product.

During the preparation of this Product, three regional meetings were held between the author team and representatives from relevant local, county, state, and federal agencies, as well non-governmental organizations. Many of the questions posed in the prospectus for SAP 4.1 were discussed in detail and the feedback has been incorporated into the Product. However, the available data are insufficient to answer all of the questions at both the local and regional scale. Therefore, the results of this Product are best used as a "starting point" for audiences seeking information about sensitivity to and implications of sea-level rise.

Many of the findings included in this Product are expressed using common terms of likelihood (*e.g.*, very likely, unlikely), similar to those used in the 2007 Intergovernmental Panel on Climate Change (IPCC) Fourth Assessment Report, *Climate Change 2007: The Physical Science Basis*. The likelihood determinations used in this Product were established by the authors and modeled after other CCSP SAPs such as CCSP SAP 1.1, *Temperature Trends in the Lower Atmosphere: Steps for Understanding and Reconciling Differences.* However, characterizations of likelihood in this Product are largely based on the judgment of the authors and uncertainties from published peer-reviewed literature (Figure P.1). Data on how coastal ecosystems and specific species may respond to climate change is limited to a small number of site-specific studies, often carried out for purposes unrelated to efforts to evaluate the potential impact of sea-level rise. Nevertheless, being able to characterize current understanding—and the uncertainty associated with that information—is important. In the main body of this Product, any use of the terms in Figure P.1 reflect qualitative assessment of potential changes based on the authors' review and understanding of available published coastal science literature and of governmental policies (the appendices do not contain findings). Statements that do not use these likelihood terms either have an insufficient basis for assessing likelihood or present information provided in the referenced literature which was not accompanied by assessments of likelihood.

The International System of Units (SI) have been used in this Product; with English units often provided in parentheses. Where conversions are not provided, some readers may wish to convert from SI to English units using the following table:

Table P.1. Conversion from the International System of Units (SI) to English units

Multiply	By	To obtain
Length		
centimeter (cm)	0.3937	inch (in)
millimeter (mm)	0.0394	inch (in)
meter (m)	3.2808	foot (ft)
kilometer (km)	0.6214	mile (mi)
meter (m)	1.0936	yard (yd)
Area		
square meter (sq m)	0.000247	acres
hectare (ha)	2.47	acres
square kilometer (sq km)	247	acres
square meter (sq m)	10.7639	square foot (sq ft)
hectare (ha)	0.00386	square mile (sq mi)
square kilometer (sq km)	0.3861	square mile (sq mi)
Rate of Change		
meters per year (m per year)	3.28084	foot per year (ft per year)
millimeters per year (mm per year)	0.03937	inch per year (in per year)
meters per second (m per sec)	1.943	knots

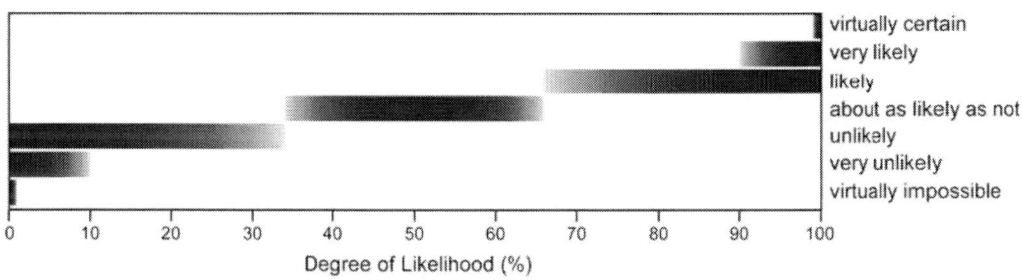

Figure P.1. Likelihood terms and related probabilities used for this Product (with the exception of Appendix 1).

FUTURE SEA-LEVEL SCENARIOS ADDRESSED IN THIS PRODUCT

In this Product, the term "sea level" refers to mean sea level or the average level of tidal waters, generally measured over a 20-year period. These measurements generally indicate the water level relative to the land, and thus incorporate changes in the elevation of the land (*i.e.*, subsidence or uplift) as well as absolute changes in sea level (*i.e.*, rise in sea level caused by increasing its volume or adding water). For clarity, scientists often use two different terms:

- "Global sea-level rise" is the average increase in the level of the world's oceans that occurs due to a variety of factors, the most significant being thermal expansion of the oceans and the addition of water by melting of land-based ice sheets, ice caps, and glaciers.

- "Relative sea-level rise" refers to the change in sea level relative to the elevation of the adjacent land, which can also subside or rise due to natural and human-induced factors. Relative sea-level changes include both global sea-level rise and changes in the vertical elevation of the land surface.

In this Product, both terms are used. Global sea-level rise is used when referring to the worldwide average increase in sea level. Relative sea-level rise, or simply sea-level rise, is used when referring to the scenarios used in this Product and effects on the coast.

This Product does not provide a forecast of future rates of sea-level rise. Rather, it evaluates the implications of three relative sea-level rise scenarios over the next century developed from a combination of the twentieth century relative sea-level rise rate and either a 2 or 7 millimeter per year increase in global sea level:

- Scenario 1: the twentieth century rate, which is generally 3 to 4 millimeters per year in the mid-Atlantic region (30 to 40 centimeters total by the year 2100);
- Scenario 2: the twentieth century rate plus 2 millimeters per year acceleration (up to 50 centimeters total by 2100);
- Scenario 3: the twentieth century rate plus 7 millimeters per year acceleration (up to 100 centimeters total by 2100).

The twentieth century rate of sea-level rise refers to the local long-term rate of relative sea-level rise that has been observed at NOAA National Ocean Service (NOS) tide gauges in the mid-Atlantic study region. Scenario 1 assesses the impacts if future sea-level rise occurs at the same rate as was observed over the twentieth century at a particular location. Scenarios 1 and 2 are within the range of those reported in the recent IPCC Report *Climate Change 2007: The Physical Science Basis,* specifically in the chapter *Observations: Oceanic Climate Change and Sea Level*, while Scenario 3 exceeds the IPCC scenario range by up to 40 centimeters by 2100. Higher estimates, as suggested by some recent publications, are the basis for Scenario 3. In addition to these three scenarios, some chapters refer to even higher sea-level rise scenarios, such as a 200 centimeter rise over the next few hundred years (a high but plausible estimate if ice sheet melting on Greenland and West Antarctica exceeds IPCC model estimates).

Product Organization

This Product is divided into four parts:

Part I first provides context and addresses the effects of sea-level rise on the physical environment. Chapter 1 provides the context for sea-level rise and its effects. Chapter 2 discusses the current knowledge and limitations in coastal elevation mapping. Chapter 3 describes the physical changes at the coast that will result in changes to coastal landforms (*e.g.*, barrier islands) and shoreline position in response to sea-level rise. Chapter 4 considers the ability of wetlands to accumulate sediments and survive in response to rising sea level. Chapter 5 examines the habitats and species that will be vulnerable to sea-level rise related impacts.

Part II describes the societal impacts and implications of sea-level rise. Chapter 6 provides a framework for assessing shoreline protection options in response to sea-level rise. Chapter 7 discusses the extent of vulnerable population and infrastructure, and Chapter 8 addresses the implications for public access to the shore. Chapter 9 reviews the impact of sea-level rise to flood hazards.

Part III examines strategies for coping with sea-level rise. Chapter 10 outlines key considerations when making decisions to reduce vulnerability. Chapter 11 discusses what organizations are currently doing to adapt to sea-level rise, and Chapter 12 examines possible institutional barriers to adaptation.

Part IV examines national implications and a science strategy for moving forward. Chapter 13 discusses sea-level rise impacts and implications at a national scale and highlights how coasts in other parts of the United States are vulnerable to sea-level rise. Chapter 14 presents opportunities for future efforts to reduce uncertainty and close gaps in scientific knowledge and understanding.

Finally, this Product also includes two appendices: Appendix 1 discusses many of the species that depend on potentially vulnerable habitat in specific estuaries, providing local elaboration of the general issues examined in Chapter 5. The Appendix also describe key statutes, regulations, and other policies that currently define how state and local governments are responding to sea-level rise, providing support for some of the observations made in Part III. This Appendix is provided as background information and does not include findings or an independent assessment of likelihood.

Appendix 2 reviews some of the basic approaches that have been used to conduct shoreline change or land loss assessments in the context of sea-level rise and some of the difficulties that arise in using these methods.

Technical and scientific terms are used throughout this Product. To aid readers with these terms, a Glossary and a list of Acronyms and Abbreviations are included at the end of the Product.

EXECUTIVE SUMMARY

Authors: K. Eric Anderson, USGS; Donald R. Cahoon, USGS; Stephen K. Gill, NOAA; Benjamin T. Gutierrez, USGS; E. Robert Thieler, USGS; James G. Titus, U.S. EPA; S. Jeffress Williams, USGS (lead authors arranged in alphabetical order).

Global sea level is rising, and there is evidence that the rate is accelerating. Increasing atmospheric concentrations of greenhouse gases, primarily from human contributions, are very likely warming the atmosphere and oceans. The warmer temperatures raise sea level by expanding ocean water, melting glaciers, and possibly increasing the rate at which ice sheets discharge ice and water into the oceans. Rising sea level and the potential for stronger storms pose an increasing threat to coastal cities, residential communities, infrastructure, beaches, wetlands, and ecosystems. The potential impacts to the United States extend across the entire country: ports provide gateways for transport of goods domestically and abroad; coastal resorts and beaches are central to the U.S. economy; wetlands provide valuable ecosystem services such as water filtering and spawning grounds for commercially important fisheries. How people respond to sea-level rise in the coastal zone will have potentially large economic and environmental costs.

This Synthesis and Assessment Product examines the implications of rising sea level, with a focus on the mid-Atlantic region of the United States, where rates of sea-level rise are moderately high, storm impacts occur, and there is a large extent of critical habitat (marshes), high population densities, and infrastructure in low-lying areas. Although these issues apply to coastal regions across the country, the mid-Atlantic region was selected as a focus area to explore how addressing both sensitive ecosystems and impacts to humans will be a challenge. Using current scientific literature and expert panel assessments, this Product examines potential risks, possible responses, and decisions that may be sensitive to sea-level rise.

The information, data, and tools needed to inform decision-making with regard to sea-level rise are evolving, but insufficient to assess the implications at scales of interest to all stakeholders. Accordingly, this Product can only provide a starting point to discuss impacts and examine possible responses at the regional scale. The Product briefly summarizes national scale implications and outlines the steps involved in providing information at multiple scales (*e.g.*, local).

ES.1. Why Is Sea Level Rising? How Much will it Rise?

During periods of climate warming, two major processes cause global mean sea-level rise: (1) as the ocean warms, the water expands and increases its volume and (2) land reservoirs of ice and water, including glaciers and ice sheets, contribute water to the oceans. In addition, the land in many coastal regions is subsiding, adding to the vulnerability to the effects of sea-level rise.

Recent U.S. and international assessments of climate change show that global average sea level rose approximately 1.7 millimeters per year through the twentieth century, after a period of little change during the previous two thousand years. Observations suggest that the rate of global sea-level rise may be accelerating. In 2007, the Intergovernmental Panel on Climate Change (IPCC) projected that global sea level will likely rise between 19 and 59 centimeters (7 and 23 inches) by the end of the century (2090 to 2099), relative to the base period (1980 to 1999), excluding any rapid changes in ice flow from Greenland and Antarctica. According to the IPCC, the average rate of global sea-level rise during the twenty-first century is *very likely* to exceed the average rate over the last four decades. Recently observed accelerated ice flow and melting in some Greenland outlet glaciers and West Antarctic ice streams could substantially increase the contribution from the ice sheets to rates of global sea-level rise. Understanding of the magnitude and timing of these processes is limited and, thus, there is currently no consensus on the upper bound of global sea-level rise. Recent studies suggest the potential for a meter or more of global sea-level rise by the year 2100, and possibly several meters within the next several centuries.

In the mid-Atlantic region from New York to North Carolina, tide-gauge observations indicate that relative sea-level rise (the combination of global sea-level rise and land subsidence) rates were higher than the global mean and generally ranged between 2.4 and 4.4 millimeters per year, or about 0.3 meters (1 foot) over the twentieth century.

ES.2. What Are the Effects of Sea-Level Rise?

Coastal environments such as beaches, barrier islands, wetlands, and estuarine systems are closely linked to sea level. Many of these environments adjust to increasing water level by growing vertically, migrating inland, or expanding laterally. If the rate of sea-level rise accelerates significantly, coastal environments and human populations will be affected. In some cases, the effects will be limited in scope and similar to those observed during the last century. In other cases, thresholds may be crossed, beyond which the impacts would be much greater. If the sea rises more rapidly than the rate with which a particular coastal system can keep pace, it could fundamentally change the state of the coast. For example, rapid sea-level rise can cause rapid landward migration or segmentation of some barrier islands, or disintegration of wetlands.

Today, rising sea levels are submerging low-lying lands, eroding beaches, converting wetlands to open water, exacerbating coastal flooding, and increasing the salinity of estuaries and freshwater aquifers. Other impacts of climate change, coastal development, and natural coastal processes also contribute to these impacts. In undeveloped or less-developed coastal areas where human influence is minimal, ecosystems and geological systems can sometimes shift upward and landward with the rising water levels. Coastal development, including buildings, roads, and other infrastructure, are less mobile and more vulnerable. Vulnerability to an accelerating rate of sea-level rise is compounded by the high population density along the coast, the possibility of other effects of climate change, and the susceptibility of coastal regions to storms and environmental stressors, such as drought or invasive species.

ES.2.1. Sea-Level Rise and the Physical Environment

The coastal zone is dynamic and the response of coastal areas to sea-level rise is more complex than simple inundation. Erosion is a natural process from waves and currents and can cause land to be lost even with a stable sea level. Sea-level rise can exacerbate coastal change due to erosion and accretion. While some wetlands can keep pace with sea-level rise due to sediment inputs, those that cannot keep pace will gradually degrade and become submerged. Shore protection and engineering efforts also affect how coasts are able to respond to sea-level rise.

For coastal areas that are vulnerable to inundation by sea-level rise, elevation is generally the most critical factor in assessing potential impacts. The extent of inundation is controlled largely by the slope of the land, with a greater area of inundation occurring in locations with more gentle gradients. Most of the currently available elevation data do not provide the degree of confidence that is needed for making quantitative assessments of the effects of sea-level rise for local planning and decision making. However, systematic collection of high-quality elevation data (*i.e.*, lidar) will improve the ability to conduct detailed assessments (Chapter 2).

Nationally, coastal erosion will probably increase as sea-level rises at rates higher than those that have been observed over the past century. The exact manner and rates at which these changes are likely to occur will depend on the character of coastal landforms (*e.g.*, barrier islands, cliffs) and physical processes (Part I). Particularly in sandy shore environments which comprise the entire mid-Atlantic ocean coast (Figure ES.1), it is *virtually certain* that coastal headlands, spits, and barrier islands will erode at a faster pace in response to future sea-level rise. For sea-level rise scenarios greater than 7 millimeters per year, it is *likely* that some barrier islands in this region will cross a threshold where rapid barrier island migration or segmentation will occur (Chapter 3).

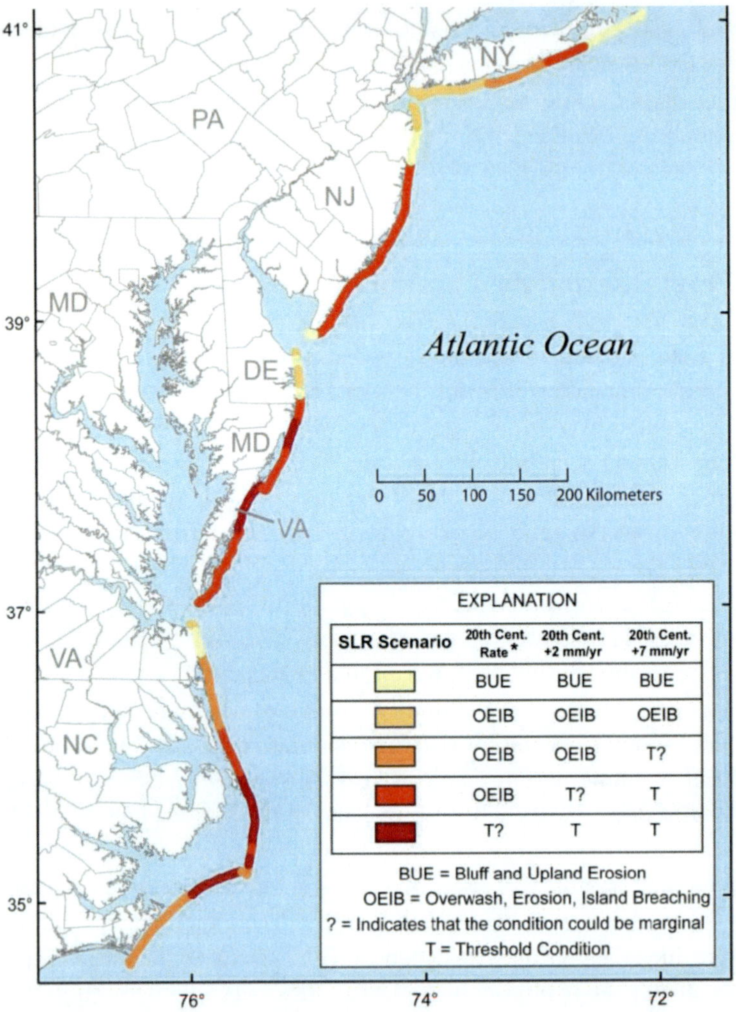

Figure ES.1. Potential mid-Atlantic coastal landform responses to three sea-level rise scenarios. Most coastal areas are currently experiencing erosion, which is expected to increase with future sea-level rise. In addition to undergoing erosion, coastal segments denoted with a "T" may also cross a threshold where rapid barrier island migration or segmentation will occur.

Tidal wetlands in the United States, such as the Mississippi River Delta in Louisiana and Blackwater River marshes in Maryland, are already experiencing submergence by relative sea-level rise and associated high rates of wetland loss.

For the mid-Atlantic region (Figure ES.2), acceleration in sea-level rise by 2 millimeters per year will cause many wetlands to become stressed; it is *likely* that most wetlands will not survive acceleration in sea-level rise by 7 millimeters per year. Wetlands may expand inland where low-lying land is available but, if existing wetlands cannot keep pace with sea-level rise, the result will be an overall loss of wetland area in the Mid-Atlantic. The loss of associated wetland ecosystem functions (*e.g.*, providing flood control, acting as a storm surge buffer, protecting water quality buffer, and serving as a nursery area) can have important societal consequences, such as was seen with the storm surge impacts associated with Hurricanes Katrina and Rita in southern Louisiana, including New Orleans, in 2005.

Nationally, tidal wetlands already experiencing submergence by sea-level rise and associated land loss (*e.g.*, Mississippi River delta in Louisiana, and Blackwater River marshes in Maryland) will continue to lose area in response to future accelerated rates of sea-level rise and changes in other climate and environmental drivers.

Terrestrial and aquatic plants and animals that rely on coastal habitat are likely to be stressed and adversely affected as sea level rises. The quality, quantity, and spatial distribution of coastal habitats will change as a result of erosion, salinity changes, and wetland loss. Depending on local conditions, habitat may be lost or migrate inland in response to sea-level rise. Loss of tidal marshes would seriously threaten coastal ecosystems, causing fish and birds to move or produce fewer offspring. Many estuarine beaches may also be lost, threatening numerous species (Chapter 5).

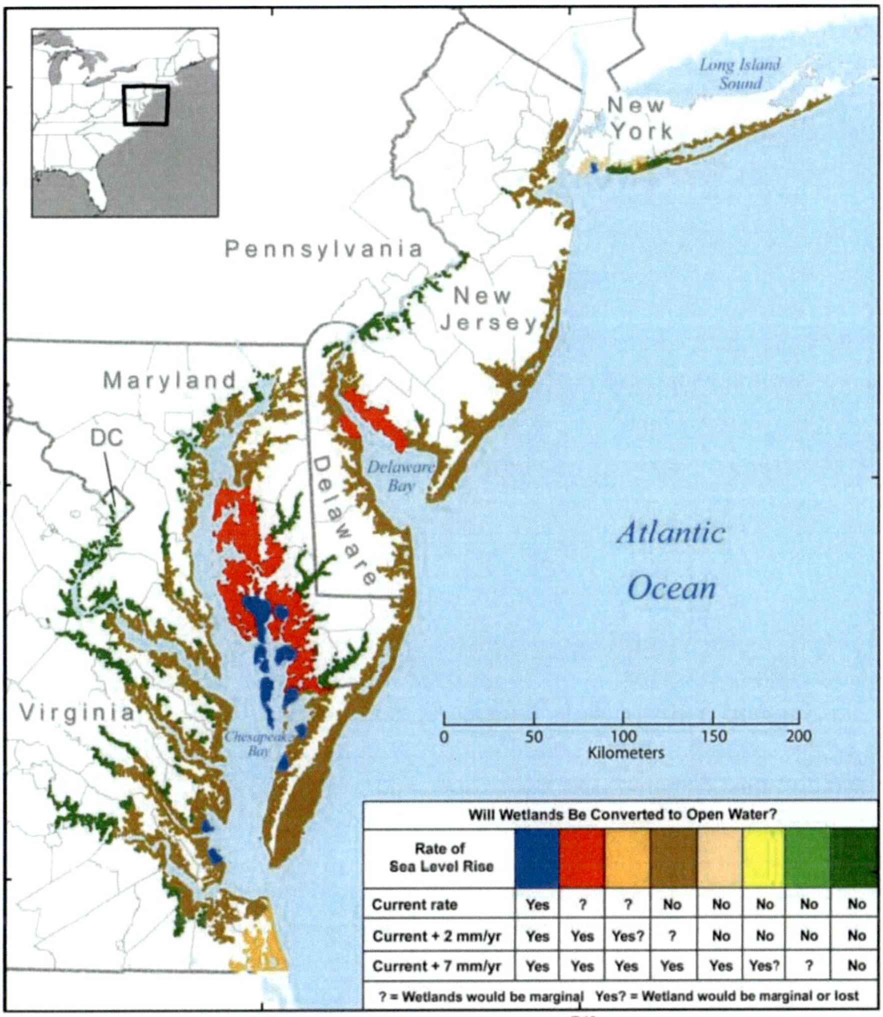

Will Wetlands Be Converted to Open Water?								
Rate of Sea Level Rise	(blue)	(red)	(orange)	(brown)	(tan)	(yellow)	(light green)	(dark green)
Current rate	Yes	?	?	No	No	No	No	No
Current + 2 mm/yr	Yes	Yes	Yes?	?	No	No	No	No
Current + 7 mm/yr	Yes	Yes	Yes	Yes	Yes	Yes?	?	No

? = Wetlands would be marginal Yes? = Wetland would be marginal or lost

Figure ES.2. Areas where wetlands would be marginal or lost (*i.e.*, converted to open water) under three sea-level rise scenarios.

Sea-level rise is just one of many factors affecting coastal habitats: sediment input, nutrient runoff, fisheries management, and other factors are also important. Under natural conditions, habitats are continually shifting, and species generally have some flexibility to adapt to varied geography and/or habitat type. Future habitat and species loss will be determined by factors that include rates of wetland submergence, coastal erosion, and whether coastal landforms and present-day habitats have space to migrate inland. As coastal development continues, the ability for habitats to change and migrate inland along the rest of the coast will not only be a function of the attributes of the natural system, but also of the coastal management policies for developed and undeveloped areas.

ES.2.2. Societal Impacts and Implications

Increasing population, development, and supporting infrastructure in the coastal zone often compete with the desire to maintain the benefits that natural ecosystems (*e.g.*, beaches, barrier islands, and wetlands) provide to humans. Increasing sea level will put additional stress on the ability to manage these competing interests effectively (Chapter 7). In the Mid-Atlantic, for example, movement to the coast and development continues, despite the growing vulnerability to coastal hazards.

Rising sea level increases the vulnerability of development on coastal floodplains. Higher sea level provides an elevated base for storm surges to build upon and diminishes the rate at which low-lying areas drain, thereby increasing the risk of flooding from rainstorms. Increases in shore erosion also contribute to greater flood damages by removing protective dunes, beaches, and wetlands and by leaving some properties closer to the water's edge (Chapter 9).

ES.3. How Can People Prepare For Sea-Level Rise?

ES.3.1. Options for Adapting to Sea-level Rise

At the current rate of sea-level rise, coastal residents and businesses have been responding by rebuilding at the same location, relocating, holding back the sea by coastal engineering, or some combination of these approaches. With a substantial acceleration of sea-level rise, traditional coastal engineering may not be economically or environmentally sustainable in some areas (Chapter 6).

Nationally, most current coastal policies do not accommodate accelerations in sea-level rise. Floodplain maps, which are used to guide development and building practices in hazardous areas, are generally based upon recent observations of topographic elevation and local mean sea-level. However, these maps often do not take into account accelerated sea-level rise or possible changes in storm intensity (Chapter 9). As a result, most shore protection structures are designed for current sea level, and development policies that rely on setting development back from the coast are designed for current rates of coastal erosion, not taking into account sea level rise.

ES.3.2. Adapting to Sea-level Rise

The prospect of accelerated sea-level rise underscores the need to rigorously assess vulnerability and examine the costs and benefits of taking adaptive actions. Determining whether, what, and when specific actions are justified is not simple, due to uncertainty in the timing and magnitude of impacts, and difficulties in quantifying projected costs and benefits. Key opportunities for preparing for sea-level rise include: provisions for preserving public access along the shore (Chapter 8); land-use planning to ensure that wetlands, beaches, and associated coastal ecosystem services are preserved (Chapter 10); siting and design decisions such as retrofitting (*e.g.*, elevating buildings and homes) (Chapter 10); and examining whether and how changing risk due to sea-level rise is reflected in flood insurance rates (Chapter 10).

However, the time, and often cultural shift, required to make change in federal, state, and local policies is sometimes a barrier to change. In the mid-Atlantic coastal zone, for example, although the management community recognizes sea-level rise as a coastal flooding hazard and state governments are starting to face the issue of sea-level rise, only a limited number of analyses and resulting statewide policy revisions to address rising sea level have been undertaken (Chapters 9, 11). Current policies in some areas are now being adapted to include the effects of sea-level rise on coastal environments and infrastructure. Responding to sea-level rise requires careful consideration regarding whether and how particular areas will be protected with structures, elevated above the tides, relocated landward, or left alone and potentially given up to the rising sea (Chapter 12).

Many coastal management decisions made today have implications for sea-level rise adaptation. Existing state policies that restrict development along the shore to mitigate hazards or protect water quality (Appendix 1) could preserve open space that may also help coastal ecosystems adapt to rising sea level. On the other hand, efforts to fortify coastal development can make it less likely that such an area would be abandoned as sea level rises (Chapter 6). A prime opportunity for adapting to sea-level rise in developed areas may be in the aftermath of a severe storm (Chapter 9).

ES.4. How Can Science Improve Understanding and Preparedness for Future Sea-Level Rise?

This Product broadly synthesizes physical, biological, social, and institutional topics involved in assessing the potential vulnerability of the mid-Atlantic United States to sea-level rise. This includes the potential for landscape changes and associated geological and biological processes; and the ability of society and its institutions to adapt to change. Current limitations in the ability to quantitatively assess these topics at local, regional, and national scales may affect whether, when, and how some decisions will be made.

Scientific syntheses and assessments such as this have different types and levels of uncertainty. Part I of this Product describes the physical settings and processes in the Mid-Atlantic and how they may be impacted by sea-level rise. There is uncertainty regarding coastal elevations and the extent to which some areas will be inundated. In some areas, coastal elevations have been mapped with great detail and accuracy, and thus the data have the requisite high degree of certainty for local decision making by coastal managers. In many other areas, the coarser resolution and limited vertical accuracy of the available elevation data preclude their use in detailed assessments, but the uncertainty can be explicitly quantified (Chapter 2). The range of physical and biological processes associated with coastal change is poorly understood at some of the time and space scales required for decision making. For example, although the scope and general nature of the changes that can occur on ocean coasts in response to sea-level rise are widely recognized, how these changes occur in response to a specific rise in sea level is difficult to predict (Chapter 3). Similarly, current model projections of wetland vulnerability on regional and national scales are uncertain due to the coarse level of resolution of landscape-scale models. While site-specific model projections are quite good where local information has been acquired on factors that control local accretionary processes in specific wetland settings, such projections cannot presently be generalized so as to apply to larger regional or national scales with high confidence (Chapter 4). The cumulative impacts of physical and biological change due to sea-level rise on the quality and quantity of coastal habitats are not well understood.

Like the uncertainties associated with the physical settings, the potential human responses to future sea-level rise described in Part II of this Product are also uncertain. Society generally responds to changes as they emerge. The decisions that people make to respond to sea-level rise could be influenced by the physical setting, the properties of the built environment, social values, the constraints of regulations and economics, as well as the level of uncertainty in the form and magnitude of future coastal change. This Product examines some of the available options and assesses actions that federal and state governments and coastal communities could take in response to sea-level rise. For example, as rising sea level impacts coastal lands, a fundamental choice is whether to attempt to hold back the sea or allow nature to takes its course. Both choices have important costs and uncertainties (Chapter 6).

Part III of this Product focuses on what might be done to prepare for sea-level rise. As discussed above, the rate, timing, and impacts of future sea-level rise are uncertain, with important implications for decision-making. For example, planning for sea-level rise requires examining the benefits and costs of such issues as coastal wetland protection, existing and planned coastal infrastructure, and management of floodplains in the context of temporal and spatial uncertainty (Chapter 10). In addition, institutional barriers can make it difficult to incorporate the potential impacts of future sea-level rise into coastal planning (Chapter 12).

ES.4.1. Enhance Understanding

An integrated scientific program of sea-level studies would reduce gaps in current knowledge and the uncertainty about the potential responses of coasts, estuaries, wetlands, and human populations to sea-level rise. This program should focus on expanded efforts to monitor ongoing physical and environmental changes, using new technologies and higher resolution elevation data as available. Insights from the historic and geologic past also provide important perspectives. A key area of uncertainty is the vulnerability of coastal landforms and wetlands to sea-level rise; therefore, it is important to understand the dynamics of barrier island processes and wetland accretion, wetland migration, and the effects of land-use change as sea-level rise continues. Understanding, predicting, and responding to the environmental and societal effects of sea-level rise would require an integrated program of research that includes both natural and social sciences. Social science research is a necessary component as sea-level rise vulnerability, sea-level rise impacts, and the success of many adaptation strategies will depend on characterizing the social, economic, and political contexts in which management decisions are made (Chapter 14).

ES.4.2. Enhance Decision Support

Decision making on regional and local levels in the coastal zone can be supported by improved understanding of vulnerabilities and risks of sea-level rise impacts. Developing tools, datasets, and other coastal management information is key to supporting and promoting sound coastal planning, policy making, and decisions. This includes providing easy access to data and information resources and applying this information in an integrated framework using such tools as geographic information systems. Integrated assessments linking physical vulnerability with economic analyses and planning options will be valuable, as will efforts to assemble and assess coastal zone planning adaptation options for federal, state, and local decision makers. Stakeholder participation in every phase of this process is important, so that decision makers and the public have access to the information that they need and can make well-informed choices regarding sea-level rise and the consequences of different management decisions. Coastal planning and policies that are consistent with the reality of a rising sea could enable U.S. coastal communities to avoid or adapt to its potential environmental, societal, and economic impacts.

PART I OVERVIEW. THE PHYSICAL ENVIRONMENT

Authors: Donald R. Cahoon, USGS; S. Jeffress Williams, USGS; Benjamin T. Gutierrez, USGS; K. Eric Anderson, USGS; E. Robert Thieler, USGS; Dean B. Gesch; USGS

The first part of this Product examines the potential physical and environmental impacts of sea-level rise on the coastal environments of the mid-Atlantic region. Rising sea level over the next century will have a range of effects on coastal regions, including land loss and shoreline retreat from erosion and inundation, an increase in the frequency of storm-related flooding, and intrusion of salt water into coastal freshwater aquifers. The sensitivity of a coastal region to sea-level rise depends both on the physical aspects (shape and composition) of a coastal landscape and its ecological setting. One of the most obvious impacts is that there will be land loss as coastal areas are inundated and eroded. Rising sea level will not only inundate the landscape but will also be a driver of change for the coastal landscape. These impacts will have large effects on natural environments such as coastal wetland ecosystems, as well as effects on human development in coastal regions (see Part II of this Product). Making long-term projections of coastal change is difficult because of the multiple, interacting factors that contribute to that change. Given the large potential impacts to human and natural environments, there is a need to improve our ability to conduct long-term projections.

Part I describes the physical settings of the mid-Atlantic coast as well as the processes that influence shoreline change and land loss in response to sea-level rise. Part I also provides an assessment of coastal changes that may occur over the twenty-first century, as well as the consequences of those changes for coastal habitats and the flora and fauna they support.

Chapter 1 provides an overview of the current understanding of climate change and sea-level rise and their potential effects on both natural environments and society, and summarizes the background information that was used to develop this Product. Sea-level rise will have a range of impacts to both natural systems and human development and infrastructure in coastal regions. A major challenge is to understand the extent of these impacts and how to develop planning and adaption strategies that address both the quality of the natural environment and human interests.

Chapter 2 highlights the important issues in analysis of sea-level rise vulnerability based on coastal elevation data. Elevation is a critical factor in determining vulnerability to inundation, which will be the primary response to sea-level rise for only some locations in the mid-Atlantic region. Because sea-level rise impact assessments often rely on elevation data, it is important to understand the inherent accuracy of the underlying data and its effects on the uncertainty of any resulting vulnerability maps and statistical summaries. The existing studies of sea-level rise vulnerability in the Mid-Atlantic based on currently available elevation data do not provide the level of confidence that is optimal for local decision making. However, recent research using newer high-resolution, high accuracy elevation data is leading toward development of improved capabilities for vulnerability assessments.

Chapter 3 summarizes the factors and processes controlling the dynamics of ocean coasts. The major factor affecting the location and shape of coasts at centennial and longer time scales is global sea-level change, which is linked to the Earth's climate. These close linkages are well documented in the scientific literature from field studies conducted over the past few decades. The details of the process-response relationships, however, are the subject of active, ongoing research. The general characteristics and shape of the coast (coastal morphology) reflects complex and ongoing interactions between changes in sea level, the physical processes that act on the coast (hydrodynamic regime, *e.g.*, waves and tidal characteristics), the availability of sediment (sediment supply) transported by waves and tidal currents at the shore, and underlying geology (the structure and composition of the landscape which is often referred to as the geologic framework). Variations in these three factors are responsible for the different coastal landforms and environments occurring in the coastal regions of the United States. Chapter 2 presents a synthesis and assessment of the potential changes that can be expected for the mid-Atlantic shores of the United States which are primarily comprised of beaches and barrier islands.

Chapter 4 describes the vulnerability of coastal wetlands in the mid-Atlantic region to current and future sea-level rise. The fate of coastal wetlands is determined in large part by the way in which wetland vertical development processes change with climate drivers. In addition, the processes by which wetlands build vertically vary by geomorphic setting.

Chapter 3 identifies those important climate drivers affecting wetland vertical development in the geomorphic settings of the mid-Atlantic region. The information on climate drivers, wetland vertical development, geomorphic settings, and local sea-level rise trends was synthesized and assessed using an expert decision process to determine wetland vulnerability for each geomorphic setting in each subregion of the mid-Atlantic region.

Chapter 5 summarizes the potential impacts to biota as a result of habitat change or loss driven by sea-level rise. Habitat quality, extent, and spatial distribution will change as a result of shore erosion, wetland loss, and shifts in estuarine salinity gradients. Of particular concern is the loss of wetland habitats and the important ecosystem functions they provide, which include critical habitat for wildlife, the trapping of sediments, nutrients, and pollutants, the cycling of nutrients and minerals, the buffering of storm impacts on coastal environments, and the exchange of materials with adjacent ecosystems.

In: Coastal Sensitivity to Sea Level Rise-Focusing on the Mid-… ISBN: 978-1-60741-440-7
Editor: Melvin C. Urajner

Chapter 1

SEA-LEVEL RISE AND ITS EFFECTS ON THE COAST

Lead Authors: S. Jeffress Williams[1], Benjamin T. Gutierrez[1], James G. Titus[2], Stephen K. Gill[3], Donald R. Cahoon[1], E. Robert Thieler[1] and K. Eric Anderson[4]

[1]USGS
[2]U.S. EPA
[3]NOAA
[4]USGS (retired)

Contributing Authors: Duncan FitzGerald[5], Virginia Burkett[1] and Jason Samenow[2]

[5]Boston University

Key Points

- Consensus in the climate science community is that the global climate is changing, mostly due to mankind's increased emissions of greenhouse gases (25 percent increase in the last century), such as carbon dioxide, methane, and nitrous oxide from burning of fossil fuels and land-use change. Warming of the climate system is unequivocal, but the effects of climate change are highly variable across regions and difficult to predict with high confidence based on limited observations over time and space. Two effects of atmospheric warming on coasts on regional, national, and global scales are sea-level rise and increase in major cyclone intensity.
- Global sea level has risen about 120 meters at highly variable rates due to natural processes since the end of the Last Glacial Maximum (i.e., last Ice Age). More recently, the sea-level rise rate has increased over natural rise due to increase in the burning of fossil fuels. In some regions, such as the mid-Atlantic region and much of the Gulf of Mexico, sea-level rise is significantly greater than the observed global sea-level rise due to sinking of the land as a result of sediment compaction processes.

- Instrumental observations over the past 15 years show that global mean sea level has been highly variable at regional scales around the world and, on average, the rate of rise appears to have accelerated over twentieth century rates, possibly due to atmospheric warming causing expansion of ocean water and ice-sheet melting.
- Results of climate model studies suggest sea-level rise in the twenty-first century will significantly exceed rates over the past century. Rates and the magnitude of rise could be much greater if warming affects dynamical processes that determine ice flow and losses in Greenland and Antarctica.
- Beyond the scope of this Product but important to consider, global sea-level elevations at the peak of the last interglacial warm cycle were 4 to 6 meters (13 to 20 feet) above present, and could be realized within the next several hundred years if warming and glacier and ice-sheet melting continue.
- Coastal regions are characterized by dynamic landforms and processes because they are the juncture between the land, oceans, and atmosphere. Features such as barrier islands, bluffs, dunes, and wetlands constantly undergo change due to driving processes such as storms, sediment supply, and sea-level change. Based on surveys over the past century, all U.S. coastal states are experiencing overall erosion at highly variable rates. Sea-level rise will have profound effects by increasing flooding frequency and inundating low-lying coastal areas, but other processes such as erosion and accretion will have cumulative effects that are profound but not yet predictable with high reliability. There is some recent scientific opinion that coastal landforms such as barrier islands and wetlands may have thresholds or tipping points with sea-level rise and storms, leading to rapid and irreversible change.
- Nearly one-half of the 6.7 billion people around the world live near the coast and are highly vulnerable to storms and sea-level rise. In the United States, coastal populations have doubled over the past 50 years, greatly increasing exposure to risk from storms and sea-level rise. Continued population growth in low-lying coastal regions worldwide and in the United States will increase vulnerability to these hazards as the effects of climate change become more pronounced.
- Most coastal regions are currently managed under the premise that sea-level rise is not significant and that shorelines are static or can be fixed in place by engineering structures. The new reality of sea-level rise due to climate change requires new considerations in managing areas to protect resources and reduce risk to humans. Long-term climate change impact data are essential for adaptation plans to climate change and coastal zone plans are most useful if they have the premise that coasts are dynamic and highly variable.

1.1. Introduction

The main objective of this Product is to review and assess the potential impacts of sea-level rise on U.S.coastal regions. Careful review and critique of sea-level and climate change science is beyond the scope of this Product; however, that information is central in assessing coastal impacts. Climate and coastal scientific disciplines are relatively recent, and while uncertainty exists in predicting quantitatively the magnitude and rates of change in sea level, a solid body of scientific evidence exists that sea level has risen over the recent geologic past, is currently rising and contributing to various effects such as coastal erosion, and has the potential to rise at an accelerated rate this century and beyond. Worldwide data also show that rates of global sea-level rise are consistent with increasing greenhouse gas concentrations and global warming (IPCC, 2001, 2007; Hansen *et al.*, 2007; Broecker and Kunzig, 2008). Global climate change is already having significant and wide ranging effects on the Earth's ecosystems and human populations (Nicholls *et al.*, 2007).

In recognition of the influence of humans on the Earth, including the global climate, the time period since the nineteenth century is being referred to by scientists as the Anthropocene Era (Pearce, 2007; Zalasiewicz, 2008). Changes to the global climate have been dramatic and the rapid rate of climate change observed over the past two decades is an increasing challenge for adaptation, by humans and animals and plants alike.

Effects from climate change are not uniform, but vary considerably from region to region and over a range of time scales (Nicholls *et al.*, 2007). These variations occur due to regional and local differences in atmospheric, terrestrial, and oceanographic processes. The processes driving climate change are complex and so-called feedback interactions between the processes can both enhance and diminish sea-level rise impacts, making prediction of long-term effects difficult. Accelerated global sea-level rise, a likely major long-term outcome of climate change, will have increasingly far-reaching impacts on coastal regions of the United States and around the world (Nicholls *et al.*, 2007). Relative sea-level rise impacts are already evident for many coastal regions and will increase significantly during this century and beyond (FitzGerald *et al*, 2008; IPCC, 2007; Nicholls *et al.*, 2007). Sea-level rise will cause significant and often dramatic changes to coastal landforms (*e.g.*, barrier islands, beaches, dunes, marshes), as well as ecosystems, estuaries, waterways, and human populations and development in the coastal zone (Nicholls *et al.*, 2007; Rosenzweig *et al.*, 2008; FitzGerald *et al.*, 2008). Low-lying coastal plain regions, particularly those that are densely populated (*e.g.*,the Mid-Atlantic, the north central Gulf of Mexico), are especially vulnerable to sea-level rise and land subsidence and their combined impacts to the coast and to development in the coastal zone (*e.g.*, McGranahan *et al.*, 2007; Day *et al.*, 2007a).

The effects of sea-level rise are not necessarily obvious in the short term, but are evident over the longer term in many ways. Arguably, the most visible effect is seen in changing coastal landscapes, which are altered through more frequent flooding, inundation, and coastal erosion as barrier islands, beaches, and sand dunes change shape and move landward in concert with sea-level rise and storm effects. In addition, the alteration or loss of coastal habitats such as wetlands, bays, and estuaries has negative impacts on many animal and plant species that depend on these coastal ecosystems.

Understanding how sea-level rise is likely to affect coastal regions and, consequently, how society will choose to address this issue in the short term in ways that are sustainable for the long term, is a major challenge for both scientists and coastal policy makers and managers. While human populations in high-risk coastal areas continue to expand rapidly, the analyses of long-term sea-level measurements show that sea level rose on average 19 centimeters (cm) (7.5 inches [in]) globally during the twentieth century (Jevrejeva *et al.*, 2008). In addition, satellite data show global sea-level rise has accelerated over the past 15 years, but at highly variable rates on regional scales. Analyses indicate that the magnitude and rate of sea-level rise for this century and beyond is likely to exceed that of the past century (Meehl *et al.*, 2007; Rahmstorf, 2007; Jevrejeva *et al.*, 2008).

Over the last century, humans have generally responded to eroding shorelines and flooding landscapes by using engineering measures to protect threatened property or by relocating development inland to higher ground. In the future, these responses will become more widespread and more expensive for society as sea-level rise accelerates (Nicholls *et al.*, 2007). Currently the world population is 6.7 billion people and is predicted to expand to 9.1 billion by the year 2042 (UN, 2005). Globally, 44 percent of the world's population lives within 150 kilometers (km) (93 miles [mi]) of the ocean (<http://www.oceansatlas.org/index.jsp>) and more than 600 million people live in low elevation coastal zone areas that are less than 10 meters (m) (33 feet [ft]) above sea level (McGranahan *et al.*, 2007), putting them at significant risk to the effects of sea-level rise. The 10 m elevation was chosen as a benchmark for providing population statistics to meet data resolution and quality needs because that elevation is a commonly used reference elevation for coastal plain regions vulnerable to coastal hazards such as storm-surge flooding and sea-level rise. Eight of the 10 largest cities in the world are sited on the ocean coast. In the United States, 14 of the 20 largest urban centers are located within 100 km of the coast and less than 10 m above sea level. Using the year 2000 census data for U.S. coastal counties as defined by the National Oceanic and Atmospheric Administration (NOAA) and excluding the Great Lakes states, approximately 126 million people resided in coastal areas (Crossett *et al.*, 2004). The Federal Emergency Management Agency (FEMA), using the same 2000 census data but different criteria for defining coastal counties, estimated the coastal population to be 86 million people (Crowell, *et al.,* 2007). Regardless, U.S. coastal populations have expanded greatly over the past 50 years, increasing exposure to risk from storms and sea-level rise. Continued population growth in low-lying coastal regions worldwide and in the United States will increase vulnerability to these hazards.

Modern societies around the world have developed and populations have expanded over the past several thousand years under a relatively mild and stable world climate and relatively stable sea level (Stanley and Warne, 2003; Day *et al.*, 2007b). However, with continued population growth, particularly in coastal areas, and the probability of accelerated sea-level rise and increased storminess, adaptation to expected changes will become increasingly challenging.

This Product reviews available scientific literature through late 2008 and assesses the likely effects of sea-level rise on the coast of the United States, with a focus on the mid-Atlantic region. An important point to emphasize is that sea-level rise impacts will be far-reaching. Coastal lands will not simply be flooded by rising seas, but will be modified by a variety of processes (*e.g.*, erosion, accretion) whose impacts will vary greatly by location and geologic setting. For example, the frequency and magnitude of flooding may change and sea-level rise can alsoaffect water table elevations, impacting fresh water supplies. These changes will have a broad range of human and environmental impacts. To effectively cope with sea-level rise and its impacts, current policies and economic considerations should be examined, and possible options for changing planning and management activities are warranted so that society and the environment are better able to adapt to potential accelerated rise in sea level. This Product examines the potential coastal impacts for three different plausible scenarios of future sea-level rise, and focuses on the potential effects to the year 2100. The effects, of course, will extend well beyond 2100, but detailed discussion of effects farther into the future is outside the scope of this Product.

1.1.1. Climate Change Basis for this Product

The scientific study of climate change and associated global sea-level rise is complicated due to differences in observations, data quality, cumulative effects, and many other factors. Both direct and indirect methods are useful for studying past climate change. Instrument records and historical documents are most accurate, but are limited to the past 100 to 150 years in the United States. Geological information from analyses of continuous cores sampled from ice sheets and glaciers, sea and lake sediments, and sea corals provide useful proxies that have allowed researchers to decipher past climate conditions and a record of climate and sea-level changes stretching back millions of years before recorded history (Miller *et al.*, 2005; Jansen *et al.*, 2007). The most precise methods have provided accurate high-resolution data on the climate (*e.g.*, global temperature, atmospheric composition) dating back more than 400,000 years.

The Intergovernmental Panel on Climate Change (IPCC) 2007 Fourth Assessment Report provides a comprehensive review and assessment of global climate change trends, expected changes over the next century, and the impacts and challenges that both humans and the natural world are likely to be confronted with during the next century (IPCC, 2007). Some key findings from this report are summarized in Box 1.1. A 2008 U.S. Climate Change Science Program (CCSP) report provides a general assessment of current scientific understanding of climate change impacts to the United States (CENR, 2008) and the recent CCSP Synthesis and Assessment (SAP) 3.4 report on Abrupt Climate Change discusses the effects of complex changes in ice sheets and glaciers on sea level (Steffen *et al.,* 2008). CCSP SAP 4.1 provides more specific information and scientific consensus on the likely effects and implications of future sea-level rise on coasts and wetlands of the United States and also includes a science strategy for improving the understanding of sea-level rise, documenting its effects, and devising robust models and methods for reliably predicting future changes and impacts to coastal regions.

Box 1.1. Selected Findings of the Intergovernmental Panel on Climate Change (IPCC) (2007a and b) on Climate and Global Sea-Level Rise

Recent Global Climate Change

Note: The likelihood scale, established by the IPCC and used throughout SAP 4.1, is described in the Preface. The terms used in that scale will be italicized when used as such in this Product

Warming of the climate system is unequivocal, as is now evident from observations of increases in global average air and ocean temperatures, widespread melting of snow and ice, and rising global average sea level.

Human-induced increase in atmospheric carbon dioxide is the most important factor affecting the warming of the Earth's climate since the start of the Industrial Era. The atmospheric concentration of carbon dioxide in 2005 exceeds by far the natural range over the last 650,000 years.

Most of the observed increase in global average temperatures since the mid-twentieth century is *very likely* due to the observed increase in human-caused greenhouse gas concentrations. Discernible human influences now extend to other aspects of climate, including ocean warming, continental-average temperatures, temperature extremes, and wind patterns.

Recent Global Sea-Level Rise

Observations since 1961 show that the average temperature of the global ocean has increased to depths of at least 3,000 meters (m) and that the ocean has been absorbing more than 80 percent of the heat added to the climate system. Such warming causes seawater to expand, contributing to global sea-level rise.

Mountain glaciers and snow cover have declined on average in both hemispheres. Widespread decreases in glaciers and ice caps have contributed to global sea-level rise.

New data show that losses from the ice sheets of Greenland and Antarctica have *very likely* contributed to global sea-level rise between 1993 and 2003.

Global average sea level rose at an average rate of 1.8 (1.3 to 2.3) millimeters (mm) per year between 1961 and 2003. The rate was faster between 1993 and 2003: about 3.1 (2.4 to 3.8) mm per year. Whether the faster rate for 1993 to 2003 reflects decadal variability or an increase in the longer term trend is unclear (see Figure 1.3).

Global average sea level in the last interglacial period (about 125,000 years ago) was *likely* 4 to 6 m higher than during the twentieth century, mainly due to the retreat of polar ice. Ice core data indicate that average polar temperatures at that time were 3 to 5°C higher than present, because of differences in the Earth's orbit. The Greenland ice sheet and other arctic ice fields *likely* contributed no more than 4 m of the observed global sea-level rise. There may also have been contributions from Antarctica ice sheet melting.

Projections of the Future

Continued greenhouse gas emissions at or above current rates would cause further warming and induce many changes in the global climate system during the twenty-first century that would *very likely* be larger than those observed during the twentieth century.

Based on a range of possible greenhouse gas emission scenarios for the next century, the IPCC estimates the global increase in temperature will likely be between 1.1 and 6.4ºC. Estimates of sea-level rise for the same scenarios are 0.18 m to 0.59 m, excluding the contribution from accelerated ice discharges from the Greenland and Antarctica ice sheets.

Extrapolating the recent acceleration of ice discharges from the polar ice sheets would imply an additional contribution up to 0.20 m. If melting of these ice caps increases, larger values of sea-level rise cannot be excluded.

In addition to global sea-level rise, the storms that lead to coastal storm surges could become more intense. The IPCC indicates that, based on a range of computer models, it is *likely* that hurricanes will become more intense, with larger peak wind speeds and more heavy precipitation associated with ongoing increases of tropical sea surface temperatures, while the tracks of "winter" or extratropical cyclones are projected to shift towards the poles along with some indications of an increase in intensity in the North Atlantic.

1.2. Why Is Global Sea Level Rising?

The elevation of global sea level is determined by the dynamic balance between the mass of ice on land (in glaciers and ice sheets) and the mass of water in ocean basins. Both of these factors are highly influenced by the Earth's atmospheric temperature. During the last 800,000 years, global sea level has risen and fallen about 120 m (400 ft) in response to the alternating accumulation and decline of large continental ice sheets about 2 to 3 km (1 to 2 mi) thick as climate warmed and cooled in naturally occurring 100,000 year astronomical cycles (Imbrie and Imbrie, 1986; Lambeck *et al.,* 2002). Figure 1.1 shows a record of large global sea-level change over the past 400,000 years during the last four cycles, consisting of glacial maximums with low sea levels and interglacial warm periods with high sea levels. The last interglacial period, about 125,000 years ago, lasted about 10,000 to 12,000 years, with average temperatures warmer than today but close to those predicted for the next century, and global sea level was 4 to 6 m (13 to 20 ft) higher than present (Imbrie and Imbrie, 1986). Following the peak of the last Ice Age about 21,000 years ago, the Earth entered the present interglacial warm period. Global sea level rose very rapidly at average rates of 10 to 20 mm per year punctuated with periodic large "meltwater pulses" with rates of more than 50 mm per year from about 21,000 to 6,000 years ago. Sea-level rise then slowed to a rate of about 0.5 mm per year from 6,000 to 3,000 years ago (Fairbanks, 1989; Rohling *et al.,* 2008). During the past 2,000 to 3,000 years the rate slowed to approximately 0.2 mm per year until an acceleration occurred in the late nineteenth century (IPCC 2001).

There is growing scientific evidence that, at the onset of the present interglacial warm period, the Earth underwent abrupt changes when the climate system crossed several thresholds or tipping points (points or levels in the evolution of the Earth's climate leading to irreversible change) that triggered dramatic changes in temperature, precipitation, ice cover, and sea level. These changes are thought to have occurred over a few decades to a century and the causes are not well understood (NRC, 2002; Alley *et al.*, 2003). One cause is thought to be disruption of major ocean currents by influxes of fresh water from glacial melt. It is not known with any confidence how anthropogenic climate change might alter the natural glacial-interglacial cycle or the forcings that drive abrupt change in the Earth's climate system. Imbrie and Imbrie (1986) surmise that the world might experience a "super-interglacial" period with mean temperatures higher than past warm periods.

At the peak of the last Ice Age, sea level was approximately 120 m lower than today and the shoreline was far seaward of its present location, at the margins of the continental shelf (Figure 1.2). As the climate warmed and ice sheets melted, sea level rose rapidly but at highly variable rates, eroding and submerging the coastal plain to create the continental shelves, drowning ancestral river valleys, and creating major estuaries such as Long Island Sound, Delaware Bay, Chesapeake Bay, Tampa Bay, Galveston Bay, and San Francisco Bay.

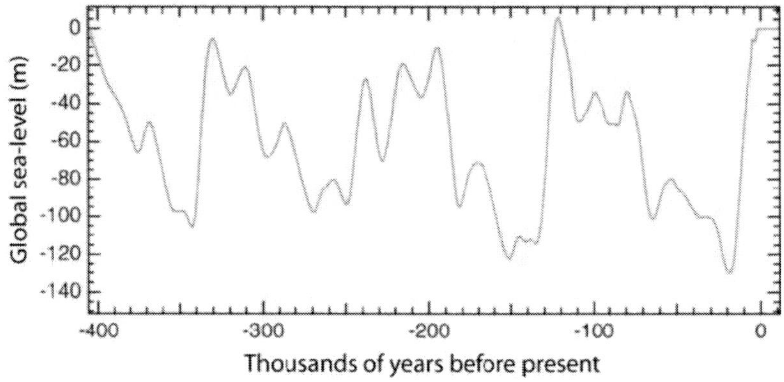

Figure 1.1. Plot of large variations in global sea level elevation over the last 400,000 years resulting from four natural glacial and interglacial cycles. Evidence suggests that sea level was about 4 to 6 meters (m) higher than present during the last interglacial warm period 125,000 years ago and 120 m lower during the last Ice Age, about 21,000 years ago (see reviews in Muhs *et al.*, 2004 and Overpeck *et al.*, 2006). (Reprinted from Quaternary Science Reviews, 21/1-3, Phillippe Huybrechts, Sea-level changes at the LGM from ice-dynamic reconstructions of the Greenland and Antarctic ice sheets during the glacial cycles, 203231, Copyright [2002], with permission from Elsevier.)

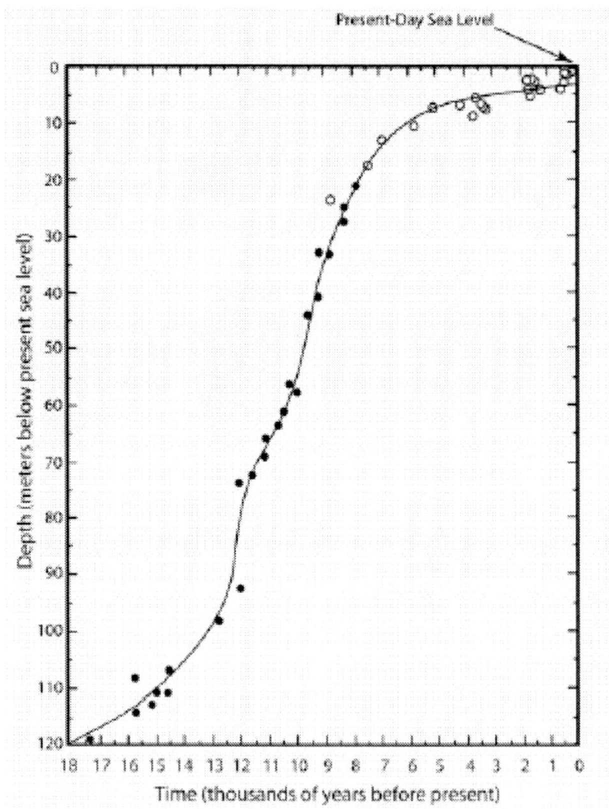

Figure 1.2. Generalized plot of the rise in global sea level at variable rates over the last 18,000 years as the Earth moved from a glacial period to the present interglacial warm period. This curve is reconstructed from geologic samples, shown as data points. Rise was rapid but highly variable for much of the time and slowed about 3,000 years ago. Recent acceleration is not shown at this scale. Reprinted

by permission from Macmillan Publishers Ltd: Nature (Fairbanks, R.G., A 17,000-year glacio-eustatic sea level record—influence of glacial melting rates on the Younger Dryas event and deep-sea circulation, 349[6250], 637-642), ©1989.

Global sea level was relatively stable with rates of rise averaging 0 to 0.2 mm per year until rates increased in the late nineteenth and early twentieth centuries (Bindoff *et al.,* 2007; Lambeck *et al.,* 2004; Gehrels *et al.*, 2008). Some studies indicate that acceleration in sea-level rise may have begun earlier, in the late eighteenth century (Jevrejeva *et al.,* 2008). Analyses of tide-gauge data indicate that the twentieth century rate of sea-level rise averaged 1.7 mm per year on a global scale (Figure 1.3) (Bindoff *et al.*, 2007), but that the rate fluctuated over decadal periods throughout the century (Church and White, 2006; Jevrejeva *et al.*, 2006, 2008). Between 1993 and 2003, both satellite altimeter and tide-gauge observations indicate that the rate of sea-level rise increased to 3.1 mm per year (Bindoff *et al.*, 2007); however, with such a short record, it is not yet possible to determine with certainty whether this is a natural decadal variation or due to human-induced climate warming (Bindoff *et al.*, 2007).

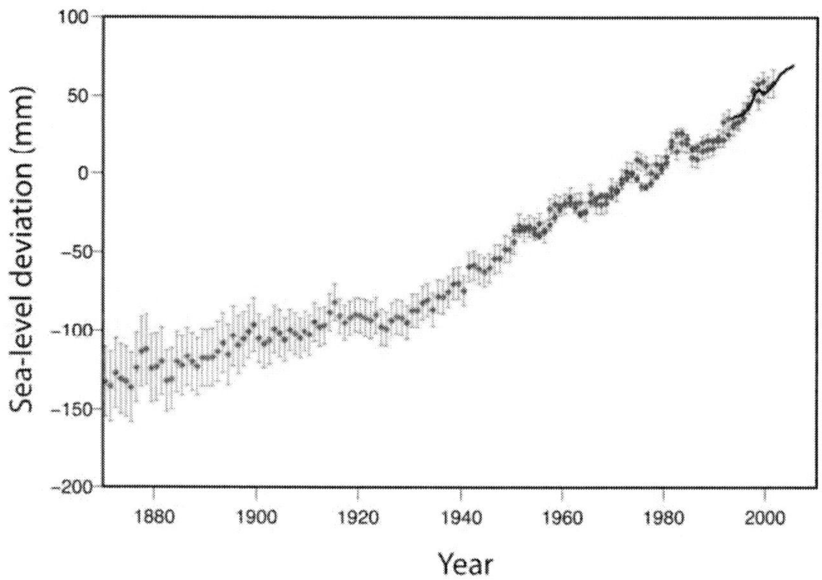

Figure 1.3. Annual averages of global mean sea level in millimeters from IPCC (2007). The red curve shows sea-level fields since 1870 (updated from Church and White, 2006); the blue curve displays tide gauge data from Holgate and Woodworth (2004), and the black curve is based on satellite observations from Leuliette *et al.* (2004). The red and blue curves are deviations from their averages for 1961 to 1990, and the black curve is the deviation from the average of the red curve for the period 1993 to 2001. Vertical error bars show 90 percent confidence intervals for the data points. Adapted from *Climate Change 2007: The Physical Science Basis*. Working Group I Contribution to the Fourth Assessment Report of the Intergovernmental Panel on Climate Change. Figure 5.13. Cambridge University Press.

Box 1.2. Relative Sea Level

"Global sea-level rise" results mainly from the worldwide increase in the volume of the world's oceans that occurs as a result of thermal expansion of warming ocean water and the addition of water to the ocean from melting ice sheets and glaciers (ice masses on land). "Relative sea-level rise" is measured directly by coastal tide gauges, which record both the movement of the land to which they are attached and changes in global sea level. Global sea-level rise can be estimated from tide gauge data by subtracting the land elevation change component. Thus, tide gauges are important observation instruments for measuring sea-level change trends. However, because variations in climate and ocean circulation can cause fluctuations over 10-year time periods, the most reliable sea level data are from tide gauges having records 50 years or longer and for which the rates have been adjusted using a global isostatic adjustment model (Douglas *et al.*, 2001)

At regional and local scales along the coast, vertical movements of the land surface can also contribute significantly to sea-level change and the combination of global sea-level and land-level change is referred to as "relative sea level" (Douglas, 2001).Thus, "relative sea-level rise" refers to the change in sea level relative to the elevation of the land, which includes both global sea-level rise and vertical movements of the land. Both terms, global sea level and relative sea level, are used throughout this Product.

Vertical changes of the land surface result from many factors including tectonic processes and subsidence (sinking of the land) due to compaction of sediments and extraction of subsurface fluids such as oil, gas, and water. A principal contributor to this change along the Atlantic Coast of North America is the vertical relaxation adjustments of the Earth's crust to reduced ice loading due to climate warming since the last Ice Age. In addition to glacial adjustments, sediment loading also contributes to regional subsidence of the land surface. Subsidence contributes to high rates of relative sea-level rise (9.9 millimeters per year) in the Mississippi River delta where thick sediments have accumulated and are compacting. Likewise, fluid withdrawal from coastal aquifers causes the sediments to compact locally as the water is extracted. In Louisiana, Texas, and Southern California, oil, gas and ground-water extraction have contributed markedly to subsidence and relative sea-level rise (Gornitz and Lebedeff, 1987; Emery and Aubrey, 1991; Nicholls and Leatherman, 1996; Galloway *et al.*, 1999; Morton *et al.*, 2004). In locations where the land surface is subsiding, rates of relative sea-level rise exceed the average rate of global rise (*e.g.*, the north central Gulf of Mexico Coast and mid-Atlantic coast).

1.3. Relative Sea-Level Rise around the United States

Geologic data from radiocarbon age-dating organic sediments in cores and coral reefs are indirect methods used for determining sea-level elevations over the past 40,000 years, but the records from long-term (more than 50 years) tide-gauge stations have been the primary direct measurements of relative sea-level trends over the past century (Douglas, 2001). Figure 1.4 shows the large variations in relative sea level for U.S. coastal regions. The majority of the Atlantic Coast and Gulf of Mexico Coast experience higher rates of sea-level rise (2 to 4 mm per year and 2 to 10 mm per year, respectively) than the current global average (1.7 mm per year).

There are large variations for relative sea-level rise (and fall) around the United States, ranging from a fall of 16.68 mm per year at Skagway in southeast Alaska due to tectonic processes and land rebound upward as a result of glacier melting (Zervas, 2001), to a rise of 9.85 mm per year at Grand Isle, Louisiana, due to land subsidence downward from natural causes and possibly oil and gas extraction.

The rate of relative sea-level rise (see Box 1.2 for definition) measured by tide gauges at specific locations along the Atlantic coast of the United States varies from 1.75 mm to as much as 4.42 mm per year (Table 1.1; Figure 1.4; Zervas, 2001). The lower rates, which occur along New England and from Georgia to northern Florida, are close to the global rate of 1.7 ±0.5 mm per year (Bindoff *et al.*, 2007). The highest rates are in the mid-Atlantic region between northern New Jersey and southern Virginia. Figure 1.5 is an example of the monthly average (mean) sea-level record and the observed relative sea-level rise trend at Baltimore, Maryland. At this location, the relative sea-level trend is 3.12 (±0.08) mm per year, almost twice the present rate of global sea-level rise. Subsidence of the land surface, attributed mainly to adjustments of the Earth's crust in response to the melting of the Laurentide ice sheet and to the compaction of sediments due to freshwater withdrawal from coastal aquifers, contributes to the high rates of relative sea-level rise observed in this region (Gornitz and Lebedeff, 1987; Emery and Aubrey, 1991; Kearney and Stevenson, 1991; Douglas, 2001; Peltier, 2001).

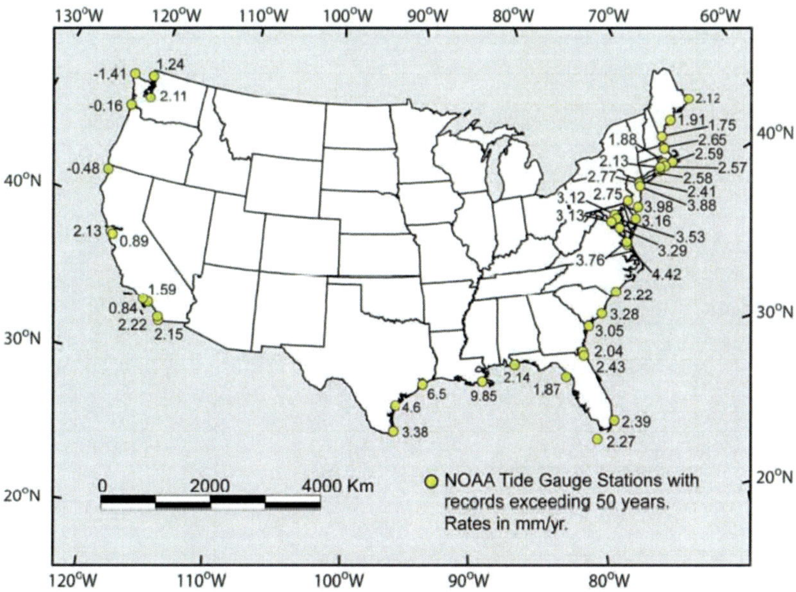

Figure 1.4. Map of twentieth century annual relative sea-level rise rates around the U.S. coast. The higher rates for Louisiana (9.85 millimeters [mm] per year) and the mid-Atlantic region (1.75 to 4.42 mm per year) are due to land subsidence. Sea level is stable or dropping relative to the land in the Pacific Northwest, as indicated by the negative values, where the land is tectonically active or rebounding upward in response to the melting of ice sheets since the last Ice Age (data from Zervas, 2001).

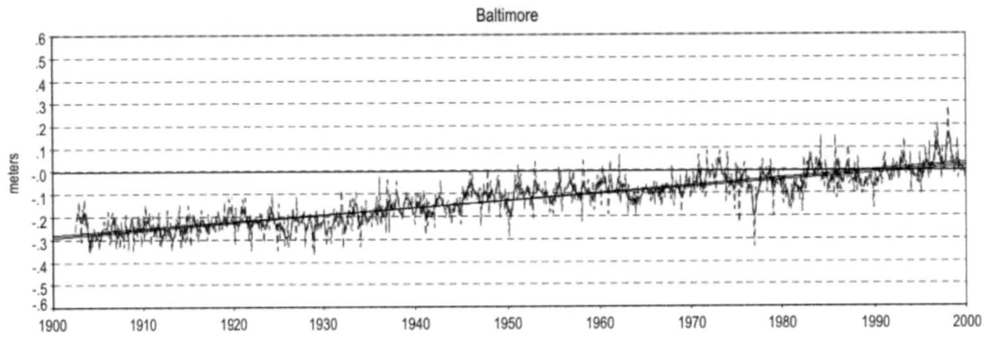

Figure 1.5. The monthly computed average sea-level record (black line) from 1900 to 2000 from the Baltimore, Maryland tide gauge. Blue line is the observed data. The zero line is the latest 19-year National Tidal Datum Epoch mean value. The rate, 3.12 millimeters (mm) per year, is nearly double the present rate (1.7 mm per year) of global sea-level rise due to land subsidence (based on Zervas, 2001).

Table 1.1. Rates of relative sea-level rise for selected long-term tide gauges on the Atlantic coast of the United States (Zervas, 2001). For comparison, the global average rate is 1.7 millimeters per year.

Station	Rate of Sea-level rise (mm per year)	Time Span of Record
Eastport, Maine	2.12 ±0.13	1929-1999
Portland, Maine	1.91 ±0.09	1912-1999
Seavey Island, Maine	1.75 ±0.17	1926-1999
Boston, Massachusetts	2.65 ±0.10	1921-1999
Woods Hole, Massachusetts	2.59 ±0.12	1932-1999
Providence, Rhode Island	1.88 ±0.17	1938-1999
Newport, Rhode Island	2.57 ±0.11	1930-1999
New London, Connecticut	2.13 ±0.15	1938-1999
Montauk, New York	2.58 ±0.19	1947-1999
Willets Point, New York	2.41 ±0.15	1931-1999
The Battery, New York	2.77 ±0.05	1905-1999
Sandy Hook, New Jersey	3.88 ±0.15	1932-1999
Atlantic City, New Jersey	3.98 ±0.11	1911-1999
Philadelphia, Pennsylvania	2.75 ±0.12	1900-1999
Lewes, Delaware	3.16 ±0.16	1919-1999
Baltimore, Maryland	3.12 ±0.08	1902-1999
Annapolis, Maryland	3.53 ±0.13	1928-1999
Solomons Island, Maryland	3.29 ±0.17	1937-1999
Washington, D.C.	3.13 ±0.21	1931-1999
Hampton Roads, Virginia	4.42 ±0.16	1927-1999
Portsmouth, Virginia	3.76 ±0.23	1935-1999
Wilmington, North Carolina	2.22 ±0.25	1935-1999
Charleston, South Carolina	3.28 ±0.14	1921-1999

Table 1.1. (Continued)

Station	Rate of Sea-level rise (mm per year)	Time Span of Record
Fort Pulaski, Georgia	3.05 ±0.20	1935-1999
Fernandina Beach, Florida	2.04 ±0.12	1897-1999
Mayport, Florida	2.43 ±0.18	1928-1999
Miami, Florida	2.39 ±0.22	1931-1999
Key West, Florida	2.27 ±0.09	1913-1999

While measuring and dealing with longer term global averages of sea-level change is useful in understanding effects on coasts, shorter term and regional-scale variations due primarily to warming and oceanographic processes can be quite different from long term averages, and equally important for management and planning. As shown in Figure 1.6 from Bindoff *et al.* (2007) based on a decade of data, some of the highest rates of rise are off the U.S. Mid-Atlantic and the western Pacific, while an apparent drop occurred off the North and South American Pacific Coast.

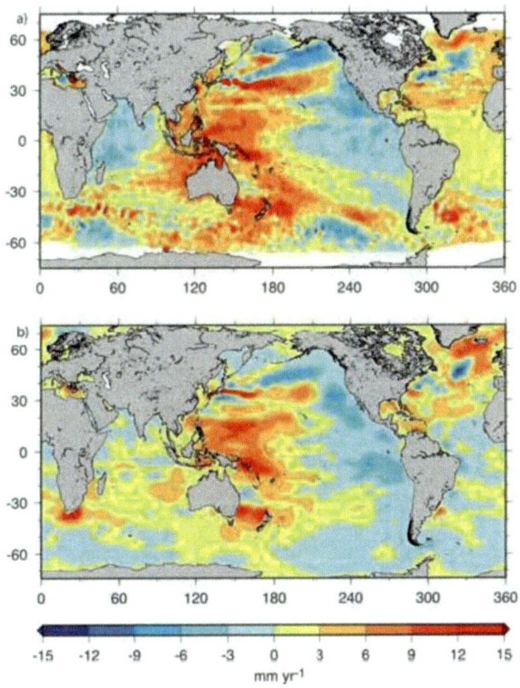

Figure 1.6. (Top) Geographic distribution of short-term linear trends in mean sea level (millimeters [mm] per year) for 1993 to 2003 based on TOPEX/Poseidon satellite altimetry (updated from Cazenave and Nerem, 2004) and (bottom) geographic distribution of linear trends in thermal expansion (mm per year) for 1993 to 2003 (based on temperature data down to 700 meters [from Ishii *et al.*, 2006]). Adapted from *Climate Change 2007: The Physical Science Basis*. Working Group I Contribution to the Fourth Assessment Report of the Intergovernmental Panel on Climate Change. Figure 5.15. Cambridge University Press.

Recently, the IPCC Fourth Assessment Report (IPCC, 2007) estimated that global sea level is likely to rise 18 to 59 cm (7 to 23 in) over the next century; however, possible increased melt water contributions from Greenland and Antarctic have been excluded (Meehl *et al.,* 2007; IPCC, 2007). The IPCC projections (Figure 1.7) represent a "likely range" which inherently allows for the possibility that the actual rise may be higher or lower. Recent observations suggest that sea-level rise rates may already be approaching the higher end of the IPCC estimates (Rahmstorf *et al.*, 2007; Jevrejeva *et al.*, 2008). This is because potentially important meltwater contributions from Greenland and Antarctica were excluded due to limited data and an inability at that time to adequately model ice flow processes. It has been suggested by Rahmstorf (2007) and other climate scientists that a global sea-level rise of 1 m (3 ft) is plausible within this century if increased melting of ice sheets in Greenland and Antarctica is added to the factors included in the IPCC estimates. Therefore, thoughtful precaution suggests that a global sea-level rise of 1 m to the year 2100 should be considered for future planning and policy discussions.

This Product focuses on the effects of sea-level rise on U.S. coasts over the next century, but climate warming and its effects are likely to continue well beyond that due to the amount of greenhouse gases already in the atmosphere. Currently, the amount of potential melting from land-based ice masses (primarily Greenland and West Antarctica) is uncertain and is therefore not fully incorporated into all sea-level rise model projections.

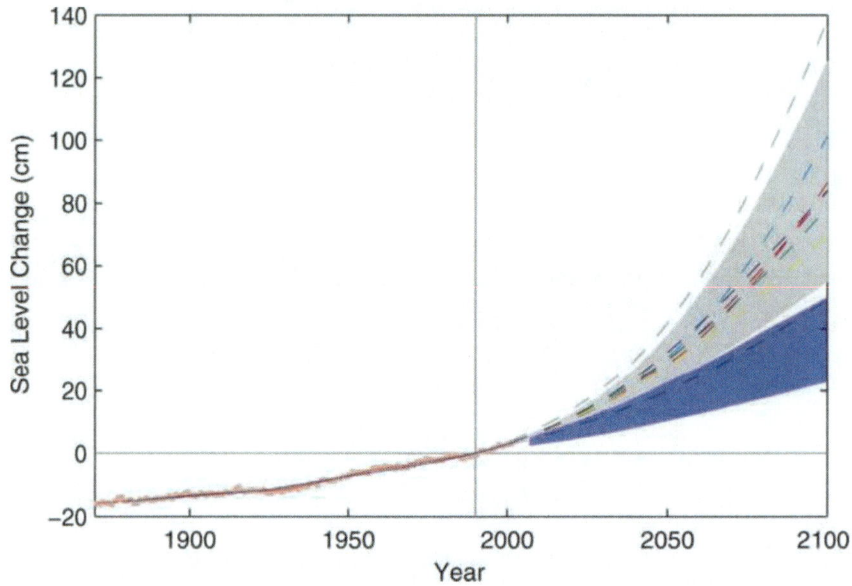

Figure 1.7. Plot in centimeters rise over time of past sea-level observations and several future sea-level projections to the year 2100 based on various computer models. The blue shaded area is the projection by Bindoff *et al.* (2007) and the basis for the IPCC (2007) estimates. The higher gray and dash line projections are from Rahmstorf (2007) considering the factors used in the IPCC estimates, and also potentially increased melting of ice sheets in Greenland and Antarctica. From: Rahmstorf, S., 2007: A semi-empirical approach to projecting future sea-level rise. *Science*, 315(5810), 368-370.Reprinted with permission from AAAS.

Recent observations of changes in ice cover and glacial melting on Greenland, West Antarctica, and smaller glaciers and ice caps around the world indicate that ice loss could be more rapid than the trends evaluated for the IPCC (2007) report (Chen *et al.*, 2006; Shepherd and Wingham, 2007; Meier *et al.*, 2007; Fettweis *et al.*, 2007). The science needed to assign probability to these high scenarios is not yet well established, but scientists agree that this topic is worthy of continued study because of the grave implications for coastal areas in the United States and around the world.

1.4. Impacts of Sea-Level Rise for the United States

1.4.1. Coastal Vulnerability for the United States

Coastal communities and habitats will be increasingly stressed by climate change impacts due to sea-level rise and storms (Field *et al.*, 2007). To varying degrees over decades, rising sea level will affect entire coastal systems from the ocean shoreline well landward. The physical and ecological changes that occur in the near future will impact people and coastal development. Impacts from sea-level rise include: land loss through submergence and erosion of lands in coastal areas; migration of coastal landforms and habitats; increased frequency and extent of storm-related flooding; wetland losses; and increased salinity in estuaries and coastal freshwater aquifers. Each of these effects can have impacts on both natural ecosystems and human developments. Often the impacts act together and the effects are cumulative. Other impacts of climate change, such as increasingly severe droughts and storm intensity—combined with continued rapid coastal development—could increase the magnitude and extent of sea-level rise impacts (Nicholls, *et al.*, 2007). To deal with these impacts, new practices in managing coasts and the combined impacts of mitigating changes to the physical system (*e.g.*, coastal erosion or migration, wetland losses) and impacts to human populations (*e.g.*, property losses, more frequent flood damage) should be considered.

Global sea-level rise, in combination with the factors above, is already having significant effects on many U.S. coastal areas. Flooding of low-lying regions by storm surges and spring tides is becoming more frequent. In certain areas, wetland losses are occurring, fringe forests are dying and being converted to marsh, farmland and lawns are being converted to marsh (*e.g.*, see Riggs and Ames, 2003; 2007), and some roads and urban centers in low elevation areas are more frequently flooded during spring high tides (Douglas, 2001). In addition, "ghost forests" of standing dead trees killed by salt water intrusion are becoming increasingly common in southern New Jersey, Maryland, Virginia, Louisiana, and North Carolina (Riggs and Ames, 2003). Relative sea level rise is causing salt water intrusion into estuaries and threatening freshwater resources in some parts of the mid-Atlantic region (Barlow, 2003).

Continued rapid coastal development exacerbates both the environmental and the human impact of rising sea level. Due to the increased human population in coastal areas, once sparsely developed coastal areas have been transformed into high-density year-round urban complexes (*e.g.*, Ocean City, Maryland; Virginia Beach, Virginia; Myrtle Beach, South Carolina). With accelerated rise in sea level and increased intensity of storms, the vulnerability of development at the coast and risks to people will increase dramatically unless new and innovative coastal zone management and planning approaches are employed.

1.4.2. Climate Change, Sea-Level Rise and Storms

Although storms occur episodically, they can have long-term impacts to the physical environment and human populations. Coupled with rise in sea level, the effects of storms could be more extensive in the future due to changes in storm character, such as intensity, frequency, and storm tracking. In addition to higher sea level, coastal storm surge from hurricanes could become higher and more intense rainfall could raise the potential for flooding from land runoff. Recent studies (*e.g.*, Emanuel, *et al.*, 2004, 2008; Emanuel, 2005; Komar and Allen, 2008; Elsner *et al.*, 2008) have concluded that there is evidence that hurricane intensity has increased during the past 30 years over the Atlantic Ocean; however, it is unknown whether these trends will continue. There is currently no scientific consensus on changes in the frequency of major storms. Emanuel *et al.* (2008) suggest that increased wind shear from global warming, which weakens hurricanes, may reduce the global frequency of hurricanes. This is in agreement with Gutowski *et al.* (2008).

Land-falling Atlantic coast hurricanes can produce storm surges of 5 m (16 ft) or more (Karl *et al.*, 2008). The power and frequency of Atlantic hurricanes has increased substantially in recent decades, though North American mainland land-falling hurricanes do not appear to have increased over the past century (Karl *et al.*, 2008). The IPCC (2007) and Karl *et al.* (2008) indicate that, based on computer models, it is likely that hurricanes will become more intense, with increases in tropical sea surface temperatures.

Although hurricane intensity is expected to increase on average, the effects on hurricane frequency in the Atlantic are still not certain and are the topic of considerable scientific study (Elsner *et al.*, 2008; Emanuel *et al.*, 2008; see also review in Karl *et al.*, 2008).

Extratropical cyclones can also produce significant storm surges. These storms have undergone a northward shift in track over the last 50 years (Karl *et al.*, 2008). This has reduced storm frequencies and intensities in the mid-latitudes and increased storm frequencies and intensities at high latitudes (Gutowski *et al.*, 2008). Karl *et al.* (2008) conclude that future intense extratropical cyclones will become more frequent with stronger winds and more extreme wave heights though the overall number of storms may decrease. So, while general storm projections are possible, specific projections for regional changes in extratropical cyclone activity, such as for the mid-Atlantic coast, are not yet available. Thus, while increased storm intensity is a serious risk in concert with sea-level rise, specific storm predictions are not so well established that planners can yet rely on them.

1.4.3. Shoreline Change and Coastal Erosion

The diverse landforms comprising more than 152,750 km (95,471 mi) of U.S. tidal coastline (<http://shoreline.noaa.gov/faqs.html>) reflect a dynamic interaction between:

(1) natural factors and physical processes that act on the coast (*e.g.*, storms, waves, currents, sand sources and sinks, relative sea level), (2) human activity (*e.g.*, dredging, dams, coastal engineering), and (3) the geological character of the coast and nearshore. Variations of these physical processes in both location and time, and the local geology along the coast, result in the majority of the U.S. coastlines undergoing overall long-term erosion at highly varying rates, as shown in Figure 1.7.

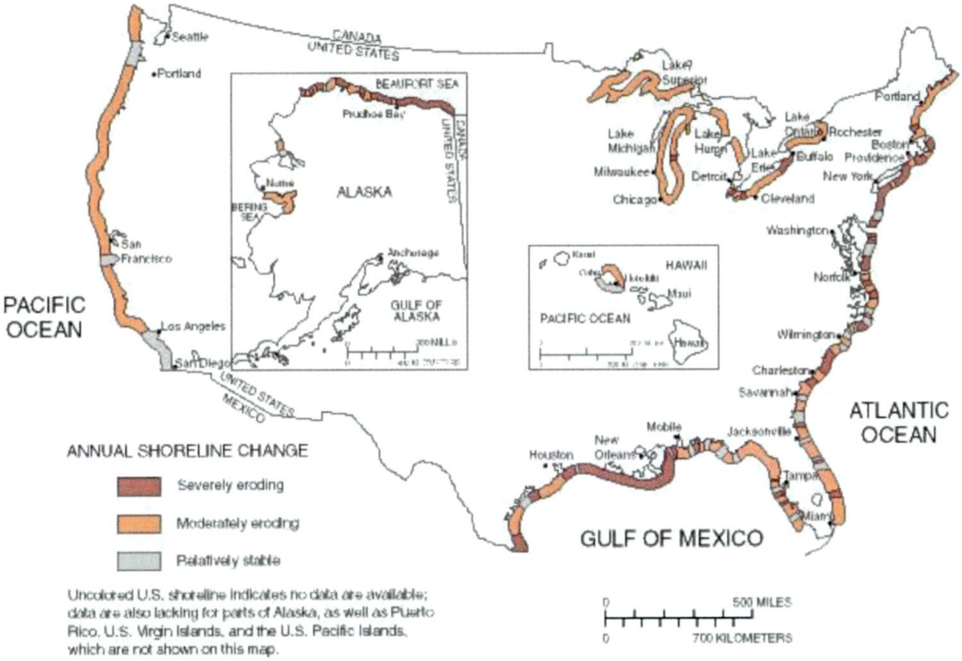

Figure 1.8. Shoreline change around the United States based on surveys over the past century. All 30 coastal states are experiencing overall erosion at highly variable rates due to natural processes (*e.g.*, storms, sea-level rise) and human activity (From USGS, 1985).

The complex interactions between these factors make it difficult to relate sea-level rise and shoreline change and to reach agreement among coastal scientists on approaches to predict how shorelines will change in response to sea-level rise. The difficulty in linking sea-level rise to coastal change stems from the fact that shoreline change is not driven solely by sea-level rise. Instead, coasts are in dynamic flux, responding to many driving forces, such as the underlying geological character, changes in tidal flow, and volume of sediment in the coastal system. For example, FitzGerald *et al.* (2008) discuss the dramatic effects that changes in tidal wetland area can have on entire coastal systems by altering tidal flow, which in turn affects the size and shape of tidal inlets, ebb and flood tide deltas, and barrier islands. Consequently, while there is strong scientific consensus that climate change is accelerating sea-level rise and affecting coastal regions, there are still considerable uncertainties predicting in any detail how the coast will respond to future sea-level rise in concert with other driving processes.

There is some scientific opinion that barrier islands, wetlands, and other parts of coastal systems might have tipping points or thresholds, such that when limits are exceeded the landforms become unstable and undergo large irreversible changes (NRC, 2002; Riggs and Ames, 2003; Nicholls *et al.*, 2007). These changes are thought to occur rapidly and are thus far unpredictable. It is possible that this is happening to barrier islands along the Louisiana coast that are subject to high rates of sea-level rise, frequent major storms over the past decade, and limited sediment supply (Sallenger *et al.*, 2007). Further deterioration of the barrier islands and wetlands may also occur in the near future along the North Carolina Outer

Banks coast as a result of increased sea-level rise and storm activity (Culver *et al.*, 2007, 2008; Riggs and Ames, 2003).

1.4.4. Managing the Coastal Zone as Sea Level Rises

A key issue for coastal zone management is how and where to adapt to the changes that will result from sea-level rise in ways that benefit or minimize impacts to both the natural environment and human populations. Shore protection policies have been developed in response to shoreline retreat problems that affect property or coastal wetland losses. While it is widely recognized that sea-level rise is an underlying cause of these changes, there are few existing policies that explicitly address or incorporate sea-level rise into decision making. Many property owners and government programs engage in coastal engineering activities designed to protect property and beaches such as beach nourishment or seawall or breakwater construction. Some of the current practices affect the natural behavior of coastal landforms and disrupt coastal ecosystems. In the short term, an acceleration of sea-level rise may simply increase the cost of current shore protection practices (Nordstrom, 2000). In the long term, policy makers might evaluate whether current approaches and justifications for coastal development and protection need to be modified to reflect the increasing vulnerability to accelerating rates of sea-level rise.

To facilitate these decisions, policy makers require credible scientific data and information. Predicting sea-level rise impacts such as shoreline changes or wetland losses with quantitative precision and certainty is often not possible. Related effects of climate change, including increased storms, precipitation, runoff, drought, and sediment supply add to the difficulty of providing accurate reliable information. Predicting future effects is challenging because the ability to accurately map and quantify the physical response of the coast to sea-level rise, in combination with the wide variety of other processes and human engineering activities along the shoreline, has not yet been well developed.

United States coastal regions are generally managed under the premise that sea level is stable, shorelines are static, and storms are regular and predictable. This Product examines how sea-level rise and changes in storm intensity and frequency due to climate change call for new considerations in managing areas to protect resources and reduce risk. This SAP 4.1 also examines possible strategies for coastal planning and management that will be effective as sea-level rise accelerates. For instance, broader recognition is needed that coastal sediments are a valuable resource, best conserved by implementing Best Coastal Sediment Management practices (see <http://www.wes.army.mil/rsm/>) on local, regional, and national levels in order to conserve sediment resources and maintain natural sediment transport processes.

This Product assesses the current scientific understanding of how sea-level rise can impact the tidal inundation of low-lying lands, ocean shoreline processes, and the vertical accretion of tidal wetlands. It also discusses the challenges that will be present in planning for future sea-level rise and adapting to these impacts. The SAP 4.1 is intended to provide information for coastal decision makers at all levels of government and society so they can better understand this topic and incorporate the effects of accelerating rates of sea-level rise into long-term management and planning.

REFERENCES

Alley, R. B., Marotzke, J., Nordhaus, W. D., Overpeck, J. T. D., Peteet, M., Pielke, R. A. Jr., Pierrehumbert, R. T., Rhines, P. B., Stocker, T. F., Talley, L. D. & Wallace, J. M. (2003). Abrupt climate change. *Science, 299(5615)*, 2005-2010.

Barlow, P.M., (2003). *Ground Water in Freshwater-Saltwater Environments of the Atlantic Coast.* US Geological Survey circular 1262. U.S. Geological Survey, Reston, VA, 113 pp.

Bindoff, N. L., Willebrand, J. Artale, V., Cazenave, A., Gregory, J., Gulev, S., Hanawa, K., Le Quéré, Levitus, S., Nojiri, Y., Shum, C. K., Talley, L. D. & Unnikrishnan, A. (2007). Observations: oceanic climate change and sea level. In: *Climate Change 2007: The Physical Science Basis.* Contribution of Working Group I to the Fourth Assessment Report of the Intergovernmental Panel on Climate [Solomon, S., D. Qin, M. Manning, Z. Chen, M. Marquis, K.B. Avery, M. Tignor, and H.L. Miller (eds.)]. Cambridge University Press, Cambridge, UK and New York, pp. 385-432.

Broecker, W.S. &Kunzig, R. (2008). *Fixing Climate: What Past Climate Changes Reveal about the Current Threat- and How to Counter It.* Hill and Wang, New York, 253 pp.

Cazenave, A. & Nerem, R. S. (2004). Present-day sea level change: observations and causes. *Reviews of Geophysics, 42(3)*, RG3001, doi:10.1029/2003RG000139.

CCSP (Climate Change Science Program), (2006). *Prospectus for Synthesis and Assessment Product 4.1: Coastal Elevations and Sensitivity to Sea-Level Rise.* U.S. Climate Change Science Program, Washington DC, 19 pp.

CENR (Committee on Environment and Natural Resources), 2008: *Scientific Assessment of the Effects of Global Change on the United States.* National Science and Technology Council, Committee on Environment and Natural Resources, Washington, DC, 261 pp.

Chen, J. L., Wilson, C. R. & Tapley, B. D. (2006). Satellite gravity measurements confirm accelerated melting of Greenland ice sheet. *Science, 313(5795),* 1958-1960.

Church, J. A. & White, N. J. (2006). A 20th century acceleration in global sea-level rise. *Geophysical Research Letters, 33(1)*, L01602, doi:10.1029/2005GL024826.

Cronin, T. M., Vogt, P. R., Willard, D.A., Thunell, R., Halka, J., Berke, M. & Pohlman, J. (2007). Rapid sea level rise and ice sheet response to 8,200-year climate event, *Geophysical Research Letters, 34*, L20603, doi:10.1029/2007GL031318.

Crossett, K., Culliton, T. J., Wiley, P. C. & Goodspeed, T. R. (2004). *Population Trends along the Coastal United States, 1980–2008.* NOAA National Ocean Service Special Projects Office, [Silver Spring, MD], 47 pp.

Crowell, M., Edelman, S., Coulton, K., & McAfee, S. (2007). How many people live in coastal areas? *Journal of Coastal Research, 23(5)*, iii-vi.

Culver, S. J., Grand Pre, C. A., Mallinson, D. J., Riggs, S. R., Corbett, D. R., Foley, J. Hale, M., Metger, L., Ricardo, J., Rosenberger, J. Smith, C. G., Smith, C.W., Synder, S. W., Twamley, D., Farrell, K. & Horton, B. (2007). Late Holocene barrier island collapse: Outer Banks, North Carolina, USA. *The Sedimentary Record, 5(4)*, 4-8.

Culver, S. J., Farrell, K. M., Mallinson, D. J., Horton, B. P., Willard, D. A., Thieler, E. R., Riggs, S. R. Snyder, S. W., Wehmiller, J. F., Bernhardt, C. E. & Hillier, C. (2008). Micropaleontologic record of late Pilocene and Quaternary paleoenvironments in the northern Albemarle Embayment, North Carolina, U.S.A. *Paleogeography, Paleoclimatology, Paleoecology, 264(1-2)*, 54-77.

Day, J. W., Jr., Biesch, D. F., Clairain, E. J., Kemp, G. P., Laska, S. B., Mitsch, W. J. Orth, K., Mashriqui, H., Reed, D. J., Shabman, L., Simenstad, C. A., Streever, B. J., Twilley, R. R., Watson, C. C., Wells, J. T. & Whigham, D. F. (2007a). Restoration of the Mississippi Delta: lessons from hurricanes Katrina and Rita. *Science, 315(5819)*, 1679-1684.

Day, J. W., Gunn, J. D., Folan, J., A Yáñez-Arancibia, & Horton, B. P. (2007b). Emergence of complex societies after sea level stabilized. *EOS Transactions of the American Geophysical Union, 88(15)*, 169-170.

Douglas, B. C., (2001). Sea level change in the era of the recording tide gauges. In: *Sea Level Rise: History and Consequences* [Douglas, B.C., M.S. Kearney, and S.P. Leatherman (eds.)]. International geophysics series v. 75. Academic Press, San Diego, pp. 37-64.

Elsner, J. B., Kossin, J. P. & Jagger, T. H. (2008). The increasing intensity of the strongest tropical cyclones. *Nature, 455(7209)*, 92-95.

Emanuel, K. A., (2005). Increasing destructiveness of tropical cyclones over the past 30 years. *Nature, 436(7051)*, 686-688.

Emanuel, K., (2008). The hurricane--climate connection. *Bulletin of the American Meteorological Society, 89(5)*, ES10-ES20.

Emanuel, K., DesAutels, C., Holloway, C. & Korty, R. (2004). Environmental control of tropical cyclone intensity. *Journal of the Atmospheric Sciences, 61(7)*, 843-858.

Emanuel, K., Sundararajan, R. & Williams, J. (2008). Hurricanes and global warming: results from downscaling IPCC AR4 simulations. *Bulletin of the American Meteorological Society, 89(3)*, 347-367.

Emery, K. O. & Aubrey, D. G. (1991). *Sea Levels, Land Levels, and Tide Gauges*. Springer-Verlag, New York, 237 pp.

Fairbanks, R. G., (1989). A 17,000-year glacio-eustatic sea level record--influence of glacial melting rates on the Younger Dryas event and deep-sea circulation. *Nature, 342(6250)*, 637-642.

Fettweis, X., J. P. van Ypersele, Gallee, H., Lefebre, F. &Lefebvre, W. (2007). The 9179-2005 Greenland ice sheet melt extent from passive microwave data using and improved version of the melt retrieval XPGR algorithm. *Geophysical Research Letters*, 34, L05502, doi:10.1029/2006GL028787.

Field, C. B., Mortsch, L. D., Brklacich, M., Forbes, D. L., Kovacs, P. Patz, J. A. Running, S.W. & Scott, M. J. (2007). North America. In: *Climate Change 2007: Impacts, Adaptation and Vulnerability*. Contribution of Working Group II to the Fourth Assessment Report of the Intergovernmental Panel on Climate Change [Parry, M.L., O.F. Canziani, J.P. Palutikof, P.J. van der Linden, and C.E. Hanson (eds.)]. Cambridge University Press, Cambridge, UK, and New York, pp. 617-652.

FitzGerald, D. M., Fenster, M. S., Argow, B. A. & Buynevich, I. V. (2008). Coastal impacts due to sea-level rise. *Annual Review of Earth and Planetary Sciences, 36*, 601-647.

Galloway, D., Jones, D. R. & Ingebritsen, S. E. (1999). *Land Subsidence in the United States*. USGS circular 1182. U.S. Geological Survey, Reston, VA, 177 pp.

Gehrels, W. R., Hayward, B. W. Newnham, R. M. & Southall, K. E. (2008). A 20th century acceleration in sea-level rise in New Zealand. *Geophysical Research Letters, 35*, L02717, doi:10.1029/2007GL032632.

Gornitz, V. & Lebedeff, S. (1987). Global sea-level changes during the past century. In: *Sea-Level Fluctuation and Coastal Evolution* [Nummedal, D., O.H. Pilkey, and J.D. Howard, (eds.)]. Special publication 41. Society of Economic Paleontologists and Mineralogists, Tulsa, OK, pp. 3-16.

Gutowski, W. J., Hegerl, G. C., Holland, G. J., Knutson, T. R., Mearns, L. O., R. J. Stouffer, Webster, P. J., Wehner, M. F. & Zwiers, F. W. (2008). Causes of observed changes in extremes and projections of future changes. In: *Weather and Climate Extremes in a Changing Climate: Regions of Focus: North America, Hawaii, Caribbean, and U.S. Pacific Islands*. [Karl, T.R., G.A. Meehl, C.D. Miller, S.J. Hassol, A.M. Waple, and W.L. Murray (eds.)]. Synthesis and Assessment Product 3.3. U.S. Climate Change Science Program, Washington, DC, pp. 81-116.

Hansen, J., Sato, M., Kharecha, P., Russell, G., Lea, D. W. & Siddall, M. (2007). Climate change and trace gases. *Philosophical Transactions of the Royal Society A, 365(1856)*, 1925-1954.

Holgate, S. J. & Woodworth, P. L., (2004). Evidence for enhanced coastal sea level rise during the 1990s. *Geophysical Research Letters*, 31, L07305, doi:10.1029/2004GL019626.

Huybrechts, P., (2002). Sea-level changes at the LGM from ice-dynamic reconstructions of the Greenland and Antarctic ice sheets during the glacial cycles. *Quaternary Science Reviews, 21(1-3)*, 203-231.

Imbrie, J. & Imbrie, K. P. (1986). *Ice Ages: Solving the Mystery*. Harvard University Press, Cambridge, MA, 224 pp.

IPCC (Intergovernmental Panel on Climate Change), (2001). *Climate Change 2001: The Scientific Basis.* Contribution of Working Group I to the Third Assessment Report of the Intergovernmental Panel on Climate Change [Houghton, J.T., Y. Ding, D.J. Griggs, M. Noguer, P. J. van der Linden, X. Dai, Maskell, K. and Johnson C. A. (eds.)]. Cambridge University Press, Cambridge, UK, and New York, 881 pp.

IPCC (Intergovernmental Panel on Climate Change), (2007). *Climate Change 2007: The Physical Science Basis.* Contribution of Working Group I to the Fourth Assessment Report of the Intergovernmental Panel on Climate Change [Solomon, S., D. Qin, M. Manning, Z. Chen, M. Marquis, K. B. Averyt, M. Tignor, and H. L. Miller (eds.)]. Cambridge University Press, Cambridge, UK, and New York, 996 pp.

Ishii, M., Kimoto, M. Sakamoto, K. & Iwasaki, S. I. (2006). Steric sea level changes estimated from historical ocean subsurface temperature and salinity analyses. *Journal of Oceanography, 62(2),* 155-170.

Jansen, E., Overpeck, J., Briffa, K. R., Duplessy, J. C., Joos, F., V. Masson-Delmotte, D. Olago, Otto-Bliesner, B., Peltier, W. R., Rahmstorf, S., Ramesh, R., Raynaud, D., D. Rind, Solomina, O. Villalba, R. & D. Zhang, (2007). Palaeoclimate. In: *Climate Change 2007: The Physical Science Basis.* Contribution of Working Group I to the Fourth Assessment Report of the Intergovernmental Panel on Climate Change [Solomon, S., D. Qin, M. Manning, Z. Chen, M. Marquis, K.B. Averyt, M. Tignor, and H.L. Miller (eds.)]. Cambridge University Press, Cambridge, UK, and New York, pp. 433-497.

Jevrejeva, S., Grinsted, A., Moore, J. C. & Holgate, S. (2006). Nonlinear trends and multiyear cycles in sea level records. *Journal of Geophysical Research, 111*, C09012, doi:10.1029/2005JC003229.

Jevrejeva, S., Moore, J. C. Grinsted, A. & Woodworth, P. L. (2008). Recent global sea level acceleration started over 200 years ago? *Geophysical Research Letters*, *35*, L08715, doi:10.1029/2008GL033611.

Karl, T.R., Meehl, G. A. Peterson, T. C., Kunkel, K. E., Gutowski, W. J., Jr. & D. R. Easterling, (2008). Executive summary. In: *Weather and Climate Extremes in a Changing Climate: Regions of Focus: North America, Hawaii, Caribbean, and U.S. Pacific Islands*. [Karl, T.R., G.A. Meehl, C.D. Miller, S.J. Hassol, A.M. Waple, and W.L. Murray (eds).]. Synthesis and Assessment Product 3.3. U.S. Climate Change Science Program, Washington, DC, pp. 1-9.

Kearney, M. S. & Stevenson, J. C. (1991). Island land loss and marsh vertical accretion rate evidence for historical sea-level changes in Chesapeake Bay. *Journal of Coastal Research*, *7(2)*, 403-415.

Komar, P.D. & Allan, J. C. (2008). Increasing hurricane-generated wave heights along the U.S. East coast and their climate controls. *Journal of Coastal Research*, *24(2)*, 479-488.

Lambeck, K., Esat, T. M. & E.-K. Potter, (2002). Links between climate and sea levels for the past three million years. *Nature*, *419 (6903)*, 199-206.

Lambeck, K., Anzidei, M., Antonioli, F., Benini, A . & Esposito, A. 2004: Sea level in Roman time in the central Mediterranean and implications for recent change. *Earth and Planetary Science Letters*, *224(3-4)*, 563-575.

Leuliette, E. W., Nerem, R. S. & Mitchum, G. T. (2004). Calibration of TOPEX/Poseidon and Jason altimeter data to construct a continuous record of mean sea level change. *Marine Geodesy*, *27(1–2)*, 79–94.

McGranahan, G., Balk, D. & Anderson, B. (2007). The rising tide: assessing the risks of climate change and human settlements in low elevation coastal zones. *Environment & Urbanization*, *19(1)*, 17-37.

Meehl, G. A., Stocker, T. F., Collins, W. D., Friedlingstein, P., Gaye, A. T., Gregory, J. M., Kitoh, A., Knutti, R., Murphy, J. M., Noda, A., Raper, S.C.B., Watterson, I.G., Weaver A.J., and Zhao, Z. C. (2007). Global climate projections. In: *Climate Change 2007: The Physical Science Basis.* Contribution of Working Group I to the Fourth Assessment Report of the Intergovernmental Panel on Climate Change [Solomon, S., D. Qin, M. Manning, Z. Chen, M. Marquis, K.B. Averyt, M. Tignor, and H.L. Miller (eds.)]. Cambridge University Press, Cambridge, UK, and New York, pp. 747-845.

Meier, M. F., Dyurgerov, M. B., Ursula, K. R., S. O' Neel, W. T. Pfeffer, Anderson, R. S., Anderson, S. P. & Glazovsky, A. F. (2007). Glaciers dominate Eustatic sea-level rise in the 21st century. *Science*, *317(5841),* 1064-1067.

Miller, K. G., Kominz, M. A., Browning, J. V. Wright, J. D., Mountain, G. S., Katz, M. E., Sugarman, P. J., Cramer, B. S. Christie-Blick, N. & Pekar, S. F. (2005). The Phanerozoic record of global sea-level change. *Science*, *310(5752)*, 1293-1298.

Morton, R. A., Miller, T. L. & L. J. Moore, (2004). *National Assessment of Shoreline Change: Part 1, Historical Shoreline Changes and Associated Coastal Land Loss along the U.S. Gulf of Mexico.* U.S. Geological Survey open file report 20041043. U.S. Geological Survey, St. Petersburg, FL, 44 pp. <http://pubs.usgs.gov/of/2004/1043>

Muhs, D. R., Wehmiller, J. F., Simmons, K. R. & York, L. L. (2004). Quaternary sealevel history of the United States. In: *The Quaternary Period of the United States* [Gillespie, A. R., Porter, S. C. & Atwater B. F. (eds.)]. Elsevier, Amsterdam, pp. 147-183.

Nicholls, R. J., Wong, P. P., Burkett, V. R., Codignotto, J. O., Hay, J. E., McLean, R. F., Ragoonaden, S. & C. D. Woodroffe, (2007). Coastal systems and low-lying areas. *Climate Change 2007: Impacts, Adaptation and Vulnerability*. Contribution of Working Group II to the Fourth Assessment Report of the Intergovernmental Panel on Climate Change [Parry, M.L., O.F. Canziani, J.P. Palutikof, P.J. van der Linden, and C.E. Hanson (eds.)]. Cambridge University Press, Cambridge, UK, and New York, pp. 315-356.

Nordstrom, K. F., (2000). *Beaches and Dunes of Developed Coasts*. Cambridge University Press, New York, 338 pp.

NRC (National Research Council) (2002). *Abrupt Climate Change: Inevitable Surprises*. National Academy Press, Washington, DC, 230 pp.

Overpeck, J. T., Otto-Bliesner, B. L., Miller, G. H., Muhs, D. R., Alley, R. B. & Keihl, J. T. (2006). Paleo-climatic evidence for the future ice-sheet instability and rapid sea level rise. *Science, 311(5768)*, 1747-1750.

Pearce, F., (2007). *With Speed and Violence: Why Scientists Fear Tipping Points in Climate Change*. Beacon Press, Boston, 278 pp.

Peltier, W. R., (2001). Global glacial isostatic adjustment and modern instrumental records of relative sea level history. In: *Sea Level Rise: History and Consequences* [Douglas, B.C., M.S. Kearney, and S.P. Leatherman (eds.)]. International geophysics series v. 75. Academic Press, San Diego, pp. 65-95.

Rahmstorf, S., (2007). A semi-empirical approach to projecting future sea-level rise. *Science, 315(5810)*, 368-370.

Rahmstorf, S., Cazenave, A., Church, J. A., Hansen, J. E., Keeling, R. F., Parker, D. E. & Somerville, R. C. J. (2007). Recent climate observations compared to projections. *Science, 316(5825)*, 709.

Riggs, S. R. & Ames, D. V. (2003). *Drowning of the North Carolina Coast: Sea-Level Rise and Estuarine Dynamics*. Publication number UNC-SG-03-04. North Carolina Sea Grant, Raleigh, NC, 152 pp.

Riggs, S. R. & Ames, D. V. (2007). *Effect of Storms on Barrier Island Dynamics, Core Banks, Cape Lookout National Seashore, North Carolina, 1960-2001*. U.S. Geological Survey scientific investigations report 2006-5309. U.S. Geological Survey, Reston, VA, 78 pp.

Rohling, E. J., Grant, K., Hemleben, Ch., Siddall, M., Hoogakker, B.A.A., Bolshaw, M. and Kucera, M. (2008). High rates of sea-level rise during the last interglacial period. *Nature Geoscience, 1(1)*, 38-42.

Rosenzweig, C., Karoly, D., Vicarelli, M., Neofotis, P., Wu, Q., Casassa, G., Menzel, A. T.L. Root, Estrella, N., Seguin, B., Tryjanowski, P., Liu, C., Rawlins, S. & Imeson, A. (2008). Attributing physical and biological impacts to anthropogenic climate change. *Naturc, 453(7193)*, 353-358.

Sallenger, A. S., C. W. Wright, and Lillycrop, J. (2007). Coastal-change impacts during Hurricane Katrina: an overview. In: *Coastal Sediments '07*, [Kraus, N.C. & J.D. Rosati (eds.)]. America Society of Civil Engineers, Reston, VA, pp. 888-896.

Shepherd, A. & D. Wingham, 2007: Recent sea-level contributions of the Antarctic and Greenland ice sheets. *Science, 315(5818)*, 1529-1532.

Stanley, D. J. & Warne, A. G. (1993). Nile delta: recent geological evolution and human impact. *Science, 260(5108)*, 628-634.

Steffen, K., Clark, P. U., Cogley, J. G., Holland, D., Marshall, S., Rignot, E. & Thomas, R. (2008). Rapid changes in glaciers and ice sheets and their impacts on sea level. In: *Abrupt Climate Change.* A Report by the U.S. Climate Science Program and the Subcommittee on Global Change Research. U.S. Geological Survey, Reston, VA, pp 60-142.

UN (United Nations), (2005). *World Population Prospects: The 2004 Revision.* Volume III: Analytical Report. United Nations Publication, Sales No. E.05.XIII.7, New York, 194 pp.

USGS (U.S. Geological Survey), (1985). *National Atlas of the United States: Coastal Erosion and Accretion.* U.S. Geological Survey, Reston, VA, 1 map.

Zalasiewicz, J., Williams, M., Smith, A., Barry, T. L., Coe, A. L., Brown, P. R., Brenchley, P. Cantrill, D., Gale, A., Gibbard, P., Gregory, F. J., Hounslow, M. W., A. C. Kerr, Pearson, P. Knox, R., Powell, J., Waters, C., Marshall, J., Oates, M., Rawson, P. & Stone, P. (2008). Are we now living in the Anthropocene? *GSA Today, 18(2),* 4-7.

Zervas, C., (2001). *Sea Level Variations of the United States 1854-1999.* NOAA technical report NOS CO-OPS 36. NOAA National Ocean Service, Silver Spring, MD, 186 pp. <http://tidesandcurrents.noaa.gov/publications/techrpt36doc.pdf>

In: Coastal Sensitivity to Sea Level Rise-Focusing on the Mid-… ISBN: 978-1-60741-440-7
Editor: Melvin C. Urajner

Chapter 2

COASTAL ELEVATIONS

Lead Author: Dean B. Gesch[1]
Contributing Authors: Benjamin T. Gutierrez[1] and Stephen K. Gill[2]
[1]USGS
[2]NOAA

Key Findings

- Coastal changes are driven by complex and interrelated processes. Inundation will be the primary response to sea-level rise in some coastal locations; yet there has been little recognition in previous studies that inundation is just one response out of a number of possible responses to sea-level rise. A challenge remains to quantify the various effects of sea-level rise and to identify the areas and settings along the coast where inundation will be the dominant coastal change process in response to rising seas.
- Sheltered, low-energy coastal areas, where sediment influx is minimal and wetlands are absent or are unable to build vertically in response to rising water levels, may be submerged. In these cases, the extent of inundation is controlled largely by the slope of the land, with a greater degree of inundation occurring in areas with more gentle gradients. In areas that are vulnerable to a simple inundation response to rising seas, elevation is a critical factor in assessing potential impacts.
- Accurate delineations of potential inundation zones are critical for meeting the challenge of fully determining the potential socioeconomic and environmental impacts of predicted sea-level rise.
- Coastal elevation data have been widely used to quantify the potential effects of predicted sea-level rise, especially the area of land that could be inundated and the affected population. Because sea-level rise impact assessments often rely on elevation data, it is critical to understand the inherent accuracy of the underlying data and its effects on the uncertainty of any resulting vulnerability maps and statistical summaries.

- The accuracy with which coastal elevations have been mapped directly affects the reliability and usefulness of sea-level rise impact assessments. Although previous studies have raised awareness of the problem of mapping and quantifying sea-level rise impacts, the usefulness and applicability of many results are hindered by the coarse resolution of available input data. In addition, the uncertainty of elevation data is often neglected.
- Existing studies of sea-level rise vulnerability based on currently available elevation data do not provide the degree of confidence that is optimal for local decision making.
- There are important technical considerations that need to be incorporated to improve future sea-level rise impact assessments, especially those with a goal of producing vulnerability maps and statistical summaries that rely on the analysis of elevation data. The primary aspect of these improvements focuses on using high-resolution, high-accuracy elevation data, and consideration and application of elevation uncertainty information in development of vulnerability maps and area statistics.
- Studies that use elevation data as an input for vulnerability maps and/or statistics need to have a clear statement of the absolute vertical accuracy. There are existing national standards for quantifying and reporting elevation data accuracy.
- Currently best available elevation data for the entire mid-Atlantic region do not support an assessment using a sea-level rise increment of 1 meter or less, using national geospatial standards for accuracy assessment and reporting. This is particularly important because the 1-meter scenario is slightly above the range of current sea-level rise estimates for the remainder of this century and slightly above the highest scenario used in this report.
- High-quality lidar elevation data, such as that which could be obtained from a national lidar data collection program, would be necessary for the entire coastal zone to complete a comprehensive assessment of sea-level rise vulnerability in the mid-Atlantic region. The availability of such elevation data will narrow the uncertainty range of elevation datasets, thus improving the ability to conduct detailed assessments that can be used in local decision making.

2.1. Introduction

Sea-level rise is a coastal hazard that can exacerbate the problems posed by waves, storm surges, shoreline erosion, wetland loss, and saltwater intrusion (NRC, 2004). The ability to identify low-lying lands is one of the key elements needed to assess the vulnerability of coastal regions to these impacts. For nearly three decades, a number of large area sea-level rise vulnerability assessments have focused mainly on identifying land located below elevations that would be affected by a given sea-level rise scenario (Schneider and Chen, 1980; U.S. EPA, 1989; Najjar *et al.*, 2000; Titus and Richman, 2001; Ericson *et al.*, 2006; Rowley *et al.*, 2007). These analyses require use of elevation data from topographic maps or digital elevation models (DEMs) to identify low-lying land in coastal regions. Recent reports have stressed that sea-level rise impact assessments need to continue to include maps of these areas subject to inundation based on measurements of coastal elevations (Coastal States

Organization, 2007; Seiden, 2008). Accurate mapping of the zones of potential inundation is critical for meeting the challenge of determining the potential socioeconomic and environmental impacts of predicted sea-level rise (FitzGerald *et al.*, 2008).

Identification of the socioeconomic impacts of projected sea-level rise on vulnerable lands and populations is an important initial step for the nation in meeting the challenge of reducing the effects of natural disasters in the coastal zone (Subcommittee on Disaster Reduction, 2008). A number of state coastal programs are using sea-level rise inundation models (including linked storm surge/sea-level rise models) to provide a basis for coastal vulnerability and socioeconomic analyses (Coastal States Organization, 2007). State coastal managers are concerned that these research efforts and those of the federal government should be well coordinated, complementary, and not redundant. Despite the common usage of elevation datasets to investigate sea-level rise vulnerability, there are limitations to elevation-based analyses. These limitations are related to the relevance of this approach in a variety of settings and to the data sources and methodologies used to conduct these analyses. Thus, an important objective of this Chapter is to review the available data and techniques, as well as the suitability of elevation-based analyses for informing sea-level rise assessments, to provide guidance for both scientists and coastal managers.

While elevation-based analyses are a critical component of sea-level rise assessments, this approach only addresses a portion of the vulnerability in coastal regions. Coastal changes are driven by complex and interrelated processes such as storms, biological processes, sea-level rise, and sediment transport, which operate over a range of time scales (Carter and Woodroffe, 1994; Brinson *et al.*, 1995; Eisma, 1995; Pilkey and Cooper, 2004; FitzGerald *et al.*, 2008). The response of a coastal region to sea-level rise can be characterized by one or more of the processes in the following broad categories (Leatherman, 2001; Valiela, 2006; FitzGerald *et al.*, 2008):

- land loss by inundation of low-lying lands;
- land loss due to erosion (removal of material from beaches, dunes, and cliffs);
- barrier island migration, breaching, and segmentation;
- wetland accretion and migration;
- wetland drowning (deterioration and conversion to open water);
- expansion of estuaries;
- salt water intrusion (into freshwater aquifers and surface waters); and
- increased frequency of storm flooding (especially of uplands and developed coastal lands).

Because large portions of the population (both in the United States and worldwide) are located in coastal regions, each of these impacts has consequences for the natural environment as well as human populations. Using elevation datasets to identify and quantify low-lying lands is only one of many aspects that need to be considered in these assessments. Nonetheless, analyses based on using elevation data to identify low-lying lands provide an important foundation for sea-level rise impact studies.

There is a large body of literature on coastal processes and their role in both shoreline and environmental change in coastal regions (Johnson, 1919; Curray, 1964; Komar, 1983; Swift *et al.*, 1985; Leatherman, 1990; Carter and Woodroffe, 1994; Brinson, 1995; Eisma, 1995;

Wright, 1995; Komar, 1998; Dean and Dalrymple, 2002; FitzGerald *et al.*, 2008). However, there is generally little discussion of the suitability of using elevation data to identify the vulnerability of coastal regions to sea-level rise. While it is straightforward to reason that low-lying lands occurring below a future sea-level rise scenario are vulnerable, it is often generally assumed that these lands will be inundated. Instead, inundation is likely only one part of the response out of a number of possible sea-level rise impacts. Despite this, some assessments have opted for inundation-based assessments due to the lack of any clear alternatives and the difficulty in accounting for complex processes such as sedimentation (Najjar *et al.*, 2000). It is plausible that extreme rates of sea-level rise (*e.g.*, 1 meter or more in a single year) could result in widespread simple coastal inundation. However, in the more common and likely case of much lower sea-level rise rates, the physical processes are more complex and rising seas do not simply flood the coastal landscape below a given elevation contour (Pilkey and Thieler, 1992). Instead, waves and currents will modify the landscape as sea level rises (Bird, 1995; Wells, 1995). Still, inundation is an important component of coastal change (Leatherman, 2001), especially in very low gradient regions such as North Carolina.

However, due to the complexity of the interrelated processes of erosion and sediment redistribution, it is difficult to distinguish and quantify the individual contributions from inundation and erosion (Pilkey and Cooper, 2004).

Inundation will be the primary response to sea-level rise only in some coastal locations. In many other coastal settings, long-term erosion of beaches and cliffs or wetland deterioration will alter the coastal landscape leading to land loss. To distinguish the term inundation from other processes, especially erosion, Leatherman (2001) offered the following important distinction:

- *erosion* involves the physical removal of sedimentary material
- *inundation* involves the permanent submergence of land.

Another term that can confuse the discussion of sea-level rise and submergence is the term *flooding* (Wells, 1995; Najjar *et al.*, 2000), which in some cases has been used interchangeably with *inundation*. *Flooding* often connotes temporary, irregular high-water conditions. The term *inundation* is used in this Chapter (but not throughout the entire Product) to refer to the permanent submergence of land by rising seas.

It is unclear whether simply modeling the inundation of the land surface provides a useful approximation of potential land areas at risk from sea-level rise. In many settings, the presence of beaches, barrier islands, or wetlands indicates that sedimentary processes (erosion, transport, or accumulation of material) are active in both the formation of and/or retreat of the coastal landscape. Sheltered, low-energy coastal areas, where sediment influx is minimal and wetlands are absent or are unable to build vertically in response to rising water levels, may be submerged. In these cases, the extent of inundation is controlled by the slope of the land, with a greater degree of inundation occurring in the areas with more gentle gradients (Leatherman, 2001). In addition, inundation is a likely response in heavily developed regions with hardened shores. The construction of extensive seawalls, bulkheads, and revetments to armor the shores of developed coasts and waterways have formed nearly immovable shorelines that may become submerged. However, the challenge remains to quantify the various effects of sea-level rise and to identify the areas and settings along the coast where inundation will be the dominant coastal change process from sea-level rise.

Despite several decades of research, previous studies do not provide the full answers about sea-level rise impacts for the mid-Atlantic region with the degree of confidence that is optimal for local decision making. Although these studies have illuminated the challenges of mapping and quantifying sea-level rise impacts, the usefulness and applicability of many results are hindered by the quality of the available input data. In addition, many of these studies have not adequately reported the uncertainty in the underlying elevation data and how that uncertainty affects the derived vulnerability maps and statistics. The accuracy with which coastal elevations have been mapped directly affects the reliability and usefulness of sea-level rise impact assessments. Elevation datasets often incorporate a range of data sources, and some studies have had to rely on elevation datasets that are poorly suited for detailed inundation mapping in coastal regions, many of which are gently sloping landscapes (Ericson *et al.*, 2006; Rowley *et al.*, 2007; McGranahan *et al.*, 2007). In addition to the limited spatial detail, these datasets have elevation values quantized only to whole meter intervals, and their overall vertical accuracy is poor when compared to the intervals of predicted sea-level rise over the next century. These limitations can undermine attempts to achieve high-quality assessments of land areas below a given sea-level rise scenario and, consequently, all subsequent analyses that rely on this foundation.

Due to numerous studies that used elevation data, but have lacked general recognition of data and methodology constraints, this Chapter provides a review of data sources and methodologies that have been used to conduct sea-level rise vulnerability assessments. New high-resolution, high-accuracy elevation data, especially lidar (light detection and ranging) data, are becoming more readily available and are being integrated into national datasets (Gesch, 2007) as well as being used in sea-level rise applications (Coastal States Organization, 2007). Research is also progressing on how to take advantage of the increased spatial resolution and vertical accuracy of the new data (Poulter and Halpin, 2007; Gesch, 2009). Still, there is a critical need to thoroughly evaluate the elevation data, determine how to appropriately utilize the data to deliver well-founded results, and accurately communicate the associated uncertainty.

The widespread use of vulnerability assessments, and the attention they receive, is likely an indication of the broad public interest in sea-level rise issues. Because of this extensive exposure, it is important for the coastal science community to be fully engaged in the technical development of elevation-based analyses. Many recent reports have been motivated and pursued from an economic or public policy context rather than a geosciences perspective. It is important for scientists to communicate and collaborate with coastal managers to actively identify and explain the applications and limitations of sea-level rise impact assessments. Arguably, sea-level rise is one of the most visible and understandable consequences of climate change for the general public, and the coastal science community needs to ensure that appropriate methodologies are developed to meet the needs for reliable information. This Chapter reviews the various data sources that are available to support inundation vulnerability assessments. In addition, it outlines what is needed to conduct and appropriately report results from elevation-based sea-level rise vulnerability analyses and discusses the context in which these analyses need to be applied.

2.2. Elevation Data

Measurement and representation of coastal topography in the form of elevation data provide critical information for research on sea-level rise impacts. Elevation data in its various forms have been used extensively for sea-level rise studies. This section reviews elevation data sources in order to provide a technical basis for understanding the limitations of past sea-level rise impact analyses that have relied on elevation data. While use of coastal elevation data is relatively straightforward, there are technical aspects that are important considerations for conducting valid quantitative analyses.

2.2.1. Topographic Maps, Digital Elevation Models, and Accuracy Standards

Topographic maps with elevation contours are perhaps the most recognized form of elevation information. The U.S. Geological Survey (USGS) has been a primary source of topographic maps for well over a century. The base topographic map series for the United States (except Alaska) is published at a scale of 1:24,000, and the elevation information on the maps is available in digital form as digital elevation models. The USGS began production of DEMs matching the 1:24,000-scale quadrangle maps in the mid-1970s using a variety of image-based (photogrammetric) and cartographic techniques (Osborn *et al.*, 2001). Coverage of the conterminous United States with 30-meter (m) (98-foot [ft]) horizontal resolution DEMs was completed in 1999, with most of the individual elevation models being derived from the elevation contours and spot heights on the corresponding topographic maps. Most of these maps have a 5-ft, 10-ft, 20-ft, or 40-ft contour interval, with 5-ft being the contour interval used in many low relief areas along the coast. About the time 30-m DEM coverage was completed, the USGS began development of a new seamless raster (gridded) elevation database known as the National Elevation Dataset (NED) (Gesch *et al.*, 2002). As the primary elevation data product produced and distributed by the USGS, the NED includes many USGS DEMs as well as other sources of elevation data. The diverse source datasets are processed to a specification with a consistent resolution, coordinate system, elevation units, and horizontal and vertical datums to provide the user with an elevation product that represents the best publicly available data (Gesch, 2007). DEMs are also produced and distributed in various formats by many other organizations, and they are used extensively for mapping, engineering, and earth science applications (Maune, 2007; Maune *et al.*, 2007a).

Because sea-level rise impact assessments often rely on elevation data, it is important to understand the inherent accuracy of the underlying data and its effects on the uncertainty of any resulting maps and statistical summaries from the assessments. For proper quantitative use of elevation data, it is important to identify and understand the vertical accuracy of the data. Vertical accuracy is an expression of the overall quality of the elevations contained in the dataset in comparison to the true ground elevations at corresponding locations. Accuracy standards and guidelines exist, in general for geospatial data, and specifically for elevation data. For topographic maps, the National Map Accuracy Standards (NMAS) issued in 1947 are the most commonly used; they state that "vertical accuracy, as applied to contour maps on all publication scales, shall be such that not more than 10 percent of the elevations tested shall be in error by more than one-half the contour interval" (USGS, 1999). An alternative way to state the NMAS vertical accuracy standard is that an elevation obtained from the topographic map will be accurate to within one-half of the contour interval 90 percent of the time. This has also been referred to as "linear error at 90 percent confidence" (LE90) (Greenwalt and Shultz, 1962). For example, on a topographic map with a 10-ft contour interval that meets NMAS, 90 percent of the elevations will be accurate to within 5 ft, or stated alternatively, any elevation taken from the map will be within 5 ft of the actual elevation with a 90-percent confidence level. Even though the NMAS was developed for printed topographic maps and it predates the existence of DEMs, it is important to understand its application because many DEMs are derived from topographic maps.

As the production and use of digital geospatial data became commonplace in the 1990s, the Federal Geographic Data Committee (FGDC) developed and published geospatial positioning accuracy standards in support of the National Spatial Data Infrastructure (Maune *et al.*, 2007b). The FGDC standard for testing and reporting the vertical accuracy of elevation data, termed the National Standard for Spatial Data Accuracy (NSSDA), states that the "reporting standard in the vertical component is a linear uncertainty value, such that the true or theoretical location of the point falls within +/- of that linear uncertainty value 95 percent of the time" (Federal Geographic Data Committee, 1998). In practice, the vertical accuracy of DEMs is often reported as the root mean square error (RMSE). The NSSDA provides the method for translating a reported RMSE to a linear error at the 95-percent confidence level. Maune *et al.* (2007b) provide a useful comparison of NMAS and NSSDA vertical accuracy measures for common contour intervals (Table 2.1) and methods to convert between the reporting standards. The NSSDA, and in some cases even the older NMAS, provides a useful approach for testing and reporting the important vertical accuracy information for elevation data used in sea-level rise assessments.

2.2.2. Lidar Elevation Data

Currently, the highest resolution elevation datasets are those derived from lidar surveys. Collected and post-processed under industry-standard best practices, lidar elevation data routinely achieve vertical accuracies on the order of 15 centimeters (cm) (RMSE). Such accuracies are well suited for analyses of impacts of sea-level rise in sub-meter increments (Leatherman, 2001). Using the conversion methods between accuracy standards documented by Maune *et al.* (2007b), it can be shown that lidar elevation data with an accuracy of equal to or better than 18.5 cm (RMSE) is equivalent to a 2-ft contour interval map meeting NMAS.

Lidar is a relatively recent remote sensing technology that has advanced significantly over the last 10 years to the point where it is now a standard survey tool used by government agencies and the mapping industry to collect very detailed, high-accuracy elevation measurements, both on land and in shallow water coastal areas. The discussion of lidar in this Chapter is limited to topographic lidar used to map land areas. Lidar measurements are acquired using laser technology to precisely measure distances, most often from an aircraft, that are then converted to elevation data and integrated with Global Positioning System (GPS) information (Fowler *et al.*, 2007). Because of their high vertical accuracy and spatial resolution, elevation data derived from lidar surveys are especially useful for applications in low relief coastal environments. The technical advantages of lidar in dynamic coastal settings, including the ability to perform repeat high-precision surveys, have facilitated successful use of the data in studies of coastal changes due to storm impacts (Brock *et al.*, 2002; Sallenger *et al.*, 2003; Stockdon *et al.*, 2007). Numerous organizations, including many state programs, have recognized the advantages of lidar for use in mapping the coastal zone. As an example, the Atlantic states of Maine, Connecticut, New Jersey, Delaware, Maryland, North Carolina, and Florida have invested in lidar surveys for use in their coastal programs (Coastal States Organization, 2007; Rubinoff, *et al.*, 2008).

2.2.3. Tides, Sea Level, and Reference Datums

Sea-level rise assessments typically focus on understanding potential changes in sea level, but elevation datasets are often referenced to a "vertical datum," or reference point, that may differ from sea level at any specific location. In any work dealing with coastal elevations, water depths, or water levels, the reference to which measurements are made must be carefully addressed and thoroughly documented. All elevations, water depths, and sea-level data are referenced to a defined vertical datum, but different datums are used depending on the data types and the original purpose of the measurements. A detailed treatment of the theory behind the development of vertical reference systems is beyond the scope of this Product. However, a basic understanding of vertical datums is necessary for fully appreciating the important issues in using coastal elevation data to assess sea-level rise vulnerability. Zilkoski (2007), Maune *et al.* (2007a), and NOAA (2001) provide detailed explanations of vertical datums and tides, and the brief introduction here is based largely on those sources.

Table 2.1. Comparison of National Map Accuracy Standards and National Standard for Spatial Data Accuracy vertical accuracy values with the equivalent common contour intervals (Maune *et al.*, 2007b).

NMAS Equivalent contour interval	NMAS 90-percent confidence level (LE90)	NSSDA RMSE	NSSDA 95-percent confidence level
1 ft	0.5 ft	0.30 ft (9.25 cm)	0.60 ft (18.2 cm)
2 ft	1 ft	0.61 ft (18.5 cm)	1.19 ft (36.3 cm)
5 ft	2.5 ft	1.52 ft (46.3 cm)	2.98 ft (90.8 cm)
10 ft	5 ft	3.04 ft (92.7 cm)	5.96 ft (1.816 m)
20 ft	10 ft	6.08 ft (1.853 m)	11.92 ft (3.632 m)

Land elevations are most often referenced to an orthometric (sea-level referenced) datum, which is based on a network of surveyed (or "leveled") vertical control benchmarks. These benchmarks are related to local mean sea level at specific tide stations along the coast. The elevations on many topographic maps, and thus DEMs derived from those maps, are referenced to the National Geodetic Vertical Datum of 1929 (NGVD 29), which uses mean sea level at 26 tide gauge sites (21 in the United States and 5 in Canada). Advances in surveying techniques and the advent of computers for performing complex calculations allowed the development of a new vertical datum, the North American Vertical Datum of 1988 (NAVD 88). Development of NAVD 88 provided an improved datum that allowed for the correction of errors that had been introduced into the national vertical control network because of crustal motion and ground subsidence. In contrast to NGVD 29, NAVD 88 is tied to mean sea level at only one tide station, located at Father Point/Rimouski, Quebec, Canada. Orthometric datums such as NGVD 29 and NAVD 88 are referenced to tide gauges, so they are sometimes informally referred to as "sea level" datums because they are inherently tied to some form of mean sea level. NAVD 88 is the official vertical datum of the United States, as stated in the Federal Register in 1993, and as such, it should serve as the reference for all products using land elevation data.

Water depths (bathymetry data) are usually referenced vertically to a tidal datum, which is defined by a specific phase of the tides. Unlike orthometric datums such as NGVD 29 and NAVD 88, which have national or international coverage, tidally referenced datums are local datums because they are relative to nearby tide stations. Determination of tidal datums in the United States is based on observations of water levels over a 19-year period, or tidal epoch. The current official tidal epoch in use is the 1983-2001 National Tidal Datum Epoch (NTDE). Averaging over this period is necessary to remove random and periodic variations caused by seasonal differences and the nearly 19-year cycle of the lunar orbit. NTDEs are updated approximately every 25 years to account for relative sea-level change (NOAA, 2001). The following are the most commonly used tidal datums:

- Mean higher high water (MHHW): the average of the higher high water levels observed over a 19-year tidal epoch (only the higher water level of the pair of high waters in a tidal day is used);
- Mean high water (MHW): the average of the high water levels observed over a 19-year tidal epoch;
- Local mean sea level (LMSL): the average of hourly water levels observed over a 19-year tidal epoch;
- Mean low water (MLW): the average of the low water levels observed over a 19year tidal epoch; and
- Mean lower low water (MLLW): the average of the lower low water levels observed over a 19-year tidal epoch (only the lower water level of the pair of low waters in a tidal day is used). MLLW is the reference chart datum used for NOAA nautical chart products.

As an illustration, Figure 2.1 depicts the relationship among vertical datums for a point located on the shore at Gibson Island, Chesapeake Bay. These elevations were calculated with use of the "VDatum" vertical datum transformation tool (Parker *et al.*, 2003; Myers, 2005), described in the following section. Sea-level rise trends at specific tide stations are generally calculated based on observed monthly mean sea level values to filter out the high frequency fluctuations in tide levels.

Based on surveys at tide stations, NAVD 88 ranges from 15 cm below to 15 cm above LMSL in the mid-Atlantic region. Due to slopes in the local sea surface from changes in tidal hydrodynamics, LMSL generally increases in elevation relative to NAVD 88 for locations increasingly farther up estuaries and tidal rivers. For smaller scale topographic maps and coarser resolution DEMs, the two datums are often reported as being equivalent, when in reality they are not. The differences should be reported as part of the uncertainty analyses. Differences between NAVD 88 and LMSL on the U.S. West Coast often exceed 100 cm and must be taken into account in any inundation mapping application. Similarly, but more importantly, many coastal projects still inappropriately use NGVD 29 as a proxy for local mean sea level in planning, designing, and reference mapping. In the Mid-Atlantic, due to relative sea level change since 1929, the elevation of NGVD 29 ranges from 15 cm to more than 50 cm below the elevation of LMSL (1983-2001 NTDE). This elevation difference must be taken into account in any type of inundation mapping. Again, because LMSL is a sloped surface relative to orthometric datums due to the complexity of tides in estuaries and inland waterways, the elevation separation between LMSL and NGVD 29 increases for locations farther up estuaries and tidal rivers.

2.2.4. Topographic/Bathymetric/Water Level Data Integration

High-resolution datasets that effectively depict elevations across the land-sea boundary from land into shallow water are useful for many coastal applications (NRC, 2004), although they are not readily available for many areas. Sea-level rise studies can benefit from the use of integrated topographic/bathymetric models because the dynamic land/water interface area, including the intertidal zone, is properly treated as one seamless entity. In addition, other coastal research topics rely on elevation data that represent near-shore topography and bathymetry (water depths), but because existing topographic, bathymetric, and water level data have been collected independently for different purposes, they are difficult to use together. The USGS and the National Oceanic and Atmospheric Administration (NOAA) have worked collaboratively to address the difficulties in using disparate elevation and depth information, initially in the Tampa Bay region in Florida (Gesch and Wilson, 2002). The key to successful integration of topographic, bathymetric, and water level data is to place them in a consistent vertical reference frame, which is generally not the case with terrestrial and marine data. A vertical datum transformation tool called VDatum developed by NOAA's National Ocean Service provides the capability to convert topographic, bathymetric and water level data to a common vertical datum (Parker *et al.*, 2003; Myers, 2005). Work was completed in mid-2008 on providing VDatum coverage for the mid-Atlantic region. VDatum uses tidal datum surfaces, derived from hydrodynamic models corrected to match observations at tide stations, to interpolate the elevation differences between LMSL and NAVD 88. An integrated uncertainty analysis for VDatum is currently underway by NOAA.

The National Research Council (NRC, 2004) has recognized the advantages of seamless data across the land/water interface and has recommended a national implementation of VDatum and establishment of protocols for merged topographic/bathymetric datasets (NOAA, 2008a). Work has continued on production of other such merged datasets for coastal locations, including North Carolina and the Florida panhandle (Feyen *et al.*, 2005, 2008). Integrated topographic/bathymetric lidar (Nayegandhi *et al.*, 2006; Guenther, 2007) has been identified as a valuable technology for filling critical data gaps at the land/water interface, which would facilitate development of more high quality datasets (NRC, 2004).

2.3. Vulnerabilty Maps and Assessments

Maps that depict coastal areas at risk of potential inundation or other adverse effects of sea-level rise are appealing to planners and land managers that are charged with communicating, adapting to, and reducing the risks (Coastal States Organization, 2007). Likewise, map-based analyses of sea-level rise vulnerability often include statistical summaries of population, infrastructure, and economic activity in the mapped impact zone, as this information is critical for risk management and mitigation efforts. Many studies have relied on elevation data to delineate potential impact zones and quantify effects. During the last 15 years, this approach has also been facilitated by the increasing availability of spatially extensive elevation, demographic, land use/land cover, and economic data and advanced geographic information system (GIS) tools. These tools have improved access to data and have provided the analytical software capability for producing map-based analyses and statistical summaries. The body of peer reviewed scientific literature cited in this Chapter includes numerous studies that have focused on mapping and quantifying potential sea-level rise impacts.

A number of terms are used in the literature to describe the adverse effects of sea-level rise, including *inundation*, *flooding*, *submergence*, and *land loss*. Likewise, multiple terms are used to refer to what this Chapter has called vulnerability, including *at risk*, *subject to*, *impacted by*, and *affected by*. Many reports do not distinguish among the range of responses to sea-level rise, as described in Section 2.1. Instead, simple inundation, as a function of increased water levels projected onto the land surface, is assumed to reflect the vulnerability.

Relationship of vertical datums for Gibson Island, Chesapeake Bay

0.72 ft	MHHW	0.219 m
0.44 ft	MHW	0.134 m
0.00 ft	NAVD 88	0.000 m
-0.04 ft	LMSL	-0.012 m
-0.53 ft	MLW	-0.163 m
-0.75 ft	MLLW	-0.229 m
-0.80 ft	NGVD 29	-0.244 m

Figure 2.1. Diagram of the VDatum derived relationship among vertical datums for a point on the shore at Gibson Island, Chesapeake Bay. The point is located between the tide stations at Baltimore and Annapolis, Maryland where datum relationships are based on observations. The numbers represent the vertical difference above or below NAVD 88. For instance, at this location in the Chesapeake Bay the estimated MLLW reference is more than 20 centimeters below the NAVD 88 zero reference, whereas local mean sea level is only about 1 centimeter below NAVD zero.

Monmonier (2008) has recognized the dual nature of sea-level rise vulnerability maps as both tools for planning and as cartographic instruments to illustrate the potential catastrophic impacts of climate change. Monmonier cites reports that depict inundation areas due to very large increases in global sea-level. Frequently, however, the sea-level rise map depictions have no time scales and no indication of uncertainty or data limitations. Presumably, these broad-scale maps are in the illustration category, and only site-specific, local scale products are true planning tools, but therein is the difficulty. With many studies it is not clear if the maps (and associated statistical summaries) are intended simply to raise awareness of potential broad impacts or if they are intended to be used in decision making for specific locations.

2.3.1. Large-Area Studies (Global and United States)

Sea-level rise as a consequence of climate change is a global concern, and this is reflected in the variety of studies conducted for locations around the world as well as within the United States. Table 2.2 summarizes the characteristics of a number of the sea-level rise assessments conducted over broad areas, with some of the studies discussed in more detail below.

Schneider and Chen (1980) presented one of the early reports on potential sea-level rise impacts along U.S. coastlines. They used the 15-ft and 25-ft contours from USGS 1:24,000-scale maps to "derive approximate areas flooded within individual counties" along the coast. As with many of the vulnerability studies, Schneider and Chen also combined their estimates of submerged areas with population and property value data to estimate socioeconomic impacts, in this case on a state-by-state basis.

Reports to Congress by the U.S. Environmental Protection Agency (EPA) and the Federal Emergency Management Agency (FEMA) contributed to the collection of broad area assessments for the United States. The EPA report (U.S. EPA, 1989; Titus *et al.*, 1991) examined several different global sea-level rise scenarios in the range of 0.5 to 2 m (1.6 to 6.6 ft), and also discussed impacts on wetlands under varying shoreline protection scenarios. For elevation information, the study used contours from USGS topographic maps supplemented with wetland delineations from Landsat satellite imagery and tide gauge data. The study found that the available data were inadequate for production of detailed maps. The FEMA (1991) report estimated the increase of land in the 100-year floodplain from sea-level rises of 1 ft (0.3 m) and 3 ft (0.9 m). FEMA also estimated the increase in annual flood damages to insured properties by the year 2100, given the assumption that the trends of development would continue.

Table 2.2. Characteristics of some sea-level rise assessments conducted over broad areas. GTOPO30 is a global raster DEM with a horizontal grid spacing of 30 arc seconds (approximately 1 kilometer). SRTM is the Shuttle Radar Topography Mission data. NED is the National Elevation Dataset.

Study	Study area	Elevation data	Sea-level rise scenario	Elevation accuracy reported?	Maps published?
Schneider and Chen (1980)	Conterminous United States	15- and 25- foot contours from USGS 1:24,000-scale maps	4.6 and 7.6 m	No	Yes
U.S. EPA (1989)	Conterminous United States	Contours from USGS maps	0.5, 1, and 2 m	No	No
Titus *et al.* (1991)	Conterminous United States	Contours from USGS maps, Wetland delineations, and tide data	0.5, 1, and 2 m	No	No
FEMA (1991)	United States	Coastal floodplain maps	1 ft and 3 ft	No	No
Small and Nicholls (2003)	Global	GTOPO30	5-m land elevation increments	Estimated a 5-Meter uncertainty for elevation data (no error metric specified)	No
Ericson *et al.* (2006)	40 deltas distributed worldwide	GTOPO30	0.5-12.5 mm per year for years 2000-2050	No	No
Rowley *et al.* (2007)	Global	GLOBE (GTOPO30)	1, 2, 3, 4, 5, and 6 m	No	Yes

Table 2.2. (Continued)

Study	Study area	Elevation data	Sea-level rise scenario	Elevation accuracy reported?	Maps published?
McGranahan *et al.* (2007)	Global	SRTM	Land elevations 0 to 10 m (to define the "low elevation coastal zone")	No, although 10-meter elevation increment was used in recognition of data limitations	Yes
Demirkesen *e tal.* (2007)	Izmir, Turkey	SRTM	2 and 5 m	Yes, but no error metric specified	Yes
Demirkesen *et al.* (2008)	Turkey	SRTM	1, 2, and 3 m	Yes, but no error metric specified	Yes
Marfai and King (2008)	Semarang, Indonesia	Local survey data	1.2 and 1.8 m	No	Yes
Kafalenos *et al.* (2008)	U.S. Gulf Coast	NED	2 and 4 ft	No	Yes

Elevation datasets with global or near-global extent have been used for vulnerability studies across broad areas. For their studies of the global population at risk from coastal hazards, Small and Nicholls (2003) and Ericson *et al.* (2006) used GTOPO30, a global 30-arc-second (about 1-kilometer [km]) elevation dataset produced by the USGS (Gesch *et al.*, 1999). Rowley *et al.* (2007) used the GLOBE 30-arc-second DEM (Hastings and Dunbar, 1998), which is derived mostly from GTOPO30. As with many vulnerability studies, these investigations used the delineations of low-lying lands from the elevation model to quantify the population at risk from sea-level rise, in one instance using increments as small as 1 m (Rowley *et al.*, 2007).

Elevation data from the Shuttle Radar Topography Mission (SRTM) (Farr *et al.*, 2007) are available at a 3-arc-second (about 90-m) resolution with near-global coverage. Because of their broad area coverage and improved resolution over GTOPO30, SRTM data have been used in several studies of the land area and population potentially at risk from sea-level rise (McGranahan *et al.*, 2007; Demirkesen *et al.*, 2007, 2008). Similar to other studies, McGranahan *et al.* (2007) present estimates of the population at risk, while Demirkesen *et al.* (2007) document the dominant land use/land cover classes in the delineated vulnerable areas.

2.3.2. Mid-Atlantic Region, States, and Localities

A number of sea-level rise vulnerability studies have been published for sites in the mid-Atlantic region, the focus area for this Product. Table 2.3 summarizes the characteristics for these reports, and important information from some of the studies is highlighted.

Table 2.3. Characteristics of some sea-level rise vulnerability studies conducted over mid-Atlantic locations. GTOPO30 is a global raster DEM with a horizontal grid spacing of 30 arc seconds (approximately 1 kilometer). SRTM is the Shuttle Radar Topography Mission data. NED is the National Elevation Dataset.

Study	Study area	Elevation data USGS DEMs	Sea-level rise scenarios	Elevation accuracy reported?	Maps published?
Titus and Richman (2001)	U.S. Atlantic and Gulf coasts	derived from 1:250,000- scale maps 30-meter	1.5- and 3.5-m land elevation increments	No	Yes
Najjar *et al.* (2000)	Delaware	USGS DEMs	2 ft	No	Yes
Kleinosky *et al.* (2007)	Hampton Roads, Virginia	10-meter and 30-meter USGS DEMs	30, 60, and 90 cm	No	Yes
Wu *et al.* (2002)	Cape May County, New Jersey	30-meter USGS DEMs	60 cm	No	Yes
Gornitz *et al.* (2002)	New York City area	30-meter USGS DEMs	5-ft land elevation increments	No, although Only qualitative results were reported	Yes
Titus and Wang (2008)	Mid-Atlantic states	Contours from USGS 1:24,000-scale maps, lidar, local data	0.5-m land elevation increments	Yes, RMSE vs. lidar for a portion of the study area	Yes
Larsen *et al.* (2004)	Blackwater National Wildlife Refuge, Maryland	Lidar	30-cm land elevation increments	No	Yes
Gesch, (2009)	North Carolina	GTOPO30, SRTM, NED, lidar	1 m	Yes, with NSSDA error metric (95% confidence)	Yes

A study by Titus and Richman (2001) is often referred to in discussions of the land in the United States that is subject to the effects of sea-level rise. The methods used to produce the maps in that report are clearly documented. However, because they used very coarse elevation data (derived from USGS 1:250,000-scale topographic maps), the resulting products are general and limited in their applicability. The authors acknowledge the limitations of their results because of the source data they used, and clearly list the caveats for proper use of the maps. As such, these maps are useful in depicting broad implications of sea-level rise, but are not appropriate for site-specific decision making.

Numerous studies have used the NED, or the underlying USGS DEMs from which much of the NED is derived, as the input elevation information. Najjar *et al.* (2000) show an example of using USGS 30-m DEMs for a simple inundation model of Delaware for a 2ft (0.6-m) sea-level rise. In another study, Kleinosky *et al.* (2007) used elevation information from USGS 10-m and 30-m DEMs to depict vulnerability of the Hampton Roads, Virginia area to storm surge flooding in addition to sea-level rise. Storm surge heights were first determined by modeling, then 30-, 60-, and 90-cm increments of sea-level rise were added to project the expansion of flood risk zones onto the land surface. In addition, Wu *et al.* (2002) conducted a study for Cape May County, New Jersey using an approach similar to Kleinosky *et al.* (2007), where they added 60 cm to modeled storm surge heights to account for sea-level rise.

More recently, Titus and Wang (2008) conducted a study of the mid-Atlantic states (New York to North Carolina) using a variety of elevation data sources including USGS 1:24,000-scale topographic maps (mostly with 5- or 10-ft contour intervals), lidar data, and some local data provided by state agencies, counties, and municipalities. They used an approach similar to that described in Titus and Richman (2001) in which tidal wetland delineations are employed in an effort to estimate additional elevation information below the first topographic map contour.

2.3.3. Other Reports

In addition to reports by federal government agencies and studies published in the peer-reviewed scientific literature, there have been numerous assessment reports issued by various non-governmental organizations, universities, state and local agencies, and other private groups (*e.g.*, Anthoff *et al.*, 2006; Dasgupta *et al.*, 2007; Stanton and Ackerman, 2007; US DOT, 2008; Mazria and Kershner, 2007; Glick *et al.*, 2008; Cooper *et al.*, 2005; Lathrop and Love, 2007; Johnson *et al.*, 2006; Bin *et al.*, 2007; Slovinsky and Dickson, 2006). While it may be difficult to judge the technical veracity of the results in these reports, they do share common characteristics with the studies reviewed in Sections 2.3.1 and 2.3.2. Namely, they make use of the same elevation datasets (GTOPO30, SRTM, NED, and lidar) to project inundation from sea-level rise onto the land surface to quantify vulnerable areas, and they present statistical summaries of impacted population and other socioeconomic variables. Many of these reports include detailed maps and graphics of areas at risk. Although some are also available in printed formats, all of the reports listed above are available online (see Chapter 2 References for website information).

This category of reports is highlighted because some of the reports have gained wide public exposure through press releases and subsequent coverage in the popular press and on Internet news sites. For example, the report by Stanton and Ackerman (2007) has been cited at least eight times by the mainstream media (see: <http://ase. tufts.edu/ gdae/Pubs/rp/ FloridaClimate.html>). The existence of this type of report, and the attention it has received, is likely an indication of the broad public interest in sea-level rise issues. These reports are often written from an economic or public policy context rather than from a geosciences perspective. Nevertheless, it is important for the coastal science community to be cognizant of them because the reports often cite journal papers and they serve as a conduit for communicating recent sea-level rise research results to less technical audiences. It is interesting to note that all of the reports listed here were produced over the last three years, thus, it is likely that that this type of outlet will continue to be used to discuss sea-level rise issues as global climate change continues to garner more public attention. Arguably, sea-level rise is among the most visible and understandable consequences of climate change for the general public, and they will continue to seek information about it from the popular press, Internet sites, and reports such as those described here.

2.3.4. Limitations of Previous Studies

It is clear from the literature reviewed in Sections 2.3.1, 2.3.2, and 2.3.3 that the development of sea-level rise impact assessments has been an active research topic for the past 25 years. However, there is still significant progress to be made in improving the physical science-based information needed for decision making by planners and land and resource managers in the coastal zone. Although previous studies have brought ample attention to the problem of mapping and quantifying sea-level rise impacts, the quality of the available input data and the common tendency to overlook the consequences of coarse data resolution and large uncertainty ranges hinder the usefulness and applicability of many results. Specifically, for this Product, none of the previous studies covering the mid-Atlantic region can be used to fully answer with high confidence the Synthesis and Assessment Product (SAP) 4.1 prospectus question (CCSP, 2006) that relates directly to coastal elevations: "Which lands are currently at an elevation that could lead them to be inundated by the tides without shore protection measures?" The collective limitations of previous studies are described in this Section, while the "lessons learned", or recommendations for required qualities of future vulnerability assessments, are discussed in Section 2.4.

Overall, there has been little recognition in previous studies that inundation is only one response out of a number of possible responses to sea-level rise (see Section 2.1). Some studies do mention the various types of coastal impacts (erosion, saltwater intrusion, more extreme storm surge flooding) (Najjar *et al.*, 2000; Gornitz *et al.*, 2002), and some studies that focus on wetland impacts do consider more than just inundation (U.S.EPA, 1989; Larsen *et al.*, 2004). However, in general, many vulnerability maps (and corresponding statistical summaries) imply that a simple inundation scenario is an adequate representation of the impacts of rising seas (Schneider and Chen, 1980; Rowley *et al.*, 2007; Demirkesen *et al.*, 2008; Najjar *et al.*, 2000).

Based on the review of the studies cited in Sections 2.3.1, 2.3.2, and 2.3.3, these general limitations have been identified:

1. *Use of lower resolution elevation data with poor vertical accuracy.* Some studies have had to rely on elevation datasets that are poorly suited for detailed inundation mapping (e.g., GTOPO30 and SRTM). While these global datasets may be useful for general depictions of low elevation zones, their relatively coarse spatial detail precludes their use for production of detailed vulnerability maps. In addition to the limited spatial detail, these datasets have elevation values quantized only to whole meter intervals, and their overall vertical accuracy is poor when compared to the intervals of predicted sea-level rise over the next century. The need for better elevation information in sea-level rise assessments has been broadly recognized (Leatherman, 2001; Marbaix, and Nicholls, 2007; Jacob et al., 2007), especially for large-scale planning maps (Monmonier, 2008) and detailed quantitative assessments (Gornitz et al., 2002).

2. *Lack of consideration of uncertainty of input elevation data.* A few studies generally discuss the limitations of the elevation data used in terms of accuracy (Small and Nicholls, 2003; McGranahan et al., 2007; Titus and Wang, 2008). However, none of these studies exhibit rigorous accuracy testing and reporting according to accepted

national standards (NSSDA and NMAS). Every elevation dataset has some vertical error, which can be tested and measured, and described by accuracy statements. The overall vertical error is a measure of the uncertainty of the elevation information, and that uncertainty is propagated to any derived maps and statistical summaries. Gesch (2009) demonstrates why it is important to account for vertical uncertainty in sea-level rise vulnerability maps and area statistics derived from elevation data (see Box 2.1).

3. *Elevation intervals or sea-level rise increments not supported by vertical accuracy of input elevation data.* Most elevation datasets, with the exception of lidar, have vertical accuracies of several meters or even tens of meters (at the 95 percent confidence level). Figure 2.2 shows a graphical representation of DEM vertical accuracy using error bars around a specified elevation. In this case, a lidar-derived DEM locates the 1-meter elevation to within ±0.3 m at 95-percent confidence. (In other words, the true elevation at that location falls within a range of 0.7 to 1.3 m.) A less accurate topographic map-derived DEM locates the 1-m elevation to within ±2.2 m at 95-percent confidence, which means the true land elevation at that location falls within a range of 0 (assuming sea level was delineated accurately on the original topographic map) to 3.2 m. Many of the studies reviewed in this Chapter use land elevation intervals or sea-level rise increments that are 1 m or less. Mapping of sub-meter increments of sea-level rise is highly questionable if the elevation data used have a vertical accuracy of a meter or more (at the 95-percent confidence level) (Gesch, 2009). For example, by definition a topographic map with a 5-ft contour interval that meets NMAS has an absolute vertical accuracy (which accounts for all effects of systematic and random errors) of 90.8 cm at the 95-percent confidence level (Maune, et al., 2007b). Likewise, a 10-ft contour interval map has an absolute vertical accuracy of 181.6 cm (1.816 m) at the 95-percent confidence level. If such maps were used to delineate the inundation zone from a 50-cm sea-level rise, the results would be uncertain because the vertical increment of rise is well within the bounds of statistical uncertainty of the elevation data.

4. *Maps without symbology or caveats concerning the inherent vertical uncertainty of input elevation data.* Some studies have addressed limitations of their maps and statistics (Titus and Richman, 2001; Najjar et al., 2000), but most reports present maps without any indication of the error associated with the underlying elevation data (see number 3 above). Gesch (2009) presents one method of spatially portraying the inherent uncertainty of a mapped sea-level rise inundation zone (see Box 2.1).

5. *Inundated area and impacted population estimates reported without a range of values that reflect the inherent vertical uncertainty of input elevation data.* Many studies use the mapped inundation zone to calculate the at-risk area, and then overlay that delineation with spatially distributed population data or other socioeconomic variables to estimate impacts. If a spatial expression of the uncertainty of the inundation zone (due to the vertical error in the elevation data) is not included, then only one total can be reported. More complete and credible information would be

provided if a second total was calculated by including the variable (area, population, or economic parameter) that falls within an additional delineation that accounts for elevation uncertainty. A range of values can then be reported, which reflects the uncertainty of the mapped inundation zone.

6. *Lack of recognition of differences among reference orthometric datums, tidal datums, and spatial variations in sea-level datums.* The vertical reference frame of the data used in a particular study needs to be specified, especially for local studies that produce detailed maps, since there can be significant differences between an orthometric datum zero reference and mean sea level (Figure 2.1; see also Section 2.2.3). As described earlier, there are important distinctions between vertical reference systems that are used for land elevation datasets and those that are used to establish the elevations of sea level. Most of the reviewed studies did not specify which vertical reference frame was used. Often, it was probably an orthometric datum because most elevation datasets are in reference to such datums. Ideally, a tool such as VDatum will be available so that data may be easily transformed into a number vertical reference frames at the discretion of the user.

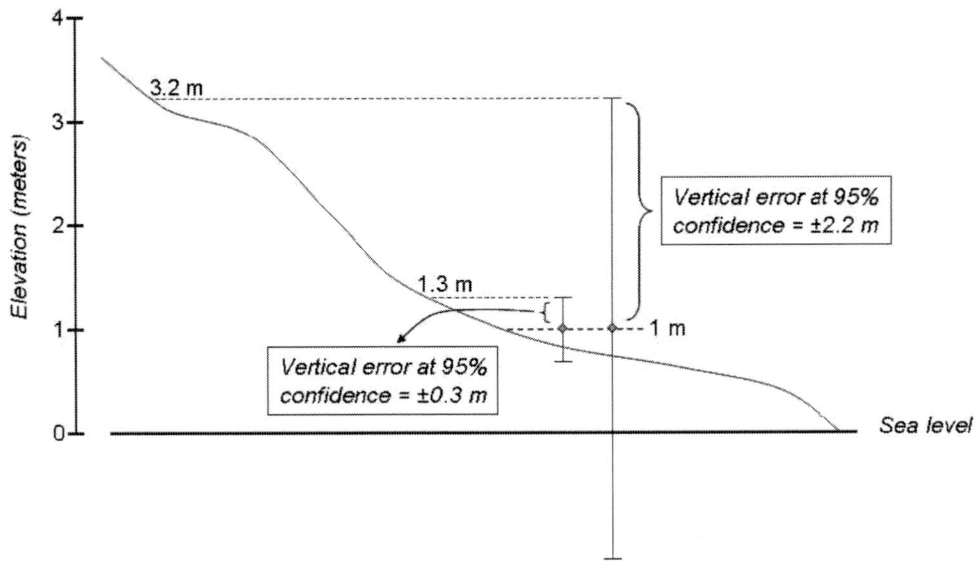

Figure 2.2. Diagram of how a sea-level rise of 1 meter is mapped onto the land surface using two digital elevation models with differing vertical accuracies. The more accurate lidar-derived DEM (±0.3 m at 95-percent confidence) results in a delineation of the inundation zone with much less uncertainty than when the less accurate topographic map-derived DEM (±2.2 m at 95-percent confidence) is used (Gesch, 2009).

BOX 2.1. A CASE STUDY USING LIDAR ELEVATION DATA

To illustrate the application of elevation uncertainty information and the advantages of lidar elevation data for sea-level rise assessment, a case study for North Carolina (Gesch, 2009) is presented and summarized here. North Carolina has a broad expanse of low-lying land (Titus and Richman, 2001), and as such is a good site for a mapping comparison. Lidar data at 1/9-arc-second (about 3 meters [m]) grid spacing were analyzed and compared to 1-arc-second (about 30 m) DEMs derived from 1:24,000-scale topographic maps. The potential inundation zone from a 1-m sea-level rise was mapped from both elevation datasets, and the corresponding areas were compared. The analysis produced maps and statistics in which the elevation uncertainty was considered. Each elevation dataset was "flooded" by identifying the grid cells that have an elevation at or below 1 m and are connected hydrologically to the ocean through a continuous path of adjacent inundated grid cells. For each dataset, additional areas were delineated to show a spatial representation of the uncertainty of the projected inundation area. This was accomplished by adding the linear error at 95-percent confidence to the 1-m sea-level increase and extracting the area at or below that elevation using the same flooding algorithm. The lidar data exhibited ±0.27 m error at 95-percent confidence based on accuracy reports from the data producer, while the topographic map-derived DEMs had ±2.21 m error at 95-percent confidence based on an accuracy assessment with high-quality surveyed control points.

Box Figure 2.1 and Box Table 2.1 show the results of the North Carolina mapping comparison. In Box Figure 2.1 the darker blue tint represents the area at or below 1-m in elevation, and the lighter blue tint represents the additional area in the vulnerable zone given the vertical uncertainty of the input elevation datasets. The more accurate lidar data for delineation of the vulnerable zone results in a more certain delineation (Box Figure 2.1B), or in other words the zone of uncertainty is small. Box Table 2.1 compares the vulnerable areas as delineated from the two elevation datasets. The delineation of the 1-m zone from the topographic map-derived DEMs more than doubles when the elevation uncertainty is considered, which calls into question the reliability of any conclusions drawn from the delineation. It is apparent that for this site the map-derived DEMs do not have the vertical accuracy required to reliably delineate a 1-m sea-level rise inundation zone. Lidar is the appropriate elevation dataset for answering the question about how much land in the study site is vulnerable to a 1-m sea-level rise, for which the answer is: "4,195 to 4,783 square kilometers (sq km) at a 95-percent confidence level". This case study emphasizes why a range of values should be given when reporting the size of the inundation area for a given sea-level rise scenario, especially for sites where high-accuracy lidar data are not available Without such a range being reported, users of an assessment report may not understand the amount of uncertainty associated with area delineations from less accurate data and the implications for any subsequent decisions based on the reported statistics.

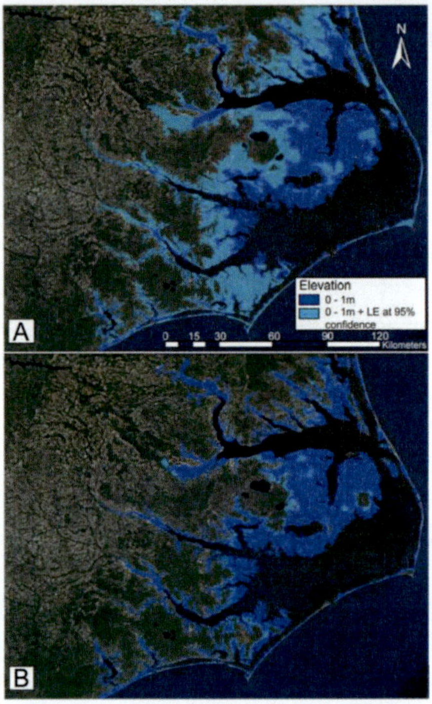

Box Figure 2.1. Lands vulnerable to a 1-meter sea-level rise, developed from topographic map-derived DEMs (A), and lidar elevation data (B) (Gesch, 2009). The background is a recent true color orthoimage.

Box Table 2.1. The area of land vulnerable to a 1-meter sea-level rise as calculated from two elevation datasets (see Box Figure 2.1), as well as the area of vulnerability when the uncertainty of the elevation data is considered (Gesch, 2009).

Elevation dataset	Area less than or equal to 1 meter in elevation (sq km)	Area less than or equal to 1 meter in elevation at 95 percent confidence (sq km)	Percent increase in vulnerable area when elevation uncertainty is included
1-arc-second (30-m) DEMs derived from 1:24,000-scale topographic maps	4,014	8,578	114%
1/9-arc-second (3-m) lidar elevation grid	4,195	4,783	14%

2.4. Future Vulnerability Assessments

To fully answer the relevant elevation question from the prospectus for this SAP 4.1 (see Section 2.3.4), there are important technical considerations that need to be incorporated to improve future sea-level rise impact assessments, especially those with a goal of producing vulnerability maps and statistical summaries of impacts. These considerations are important for both the researchers who develop impact assessments, as well as the users of those assessments who must understand the technical issues to properly apply the information. The recommendations for improvements described below are based on the review of the previous studies cited in Sections 2.3.1, 2.3.2, 2.3.3, and other recent research:

1. *Determine where inundation will be the primary response to sea-level rise.* Inundation (submergence of the uplands) is only one of a number of possible responses to sea-level rise (Leatherman, 2001; Valiela, 2006; FitzGerald et al., 2008). If the complex nature of coastal change is not recognized up front in sea-level rise assessment reports, a reader may mistakenly assume that all stretches of the coast that are deemed vulnerable will experience the same "flooding" impact, as numerous reports have called it. For the coastal settings in which inundation is the primary vulnerability, elevation datasets should be analyzed as detailed below to produce comprehensive maps and statistics.

2. *Use lidar elevation data (or other high-resolution, high-accuracy elevation source).* To meet the need for more accurate, detailed, and up-to-date sea-level rise vulnerability assessments, new studies should be based on recently collected high-resolution, high-accuracy, lidar elevation data. Other mapping approaches, including photogrammetry and ground surveys, can produce high-quality elevation data suitable for detailed assessments, but lidar is the preferred approach for cost-effective data collection over broad coastal areas. Lidar has the added advantage that, in addition to high-accuracy measurements of ground elevation, it also can be used to produce information on buildings, infrastructure, and vegetation, which may be important for sea-level rise impact assessments. As Leatherman (2001) points out, inundation is a function of slope. The ability of lidar to measure elevations very precisely facilitates the accurate determination of even small slopes, thus it is quite useful for mapping low relief coastal landforms. The numerous advantages of lidar elevation mapping in the coastal zone have been widely recognized (Leatherman, 2001; Coastal States Organization, 2007; Monmonier, 2008; Subcommittee on Disaster Reduction, 2008; Feyen et al., 2008; Gesch, 2009). A recent study by the National Research Council (NRC, 2007) concluded that FEMA's requirements for floodplain mapping would be met in all areas by elevation data with 1-ft to 2-ft equivalent contour accuracy, and that a national lidar program called "Elevation for the Nation" should be carried out to create a new national DEM. Elevation data meeting 1-ft contour interval accuracy (NMAS) would allow effective sea-level rise inundation modeling for increments in the 0.35 m range, while data with 2-ft contour interval accuracy would be suitable for increments of about 0.7 m.

3. *Test and report absolute vertical accuracy as a measure of elevation uncertainty.* Any studies that use elevation data as an input for vulnerability maps and/or statistics need to have a clear statement of the absolute vertical accuracy (in reference to true ground elevations). The NSSDA vertical accuracy testing and reporting methodology (Federal Geographic Data Committee, 1998), which uses a metric of linear error at

95-percent confidence, is the preferred approach. Vertical accuracy may be reported with other metrics including RMSE, standard deviation (one sigma error), LE90, or three sigma error. Maune *et al.* (2007b) and Greenwalt and Shultz (1962) provide methods to translate among the different error metrics. In any case, the error metric must be identified because quoting an accuracy figure without specifying the metric is meaningless. For lidar elevation data, a specific testing and reporting procedure that conforms to the NSSDA has been developed by the National Digital Elevation Program (NDEP) (2004). The NDEP guidelines are useful because they provide methods for accuracy assessment in "open terrain" *versus* other land cover categories such as forest or urban areas where the lidar sensor may not have detected ground level. NDEP also provides guidance on accuracy testing and reporting when the measured elevation model errors are from a non-Gaussian (non-normal) distribution.

4. *Apply elevation uncertainty information in development of vulnerability maps and area statistics.* Knowledge of the uncertainty of input elevation data should be incorporated into the development of sea-level rise impact assessment products. In this case, the uncertainty is expressed in the vertical error determined through accuracy testing, as described above. Other hydrologic applications of elevation data, including rainfall runoff modeling (Wu et al., 2008) and riverine flood inundation modeling (Yilmaz et al., 2004, 2005), have benefitted from the incorporation of elevation uncertainty. For sea-level rise inundation modeling, the error associated with the input elevation dataset is used to include a zone of uncertainty in the delineation of vulnerable land at or below a specific elevation. For example, assume a map of lands vulnerable to a 1-m sea-level rise is to be developed using a DEM. That DEM, similar to all elevation datasets, has an overall vertical error. The challenge, then, is how to account for the elevation uncertainty (vertical error) in the mapping of the vulnerable area. Figure 2.2 (Gesch, 2009) shows how the elevation uncertainty associated with the 1-m level, as expressed by the absolute vertical accuracy, is projected onto the land surface. The topographic profile diagram shows two different elevation datasets with differing vertical accuracies depicted as error bars around the 1-m elevation. One dataset has a vertical accuracy of ±0.3 m at the 95-percent confidence level, while the other has an accuracy of ±2.2 m at the 95-percent confidence level. By adding the error to the projected 1-m sea-level rise, more area is added to the inundation zone delineation, and this additional area is a spatial representation of the uncertainty. The additional area is interpreted as the region in which the 1-m elevation may actually fall, given the statistical uncertainty of the DEMs.

5. Recognizing that elevation data inherently have vertical uncertainty, vulnerability maps derived from them should include some type of indication of the area of uncertainty. This could be provided as a caveat in the map legend or margin, but a spatial portrayal with map symbology may be more effective. Merwade *et al.* (2008) have demonstrated this approach for floodplain mapping where the modeled inundation area has a surrounding uncertainty zone depicted as a buffer around the flood boundary. Gesch (2009) used a similar approach to show a spatial

representation of the uncertainty of the projected inundation area from a 1-m sea-level rise, with one color for the area below 1-m in elevation and another color for the adjacent uncertainty zone (see Box 2.1).

6. As with vulnerability maps derived from elevation data, statistical summaries of affected land area, population, land use/land cover types, number of buildings, infrastructure extent, and other socioeconomic variables should include recognition of the vertical uncertainty of the underlying data. In many studies, the delineated inundation zone is intersected with geospatial representations of demographic or economic variables in order to summarize the quantity of those variables within the potential impact zone. Such overlay and summarizing operations should also include the area of uncertainty associated with the inundation zone, and thus ranges of the variables should be reported. The range for a particular variable would increase from the total for just the projected inundation zone up to the combined total for the inundation zone plus the adjacent uncertainty zone. Additionally, because the combined area of the inundation zone and its adjacent uncertainty zone has a known confidence level, the range can be reported with that same confidence level. Merwade *et al.* (2008) have recommended such an approach for floodplain mapping when they state that the flood inundation extent should be reported as being "in the range from *x* units to *y* units with a *z*-% confidence level".

7. An important use of elevation data accuracy information in an assessment study is to guide the selection of land elevation intervals or sea-level rise increments that are appropriate for the available data. Inundation modeling is usually a simple process wherein sea level is effectively raised by delineating the area at and below a specified land elevation to create the inundation zone. This procedure is effectively a contouring process, so the vertical accuracy of a DEM must be known to determine the contour interval that is supported. DEMs can be contoured at any interval, but, just by doing so, it does not mean that the contours meet published accuracy standards. Likewise, studies can use small intervals of sea-level rise, but the underlying elevation data must have the vertical accuracy to support those intervals. The intervals must not be so small that they are within the bounds of the statistical uncertainty of the elevation data.

8. *Produce spatially explicit maps and detailed statistics that can be used in local decision making.* The ultimate use of a sea-level rise assessment is as a planning and decision-making tool. Some assessments cover broad areas and are useful for scoping the general extent of the area of concern for sea-level rise impacts. However, the smaller-scale maps and corresponding statistics from these broad area assessments cannot be used for local decision making, which require large-scale map products and site specific information. Such spatially explicit planning maps require high-resolution, high-accuracy input data as source information. Monmonier (2008) emphasizes that "reliable large-scale planning maps call for markedly better elevation data than found on conventional topographic maps". Even with source data that supports local mapping, it is important to remember, as Frumhoff *et al.* (2007) point out, due to the complex nature of coastal dynamics that "projecting the impacts

> of rising sea level on specific locations is not as simple as mapping which low-lying areas will eventually be inundated".

Proper treatment of elevation uncertainty is especially important for development of large-scale maps that will be used for planning and resource management decisions. Several states have realized the advantages of using high-accuracy lidar data to reduce uncertainty in sea-level rise studies and development of local map products (Rubinoff, *et al.*, 2008). Accurate local-scale maps can also be generalized to smaller-scale maps for assessments over larger areas. Such aggregation of detailed information benefits broad area studies by incorporating the best available, most detailed information.

Development of large-scale spatially explicit maps presents a new set of challenges. At scales useful for local decision making, the hydrological connectivity of the ocean to vulnerable lands must be mapped and considered. In some vulnerable areas, the drainage network has been artificially modified with ditches, canals, dikes, levees, and seawalls that affect the hydrologic paths rising water can traverse (Poulter and Halpin, 2007; Poulter *et al.*, 2008). Fortunately, lidar data often include these important features, which are important for improving large-scale inundation modeling (Coastal States Organization, 2007). Older, lower resolution elevation data often do not include these fine-scale manmade features, which is another limitation of these data for large-scale maps.

Other site-specific data should be included in impact assessments for local decision making, including knowledge of local sea-level rise trends and the differences among the zero reference for elevation data (often an orthometric datum), local mean sea level, and high water (Marbaix, and Nicholls, 2007; Poulter and Halpin, 2007). The high water level is useful for inundation mapping because it distinguishes the area of periodic submergence by tides from those areas that may become inundated as sea-level rises (Leatherman, 2001). The importance of knowing the local relationships of water level and land vertical reference systems emphasizes the need for a national implementation of VDatum (Parker *et al.*, 2003; Myers, 2005) so that accurate information on tidal dynamics can be incorporated into local sea-level rise assessments.

Another useful advance for detailed sea-level rise assessments can be realized by better overlay analysis of a delineated vulnerability zone and local population data. Population data are aggregated and reported in census blocks and tracts, and are often represented in area-based statistical thematic maps, also known as choropleth maps. However, such maps usually do not represent actual population density and distribution across the landscape because census units include both inhabited and uninhabited land. Dasymetric mapping (Mennis, 2003) is a technique that is used to disaggregate population density data into a more realistic spatial distribution based on ancillary land use/land cover information or remote sensing images (Sleeter and Gould, 2008; Chen, 2002). This technique holds promise for better analysis of population, or other socioeconomic data, to report statistical summaries of sea-level rise impacts within vulnerable zones.

2.5. Summary, Conclusions, and Future Directions

The topic of coastal elevations is most relevant to the first SAP 4.1 prospectus question (CCSP, 2006): "Which lands are currently at an elevation that could lead them to be inundated by the tides without shore protection measures?" The difficulty in directly answering this question for the mid-Atlantic region with a high degree of confidence was recognized. Collectively, the available previous studies do not provide the full answer for this region with the degree of confidence that is optimal for local decision making. Fortunately, new elevation data, especially lidar, are becoming available and are being integrated into the USGS NED (Gesch, 2007) as well as being used in sea-level rise applications (Coastal States Organization, 2007). Also, research is progressing on how to take advantage of the increased spatial resolution and vertical accuracy of new data (Poulter and Halpin, 2007; Gesch, 2009).

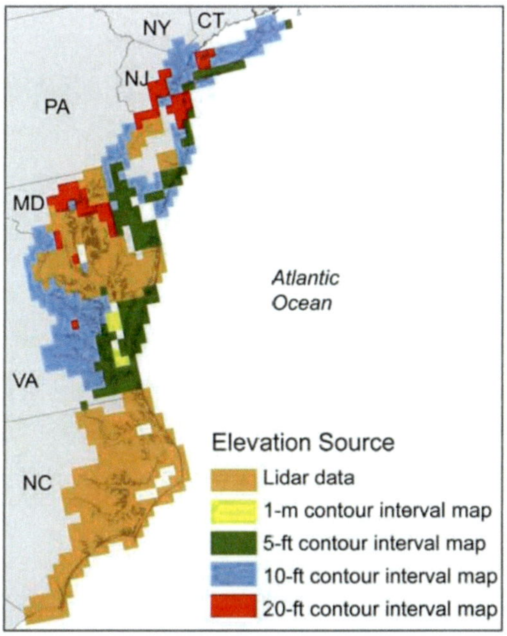

Figure 2.3. The current best available elevation source data (as of August 2008) for the National Elevation Dataset over the mid-Atlantic region.

Using national geospatial standards for accuracy assessment and reporting, the currently best available elevation data for the entire mid-Atlantic region do not support an assessment using a sea-level rise increment of 1-m or less, which is slightly above the range of current estimates for the remainder of this century and the high scenario used in this Product. Where lidar data meeting current industry standards for accuracy are available, the land area below the 1-m contour (simulating a 1-m sea-level rise) can be estimated for those sites along the coast at which inundation will be the primary response. The current USGS holdings of the best available elevation data include lidar for North Carolina, parts of Maryland, and parts of New Jersey (Figure 2.3). Lidar data for portions of Delaware and more of New Jersey and Maryland will be integrated into the NED in 2009. However, it may be some time before the full extent of the mid-Atlantic region has sufficient coverage of elevation data that are suitable for detailed assessments of sub-meter increments of sea-level rise and development of spatially explicit local planning maps.

Given the current status of the NED for the mid-Atlantic region (Figure 2.3), the finest increment of sea-level rise that is supported by the underlying elevation data varies across the area (Table 2.4 and Figure 2.4). At a minimum, a sea-level rise increment used for inundation modeling should not be smaller than the range of statistical uncertainty of the elevation data. For instance, if an elevation dataset has a vertical accuracy of ±1 m at 95percent confidence, the smallest sea-level rise increment that should be considered is 1 m.

Even then, the reliability of the vulnerable area delineation would not be high because the modeled sea-level rise increment is the same as the inherent vertical uncertainty of the elevation data. Thus, the reliability of a delineation of a given sea-level rise scenario will be better if the inherent vertical uncertainty of the elevation data is much less than the modeled water level rise For example, a sea-level rise of 0.5 m is reliably modeled with elevation data having a vertical accuracy of ±0.25 m at 95-percent confidence. This guideline, with the elevation data being at least twice as accurate as the modeled sea-level rise, was applied to derive the numbers in Table 2.4.

Table 2.4 Minimum sea-level rise scenarios for vulnerability assessments supported by elevation datasets of varying vertical accuracy.

Elevation data source	Vertical accuracy: RMSE	Vertical accuracy: linear error at 95-percent confidence	Minimum sea-level rise increment for inundation modeling
1-foot contour interval map	9.3 cm	18.2 cm	36.4 cm
Lidar	15.0 cm	29.4 cm	58.8 cm
2-foot contour interval map	18.5 cm	36.3 cm	72.6 cm
1-meter contour interval map	30.4 cm	59.6 cm	1.19 m
5-foot contour interval map	46.3 cm	90.7 cm	1.82 m
10-foot contour interval map	92.7 cm	1.82 m	3.64 m
20-foot contour interval map	1.85 m	3.63 m	7.26 m

High-quality lidar elevation data, such as that which could be collected in a national lidar survey, would be necessary for the entire coastal zone to complete a comprehensive assessment of sea-level rise vulnerability in the mid-Atlantic region. Lidar remote sensing has been recognized as a means to provide highly detailed and accurate data for numerous applications, and there is significant interest from the geospatial community in developing an initiative for a national lidar collection for the United States (Stoker *et al.*, 2007, 2008). If such an initiative is successful, then a truly national assessment of potential sea-level rise impacts could be realized. A U.S. national lidar dataset would facilitate consistent assessment of vulnerability across state or jurisdictional boundaries, an approach for which coastal states have voiced strong advocacy (Coastal States Organization, 2007). Even with the current investment in lidar by several states, there is a clear federal role in the development of a national lidar program (NRC, 2007; Monmonier, 2008; Stoker *et al.*, 2008).

Use of recent, high-accuracy lidar elevation data, especially with full consideration of elevation uncertainty as described in Section 2.4, will result in a new class of vulnerability maps and statistical summaries of impacts. These new assessment products will include a specific level of confidence, with ranges of variables reported. The level of statistical confidence could even be user selectable if assessment reports publish results at several confidence levels.

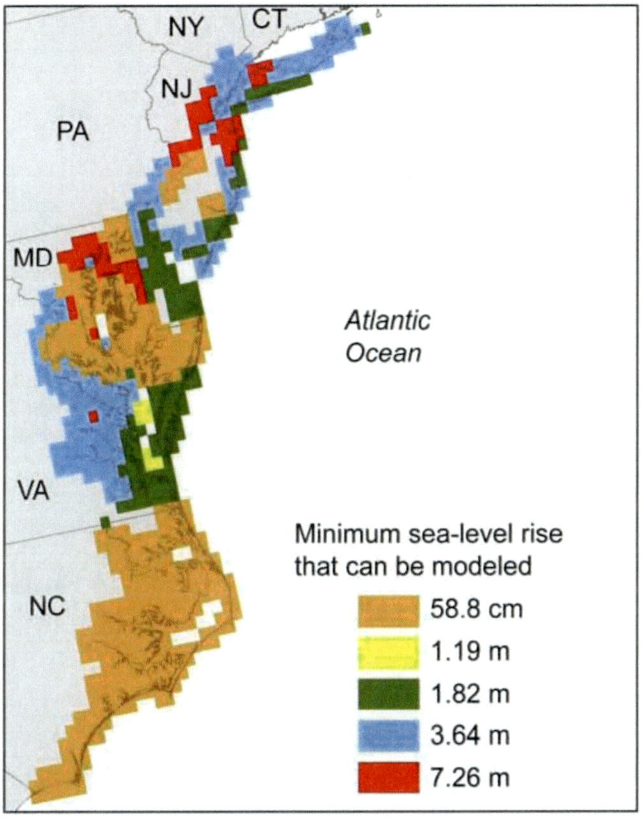

Figure 2.4. The estimated minimum sea-level rise scenarios for inundation modeling in the mid-Atlantic region given the current best available elevation data.

It is clear that improved elevation data and analysis techniques will lead to better sea-level rise impact assessments. However, new assessments must include recognition that inundation, defined as submergence of the uplands, is the primary response to rising seas in only some areas. In other areas, the response may be dominated by more complex responses such as those involving shoreline erosion, wetland accretion, or barrier island migration. These assessments should first consider the geological setting and the dominant local physical processes at work to determine where inundation might be the primary response. Analysis of lidar elevation data, as outlined above, should then be conducted in those areas.

Investigators conducting sea-level rise impact studies should strive to use approaches that generally follow the guidelines above so that results can be consistent across larger areas and subsequent use of the maps and data can reference a common baseline. Assessment results, ideally with spatially explicit vulnerability maps and summary statistics having all the qualities described in Section 2.4, should be published in peer-reviewed journals so that decision makers can be confident of a sound scientific base for their decisions made on the basis of the findings. If necessary, assessment results can be reformatted into products that are more easily used by local planners and decision makers, but the scientific validity of the information remains.

REFERENCES

* Indicates non-peer reviewed literature; available upon request

* Anthoff, D., Nicholls, R. J., Tol, R. S. J. & Vafeidis, A. T. (2006): *Global and Regional Exposure to Large Rises in Sea-level: A Sensitivity Analysis*. Working paper 96. Tyndall Centre for Climate Change Research, Southampton, UK, 31 pp. <http://www.tyndall.ac.uk/publications/working_papers/twp96.pdf>

* Bin, O., Dumas, C. B., Poulter and Whitehead, J. (2007). *Measuring the Impacts of Climate Change on North Carolina Coastal Resources*. Department of Economics, Appalachian State University, Boone, NC, 91 pp. <http://econ.appstate.edu/climate/>

Bird, E.C.F., (1995). Present and future sea level: the effects of predicted global changes. In: *Climate Change: Impact on Coastal Habitation* [Eisma, D. (ed.)]. CRC Press, Boca Raton, FL, pp. 29-56.

Brock, J. C., Wright, Sallenger, C. W., Krabill, A. H., W. B. & R. N. (2002). Basis and methods of NASA Airborne Topographic Mapper lidar surveys for coastal studies. *Journal of Coastal Research*, *18(1)*, 1-13.

Brinson, M. M., Christian, R. R. & Blum, L. K.(1995). Multiple states in the sea-level induced transition from terrestrial forest to estuary. *Estuaries*, *18(4)*, 648-659.

Carter, R. W. G. & Woodroffe, C. D. (1994). *Coastal Evolution: Late Quaternary Shoreline Morphodynamics*. Cambridge University Press, Cambridge, UK, 517 pp.

CCSP (Climate Change Science Program), (2006). *Prospectus for Synthesis and Assessment Product 4.1: Coastal Elevations and Sensitivity to Sea-level Rise*. U.S. Climate Change Science Program, Washington, DC, 19 pp. http://www.climatescience.gov/ Library/sap/sap4-1/SAP4-1prospectus-final.pdf

Chen, K., (2002). An approach to linking remotely sensed data and areal census data. *International Journal of Remote Sensing*, *23(1)*, 37-48.

* Coastal States Organization[†], 2007: *The Role of Coastal Zone Management Programs in Adaptation to Climate Change*. CSO Climate Change Work Group, Washington, DC, 27 pp.

* Cooper, M. J. P., Beevers, M. D. & Oppenheimer, M. (2005). *Future Sea Level Rise and the New Jersey Coast: Assessing Potential Impacts and Opportunities*. Woodrow Wilson School of Public and International Affairs, Princeton University, Princeton, NJ, 36 pp. <http://www.princeton.edu/~cmi/news/Future%20of%20Sea%20Level%20Rise%20and%20the%20New%20Jersey%20Coast.pdf>

Curray, J. R., (1964). Transgression and regression. In: *Papers in Marine Geology* [Miller, R.L. (ed.)]. McMillan, New York, pp. 175-203.

* Dasgupta, S., Laplante, B., Meisner, C., Wheeler, D. & Yan, J. (2007). *The Impact of Sea Level Rise on Developing Countries: A Comparative Analysis*. World Bank policy research working paper 4136. World Bank, Washington, DC, 51 pp. <http://www-wds.worldbank.org/external/default/WDSContentServer/IW3P/IB/2007/02/09/000016406_20070209161430/Rendered/PDF/wps4136.pdf>

Dean, R. G. & Dalrymple, R. A. (2002). *Coastal Processes with Engineering Applications*. Cambridge University Press, New York, 475 pp.

Demirkesen, A.C., Evrendilek, F., Berberoglu, S. & Kilic, S. (2007). Coastal flood risk analysis using landsat-7 ETM+ imagery and SRTM DEM: A case study of Izmir, Turkey. *Environmental Monitoring and Assessment*, 131(1-3), 293-300.

Demirkesen, A. C., Evrendilek F. & Berberoglu, S. (2008). Quantifying coastal inundation vulnerability of Turkey to sea-level rise. *Environmental Monitoring and Assessment, 138(1-3)*, 101-106.

Eisma, D., (1995). *Climate Change: Impact on Coastal Habitation*. CRC Press, Boca Raton, FL, 260 pp.

Ericson, J. P., Vorosmarty, C. J. Dingman, S. L., Ward, L. G. & Meybeck, M. (2006). Effective sea-level rise and deltas: causes of change and human dimension implications. *Global and Planetary Change, 50(1-2),* 63-82.

Farr, T. G., Rosen, P. A., Caro, E., Crippen, R., Duren, R., Hensley, Kobrick, S. M., Paller, M., E. Rodriguez, Roth, L., Seal, D., Shaffer, S., Shimada, Umland, J., Werner, M., Oskin, M., Burbank, D. & Alsdorf, D. (2007). The Shuttle Radar Topography Mission. *Reviews of Geophysics*, *45*, RG2004, doi:10.1029/2005RG000183.

Federal Geographic Data Committee, (1998). *Geospatial Positioning Accuracy Standards Part 3: National Standard for Spatial Data Accuracy*. FGDC-STD007.3-1998. Federal Geographic Data Committee, Reston, VA, [25 pp.] <http://www.fgdc.gov/standards/projects/FGDC-standardsprojects/accuracy/part3/chapter3>

FEMA (Federal Emergency Management Agency), (1991). *Projected Impact of Relative Sea Level Rise on the National Flood Insurance Program: Report to Congress*. Federal Insurance Administration, Washington, DC, 61 pp. <http://www.epa.gov/climatechange/effects/downloads/flood_insurance.pdf>

* Feyen, J., Hess, K., Spargo, E. Wong, A., White, S. Sellars, J. & S. Gill, (2005). Development of a continuous bathymetric/topographic unstructured coastal flooding model to study sea level rise in North Carolina. In: *Proceedings of the 9th International Conference on Estuarine and Coastal Modeling*, Charleston, South Carolina, October 31-November 2, (2005). America Society of Civil Engineers, Reston, VA, pp. 338-356.

* Feyen, J. C., Brooks, B., Marcy, and Aikman, F. (2008). Advanced inundation modeling and decision-support tools for Gulf coast communities. In: *Proceedings of Solutions to Coastal Disasters 2008*, Turtle Bay, Oahu, Hawaii, April 13-16, 2008. America Society of Civil Engineers, Reston, VA, pp. 361-372.

FitzGerald, D. M., Fenster, M. S., Argow, B. A., and I.V. Buynevich, (2008). Coastal impacts due to sea-level rise. *Annual Review of Earth and Planetary Sciences*, *36*, 601-647.

Fowler, R. A., Samberg, A., Flood, M. J. & Greaves, T. J. (2007). Topographic and terrestrial lidar. In: *Digital Elevation Model Technologies and Applications: The DEM Users*

Manual [Maune, D. (ed.)]. American Society for Photogrammetry and Remote Sensing, Bethesda, MD, 2nd edition, pp. 199-252.

Frumhoff, P. C., McCarthy, J. J., Melillo, J. M., Moser, S. C. & Wuebbles, D. J. (2007). *Confronting Climate Change in the U.S. Northeast: Science, Impacts, and Solutions.* Synthesis report of the Northeast Climate Impacts Assessment. Union of Concerned Scientists, Cambridge, MA, 146 pp. <http://www.climatechoices.org/assets/documents/climatechoices/confrontingclimate-change-in-the-u-s-northeast.pdf>

Gesch, D. B., (2007). The National Elevation Dataset. In: *Digital Elevation Model Technologies and Applications: The DEM Users Manual* [Maune, D. (ed.)]. American Society for Photogrammetry and Remote Sensing, Bethesda, MD, 2nd edition, pp. 99-118.

Gesch, D. B., (2009). Analysis of lidar elevation data for improved identification and delineation of lands vulnerable to sea level rise. *Journal of Coastal Research* (in press).

Gesch, D. & Wilson, R. (2002). Development of a seamless multisource topographic/bathymetric elevation model of Tampa Bay. *Marine Technology Society Journal, 35(4),* 58-64.

Gesch, D. B., Verdin, K. L., and Greenlee, S. K. (1999). New land surface digital elevation model covers the earth. *Eos, Transactions of the American Geophysical Union*, *80(6)*, 69-70.

Gesch, D., Oimoen, M., Greenlee, S., Nelson, C., Steuck, M. & Tyler, D. (2002). The National Elevation Dataset. *Photogrammetric Engineering and Remote Sensing, 68(1),* 5-11.

Glick, P., Clough, J., and Nunley, B. (2008). *Sea-level Rise and Coastal Habitats in the Chesapeake Bay Region*. National Wildlife Federation, [Washington, DC], 121 pp. <http://www.nwf.org/sealevelrise/pdfs/SeaLevelRiseandCoastalHabitats_ChesapeakeRegion.pdf>

Gornitz, V., Couch, S. & Hartig, E. K. (2002). Impacts of sea level rise in the New York City metropolitan area. *Global and Planetary Change, 32(1)*, 61-88.

Greenwalt, C. R. & Shultz, M. E. (1962). *Principles of Error Theory and Cartographic Applications.* ACIC technical report no. 96. United States Air Force, Aeronautical Chart and Information Center, St. Louis, MO, 60 pp.

Guenther, G. C., (2007). Airborne lidar bathymetry. In: *Digital Elevation Model Technologies and Applications: The DEM Users Manual* [Maune, D. (ed.)]. American Society for Photogrammetry and Remote Sensing, Bethesda, MD, 2nd edition, pp. 253-320.

Hastings, D. A. & Dunbar, P. K. (1998). Development and assessment of the Global Land One-km Base Elevation digital elevation model (GLOBE). *International Archives of Photogrammetry and Remote Sensing*, *32(4),* 218-221.

Jacob, K., V. Gornitz, and Rosenzweig, C. (2007). Vulnerability of the New York City metropolitan area to coastal hazards, including sea-level rise: inferences for urban coastal risk management and adaptation policies. In: *Managing Coastal Vulnerability* [McFadden, L., R. Nicholls, and E. Penning-Rowsell (eds.)]. Elsevier, Amsterdam and Oxford, pp. 139-156.

Johnson, D. W., 1919: *Shoreline Processes and Shoreline Development.* John Wiley, New York, 584 pp.

Johnson, Z., Barlow, R., Clark, I., Larsen, C. & Miller, K. (2006). *Worcester County Sea Level Rise Inundation Model.* Technical report DNR publication no. 14-982006-166.

Maryland Department of Natural Resources, Annapolis, MD, and U.S. Geological Survey, Reston, VA, 15 pp. <http://www.dnr.state.md.us/bay/czm/wcslrreport.html>

Kafalenos, R. S., Leonard, K. J., Beagan, D. M., Burkett, V. R., Keim, B. D. Meyers, A., Hunt, D. T., Hyman, R. C., Maynard, M. K., Fritsche, B. Henk, R.H., E. J. Seymour, L. E. Olson, J. R. Potter, and Savonis, M. J. (2008). What are the implications of climate change and variability for Gulf coast transportation? In: *Impacts of Climate Change and Variability on Transportation Systems and Infrastructure: Gulf Coast Study, Phase I.* A Report by the U.S. Climate Change Science Program and the Subcommittee on Global Change Research. [Savonis, M.J., V.R. Burkett, and J.R. Potter (eds.)]. Department of Transportation, Washington, DC, 104 pp. <http://www.climatescience.gov/ Library/sap/sap4-7/final-report/sap4-7-finalch4.pdf>

Kleinosky, L. R., Yarnal, B. & Fisher, A. (2007). Vulnerability of Hampton Roads, Virginia to storm-surge flooding and sea-level rise. *Natural Hazards, 40(1),* 4370.

Komar, P. D., (1983). *Handbook of Coastal Processes and Erosion.* CRC Press, Boca Raton, FL, 305 pp.

Komar, P. D., (1998). *Beach Processes and Sedimentation.* Prentice Hall, New Jersey, 2nd edition, 544 pp.

Larsen, C., Clark, I., Guntenspergen, G. R., Cahoon, D. R., Caruso, V., Hupp, C. & Yanosky, T. (2004). *The Blackwater NWR Inundation Mode. Rising Sea Level on a Low-lying Coast: Land Use Planning for Wetlands.* U.S. Geological Survey open file report 04–1302. U.S. Geological Survey, Reston, VA. <http://pubs.usgs.gov/of/2004/1302/>

* Lathrop, R. G., Jr., and Love, A. (2007). *Vulnerability of New Jersey's Coastal Habitats to Sea Level Rise.* Grant F. Walton Center for Remote Sensing and Spatial Analysis, Rutgers University, New Brunswick, NJ, and American Littoral Society, Highlands, NJ, 17 pp. <http://crssa.rutgers.edu/projects/coastal/sealevel/>

Leatherman, S. P., (1990). Modeling shore response to sea-level rise on sedimentary coasts. *Progress in Physical Geography, 14(4)*, 447-464.

Leatherman, S. P., (2001). Social and economic costs of sea level rise. In: *Sea Level Rise: History and Consequences.* [Douglas, B.C., M.S. Kearney, and S.P. Leatherman (eds.)]. Academic Press, San Diego, pp. 181-223.

Marbaix, P. & Nicholls, R. J. (2007). Accurately determining the risks of rising sea level. *Eos, Transactions of the American Geophysical Union, 88(43)*, 441, 442.

Marfai, M. A. & King, L. (2008). Potential vulnerability implications of coastal inundation due to sea level rise for the coastal zone of Semarang city, Indonesia. *Environmental Geology, 54(6)*, 1235-1245.

Maune, D. F., (2007). DEM user applications. In: *Digital Elevation Model Technologies and Applications: The DEM Users Manual* [Maune, D. (ed.)]. American Society for Photogrammetry and Remote Sensing, Bethesda, MD, 2nd edition, pp. 391423.

Maune, D. F., Kopp, S. M., Crawford, C. A. & Zervas, C. E. (2007a). Introduction. In: *Digital Elevation Model Technologies and Applications: The DEM Users Manual* [Maune, D. (ed.)]. American Society for Photogrammetry and Remote Sensing, Bethesda, MD, 2nd edition, pp. 1-35.

Maune, D. F., Maitra, J. B. & McKay, E. J. (2007b). Accuracy standards & guidelines. In: *Digital Elevation Model Technologies and Applications: The DEM Users Manual* [Maune, D. (ed.)]. American Society for Photogrammetry and Remote Sensing, Bethesda, MD, 2nd edition, pp. 65-97.

* Mazria, E. & Kershner, K. (2007). *Nation Under Siege: Sea Level Rise at Our Doorstep.* The 2030 Research Center, Santa Fe, NM, 34 pp. <http://www.architecture2030.org/pdfs/nation_under_siege.pdf>

McGranahan, G., Balk, D. & Anderson, B. (2007). The rising tide: assessing the risks of climate change and human settlements in low elevation coastal zones. *Environment & Urbanization, 19(1),* 17-37.

Mennis, J., (2003). Generating surface models of population using dasymetric mapping. *The Professional Geographer, 55(1),* 31-42.

Merwade, V., Olivera, F., Arabi, M., and Edleman, S. (2008). Uncertainty in flood inundation mapping: current issues and future directions. *Journal of Hydrologic Engineering, 13(7),* 608-620.

Monmonier, M., (2008). High-resolution coastal elevation data: The key to planning for storm surge and sea level rise. In: *Geospatial Technologies and Homeland Security: Research Frontiers and Future Challenges.* [Sui, D.Z. (ed.)]. Springer, Dordrecht and London, pp. 229-240.

* Myers, E. P., (2005). Review of progress on VDatum, a vertical datum transformation tool. In: *Oceans 2005: Proceedings of the MTS/IEEE "One Ocean" Conference*, Washington, DC, September 18–23, 2005. IEEE, Piscataway, NJ, v. 2, pp. 974 980.

Najjar, R. G., Walker, H. A., Anderson, H. A., Barron, E. J., Bord, R. J., Gibson, J. R., Kennedy, V. S., Knight, C. G., Megonigal, J. P., O'Connor, R.E., Polsky, C. D., N. P. Psuty, B.A. Richards, Sorenson, L. G. Steele, E.M. & Swanson, R.S. (2000). The potential impacts of climate change on the mid-Atlantic coastal region. *Climate Research, 14(3)*, 219-233.

National Digital Elevation Program, (2004). *Guidelines for Digital Elevation Data – Version 1.* National Digital Elevation Program, [Reston, VA], 93 pp.

Nayegandhi, A., Brock, J. C., Wright, C. W. & O'Connell, M. J. (2006). Evaluating a small footprint, waveform-resolving lidar over coastal vegetation communities. *Photogrammetric Engineering and Remote Sensing, 72(12)*, 1407-1417.

NOAA (National Oceanic and Atmospheric Administration), (2001). *Tidal Datums and Their Applications.* NOAA special publication NOS CO-OPS 1. NOAA National Ocean Service, Silver Spring, MD, 112 pp. <http://tidesandcurrents.noaa.gov/publications/tidal_datums_ and_their_applicatio ns.pdf>

NOAA (National Oceanic and Atmospheric Administration), (2008). *Topographic and Bathymetric Data Considerations: Datums, Datum Conversion Techniques, and Data Integration.* Technical report NOAA/CSC/20718-PUB. National Oceanic and Atmospheric Administration, Charleston, SC, 18 pp. <http://www.csc.noaa.gov/topobathy/>

NRC (National Research Council), (2004). *A Geospatial Framework for the Coastal Zone: National Needs for Coastal Mapping and Charting.* National Academies Press, Washington, DC, 149 pp.

NRC (National Research Council), (2007). *Elevation Data for Floodplain Mapping.* National Academies Press, Washington, DC, 151 pp.

Osborn, K., J. List, Gesch, Crowe, D. J., Merrill, G., Constance, E., Mauck, J., Lund, C., Caruso, V. & Kosovich, J. (2001). National digital elevation program (NDEP). In: *Digital Elevation Model Technologies and Applications: The DEM Users Manual* [Maune, D.

(ed.)]. American Society for Photogrammetry and Remote Sensing, Bethesda, MD, 2nd edition, pp. 83-120.

Parker, B., Hess, K., Milbert, D. & Gill, S. (2003). A national vertical datum transformation tool. *Sea Technology*, *44(9)*, 10-15.

Pilkey, O. H. & Cooper, J. A. G. (2004). Society and sea level rise. *Science*, *303(5665)*, 1781-1782.

Pilkey, O. H. & Thieler, E. R. (1992). Erosion of the U.S. shoreline. In: *Quaternary Coasts of the United States: Marine and Lacustrine Systems*. [Fletcher, C.H. & J.F. Wehmiller (eds.)]. Special publication no. 48. Society of Economic Paleontologists and Mineralogists, Tulsa, OK, pp. 3-8.

Poulter, B. & Halpin, P. N. (2007). Raster modelling of coastal flooding from sea-level rise. *International Journal of Geographical Information Science*, *22(2)*, 167-182.

Poulter, B., Goodall, J. L. & Halpin, P. N. (2008). Applications of network analysis for adaptive management of artificial drainage systems in landscapes vulnerable to sea level rise. *Journal of Hydrology*, *357(3-4)*, 207-217.

Rowley, R. J., Kostelnick, J. C., Braaten, D. Li, X. & Meisel, J. (2007). Risk of rising sea level to population and land area. *Eos, Transactions of the American Geophysical Union*, *88(9)*, 105, 107.

* Rubinoff, P., Vinhateiro, N. D. & Piecuch, C. (2008). *Summary of Coastal Program Initiatives that Address Sea Level Rise as a Result of Global Climate Change*. Rhode Island Sea Grant/Coastal Resources Center, University of Rhode Island, Narragansett, RI, 50 pp. <http://seagrant.gso.uri.edu/ccd/slr/SLR_policies_summary_Mar6_final.pdf>

Sallenger Jr., A. H., Krabill, W. B., Swift, R. N., Brock, J., List, J., Hansen, M., Holman, R. A., Manizade, S. Sontag, J. Meredith, A., Morgan, K., Yunkel, J.K., Frederick, E.B. & Stockdon, H. (2003). Evaluation of airborne topographic lidar for quantifying beach changes. *Journal of Coastal Research*, *19(1)*, 125-133.

Schneider, S. H. & Chen, R. S., (1980). Carbon dioxide warming and coastline flooding: physical factors and climatic impact. *Annual Review of Energy*, 5, 107-140.

* Seiden, E. (ed.), 2008: *Climate Change: Science, Education and Stewardship for Tomorrow's Estuaries*. National Estuarine Research Reserve System, Silver Spring, MD, 16 pp. <http://nerrs08.elkhornslough.org/files/NERRS_Climate_Change_Strategy_Paper_7.30.08.pdf>

Sleeter, R. & Gould, M. (2008). *Geographic Information System Software to Remodel Population Data Using Dasymetric Mapping Methods*. U.S. Geological Survey techniques and methods report 11-C2. U.S. Geological Survey, Reston, VA, 15 pp. <http://pubs.usgs.gov/tm/tm11c2/tm11c2.pdf>

Slovinsky, P. A. & Dickson, S. M. (2006). *Impacts of Future Sea Level Rise on the Coastal Floodplain*. MGS open-file 06-14. Maine Geological Survey, Augusta, 26 pp. <http://maine.gov/doc/nrimc/mgs/explore/marine/sea-level/mgs-open-file06-14.pdf>

Small, C. & R.J. Nicholls, 2003: A global analysis of human settlement in coastal zones. *Journal of Coastal Research*, *19(3)*, 584-599.

* Stanton, E.A. & Ackerman, F. (2007). *Florida and Climate Change: The Costs of Inaction*. Tufts University, Medford, MA, 91 pp. <http://www.ase.tufts.edu/gdae/Pubs/rp/Florida_hr.pdf>

Stockdon, H. F., Lillycrop, W. J., Howd, P. A. & Wozencraft, J. A. (2007). The need for sustained and integrated high-resolution mapping of dynamic coastal landforms. *Marine Technology Society Journal, 40(4),* 90-99.

Stoker, J., Parrish, J., Gisclair, D., Harding, Haugerud, R., Flood, M., Andersen, H., Schuckman, K., Maune, D. P. Rooney, K. Waters, A. Habib, E. Wiggins, B. Ellingson, Jones, B., Nechero, S., A. Nayegandhi, T. Saultz, and Lee, G. (2007). *Report of the First National Lidar Initiative Meeting, February 14-16, 2007, Reston, VA.* U.S. Geological Survey open-file report 2007-1189. U.S. Geological Survey, Reston, VA, 64 p.

Stoker, J., Harding, D. & Parrish, J. (2008). The need for a national lidar dataset. *Photogrammetric Engineering and Remote Sensing*, *74(9)*, 1065-1067.

Subcommittee on Disaster Reduction, (2008). *Coastal Inundation: Grand Challenges for Disaster Reduction Implementation Plans.* Committee on Environment and Natural Resources, Office of Science and Technology Policy, Executive Office of the President, Washington, DC, 4 pp. <http://www.sdr.gov/185820_Coastal_FINAL.pdf>

Swift, D. J. P., Niederoda, A. W., Vincent, C. E. & Hopkins, T.S. (1985). Barrier island evolution, middle Atlantic shelf, USA, Part I: shoreface dynamics. *Marine Geology*, *63(1-4)*, 331-361.

Titus, J. G. & Richman, C. (2001). Maps of lands vulnerable to sea level rise: modeled elevations along the US Atlantic and Gulf coasts. *Climate Research*, *18(3)*, 205-228.

Titus, J.G. & Wang, J. (2008). Maps of lands close to sea level along the middle Atlantic coast of the United States: an elevation data set to use while waiting for LIDAR. Section 1.1 in: *Background Documents Supporting Climate Change Science Program Synthesis and Assessment Product 4.1: Coastal Elevations and Sensitivity to Sea Level Rise* [Titus, J.G. & E.M. Strange (eds.)]. EPA 430R07004. U.S. Environmental Protection Agency, Washington, DC, pp. 2-44.

Titus, J. G., Park, R. A., Leatherman, S. P., Weggel, J. R., Greene, M. S., Mausel, P. W., Brown, S., Gaunt, G., Threhan, M. & Yohe, G. (1991). Greenhouse effect and sea level rise: the cost of holding back the sea. *Coastal Management*, *19(2),* 171-204.

US DOT, (2008). *The Potential Impacts of Global Sea Level Rise on Transportation Infrastructure, Phase 1 - Final Report: the District of Columbia, Maryland, North Carolina and Virginia.* Center for Climate Change and Environmental Forecasting, U.S. Department of Transportation, Washington DC.

U.S. EPA (Environmental Protection Agency), (1989). *The Potential Effects of Global Climate Change on the United States: Report to Congress.* EPA 230-05-89-050. U.S. Environmental Protection Agency, Washington, DC. <http://yosemite.epa.gov/oar/globalwarming.nsf/UniqueKeyLookup/RAMR5CK NNG/$File/potential_effects.pdf>

USGS (U.S. Geological Survey), (1999). *Map Accuracy Standards.* U.S. Geological Survey fact sheet FS-171-99. [U.S. Geological Survey, Reston, VA], 2 pp. <http://edc2.usgs.gov/pubslists/factsheets/fs17199.pdf>

Valiela, I., (2006). *Global Coastal Change.* Blackwell Publishing, Oxford, UK, 376 pp.

Wells, J. T., (1995). Effects of sea level rise on coastal sedimentation and erosion. In: *Climate Change: Impact on Coastal Habitation* [Eisma, D. (ed.)]. CRC Press, Boca Raton, FL, pp. 111-136.

Wright, L. D., (1995). *Morphodynamics of Inner Continental Shelves.* CRC Press, Boca Raton, FL, 241 pp.

Wu, S.-Y., Yarnal, B. & Fisher, A. (2002). Vulnerability of coastal communities to sea-level rise: a case study of Cape May County, New Jersey, USA. *Climate Research, 22(3)*, 255-270.

Wu, S., J. Li, and Huang, G. H. (2008). Characterization and evaluation of elevation data uncertainty in water resources modeling with GIS. *Water Resources Management, 22(8)*, 959-972.

* Yilmaz, M., Usul, N. & Akyurek, Z. (2004). Modeling the propagation of DEM uncertainty in flood inundation. In: *Proceedings of the 24th Annual ESRI International User Conference*, August 9–13, 2004, San Diego, CA. 10 pp. <http://gis.esri.com/library/userconf/proc04/docs/pap1039.pdf>

* Yilmaz, M., Usul, N. & Akyurek, Z. (2005). Modeling the propagation of DEM uncertainty on flood inundation depths. In: *Proceedings of the 25th Annual ESRI International User Conference*, July 25-29, 2005, San Diego, CA. 8 pp. <http://gis.esri.com/library/userconf/proc05/papers/pap1996.pdf>

Zilkoski, D. B., (2007). Vertical datums. In: *Digital Elevation Model Technologies and Applications: The DEM Users Manual* [Maune, D. (ed.)]. American Society for Photogrammetry and Remote Sensing, Bethesda, MD, 2nd edition, pp. 37-64.

In: Coastal Sensitivity to Sea Level Rise-Focusing on the Mid-... ISBN: 978-1-60741-440-7
Editor: Melvin C. Urajner

Chapter 3

OCEAN COASTS

Benjamin T. Gutierrez, S. Jeffress Williams and E. Robert Thieler
USGS

Key Findings

- Along the ocean shores of the Mid-Atlantic, which are comprised of headlands, barrier islands, and spits, it is *virtually certain* that erosion will dominate changes in shoreline position in response to sea-level rise and storms over the next century.
- It is *very likely* that landforms along the mid-Atlantic coast of the United States will undergo large changes if the higher sea-level rise scenarios occur. The response will vary depending on the type of coastal landforms and the local geologic and oceanographic conditions, and could be more variable than the changes observed over the last century.
- For higher sea-level rise scenarios, it is *very likely* that some barrier island coasts will cross a threshold and undergo significant changes. These changes include more rapid landward migration or segmentation of some barrier islands.

3.1. Introduction

The general characteristics of the coast, such as the presence of beaches *versus* cliffs, reflects a complex and dynamic interaction between physical processes (*e.g.,* waves and tidal currents) that act on the coast, availability of sediment transported by waves and tidal currents, underlying geology, and changes in sea level (see review in Carter and Woodroffe, 1994a). Variations in these factors from one region to the next are responsible for the different coastal landforms, such as beaches, barrier islands, and cliffs that are observed along the coast today. Based on studies of the geologic record, the scope and general nature of the changes that can occur in response to sea-level rise are widely recognized (Curray, 1964; Carter and Woodroffe, 1994a; FitzGerald *et al.*, 2008). On the other hand, determining precisely how these changes occur in response to a specific rise in sea level has been difficult.

Part of the complication arises due to the range of physical processes and factors that modify the coast and operate over a range of time periods (*e.g.,* from weeks to centuries to thousands of years) (Cowell and Thom, 1994; Stive *et al.*, 2002; Nicholls *et al.*, 2007). Because of the complex interactions between these factors and the difficulty in determining their exact influence, it has been difficult to resolve a quantitative relationship between sea-level rise and shoreline change (*e.g.*, Zhang *et al.*, 2004; Stive, 2004). Consequently, it has been difficult to reach a consensus among coastal scientists as to whether or not sea-level rise can be quantitatively related to observed shoreline changes and determined using quantitative models (Dubois, 2002; Stive, 2004; Pilkey and Cooper, 2004; Cowell *et al.*, 2006).

Along many U.S. shores, shoreline changes are related to changes in the shape of the landscape at the water's edge (*e.g.*, the shape of the beach). Changes in beach dimensions, and the resulting shoreline changes, do not occur directly as the result of sea-level rise but are in an almost continual state of change in response to waves and currents as well as the availability of sediment to the coastal system (see overviews in Carter and Woodroffe, 1994b; Stive *et al.*, 2002; Nicholls *et al.*, 2007). This is especially true for shoreline changes observed over the past century, when the increase in sea level has been relatively small (about 30 to 40 centimeters, or 12 to16 inches, along the mid-Atlantic coast). During this time, large storms, variations in sediment supply to the coast, and human activity have had a more obvious influence on shoreline changes. Large storms can cause changes in shoreline position that persist for weeks to a decade or more (Morton *et al.*, 1994; Zhang *et al.*, 2002, 2004; List *et al.*, 2006; Riggs and Ames, 2007). Complex interactions with nearshore sand bodies and/or underlying geology (the geologic framework), the mechanics of which are not yet clearly understood, also influence the behavior of beach morphology over a range of time periods (Riggs *et al.*, 1995; Honeycutt and Krantz, 2003; Schupp *et al.*, 2006; Miselis and McNinch, 2006). In addition, human actions to control changes to the shore and coastal waterways have altered the behavior of some portions of the coast considerably (*e.g.*, Assateague Island, Maryland, Dean and Perlin, 1977; Leatherman, 1984; also see reviews in Nordstrom, 1994, 2000; Nicholls *et al.*, 2007).

It is even more difficult to develop quantitative predictions of how shorelines may change in the future (Stive, 2004; Pilkey and Cooper, 2004; Cowell *et al.*, 2006). The most easily applied models incorporate relatively few processes and rely on assumptions that do not always apply to real-world settings (Thieler *et al.*, 2000; Cooper and Pilkey, 2004). In addition, model assumptions often apply best to present conditions, but not necessarily to future conditions. Models that incorporate more factors are applied at specific locations and require precise knowledge regarding the underlying geology or sediment budget (*e.g.*, GEOMBEST, Stolper *et al.*, 2005), and it is therefore difficult to apply these models over larger coastal regions. Appendix 2 presents brief summaries of a few basic methods that have been used to predict the potential for shoreline changes in response to sea-level rise.

As discussed in Chapter 2, recent and ongoing assessments of sea-level rise impacts commonly examine the vulnerability of coastal lands to inundation by specific sea-level rise scenarios (*e.g.,* Najjar *et al.*, 2000; Titus and Richman, 2001; Rowley *et al.*, 2007). This approach provides an estimate of the land area that may be vulnerable, but it does not incorporate the processes (*e.g.*, barrier island migration) nor the environmental changes (*e.g.*, salt marsh deterioration) that may occur as sea level rises. Because of these complexities, inundation can be used as a basic approach to approximate the extent of land areas that could be affected by changing sea level. Because the majority of the U.S. coasts, including those along the Mid-Atlantic, consist of sandy shores, inundation alone is unlikely to reflect the potential consequences of sea-level rise. Instead, long-term shoreline changes will involve contributions from both inundation and erosion (Leatherman, 1990, 2001) as well as changes to other coastal environments such as wetland losses.

Most portions of the open coast of the United States will be subject to significant physical changes and erosion over the next century because the majority of coastlines consist of sandy beaches which are highly mobile and in a continual state of change. This Chapter presents an overview and assessment of the important factors and processes that influence potential changes to the mid-Atlantic ocean coast due to sea-level rise expected by the end of this century. This overview is based in part on a panel assessment (*i.e.*, expert judgement) that was undertaken to address this topic for this Product (Gutierrez *et al.,* 2007). The panel assessment process is described in Section 3.2 and Box 3.1. Section 3.3 reviews the geological characteristics of the mid-Atlanic coast. Section 3.4 provides an overview of the basic factors that influence sea-level rise-driven shoreline changes. Sections 3.5 and 3.6 describe the coastal landforms of the mid-Atlantic coast of the United States and what is known regarding how these landforms respond to changes in sea-level based on a literature review included as part of the panel assessment (Gutierrez *et al.*, 2007). The potential responses of mid-Atlantic coastal landforms to sea-level rise, which were defined in the panel assessment, are presented in Section 3.7 and communicated using the likelihood terms specified in the Preface (see Figure P.1).

3.2. Assessing the Potential Impact of Sea-Level Rise on the Ocean Coasts of the Mid-Atlantic

Lacking a single agreed-upon method or scientific consensus view about shoreline changes in response to sea-level rise at a regional scale, a panel was consulted to address the key question that guided this Chapter (Gutierrez *et al.*, 2007). The panel consisted of coastal scientists whose research experiences have focused on the mid-Atlantic region and have been involved with coastal management in the mid-Atlantic region[1]. The panel discussed the changes that might be expected to occur to the ocean shores of the U.S. mid-Atlantic coast in response to predicted accelerations in sea-level rise over the next century, and considered the important geologic, oceanographic, and anthropogenic factors that contribute to shoreline changes in this region. The assessment presented here is based on the professional judgment of the panel. This qualitative assessment of potential changes that was developed by the panel is based on an understanding of both coastal science literature and their personal field observations.

This assessment focuses on four sea-level rise scenarios. As defined in the Preface and Chapter 1, the first three sea-level rise scenarios (Scenarios 1 through 3) assume that: (1) the sea-level rise rate observed during the twentieth century will persist through the twenty-first century; (2) the twentieth century rate will increase by 2 millimeters (mm) per year, and (3) the twentieth century rate will increase by 7 mm per year. Lastly, a fourth scenario is discussed, which considers a 2-meter (6.6-foot) rise over the next few hundred years. In the following discussions, sea-level change refers to the relative sea-level change, which is the combination of global sea-level change and local change in land elevation. Using these scenarios, this assessment focuses on:

- Identifying important factors and processes contributing to shoreline change over the next century;
- Identifying key geomorphic settings along the coast of the mid-Atlantic region;
- Defining potential responses of shorelines to sea-level rise; and
- Assessing the likelihood of these responses.

Box 3.1. The Panel Assessment Process Used in SAP 4.1, Chapter 3

As described in this Product, there is currently a lack of scientific consensus regarding local-, regional-, and national-scale coastal changes in response to sea-level rise, due to limited elevation and observational data and lack of adequate scientific understanding of the complex processes that contribute to coastal change. To address the question of potential future changes to the mid-Atlantic coast posed in the SAP 4.1 Prospectus, the authors assembled 13 coastal scientists for a meeting to evaluate the potential outcomes of the sea-level rise scenarios used in this Product. These scientists were chosen on the basis of their technical expertise and experience in the coastal research community, and also their involvement with coastal management issues in the mid-Atlantic region. Prior to the meeting, the scientists were provided with documents describing the Climate Change Science Program, and the Prospectus for this Product. The Prospectus included key questions and topics that the panel was charged to address. The panel was also provided a draft version of the report by Reed et al. (2008), which documented a similar panel-assessment approach used in developing Chapter 4 of this Product.

The sea-level rise impact assessment effort was conducted as an open discussion facilitated by the USGS authors over a two-day period. The main topics that the panel discussed were:

9. 1) approaches that can be used to conduct long-term assessments of coastal change;
10. 2) key geomorphic environments in the mid-Atlantic region from Long Island, NY to North Carolina;
11. 3) potential responses of these environments to sea-level rise based on an understanding of important factors and processes contributing to coastal change; and
12. 4) the likelihood of these responses to the sea-level rise scenarios used in this Product (see Section 3.7).

The qualitative, consensus-based assessment of potential changes and their likelihood developed by the panel was based on their review and understanding of peer reviewed published coastal science literature, as well as field observations drawn from other studies conducted in the mid-Atlantic region. The likelihood statements reported in Section 3.7 were determined based on the results of the discussion during the two-day meeting and revised according comments from panelists during the drafting of a summary report. The

USGS report (Gutierrez *et al.*, 2007) summarizing the process used, the basis in the published literature, and a synthesis of the resulting assessment was produced based on results of the meeting, reviewed as part of the USGS peer review process, and approved by members of the panel.

3.3. Geological Character of the Mid-Atlantic Coast

The mid-Atlantic margin of the United States is a gently sloping coastal plain that has accumulated over millions of years in response to the gradual erosion of the Appalachian mountain chain. The resulting sedimentation has constructed a broad coastal plain and a continental shelf that extends almost 300 kilometers (approximately 185 miles) seaward of the present coast (Colquhoun *et al.*, 1991). The current morphology of this coastal plain has resulted from the incision of rivers that drain the region and the construction of barrier islands along the mainland occurring between the river systems. Repeated ice ages, which have resulted in sea-level fluctuations up to 140 meters (460 feet) (Muhs *et al.*, 2004), caused these rivers to erode large valleys during periods of low sea level that then flooded and filled with sediments when sea levels rose. The northern extent of the mid-Atlantic region considered in this Product, Long Island, New York, was also shaped by the deposition of glacial outwash plains and moraines that accumulated from the retreat of the Laurentide ice sheet, which reached its maximum extent approximately 21,000 years ago. This sloping landscape that characterizes entire mid-Atlantic margin, in combination with slow rates of sea-level rise over the past 5,000 years and sufficient sand supply, is also thought to have enabled the formation of the barrier islands that comprise the majority of the Atlantic Coast (Walker and Coleman, 1987; Psuty and Ofiara, 2002).

The mid-Atlantic coast is generally described as a sediment-starved coast (Wright, 1995). Presently, sediments from the river systems of the region are trapped in estuaries and only minor amounts of sediment are delivered to the open ocean coast (Meade, 1969, 1972). In addition, these estuaries trap sandy sediment from the continental shelf (Meade, 1969). Consequently, the sediments that form the mainland beach and barrier beach environments are thought to be derived mainly from the wave-driven erosion of the mainland substrate and sediments from the seafloor of the continental shelf (Niedoroda *et al.*, 1985; Swift *et al.*, 1985; Wright, 1995). Since the largest waves and associated currents occur during storms along the Atlantic Coast, storms are often thought to be significant contributors to coastal changes (Niedoroda *et al.*, 1985; Swift *et al.*, 1985; Morton and Sallenger, 2003).

The majority of the open coasts along the mid-Atlantic region are sandy shores that include the beach and barrier environments. Although barriers comprise only 15 percent of the world coastline (Glaeser, 1978), they are the dominant shoreline type along the Atlantic Coast. Along the portion of the mid-Atlantic coast examined here, which ranges between Montauk, New York and Cape Lookout, North Carolina, barriers line the majority of the open coast. Consequently, scientific investigations exploring coastal geology of this portion of North America have focused on understanding barrier island systems (Fisher, 1962, 1968; Pierce and Colquhoun, 1970; Kraft, 1971; Leatherman, 1979; Moslow and Heron, 1979, 1994; Swift, 1975; Nummedal, 1983; Oertel, 1985; Belknap and Kraft, 1985; Hine and Snyder, 1985; Davis, 1994).

3.4. Important Factors for Mid-Atlantic Shoreline Change

Several important factors influence the evolution of the mid-Atlantic coast in response to sea-level rise including: (1) the geologic framework, (2) physical processes, (3) the sediment supply, and (4) human activity. Each of these factors influences the response of coastal landforms to changes in sea level. In addition, these factors contribute to the local and regional variations of sea-level rise impacts that are difficult to capture using quantitative prediction methods.

3.4.1. Geologic Framework

An important factor influencing coastal morphology and behavior is the underlying geology of a setting, which is also referred to as the geological framework (Belknap and Kraft, 1985; Demarest and Leatherman, 1985; Schwab *et al.*, 2000). On a large scale, an example of this is the contrast in the characteristics of the Pacific Coast *versus* the Atlantic Coast of the United States. The collision of tectonic plates along the Pacific margin has contributed to the development of a steep coast where cliffs line much of the shoreline (Inman and Nordstrom, 1971; Muhs *et al.*, 1987; Dingler and Clifton, 1994; Griggs and Patsch, 2004; Hapke *et al.*, 2006; Hapke and Reid, 2007). While common, sandy barriers and beaches along the Pacific margin are confined to river mouths and low-lying coastal plains that stretch between rock outcrops and coastal headlands. On the other hand, the Gulf of Mexico and Atlantic coasts of the United States are situated on a passive margin where tectonic activity is minor (Walker and Coleman, 1987). As a result, these coasts are composed of wide coastal plains and wide continental shelves extending far offshore. The majority of these coasts are lined with barrier beaches and lagoons, large estuaries, isolated coastal capes, and mainland beaches that abut high grounds in the surrounding landscape.

From a smaller-scale perspective focused on the mid-Atlantic region, the influence of the geological framework involves more subtle details of the regional geology. More specifically, the distribution, structure, and orientation of different rock and sediment units, as well as the presence of features such as river and creek valleys eroded into these units, provides a structural control on a coastal environment (*e.g.*, Kraft, 1971; Belknap and Kraft, 1985; Demarest and Leatherman, 1985; Fletcher *et al.*, 1990; Riggs *et al.*, 1995; Schwab *et al.*, 2000; Honeycutt and Krantz, 2003). Moreover, the framework geology can control (1) the location of features, such as inlets, capes, or sand-ridges, (2) the erodibility of sediments, and (3) the type and abundance of sediment available to beach and barrier island settings. In the mid-Atlantic region, the position of tidal inlets, estuaries, and shallow water embayments can be related to the existence of river and creek valleys that were present in the landscape during periods of lower sea level in a number of cases (*e.g.*, Kraft, 1971; Belknap and Kraft, 1985; Fletcher *et al.*, 1990). Elevated regions of the landscape, which can often be identified by areas where the mainland borders the ocean coast, form coastal headlands. The erosion of these features supplies sand to the nearshore system. Differences in sediment composition (*e.g.*, sediment size or density), can sometimes be related to differences in shoreline retreat rates (*e.g.*, Honeycutt and Krantz, 2003). In addition, the distribution of underlying geological units (rock outcrops, hard-grounds, or sedimentary strata) in shallow regions offshore of the coast can modify waves and currents and influencing patterns of sediment erosion, transport, and deposition on the adjacent shores (Riggs *et al.*, 1995; Schwab *et al.*, 2000). These complex interactions with nearshore sand bodies and/or underlying geology can also influence the behavior of beach morphology over a range of time scales (Riggs *et al.*, 1995; Honeycutt and Krantz, 2003; Schupp *et al.*, 2006; Miselis and McNinch, 2006).

3.4.2. Physical Processes

The physical processes acting on the coast are a principal factor shaping coastal landforms and consequently changes in shoreline position (see reviews in Davis, 1987; Komar, 1998). Winds, waves, and tidal currents continually erode, rework, winnow, redistribute, and shape the sediments that make up these landforms. As a result, these forces also have a controlling influence on the composition and morphology of coastal landforms such as beaches and barrier islands.

Winds have a range of effects on coastal areas. They are the main cause of waves and also generate currents that transport sediments in shallow waters. In addition, winds are a significant mechanism transporting sand along beaches and barrier islands that generate and sustain coastal dunes.

Waves are either generated by local winds or result from far-away disturbances such as large storms out at sea. As waves propagate into shallow water, their energy decreases but they are also increasingly capable of moving the sediment on the seabed. Close to shore each passing wave or breaking wave suspends sediments off the seabed. Once suspended above the bottom, these sediments can be carried by wave- or tide-generated currents.

Wave-generated currents are important agents of change on sandy shores. The main currents that waves generate are longshore currents, rip currents, and onshore and offshore directed currents that accompany the surge and retreat of breaking waves. Longshore currents are typically the most important for sediment transport that influences changes in shoreline position. Where waves approach the coast at an angle, longshore currents are generated. The speed of these currents varies, depending on the wave climate (*e.g.*, average wave height and direction) and more specifically, on the power and angle of approach of the waves (*e.g.*, high waves during storms, low waves during fair weather). These currents provide a mechanism for sand transport along the coast, referred to as littoral transport, longshore drift, or longshore transport. During storms, high incoming waves can generate longshore currents exceeding 1 meter (3 feet) per second and storm waves can transport thousands of cubic meters of sand in a relatively short time period, from hours to days. During calm conditions, waves are weaker but can still gradually transport large volumes of sand over longer time periods, ranging from weeks to months. Where there are changes in coastal orientation, the angle at which waves approach the coast changes and can lead to local reversals in longshore sediment transport. These variations can result in the creation of abundances or deficits of longshore sediment transport and contribute to the seaward growth or landward retreat of the shoreline at a particular location (*e.g.*, Cape Lookout, North Carolina, McNinch and Wells [1999]).

The effect of tidal currents on shores is more subtle except for regions near the mouths of inlets, bays, or areas where there is a change in the orientation of the shore. The rise and fall of the water level caused by tides moves the boundary between the land and sea (the shoreline), causing the level that waves act on a shore to move as well. In addition, this controls the depth of water which influences the strength of breaking waves. In regions where there is a large tidal range, there is a greater area over which waves can act on a shore. The rise and fall of the water level also generates tidal currents. Near the shore, tidal currents are small in comparison to wave-driven currents. Near tidal inlets and the mouths of bays or estuaries, tidal currents are strong due to the large volumes of water that are transported through these conduits in response to changing water levels. In these settings, tidal currents transport sediment from ocean shores to back-barrier wetlands, inland waterways on flood tides and vice versa on ebb tides. Aside from these settings, tidal currents are generally small along the mid-Atlantic region except near changes in shoreline orientation or sand banks (*e.g.*, North Carolina Capes, Cape Henlopen, Delaware). In these settings, the strong currents generated can significantly influence sediment transport pathways and the behavior of adjacent shores.

3.4.3. Sediment Supply

The availability of sediments to a coastal region also has important effects on coastal landforms and their behavior (Curray, 1964). In general, assuming a relatively stable sea level, an abundance of sediment along the coast can cause the coast to build seaward over the long term if the rate of supply exceeds the rate at which sediments are eroded and transported by nearshore currents. Conversely, the coast can retreat landward if the rate of erosion exceeds the rate at which sediment is supplied to a coastal region. One way to evaluate the role of sediment supply in a region or specific location is to examine the amount of sediment being gained or lost along the shore. This is often referred to as the sediment budget (Komar, 1996; List, 2005; Rosati, 2005). Whether or not there is an overall sediment gain or loss from a coastal setting is a critical determinant of the potential response to changes in sea level; however, it is difficult if to quantify with high confidence the sediment budget over time periods as long as a century or its precise role in influencing shoreline changes.

The recent Intergovernmental Panel on Climate Change (IPCC) chapter on coastal systems and low-lying regions noted that the availability of sediment to coastal regions will be a key factor in future shoreline changes (Nicholls *et al.*, 2007). In particular, the deposition of sediments in coastal embayments (*e.g.*, estuaries and lagoons) may be a significant sink for sediments as they deepen in response to sea-level rise and are able to accommodate sediments from coastal river systems and adjacent open ocean coasts. For this reason, it is expected that the potential for erosion and shoreline retreat will increase, especially in the vicinity of tidal inlets (see Nicholls *et al.*, 2007). In addition, others have noted an important link between changes in the dimension of coastal embayments, the sediment budget, and the potential for shoreline changes (FitzGerald *et al.*, 2006, 2008). In the mid-Atlantic region, coastal sediments generally come from erosion of both the underlying coastal landscape and the continental shelf (Swift *et al.*, 1985; Niedoroda *et al.*, 1985). Sediments delivered through coastal rivers in the mid-Atlantic region, are generally captured in estuaries contributing minor amounts of sediments to the open-ocean coast (Meade, 1969).

3.4.4 Human Impacts

The human impact on the coast is another important factor affecting shoreline changes. A variety of erosion control practices have been undertaken over the last century along much of the mid-Atlantic region, particularly during the latter half of the twentieth century (see reviews in Nordstrom, 1994; 2000). As discussed later in Chapter 6, shoreline engineering structures such as seawalls, revetments, groins, and jetties have significantly altered sediment transport processes, and consequently affect the availability of sediment (*e.g.*, sediment budget) to sustain beaches and barriers and the potential to exacerbate erosion on a local level (see discussion on Assateague Island in Box 3.2). Beach nourishment, a commonly used approach, has been used on many beaches to temporarily mitigate erosion and provide storm protection by adding to the sediment budget.

The management of tidal inlets by dredging has had a large impact to the sediment budget particularly at local levels (see review in Nordstrom, 1994; 2000). In the past, sand removed from inlet shoals has been transferred out to sea, thereby depleting the amount of sand available to sustain portions of the longshore transport system and, consequently, adjacent shores (Marino and Mehta, 1988; Dean, 1988). More recently, inlet management efforts have attempted to retain this material by returning it to adjacent shores or other shores where sand is needed.

A major concern to coastal scientists and managers is whether or not erosion management practices are sustainable for the long term, and whether or how these shoreline protection measures might impede the ability of natural processes to respond to future sea-level rise, especially at accelerated rates. It is also uncertain whether beach nourishment will be continued into the future due to economic constraints and often limited supplies of suitable sand resources. Chapter 6 describes some of these erosion control practices and their management and policy implications further. In addition, Chapter 6 also describes the important concept of "Regional Sediment Management" which is used to guide the management of sediment in inlet dredging, beach nourishment, or other erosion control activities.

3.5. Coastal Landforms of the Mid-Atlantic

For this assessment, the coastal landforms along the shores of the mid-Atlantic region are classified using the criteria developed by Fisher (1967; 1982), Hayes (1979), and Davis and Hayes (1984). Four distinct geomorphic settings, including spits, headlands, and wave-dominated and mixed-energy barrier islands, occur in the mid-Atlantic region, as shown in Figure 3.1 and described below.

3.5.1. Spits

The accumulation of sand from longshore transport has formed large spits that extend from adjacent headlands into the mouths of large coastal embayments (Figure 3.1, Sections 4, 9, and 15). Outstanding examples of these occur at the entrances of Raritan Bay (Sandy Hook, New Jersey) and Delaware Bay (Cape Henlopen, Delaware). The evolution and existence of these spits results from the interaction between alongshore transport driven by incoming waves and the tidal flow through the large embayments. Morphologically, these areas can evolve rapidly. For example, since 1842 Cape Henlopen (Figure 3.1, Section 9) has extended almost 1.5 kilometers (0.9 miles) to the north into the mouth of Delaware Bay as the northern Delaware shoreline has retreated and sediment has been transported north by longshore currents (Kraft, 1971; Kraft *et al.*, 1978; Ramsey *et al.*, 2001).

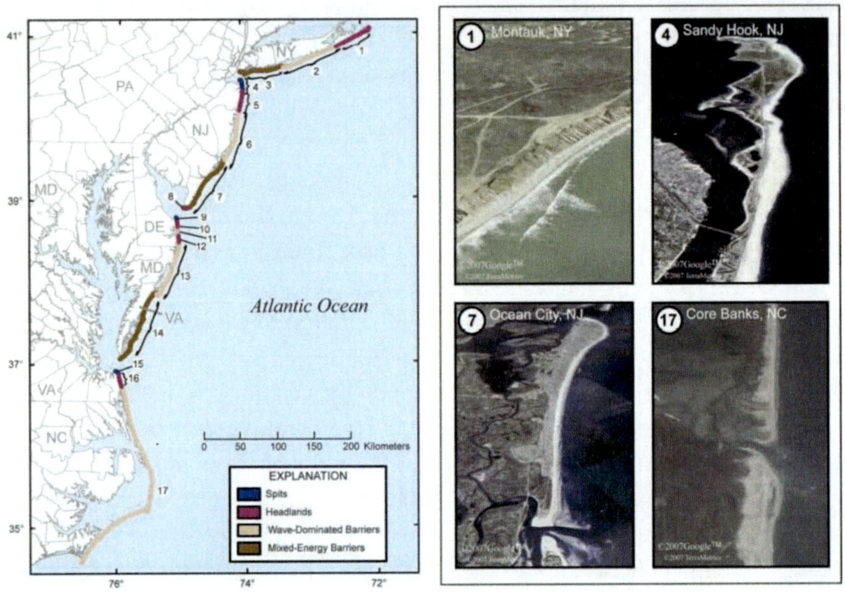

Figure 3.1. Map of the mid-Atlantic coast of the United States showing the occurrence of the four coastal landform types (geomorphic settings). Numbers on the map designate distinct portions of the coast divided by landform type and refer to the discussions in Sections 3.5 and 3.7. Numbers on the photographs refer to specific sections of the coast that are depicted on the map. Images from Google Earth. (Gutierrez *et. al.*, 2007).

3.5.2. Headlands

Along the shores of the mid-Atlantic region, coastal headlands typically occur where elevated regions of the landscape intersect the coast. These regions are often formed where drainage divides that separate creeks and rivers from one another occur in the landscape, or where glacial deposits create high grounds (Taney, 1961; Kraft, 1971; Nordstrom *et al.*, 1977). The erosion of headlands provides a source of sediment that is incorporated into the longshore transport system that supplies and maintains adjacent beaches and barriers. Coastal headlands are present on Long Island, New York (see Figure 3.1), from Southampton to Montauk (Section 1), in northern New Jersey from Monmouth to Point Pleasant (Section 5; Oertel and Kraft, 1994), in southern New Jersey at Cape May (Section 8), on Delaware north and south of Indian River and Rehoboth Bays (Sections 10 and 12; Kraft, 1971; Oertel and Kraft, 1994; Ramsey *et al.*, 2001), and on the Virginia Coast, from Cape Henry to Sandbridge (Section 16).

3.5.3. Wave-Dominated Barrier Islands

Wave-dominated barrier islands occur as relatively long and thin stretches of sand fronting shallow estuaries, lagoons, or embayments that are bisected by widely-spaced tidal inlets (Figure 3.1, Sections 2, 6, 10, 13, and 17). These barriers are present in regions where wave energy is large relative to tidal energy, such as in the mid-Atlantic region (Hayes, 1979; Davis and Hayes, 1984). Limited tidal ranges result in flow-through tidal inlets that are marginally sufficient to flush the sediments that accumulate from longshore sediment transport. In some cases, this causes the inlet to migrate over time in response to a changing balance between tidal flow through the inlet and wave-driven longshore transport. Inlets on wave-dominated coasts often exhibit large flood-tidal deltas and small ebb-tidal deltas as tidal currents are often stronger during the flooding stage of the tide.

In addition, inlets on wave-dominated barriers are often temporary features. They open intermittently in response to storm-generated overwash and migrate laterally in the overall direction of longshore transport. In many cases, these inlets are prone to filling with sands from alongshore sediment transport (*e.g.*, McBride, 1999).

Overwash produced by storms is common on wave-dominated barriers (*e.g.*, Morton and Sallenger, 2003; Riggs and Ames, 2007). Overwash erodes low-lying dunes into the island interior. Sediment deposition from overwash adds to the island's elevation. Overwash deposits (washover fans) that extend into the back-barrier waterways form substrates for back-barrier marshes and submerged aquatic vegetation.

The process of overwash is an important mechanism by which some types of barriers migrate landward and upward over time. This process of landward migration has been referred to as "roll-over" (Dillon, 1970; Godfrey and Godfrey, 1976; Fisher, 1982; Riggs and Ames, 2007). Over decades to centuries, the intermittent processes of overwash and inlet formation enable the barrier to migrate over and erode into back-barrier environments such as marshes as relative sea-level rise occurs over time. As this occurs, back-barrier environments such as marshes are eroded and buried by barrier beach and dune sands.

3.5.4. Mixed-Energy Barrier Islands

The other types of barrier islands present along the U.S. Atlantic coast are mixed-energy barrier islands, which are shorter and wider than their wave-dominated counterparts (Hayes, 1979; Figure 3.1, Sections 3, 4, 7, and 14). The term "mixed-energy" refers to the fact that both waves and tidal currents are important factors influencing the morphology of these systems. Due to the larger tidal range and consequently stronger tidal currents, mixed energy barriers are shorter in length and well-developed tidal inlets are more abundant than for wave-dominated barriers. Some authors have referred to the mixed-energy barriers as tide-dominated barriers along the New Jersey and Virginia coasts (*e.g.*, Oertel and Kraft, 1994).

The large sediment transport capacity of the tidal currents within the inlets of these systems maintains large ebb-tidal deltas seaward of the inlet mouth. The shoals that comprise ebb-tidal deltas cause incoming waves to refract around the large sand body that forms the delta such that local reversals of alongshore currents and sediment transport occur downdrift of the inlet. As a result, portions of the barrier downdrift of inlets accumulate sediment which form recurved sand ridges and give the barrier islands a 'drumstick'-like shape (Hayes 1979; Davis, 1994).

3.6. Potential Responses to Future Sea-Level Rise

Based on current understanding of the four landforms discussed in the previous section, three potential responses could occur along the mid-Atlantic coast in response to sea-level rise over the next century.

3.6.1. Bluff and Upland Erosion

Shorelines along headland regions of the coast will retreat landward with rising sea level. As sea level rises over time, uplands will be eroded and the sediments incorporated into the beach and dune systems along these shores. Along coastal headlands, bluff and upland erosion will persist under all four of the sea-level rise scenarios considered in this Product. A possible management reaction to bluff erosion is shore armoring (e.g. Nordstrom, 2000; Psuty and Ofiara, 2002; see Chapter 6). This may reduce bluff erosion in the short term but could increase long-term erosion of the adjacent coast by reducing sediment supplies to the littoral system.

3.6.2. Overwash, Inlet Processes, and Barrier Island Morphologic Changes

For barrier islands, three main processes are agents of change as sea level rises. First, with higher sea level, storm overwash may occur more frequently. This is especially critical if the sand available to the barrier, such as from longshore transport, is insufficient to allow the barrier to maintain its width and/or build vertically over time in response to rising water levels. If sediment supplies or the timing of the barrier recovery are insufficient, storm surges coupled with breaking waves will affect increasingly higher elevations of the barrier systems as mean sea level increases, possibly causing more extensive erosion and overwash. In addition, it is possible that future hurricanes may become more intense, possibly increasing the potential for episodic overwash, inlet formation, and shoreline retreat. The topic of recent and future storm trends has been debated in the scientific community, with some researchers suggesting that other climate change impacts such as strengthening wind shear may lead to a decrease in future hurricane frequency (see Chapter 1 and reviews in Meehl *et al.*, 2007; Karl *et al.*, 2008; Gutowski *et al.*, 2008). It is also expected that extratropical storms will be more frequent and intense in the future, but these effects will be more pronounced at high latitudes (60° to 90°N) and possibly decreased at midlatitudes (30° to 60°N) (Meehl *et al.*, 2007; Karl *et al.*, 2008; and Gutowski *et al.*, 2008).

Second, tidal inlet formation and migration will contribute to important changes in future shoreline positions. Storm surges coupled with high waves can cause not only barrier island overwash but also breach the barriers and create new inlets. In some cases, breaches can be large enough to form inlets that persist for some time until the inlet channels fill with sediments accumulated from longshore transport. Numerous deposits have been found along the shores of the mid-Atlantic region, indicating former inlet positions (North Carolina: Moslow and Heron, 1979 and Everts *et al.*, 1983; Fire Island, New York: Leatherman, 1985). Several inlets along the mid-Atlantic coast were formed by the storm surges and breaches from an unnamed 1933 hurricane, including Shackleford Inlet in North Carolina; Ocean City inlet in Maryland; Indian River Inlet in Delaware; and Moriches Inlet in New York. Recently, tidal inlets were formed in the North Carolina Outer Banks in response to Hurricane Isabel in 2003. While episodic inlet formation and migration are natural processes and can occur independently of long-term sea-level rise, a long-term increase in sea level coupled with limited sediment supply and increases in storm frequency and/or intensity could increase the likelihood for future inlet breaching.

Third, the combined effect of rising sea level and stronger storms could accelerate barrier island shoreline changes. These will involve both changes to the seaward facing and landward facing shores of some barrier islands. Assessments of shoreline change on barrier islands indicate that barriers have thinned in some areas over the last century (Leatherman, 1979; Jarrett, 1983; Everts *et al.*, 1983; Penland *et al.*, 2005). Evidence of barrier migration is not widespread on the mid-Atlantic coast (Morton *et al.*, 2003), but is documented at northern Assateague Island in Maryland (Leatherman, 1979) and Core Banks, North Carolina (Riggs and Ames, 2007).

3.6.3. Threshold Behavior

Barrier islands are dynamic environments that are sensitive to a range of physical and environmental factors. Some evidence suggests that changes in some or all of these factors can lead to conditions where a barrier system becomes less stable and crosses a geomorphic threshold. Once a threshold is crossed, the potential for significant and irreversible changes to the barrier island is high. These changes can involve landward migration or changes to the barrier island dimensions such as reduction in size or an increased presence of tidal inlets. Although it is difficult to precisely define an unstable barrier, indications include:

- Rapid landward migration of the barrier;
- Decreased barrier width and height, due to a loss of sand eroded from beaches and dunes;
- Increased frequency of overwash during storms;
- Increased frequency of barrier breaching and inlet formation; and
- Segmentation of the barrier.

Given the unstable state of some barrier islands under current rates of sea-level rise and climate trends, it is very likely that conditions will worsen under accelerated sea-level rise rates. The unfavorable conditions for barrier maintenance could result in significant changes, for example, to barrier islands as observed in coastal Louisiana (further discussed in Box 3.2; McBride *et al.*, 1995; McBride and Byrnes, 1997; Penland *et al.*, 2005; Day *et al.*, 2007; Sallenger *et al.*, 2007; FitzGerald *et al.*, 2008). In one case, recent observations indicate that the Chandeleur Islands are undergoing a significant land loss due to several factors which include: (1) limited sediment supply by longshore or cross-shore transport, (2) accelerated rates of sea-level rise, and (3) permanent sand removal from the barrier system by storms such as Hurricanes Camille, Georges, and Katrina. Likewise, a similar trend has been observed for Isle Dernieres, also on the Louisiana coast (see review in FitzGerald *et al.*, 2008). In addition, recent studies from the North Carolina Outer Banks indicate that there have been at least two periods during the past several thousand years where fully open-ocean conditions have occurred in Albemarle and Pamlico Sounds, which are estuaries fronted by barrier islands at the present time (Mallinson *et al.*, 2005; Culver *et al.*, 2008). This indicates that portions of the North Carolina barrier island system may have segmented or become less continuous than the present time for periods of a few hundred years, and later reformed. Given future increases in sea level and/or storm activity, the potential for a threshold crossing exists, and portions of these barrier islands could once again become segmented.

Changes in sea level coupled with changes in the hydrodynamic climate and sediment supply in the broader coastal environment contribute to the development of unstable barrier island behavior. The threshold behavior of unstable barriers could result in: barrier segmentation, barrier disintegration, or landward migration and roll-over. If the barrier were to disintegrate, portions of the ocean shoreline could migrate or back-step toward and/or merge with the mainland.

The mid-Atlantic coastal regions most vulnerable to threshold behavior can be estimated based on their physical dimensions. During storms, large portions of low-elevation, narrow barriers can be inundated under high waves and storm surge. Narrow, low-elevation barrier islands, such as the northern portion of Assateague Island, Maryland are most susceptible to storm overwash, which can lead to landward migration and the formation of new tidal inlets (*e.g.*, Leatherman, 1979; see also Box 3.2).

The future evolution of some low-elevation, narrow barriers could depend in part on the ability of salt marshes in back-barrier lagoons and estuaries to keep pace with sea-level rise (FitzGerald *et al.*, 2006, 2008; Reed *et al.*, 2008). A reduction of salt marsh in back- barrier regions could increase the volume of water exchanged with the tides (*e.g.,* the tidal prism) of back-barrier systems, altering local sediment budgets and leading to a reduction in sandy materials available to sustain barrier systems (FitzGerald *et al.*, 2006, 2008).

Box 3.2. Evidence for Threshold Crossing of Coastal Barrier Landforms

Barrier islands change and evolve in subtle and somewhat predictable ways over time in response to storms, changing sediment supply, and changes in sea level. Recent field observations suggest that some barrier islands can reach a "threshold" condition: that is, a point where they become unstable and disintegrate. Two sites where barrier island disintegration is occurring and may continue to occur are along the 72 kilometers (about 45 miles) long Chandeleur Islands in Louisiana, east of the Mississippi River Delta, due to impacts of Hurricane Katrina in September 2005; and the northern 10 kilometers (6 miles) of Assateague Island National Seashore, Maryland due to 70 years of sediment starvation caused by the construction of jetties to maintain Ocean City Inlet.

Chandeleur Islands, Louisiana

In the Chandeleur Islands, the high storm surge (about 4 meters or 13 feet) and waves associated with Hurricane Katrina in 2005 completely submerged the islands and eroded about 85 percent of the sand from the beaches and dunes (Sallenger *et al.*, 2007). Box Figure 3.2a (UTM Northing) shows the configuration of the barriers in 2002, and in 2005 after Katrina's passage. Follow-up aerial surveys by the U.S. Geological Survey indicate that erosion has continued since that time. When the Chandeleur Islands were last mapped in the late 1980s and erosion rates were calculated from the 1850s, it was estimated that the Chandeleurs would last approximately 250 to 300 years (Williams *et al.*, 1992). The results from post-Katrina studies suggest that a threshold has been crossed such that conditions have changed and natural processes may not contribute to the rebuilding of the barrier in the future.

Assateague Island National Seashore, Maryland

An example of one shoreline setting where human activity has increased the vulnerability of the shore to sea-level rise is Assateague Island, Maryland. Prior to a hurricane in 1933, Assateague Island was a continuous, straight barrier connected to

Fenwick Island (Dolan *et al.*, 1980). An inlet that formed during the storm separated the island into two sections at the southern end of Ocean City, Maryland. Subsequent construction of two stone jetties to maintain the inlet for navigation interrupted the longshore transport of sand to the south. Since then, the jetties have trapped sand, building the Ocean City shores seaward by 250 meters (820 feet) by the mid-1970s (Dean and Perlin, 1977). In addition, the development of sand shoals (ebb tidal deltas) around the inlet mouth has sequestered large volumes of sand from the longshore transport system (Dean and Perlin, 1977; FitzGerald, 1988). South of the inlet, the opposite has occurred. The sand starvation on the northern portion of Assateague Island has caused the shore to migrate almost 700 meters (2,300 feet) landward and transformed the barrier into a low-relief, overwash-dominated barrier (Leatherman, 1979; 1984). This extreme change in barrier island sediment supply has caused a previously stable segment of the barrier island to migrate. To mitigate the effects of the jetties, and to restore the southward sediment transport that was present prior to the existence of Ocean City inlet, the U.S. Army Corps of Engineers and National Park Service mechanically transfer sand from the inlet and the ebb and flood tidal deltas, where the sand is now trapped, to the shallow nearshore regions along the north end of the island. Annual surveys indicate that waves successfully transport the sediment alongshore and have slowed the high shoreline retreat rates present before the project began (Schupp *et al.*, 2007). Current plans call for continued biannual transfer of sand from the tidal deltas to Assateague Island to mitigate the continued sediment starvation by the Ocean City inlet jetties.

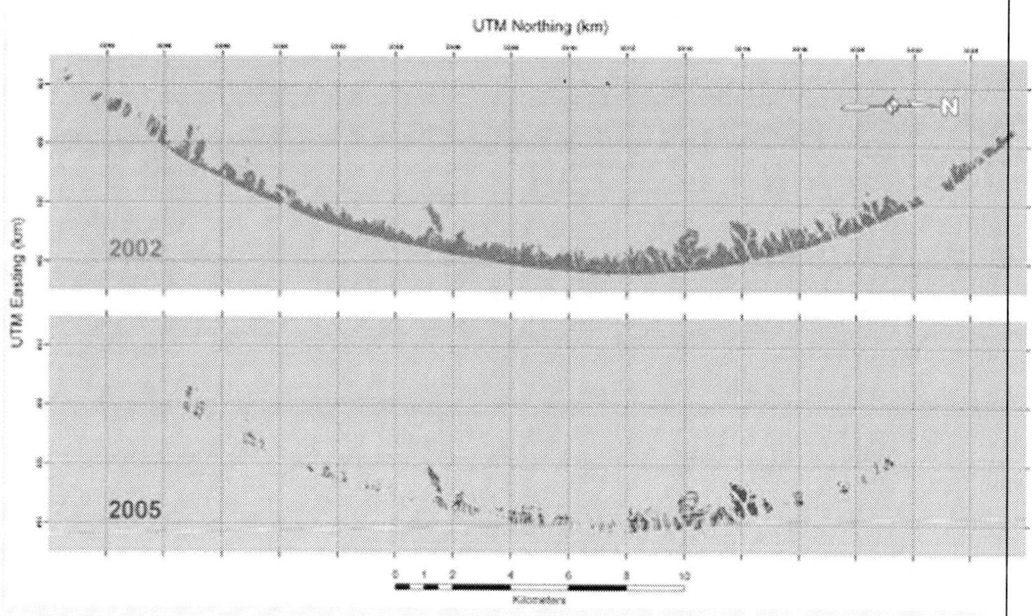

Box Figure 3.2a. Maps showing the extent of the Chandeleur Islands in 2002, three years before Hurricane Katrina and in 2005, after Hurricane Katrina. Land area above mean high water. Source: A. Sallenger, USGS.

Box Figure 3.2b. Aerial Photo of northern Assateague Island and Ocean City, Maryland showing former barrier positions. Note that in 1850, a single barrier island, shown in outlined in yellow, occupied this stretch of coast. In 1933, Ocean City inlet was created by a hurricane. The inlet improved accessibility to the ocean and was stabilized by jetties soon after. By 1942, the barrier south of the inlet had migrated landward (shown as a green shaded region). Shorelines acquired from the State of Maryland Geological Survey. Photo source: NPS.

Box Figure 3.2c. North oblique photographs of northern Assateague Island in 1998 after a severe winter storm. The left photo of Assateague Island barrier shows clear evidence of overwash. The right 2006 photo shows a more robust barrier that had been augmented by recent beach nourishment. The white circles in the photos specify identical locations on the barrier. The offset between Fenwick Island (north) and Assateague Island due to Ocean City inlet and jetties can be seen at the top of the photo. Sources: a) National Park Service, b) Jane Thomas, IAN Photo and Video Library.

3.7. Potential Changes to the Mid-Atlantic Ocean Coast Due to Sea-Level Rise

In this Section, the responses to the four sea-level rise scenarios considered in this Chapter are described according to coastal landform types (Figure 3.2). The first three sea-level rise scenarios (Scenarios 1 through 3) are: (1) a continuation of the twentieth century rate, (2) the twentieth century rate plus 2 mm per year, and (3) the twentieth century rate plus 7 mm per year. Scenario 4 specifies a 2-meter rise (6.6-foot) over the next few hundred years. Because humans have a significant impact on portions of the mid-Atlantic coast, this assessment focuses on assessing the vulnerability of the coastal system as it currently exists (see discussion in Section 3.4). However, there are a few caveats to this approach:

- This is a regional-scale assessment and there are local exceptions to these geomorphic classifications and potential outcomes;
- Given that some portions of the mid-Atlantic coast are heavily influenced by development and erosion mitigation practices, it cannot be assumed that current practices will continue into the future given uncertainties regarding the decision-making process that occurs when these practices are pursued; but,
- At the same time, there are locations where some members of the panel believe that erosion mitigation will be implemented regardless of cost.

To express the likelihood of a given outcome for a particular sea-level rise scenario, the terminology advocated by ongoing CCSP assessments was used (see Preface, Figure P.1; CCSP, 2006). This terminology is used to quantify and communicate the degree of likelihood of a given outcome specified by the assessment. These terms should not be construed to represent a quantitative relationship between a specific sea-level rise scenario and a specific dimension of coastal change, or rate at which a specific process operates on a coastal geomorphic compartment. The potential coastal responses to the sea-level rise scenarios are described below according to the coastal landforms defined in Section 3.5.

3.7.1. Spits

For sea-level rise Scenarios 1 through 3, it is *virtually certain* that the coastal spits along the mid-Atlantic coast will be subject to increased storm overwash, erosion, and deposition over the next century (see Figure 3.2, Sections 4, 9, 15). It is *virtually certain* that some of these coastal spits will continue to grow though the accumulation of sediments from longshore transport as the erosion of updrift coastal compartments occurs. For Scenario 4, it is *likely* that threshold behavior could occur for this type of coastal landform (rapid landward and/or alongshore migration).

3.7.2. Headlands

Over the next century, it is *virtually certain* that these headlands along the mid-Atlantic coast will be subject to increased erosion for all four sea-level rise scenarios (see Figure 3.2, Sections 1, 5, 8, 10, 12, and 16). It is *very likely* that shoreline and upland (bluff) erosion will accelerate in response to projected increases in sea level.

3.7.3. Wave-Dominated Barrier Islands

Potential sea-level rise impacts on wave-dominated barriers in the Mid-Atlantic vary by location and depend on the sea-level rise scenario (see Figure 3.2, Sections 2, 6, 11, 13, 17). For Scenario 1, it is *virtually certain* that the majority of the wave-dominated barrier islands along the mid-Atlantic coast will continue to experience morphological changes through erosion, overwash, and inlet formation as they have over the last several centuries, except for the northern portion of Assateague Island (Section 13). In this area, the shoreline exhibits high rates of erosion and large portions of this barrier are submerged during moderate storms. In the past, large storms have breached and segmented portions of northern Assateague Island (Morton *et al.*, 2003). Therefore, it is possible that these portions of the coast are already at a geomorphic threshold. With any increase in the rate of sea-level rise, it is *virtually certain* that this barrier island will exhibit large changes in morphology, ultimately leading to the degradation of the island. At this site, however, periodic transfer of sand from the shoals of Ocean City Inlet appear to be reducing erosion and shoreline retreat in Section13 (see Box 3.2). Portions of the North Carolina Outer Banks (Figure 3.2) may similarly be nearing a geomorphic threshold.

For Scenario 2, it is *virtually certain* that the majority of the wave-dominated barrier islands in the mid-Atlantic region will continue to experience morphological changes through overwash, erosion, and inlet formation as they have over the last several centuries. It is also *about as likely as not* that a geomorphic threshold will be reached in a few locations, resulting in rapid morphological changes in these barrier systems. Along the shores of northern Assateague Island (Section 13) and a substantial portion of Section 17 it is *very likely* that the barrier islands could exhibit threshold behavior (barrier segmentation). For this scenario, the ability of wetlands to maintain their elevation through accretion at higher rates of sea-level rise may be reduced (Reed *et al.*, 2008). It is *about as likely as not* that the loss of back-barrier marshes will lead to changes in hydrodynamic conditions between tidal inlets and back-barrier lagoons, thus affecting the evolution of barrier islands (*e.g.*, FitzGerald *et al.*, 2006; FitzGerald *et al.,* 2008).

For Scenario 3, it is *very likely* that the potential for threshold behavior will increase along many of the mid-Atlantic barrier islands. It is *virtually certain* that a 2-meter (6.6- foot) sea-level rise will lead to threshold behavior (segmentation or disintegration) for this landform type.

3.7.4. Mixed-Energy Barrier Islands

The response of mixed-energy barrier islands will vary (see Figure 3.2, Sections 3, 7, 14). For Scenarios 1 and 2, the mixed-energy barrier islands along the mid-Atlantic will be subject to processes much as have occurred over the last century such as storm overwash and shoreline erosion. Given the degree to which these barriers have been developed, it is difficult to determine the likelihood of future inlet breaches, or whether these would be allowed to persist due to common management decisions to repair breaches when they occur. In addition, changes to the back-barrier shores are uncertain due to the extent of coastal development.

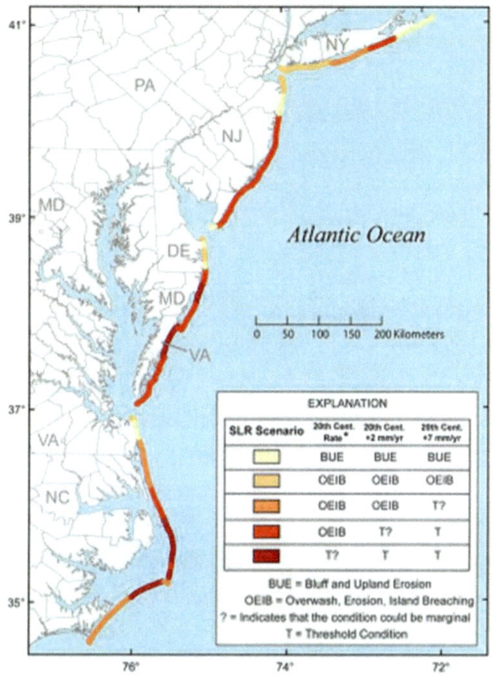

Figure 3.2. Map showing the potential sea-level rise responses for each coastal compartment. Colored portions of the coastline indicates the potential response for a given sea-level rise scenario according to the inset table (Gutierrez *et. al.*, 2007). The color scheme was created using ColorBrewer by Cindy Brewer and Mark Harrower.

It is *about as likely as not* that four of the barrier islands along the Virginia Coast (Wallops, Assawoman, Metompkin, and Cedar Islands) are presently at a geomorphic threshold. Thus, it, it is *very likely* that further sea-level rise will contribute to significant changes resulting in the segmentation, disintegration and/or more rapid landward migration of these barrier islands.

For the higher sea-level rise scenarios (Scenarios 3 and 4), it is *about as likely as not* that these barriers could reach a geomorphic threshold. This threshold is dependent on the availability of sand from the longshore transport system to supply the barrier. It is *virtually certain* that a 2-meter (6.6-foot) sea-level rise will have severe consequences along the shores of this portion of the coast, including one or more of the extreme responses described above. For Scenario 4, the ability of wetlands to maintain their elevation through accretion at higher rates of sea-level rise may be reduced (Reed *et al.*, 2008). It is *about as likely as not* that the loss of back-barrier marshes could lead to changes in the hydrodynamic conditions between tidal inlets and back-barrier lagoons, affecting the evolution of barrier islands (FitzGerald *et al.*, 2006, 2008).

REFERENCES

* Indicates non-peer reviewed literature; available upon request

Belknap, D. F. & Kraft, J. C. (1985). Influence of antecedent geology on stratigraphic preservation potential and evolution of Delaware's barrier system. *Marine Geology, 63(1-4)*, 235-262.

Carter, R. W. G. & Woodroffe C. D.(eds.), (1994a). *Coastal Evolution: Late Quaternary Shoreline Morphodynamics*. Cambridge University Press, Cambridge, UK, 517 pp.

Carter, R. W. G. & Woodroffe, C. D. (1994b). Coastal evolution: an introduction. In: *Coastal Evolution: Late Quaternary Shoreline Morphodynamics* [Carter, R.W.G. & C.D. Woodroffe (eds.)]. Cambridge University Press, Cambridge, UK, pp. 1- 32.

CCSP (Climate Change Science Program), (2006). *Recommendations for Implementingthe CCSP Synthesis and Assessment Guidelines*. Climate Change Science Program, Washington DC.

Colquhoun, D. J., Johnson, G. H., Peebles, P. C., Huddleston, P. F. & Scott, T. (1991). Quaternary geology of the Atlantic Coastal Plain. In: *Quaternary Nonglacial Geology: Conterminous U.S.* [Morrison, R.B. (ed.)]. The Geology of North America v.K-2. Geological Society of America, Boulder, CO, pp. 629-650.

Cooper, J. A. G. & Pilkey, O. H. (2004). Sea-level rise and shoreline retreat: time to abandon the Bruun Rule. *Global and Planetary Change, 43(3-4)*, 157-171.

Cowell, P. J. & Thom, B. G. (1994). Morphodynamics of coastal evolution. In: *Coastal Evolution: Late Quaternary Shoreline Morphodynamics* [Carter, R.W.G. & C.D. Woodroffe (eds.)]. Cambridge University Press, Cambridge, UK, pp. 3 3-86.

Cowell, P. J., Thom, B. G., Jones, R. A., Everts, C. H. & D. Simanovic, (2006). Management uncertainty in predicting climate-change impacts on beaches. *Journal of Coastal Research, 22(1)*, 232-245.

Culver, S. J., Farrell, K. M., Mallinson, D .J., Horton, D.A. Willard, E.R. Thieler, S. R. Riggs, S.W. Snyder, J.F. Wehmiller, C.E. Bernhardt, and C. Hillier, (2008). Micropaleontologic record of late Pliocene and Quaternary paleoenvironments in the northern Albemarle Embayment, North Carolina, USA. *Palaeogeography, Palaeoclimatology, Palaeoecology, 264(1-2)*, 54-77.

Curray, J. R., (1964). Transgression and regression. In: *Papers in Marine Geology* [Miller, R.L. (ed.)]. McMillan, New York, pp. 175-203.

Davis, R. A., (1987). *Coasts*. Prentice Hall, Upper Saddle River, NJ, 274 pp.

Davis, R. A., (1994). Barrier island systems-a geologic overview. In: *Geology of Holocene Barrier Island Systems* [Davis, R.A. (ed.)]. Springer-Verlag, New York, pp. 435- 456.

Davis, R. A. & Hayes, M. O. (1984). What is a wave-dominated coast? *Marine Geology, 60(1-4)*, 3 13-329.

Day, J. W. J., Biesch, D. F., Clairain, E. J., Kemp, G. P., Laska, S. B., Mitsch, W. J., Orth, K. H. Mashriqui, Reed, D. J., Shabman, L., Simenstad, C. A., Streever, B. J., Twilley, R. R., C. C. Watson, Wells, J. T. & Whigham, D. F. (2007). Restora tion of the Mississippi Delta: lessons from hurricanes Katrina and Rita. *Science, 315(5819)*, 1679-1684.

Dean, R. G., (1988). Sediment interaction at modified coastal inlets. In: *Hydrodynamics and Sediment Dynamics of Tidal Inlets* [Aubrey, D.G. & L. Weishar (eds.)]. Springer-Verlag, New York, pp. 412-439.

Dean, R. G. & Perlin, M. (1977). A coastal engineering study of Ocean City Inlet. In: *Coastal Sediments '77*. American Society of Civil Engineers, Reston, VA, pp. 520-540.

Demarest, J. M. & Leatherman, S. P. (1985). Mainland influence on coastal transgression: Delmarva Peninsula. *Marine Geology, 63(1-4)*, 19-33.

Dillon, W. P., (1970). Submergence effects on Rhode Island barrier and lagoon and influences on migration of barriers. *Journal of Geology, 78*, 94-106.

Dingler, J. R. & Clifton, H. E. (1994). Barrier systems of California, Oregon, and Washington. In: *Geology of Holocene Barrier Island Systems* [Davis, R.A. (ed.)]. Springer-Verlag, New York, pp. 115-165.

Dolan, R., Lins, H. F. & Stewart, J. (1980). *Geographical Analysis of Fenwick Island, Maryland, a Middle Atlantic Coast Barrier Island.* Geological Survey professional paper 1177-A. U.S. Government Printing Office, Washington, DC, 24 pp.

Dubois, R. N., (2002). How does a barrier shoreface respond to a sea-level rise? *Journal of Coastal Research, 18(2)*, iii-v.

Everts, C. H., Battley J. P. Jr., and Gibson, P.N. (1983). *Shoreline Movements:Report 1, Cape Henry, Virginia to Cape Hatteras, North Carolina, 1849-1980.* Technical report CERC-83-1. U.S. Army Corps of Engineers, Washington, DC, and National Oceanic and Atmospheric Administration, Rockville, MD, 111 pp.

* Fisher, J. J., (1962). *Geomorphic Expression of Former Inlets along the Outer Banks of North Carolina.* M.S. thesis, Department of Geology and Geography. University of North Carolina, Chapel Hill, 120 pp.

* Fisher, J. J., (1967). Origin of barrier island chain shoreline, Middle Atlantic states. In: *Abstract with Programs Annual Meeting of the Geological Society of America*, New Orleans, pp. 66-67.

Fisher, J. J., (1968). Barrier island formation: discussion. *Geological Society of America Bulletin, 79(10)*, 1421-1426.

Fisher, J. J., (1982). Barrier islands. In: *The Encyclopedia of Beaches and Coastal Environments* [Schwartz, M.L. (ed.)]. Hutchinson Ross Publishing Company, Stroudsburg, PA, volume XV, pp. 124-133. FitzGerald, D.M., 1988: Shoreline erosional-depositional processes associated with tidal inlets. In: *Hydrodynamics and Sediment Dynamics of Tidal Inlets* [Aubrey, D.G. & L. Weishar (eds.)]. Springer-Verlag, New York, pp. 186-225.

FitzGerald, D.M., I.V. Buynevich, and Argow, B. A. (2006). Model of tidal inlet and barrier island dynamics in a regime of accelerated sea-level rise. *Journal of Coastal Research*, Special Issue *39*, 789-795.

FitzGerald, D. M., Fenster, M. S., Argow, B. A. & Buynevich, I.V. (2008). Coastal impacts due to sea-level rise. *Annual Reviews of Earth and Planetary Sciences, 36*, 601-647.

Fletcher, C. H., Knebel, H. J. & Kraft, J. C. (1990). Holocene evolution of an estuarine coast and tidal wetlands. *Geological Society of America Bulletin, 102(3),* 283-297.

Glaeser, J. D., (1978). Global distribution of barrier islands in terms of tectonic setting. *Journal of Geology, 86*, 283-297.

Godfrey, P. J. & Godfrey, M. M. (1976). *Barrier Island Ecology of Cape Lookout National Seashore and Vicinity, North Carolina.* National Park Service monograph series no. 9. U.S. Government Printing Office, Washington, DC, 160 pp.

Griggs, G. B. & Patsch, K. B. (2004). California's coastal cliffs and bluffs. In: *Formation, Evolution, and Stability of Coastal Cliffs-Status and Trends* [Hampton, M. (ed.)]. U.S. Geological Survey professional paper 1693. U.S. Geological Survey, Reston, VA, pp. 53-64.

Gutierrez, B. T., Williams, S. J. & Thieler, E. R. (2007). *Potential for Shoreline Changes Due to Sea-level Rise along the U.S. Mid-Atlantic Region.* U.S. Geological Survey open file report 2007-1278. U.S. Geological Survey, Reston, VA, 26 pp. <http://pubs.usgs.gov/of/2007/1 278/>.

Gutowski, W. J., Hegerl, G. C., Holland, G. J., Knutson, T. R., Mearns, L. O., Stouffer, R. J., Webster, P. J., Wehner, M. F. & Zwiers, F. W. (2008). Causes of observed changes in extremes and projections of future changes. In: *Weather and Climate Extremes in a Changing Climate. Regions of Focus: North America, Hawaii, Caribbean, and U.S. Pacific Islands* [Karl, T.R., G.A. Meehl, C.D. Miller, S.J. Hassol, A.M. Waple, and W.L. Murray (eds.)]. Synthesis and Assessment Product 3.3. U.S. Climate Change Science Program, Washington, DC, pp. 81-116.

Hapke, C. J. & Reid, D. (2007). *The National Assessment of Shoreline Change: Part 4, Historical Coastal Cliff Retreat along the California Coast.* U.S. Geological Survey open-file report 2007-1133. U.S. Geological Survey, Reston, VA, 51 pp.

Hapke, C. J., Reid, D., Richmond, B.M., Ruggiero P. & List, J. (2006). *National Assessment of Shoreline Change: Part 3, Historical Shoreline Change and Associated Coastal Land Loss along Sandy Shorelines of the California Coast.* U.S. Geological Survey open-file report 2006-1219. U.S. Geological Survey, Reston, VA, 79 pp.

Hayes, M. O., (1979). Barrier island morphology as a function of tidal and wave regime. In: *Barrier Islands from the Gulf of St. Lawrence to the Gulf of Mexico* [Leatherman, S. (ed.)]. Academic Press, New York, pp. 211-236.

Hine, A. C. & Snyder, S. W. (1985). Coastal lithosome preservation: evidence from the shoreface and inner continental shelf off Bogue Banks, North Carolina. *Marine Geology, 63(1-4)*, 307-330.

Honeycutt, M. G. & Krantz, D. E. (2003). Influence of geologic framework on spatial variability in long-term shoreline change, Cape Henlopen to Rehoboth Beach, Delaware. *Journal of Coastal Research*, Special Issue *38*, 147-167.

Inman, D.L. & Nordstrom, C. E. (1971). On the tectonic and morphologic classification of coasts. *Journal of Geology*, *79*, 1-21.

* Jarrett, J. T., (1983). Changes of some North Carolina barrier islands since the mid-19th century. In: *Coastal Zone '83*, Proceedings of the 3rd Symposium on Coastal and Ocean Management, San Diego, 1-4 June 1983. American Society of Civil Engineers, New York, pp. 641-661.

Karl, T. R., Meehl, G. A., Peterson, T. C., Kunkel, K. E., Gutowski W. J. Jr., and Easterling, D. R. (2008). Executive summary. In: *Weather and Climate Extremes in a Changing Climate. Regions of Focus: North America, Hawaii, Caribbean, and U.S. Pacific Islands.* [Karl, T.R., G.A. Meehl, C.D. Miller, S.J. Hassol, A.M. Waple, and W.L. Murray, (eds.)]. Synthesis and Assessment Product 3.3. U.S. Climate Change Science Program, Washington, DC, pp. 1-9.

Komar, P. D., (1996). The budget of littoral sediments concepts and applications. *Shore and Beach, 64(3),* 18-26.

Komar, P. D., (1998). *Beach Processes and Sedimentation.* Prentice Hall, New Jersey, 2nd edition, 544 pp.

Kraft, J. C., (1971). Sedimentary facies patterns and geologic history of a Holocene marine transgression. *Geological Society of America Bulletin, 82(8)*, 2131-2158.

Kraft, J. C., Allen, E. A. & Maurmeyer, E. M. (1978). The geological and paleogeomorphological evolution of a spit system and its associated shoal environments: Cape Henlopen Spit, Delaware. *Journal of Sedimentary Petrology*, 48(1), 211-226.

Leatherman, S. P., (1979). Migration of Assateague Island, Maryland, by inlet and overwash processes. *Geology, 7(2),* 104-107.

Leatherman, S. P., (1984). Shoreline evolution of north Assateague Island, Maryland. *Shore and Beach, 52(3),* 3-10.

Leatherman, S. P., (1985). Geomorphic and sedimentary analysis of Fire Island, New York. *Marine Geology, 63(1-4)*, 173-195.

Leatherman, S. P., (1990). Modeling shore response to sea-level rise on sedimentary coasts. *Progress in Physical Geography*, 14(4), 447-464. Leatherman, S.P., 2001: Social and economic costs of sea level rise. In: *Sea Level Rise: History and Consequences* [Douglas, B.C., M.S. Kearney, and S.P. Leatherman (eds.)]. Academic Press, San Diego, pp. 181-223.

List, J. H., (2005). The sediment budget. In: *Encyclopedia of Coastal Science* [Schwartz, M.L. (ed.)]. Springer, Dordrecht, the Netherlands, pp. 846-850.

List, J. H., Farris, A. S. & Sullivan, C. (2006). Reversing storm hotspots on sandy beaches: spatial and temporal characteristics. *Marine Geology, 226(3-4)*, 261-279.

Mallinson, D., Riggs, S., Thieler, E. R., Culver, S., Farrell, K., Foster, D. S. Corbett, D. R. B. Horton, and Wehmiller, J. F. (2005). Late Neogene and Quaternary evolution of the northern Albemarle Embayment (mid-Atlantic continental margin, USA). *Marine Geology, 217(1-2)*, 97-117.

Marino, J. N. & Mehta, A. J. (1988). Sediment trapping at Florida's east coast inlets. In: *Hydrodynamics and Sediment Dynamics of Tidal Inlets* [Aubrey, D.G. & L. Weishar (eds.)]. Springer-Verlag, New York, pp. 284-296.

McBride, R. A., (1999). Spatial and temporal distribution of historical and active tidal inlets: Delmarva Peninsula and New Jersey, USA. In: *Coastal Sediments '99* [Kraus, N.C. & W.G. McDougal (eds.)]. America Society of Civil Engineers, Reston, VA, volume 2, pp. 1505-1521.

McBride, R. A. & Byrnes, M. R. (1997). Regional variations in shore response along barrier island systems of the Mississippi River delta plain: Historical change and future prediction. *Journal of Coastal Research, 13(3)*, 628-655.

McBride, R. A., Byrnes, M. R. & Hiland, M. W. (1995). Geomorphic response-type model for barrier coastlines: a regional perspective. *Marine Geology, 126(1-4),* 143-159.

McNinch, J. E. & Wells, J. T. (1999). Sedimentary processes and depositional history of a cape-associated shoal: Cape Lookout, North Carolina. *Marine Geology, 158(1-4)*, 233-252.

Meade, R. H., (1969). Landward transport of bottom sediments in estuaries of the Atlantic coastal plain. *Journal of Sedimentary Petrology, 39(1)*, 222-234.

Meade, R. H., (1972). Transport and deposition of sediments in estuaries. *Geological Society of America, 133(1)*, 91-120.

Meehl, G. A., Stocker, T. F., Collins, W. D., Friedlingstein, P., Gaye, A. T., J. M. Gregory, A. Kitoh, R. Knutti, J. M. Murphy, A. Noda, S. C.B. Raper, I. G. Watterson, A.J. Weaver and Zhao, Z. C. (2007). Global climate projections. In: *Climate Change 2007: The Physical Science Basis.* Contribution of Working Group I to the Fourth Assessment Report of the Intergovernmental Panel on Climate Change [Solomon, S., D. Qin, M. Manning, Z. Chen, M. Marquis, K.B. Averyt, M. Tignor, and H.L. Miller (eds.)]. Cambridge University Press, Cambridge, UK, and New York, pp. 747-845.

Miselis, J. L. & McNinch, J. E. (2006). Calculating shoreline erosion potential using nearshore stratigraphy and sediment volume, Outer Banks, North Carolina. *Journal of Geophysical Research, 111*, F020 19, doi: 10.1 029/2005JF0003 89.

Morton, R.A. & Sallenger, A. H., Jr., (2003). Morphological impacts of extreme storms on sandy beaches and barriers. *Journal of Coastal Research, 19(3)*, 560-573.

Morton, R. A., Paine, J. G. & Gibeaut, J.C. (1994). Stages and durations of post-storm beach recovery, southeastern Texas coast, USA. *Journal of Coastal Research, 10(4)*, 884-908.

Morton, R. A., Guy, K. K. Hill, H. W. and Pascoe, T. (2003). Regional morphological responses to the March 1962 Ash Wednesday Storm. In: *Proceedings CoastalSediments '03* [Davis, R. A., Sallenger, A. H. & Howd P. (eds.)]. America Society of Civil Engineers, Reston, VA.

Moslow, T. F. & Heron, S. D. (1979). Quaternary evolution of Core Banks, North Carolina: Cape Lookout to New Drum inlet. In: *Barrier Islands, from the Gulf of Saint Lawrence to the Gulf of Mexico* [Leatherman, S.P. (ed.)]. Academic Press, New York, pp. 211-236.

Moslow, T. F. & S. D. Heron, 1994: The Outer Banks of North Carolina. In: *Geology of Holocene Barrier Island Systems* [Davis, R.A. (ed.)]. Springer-Verlag, Berlin, pp. 47-74.

Muhs, D. R., Thorson, R. M., Clague, J. J., Mathews, W. H., McDowell, P. F. & Kelsey, H. M. (1987). Pacific coast and mountain system. In: Geomorphic Systems of North America [Graf, W.L. (ed.)]. *Geological Society of America*, Boulder CO, pp. 517-582.

Muhs, D. R., Wehmiller, J. F., Simmons, K. R. & York, L. L. (2004). Quaternary sea-level history of the United States. In: *The Quaternary Period of the United States.* [Gillespie, A.R., S.C. Porter and B.F. Atwater (eds.)]. Elsevier, Amsterdam, pp. 147-183.

Najjar, R.G., H.A. Walker, P.J. &erson, E.J. Barron, R.J. Brod, J.R. Gibson, V.S. Kennedy, C.G. Knight, J.P. Megonigal, R.E. O'Connor, C.D. Polsky, N.P. Psuty, B.A. Richards, Sorenson, L. G. Steele, E. M. & Swanson, R. S. (2000). The potential impacts of climate change on the mid-Atlantic coastal region. *Climate Research, 14(3)*, 219-233.

Nicholls, R. J., Wong, P. P. Burkett, V. R., Codignotto, J. O., Hay, J. E., McLean, R. F. Ragoonaden, S. & Woodroffe, C. D. (2007). Coastal systems and low-lying areas. In: *Climate Change 2007: Impacts, Adaptation and Vulnerability.* Contribution of Working Group II to the Fourth Assessment Report of the Intergovernmental Panel on Climate Change [Parry, M.L., O.F. Canziani, J.P. Palutikof, P.J. van der Linden, and C.E. Hanson (eds.)]. Cambridge University Press, Cambridge, UK, and New York, pp. 3 15-356.

Niedoroda, A. W., Swift, D. J. P., Figueiredo, A. G. & Freeland, G. L. (1985). Barrier island evolution, middle Atlantic shelf, USA. Part II: Evidence from the shelf floor. *Marine Geology, 63(1-4)*, 363-396.

Nordstrom, K. F., (1994). Developed coasts. In: *Coastal Evolution: Late Quaternary Shoreline Morphodynamics* [Carter, R.W.G. & C.D. Woodroffe (eds.)]. Cambridge University Press, Cambridge, UK, pp. 477-5 10.

Nordstrom, K. F., (2000). *Beaches and Dunes of Developed Coasts*. Cambridge University Press, New York, 338pp.

Nordstrom, K., Fisher, S. Burr, M., Frankel, E. Buckalew, T. & Kucma, G. (1977). *Coastal Geomorphology of New Jersey, Volumes I and II*. Tech report 77-1. Center for Coastal and Environmental Studies, Rutgers University, New Brunswick, NJ.

Nummedal, D., (1983). Barrier islands. In: *Handbook of Coastal Processes and Erosion* [Komar, P.D. (ed.)]. CRC Press, Boca Raton, FL, pp. 77-122.

Oertel, G. F., (1985). The barrier island system. *Marine Geology*, *63(1-4)*, 1-18.

Oertel, G. F. & Kraft, J. C. (1994). New Jersey and Delmarva barrier islands. In: *Geology of Holocene Barrier Island Systems* [Davis, R.A. (ed.)]. Springer-Verlag, New York, pp. 207-232.

Penland, S., Connor, P. F., Beall, A., Fearnley, S. & Williams, S. J. (2005). Changes in Louisiana's shoreline: 1855-2002. *Journal of Coastal Research*, Special Issue *44*, 7-39.

Pierce, J. W. & Colquhoun, (1970). Holocene evolution of a portion of the North Carolina coast. *Geological Society of America Bulletin*, *81(12)*, 3697-37 14.

Pilkey, O. H. & Cooper, J. A. G (2004). Society and sea-level rise. *Science*, *303(5665)*, 178 1-1782.

Psuty, N. P. & Ofiara, D. D. (2002). *Coastal Hazard Management: Lessons and Future Directions from New Jersey*. Rutgers University Press, New Brunswick, NJ, 429 pp.

Ramsey, K.W., Schenck, and L.T. Wang, 2001: *Physiographic Regions of the Delaware Atlantic Coast*. Delaware Geological Survey special publication 25. University of Delaware, Lewes, 1 map.

Reed, D. J., Bishara, D., Cahoon, D., Donnelly, J., Kearney, M., Kolker, A., Leonard, L. R.A. Orson, and J.C. Stevenson, (2008) Site-specific scenarios for wetlands accretion as sea level rises in the mid-Atlantic region. Section 2.1 in: *Background Documents Supporting Climate Change Science Program Synthesis and Assessment Product 4.1: Coastal Elevations and Sensitivity to Sea Level Rise* [Titus, J.G. & E.M. Strange (eds.)]. EPA 430R07004. U.S. Environmental Protection Agency, Washington, DC, pp. 134-174. <http://epa.gov/climatechange/ effects/coastal /background.html>

Riggs, S.R. & Ames, D.V. (2007). *Effect of Storms on Barrier Island Dynamics, Core Banks, Cape Lookout National Seashore, North Carolina, 1960-2001*. U.S. Geological Survey scientific investigations report 2006-5309. U.S. Geological Survey, Reston, VA, 78 pp.

Riggs, S.R., Cleary, W. J. and Snyder, S. W. (1995). Influence of inherited geologic framework upon barrier beach morphology and shoreface dynamics. *Marine Geology*, *126(1-4)*, 213-234.

Rosati, J. D., (2005). Concepts in sediment budgets. *Journal of Coastal Research*, *21(2)*, 307-322.

Rowley, R. J., Kostelnick, J. C., Braaten, D. X. Li, and Meisel, J. (2007). Risk of rising sea level to population and land area. *EOS, Transactions of the American Geophysical Union*, *88(9)*, 105, 107.

Sallenger, A.S., Wright, C.W. & Lillycrop, J. (2007). Coastal-change impacts during Hurricane Katrina: an overview. In: *Coastal Sediments '07* [Kraus, N.C. and J.D. Rosati (eds.)]. America Society of Civil Engineers, Reston, VA, pp. 888-896.

Schupp, C.A., McNinch, J.E. & List, J.H. (2006). Shore-oblique bars, gravel outcrops and correlation to shoreline hotspots. *Marine Geology, 233(1-4)*, 63-79.

Schupp, C.A., Bass, G.P. & Grosskopf, W.G. (2007). Sand bypassing restores natural processes to Assateague Island, Maryland. In: *Coastal Sediments '07* [Kraus, N.C. & J.D. Rosati (eds.)]. America Society of Civil Engineers, Reston, VA, pp. 1340-1353.

Schwab, W.C., Thieler, E.R., Allen, J.R., Foster, D.S., Swift, B.A. & Denny, J.F. (2000). Influence of inner-continental shelf geologic framework on the evolution and behavior of the barrier island system between Fire Island inlet and Shinnecock inlet, Long Island, New York. *Journal of Coastal Research, 16(2)*, 408-422.

Stive, M.J.F. (2004). How important is global warming for coastal erosion? An editorial comment. *Climatic Change, 64(1-2)*, 27-39.

Stive, M.J.F., Aarninkhof, S.G.J., Hamm, L., Hanson, H., Larson, M., Wijnberg, K.M., Nicholls, R.J. & Capohianco, M. (2002). Variability of shore and shoreline evolution. *Coastal Engineering, 47(2)*, 211-235.

Stolper, D., List, J.H. & Thieler, E.R. (2005). Simulating the evolution of coastal morphology and stratigraphy with a new morphological-behavior model (GEOMBEST). *Marine Geology*, 218(1-4), 17-36.

Swift, D.J.P. (1975). Barrier island genesis; evidence from the central Atlantic shelf, eastern USA. *Sedimentary Geology, 14(1)*, 1-43.

Swift, D.J.P., Niederoda, A.W., Vincent, C.E. & Hopkins, T.S. (1985). Barrier island evolution, middle Atlantic shelf, USA. Part I: shoreface dynamics. *Marine Geology, 63(1-4)*, 331-361.

Taney, N.E. (1961). *Geomorphology of the South Shore of Long Island, New York.* Technical memorandum no. 128. U.S. Beach Erosion Board, Washington, DC, 67 pp.

Thieler, E.R., Pilkey, O.H., Young, R.S. Bush, D.M. & Chai, F. (2000). The use of mathematical models to predict beach behavior for coastal engineering: a critical review. *Journal of Coastal Research, 16(1)*, 48-70.

Titus, J.G. & Richman, C. (2001). Maps of lands vulnerable to sea level rise: modeled elevations along the U.S. Atlantic and Gulf coasts. *Climate Research, 18(3)*, 205- 228.

Walker, H.J. & Coleman, J.M. (1987). Atlantic and Gulf Coast province. In: *Geomorphic Systems of North America* [Graf, W.L. (ed.)]: Geological Society of America, Boulder CO, pp. 51-110.

Williams, S.J., Penland, S. & Sallenger, A.H. (1992). *Atlas of Shoreline Changes in Louisiana from 1853 to 1989.* U.S. Geological Survey miscellaneous investigation series I-2 150-A; Louisiana Barrier Island Erosion Study. U.S.Geological Survey, Reston, VA and Louisiana Geological Survey, Baton Rouge, 107 pp.

Wright, L.D. (1995). *Morphodynamics of Inner Continental Shelves*. CRC Press, Boca Raton, FL, 241 pp.

Zhang, K., Douglas, B.C. & Leatherman, S.P. (2002). Do storms cause long-term beach erosion along the U.S. east barrier coast? *Journal of Geology, 110(4)*, 493-502.

Zhang, K., Douglas, B.C. & Leatherman, S.P. (2004). Global warming and coastal erosion. *Climatic Change, 64(1-2)*, 41-58.

In: Coastal Sensitivity to Sea Level Rise-Focusing on the Mid-… ISBN: 978-1-60741-440-7
Editor: Melvin C. Urajner

Chapter 4

COASTAL WETLAND SUSTAINABILITY

Lead Authors: Donald R. Cahoon[1], Denise J. Reed[2], Alexander S. Kolker[3] and Mark M. Brinson[4]
Contributing Authors: J. Court Stevenson[5], Stanley Riggs[4], Robert Christian[4], Enrique Reyes[4], Christine Voss[4] and David Kunz[4]
[1]USGS
[2]University of New Orleans
[3]Louisiana Universities Marine Consortium
[4]East Carolina University
[5] University of Maryland

Key Findings

- It is *virtually certain* that tidal wetlands already experiencing submergence by sea-level rise and associated high rates of loss (*e.g.*, Mississippi River Delta in Louisiana, Blackwater River marshes in Maryland) will continue to lose area in response to future accelerated rates of sea-level rise and changes in other climate and environmental drivers (factors that cause measurable changes).
- It is *very unlikely* that there will be an overall increase in tidal wetland area in the United States over the next 100 years, given current wetland loss rates and the relatively minor accounts of new tidal wetland development (*e.g.*, Atchafalaya Delta in Louisiana).
- Current model projections of wetland vulnerability on regional and national scales are uncertain due to the coarse level of resolution of landscape-scale models. In contrast, site-specific model projections are quite good where local information has been acquired on factors that control local accretionary processes in specific wetland settings. However, the authors have low confidence that site-specific model

simulations can be successfully generalized so as to apply to larger regional or national scales.

- An assessment of the mid-Atlantic region based on an opinion approach by scientists with expert knowledge of wetland accretionary dynamics projects with a moderate level of confidence that those wetlands keeping pace with twentieth century rates of sea-level rise (Scenario 1) would survive a 2 millimeter per year acceleration of sea-level rise (Scenario 2) only under optimal hydrology and sediment supply conditions, and would not survive a 7 millimeter per year acceleration of sea-level rise (Scenario 3). There may be localized exceptions in regions where sediment supplies are abundant, such as at river mouths and in areas where storm overwash events are frequent.
- The mid-Atlantic regional assessment revealed a wide variability in wetland responses to sea-level rise, both within and among subregions and for a variety of wetland geomorphic settings. This underscores both the influence of local processes on wetland elevation and the difficulty of generalizing from regional/national scale projections of wetland sustainability to the local scale in the absence of local accretionary data. Thus, regional or national scale assessments should not be used to develop local management plans where local accretionary dynamics may override regional controls on wetland vertical development.
- Several key uncertainties need to be addressed in order to improve confidence in projecting wetland vulnerability to sea-level rise, including: a better understanding of maximum rates at which wetland vertical accretion can be sustained; interactions and feedbacks among wetland elevation, flooding, and soil organic matter accretion;
- broad-scale, spatial variability in accretionary dynamics; land use change effects (*e.g.*, freshwater runoff, sediment supply, barriers to wetland migration) on tidal wetland accretionary processes; and local and regional sediment supplies, particularly fine-grain cohesive sediments needed for wetland formation.

4.1. Introduction

Given an expected increase in the rate of sea-level rise in the next century, effective management of highly valuable coastal wetland habitats and resources in the United States will be improved by an in-depth assessment of the effects of accelerated sea-level rise on wetland vertical development (*i.e.*, vertical accretion), the horizontal processes of shore erosion and landward migration affecting wetland area, and the expected changes in species composition of plant and animal communities (Nicholls *et al.*, 2007). This Chapter assesses current and projected future rates of vertical buildup of coastal wetland surfaces and wetland sustainability during the next century under the three sea-level rise scenarios, as described briefly above, and in greater detail in Chapter 1.

Many factors must be considered in such an assessment, including: the interactive effects of sea-level rise and other environmental drivers, (*e.g.,* changes in sediment supplies related to altered river flows and storms); local processes controlling wetland vertical and horizontal development and the interaction of these processes with the array of environmental drivers; geomorphic setting; and limited opportunities for landward migration (*e.g.*, human development on the coast, or steep slopes) (Figures 4.1 and 4.2). Consequently, there is no simple, direct answer on national or regional scales to the key question facing coastal wetland managers today, namely, "Are wetlands building vertically at a pace equal to current sea-level rise, and will they build vertically at a pace equal to future sea-level rise?" This is a difficult question to answer because of the various combinations of local drivers and processes controlling wetland elevation across the many tidal wetland settings found in North America, and also due to the lack of available data on the critical drivers and local processes across these larger landscape scales.

The capacity of wetlands to keep pace with sea-level rise can be more confidently addressed at the scale of individual sites where data are available on the critical drivers and local processes. However, scaling up from the local to the national perspective is difficult, and is rarely done, because of data constraints and because of variations in climate, geology, species composition, and human-induced stressors that become influential at larger scales. Better estimates of coastal wetland sustainability under rising sea levels and the factors influencing future sustainability are needed to inform coastal management decision making. This Chapter provides an overview of the factors influencing wetland sustainability (*e.g.*, environmental drivers, accretionary processes, and geomorphic settings), the state of knowledge of current and future wetland sustainability, including a regional case study analysis of the mid-Atlantic coast of the United States, and information needed to improve projections of future wetland sustainability at continental, regional, and local scales.

4.2. Wetland Settings of the Mid-Atlantic Region

Coastal wetlands in the continental United States occur in a variety of physical settings (Table 4.1). The geomorphic classification scheme presented in Table 4.1, developed by Reed *et al.* (2008) (based on Woodroffe, 2002 and Cahoon *et al.*, 2006), provides a useful way of examining and comparing coastal wetlands on a regional scale. Of the geomorphic settings described in Table 4.1, saline fringe marsh, back-barrier lagoon marsh, estuarine brackish marsh, tidal fresh marsh, and tidal fresh forest are found in the mid-Atlantic region of the United States. Back-barrier lagoon salt marshes are either attached to the backside of the barrier island, or are islands either landward of a tidal inlet or behind the barrier island. Saline fringe marshes are located on the landward side of lagoons where they may be able to migrate upslope in response to sea-level rise (see Section 4.3 for a description of the wetland migration process). Estuarine marshes are brackish (a mixture of fresh and salt water) and occur along channels rather than open coasts, either bordering tidal rivers or embayments; or as islands within tidal channels. Tidal fresh marshes and tidal fresh forests occur along river channels, usually above the influence of salinity but not of tides. These wetlands can be distinguished based on vegetative type (species composition; herbaceous *versus* forested) and the salinity of the area. Given the differing hydrodynamics, sediment sources, and vegetative community characteristics of these geomorphic settings, the relationship between sea-level rise and wetland response will also differ.

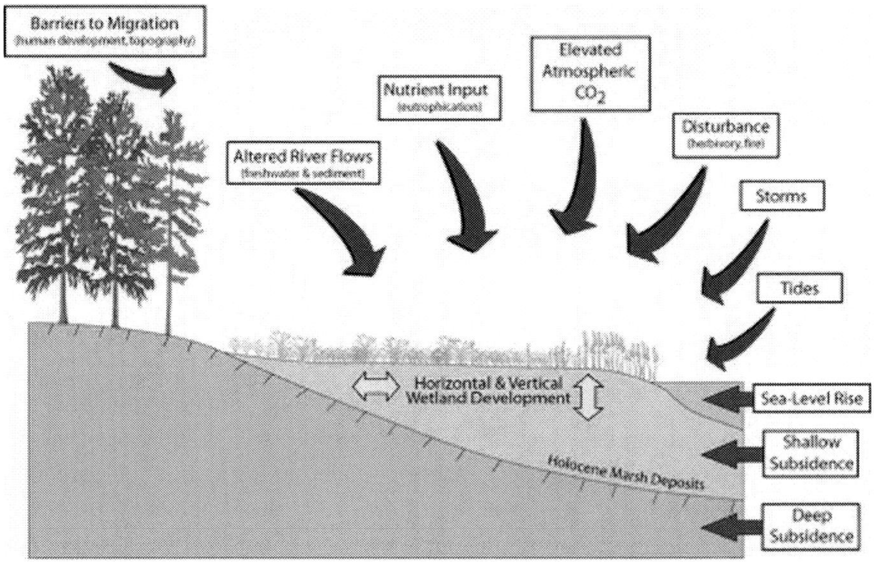

Figure 4.1. Climate and environmental drivers influencing vertical and horizontal wetland development.

Table 4.1. Wetland types and their characteristics as they are distributed within geomorphic settings in the continental United States.

Geomorphic Setting	Description	Sub-settings	Dominant accretion processes	Example Site	Dominant vegetation
Open Coast	Areas sheltered from waves and currents due to coastal topography or bathymetry		Storm sedimentation Peat accumulation	Appalachee Bay, Florida	smooth cordgrass ***(Spartina alterniflora)*** black needlerush ***(Juncus roemerianus)*** spike grass *(Distichlis spicata)* salt hay *(Spartina patens)* glasswort *(Salicornia* spp.) saltwort *(Batis maritima)*
Back-Barrier Lagoon Marsh (BB)	Occupies fill within transgressive back-barrier lagoons	Back-barrier Active flood tide delta Lagoonal fill	Storm sedimentation (including barrier overwash) Peat accumulation Oceanic inputs via inlets	Great South Bay, New York; Chincoteague Bay, Maryland, Virginia	smooth cordgrass *(Spartina alterniflora)* black needlerush (***Juncus roemerianus***) spike grass *(Distichlis spicata)* salt hay *(Spartina patens)* glasswort *(Salicornia* spp.) saltwort *(Batis maritima)*
Estuarine Embayment	Shallow coastal embayments with some river discharge, frequently drowned river valleys			Chesapeake Bay, Maryland, Virginia; Delaware Bay, New Jersey, Pennsylvania, Delaware,	

Table 4.1. (Continued)

Geomorphic Setting	Description	Sub-settings	Dominant accretion processes	Example Site	Dominant vegetation
Estuarine Embayment a. Saline Fringe Marsh (SF)	Transgressive Marshes bordering uplands at the lower end of estuaries (can also be found in back-barrier lagoons)		Storm sedimentation Peat accumulation	Peconic Bay, New York; Western Pamlico Sound, North Carolina	smooth cordgrass *(Spartina alterniflora)* black needlerush ***(Juncus roemerianus)*** spike grass *(Distichlis spicata)* salt hay *(Spartina patens)* glasswort *(Salicornia* spp.) saltwort *(Batis maritima)*
Estuarine Embayment b. Stream Channel Wetlands	Occupy estuarine/ alluvial channels rather than open coast			Dennis Creek, New Jersey; Lower Nanticoke River, Maryland	
Estuarine Brackish Marshes (ES)	Located in vicinity of turbidity maxima zone	Meander Fringing Island	Alluvial and tidal inputs Peat accumulation	Lower James River, Virginia; Lower Nanticoke River, Maryland; Neuse River Estuary, North Carolina	smooth cordgrass*(Spartina alterniflora)* salt hay *(Spartina patens)* spike grass *(Distichlis spicata)* black grass *(Juncus gerardi)* black needlerush *(Juncus roemerianus)* sedges *(Scirpus olneyi)* cattails *(Typha spp.)* big cordgrass *(Spartina cynosuroides)* pickerelweed *(Pontederis cordata)*
Tidal Fresh Marsh (FM)	Located above turbidity maxima zone; develop in drowned river valleys as filled with sediment		Alluvial and tidal inputs Peat accumulation	Upper Nanticoke River, Maryland; Anacostia River, Washington, DC	arrow arum *(Peltandra virginica)* pickerelweed *(Pontederis cordata)* arrowhead *(Sagitarria spp.)* bur-marigold *(Bidens laevis)* halberdleaf tearthumb *(Polygonum arifolium)* scarlet rose-mallow *(Hibiscus coccineus)* wild-rice *(Zizannia aquatica)* cattails *(Typha spp.)* giant cut grass *(Zizaniopsis miliacea)* big cordgrass *(Spartina cynosuroides)*
Tidal Fresh Forests (FF)	Develop in riparian zone along rivers and backwater areas beyond direct influence of seawater	Deep-water Swamps (permanently flooded) Bottomland Hard-wood Forests (seasonally flooded)	Alluvial input Peat accumulation	Upper Raritan Bay, New Jersey; Upper Hudson River, New York	bald cypress *(Taxodium distichum)* blackgum *(Nyssa sylvatica)* oak *(Quercus* spp.) green ash *(Fraxinus pennsylvanica)* (var. *lanceolata*)

Table 4.1. (Continued)

Geomorphic Setting	Description	Sub-settings	Dominant accretion processes	Example Site	Dominant vegetation
Nontidal Brackish Marsh	Transgressive marshes bordering uplands in estuaries with restricted tidal signal		Alluvial input Peat accumulation	Pamlico Sound, North Carolina	black needlerush *(Juncus roemerianus)* smooth cordgrass *(Spartina alterniflora)* spike grass *(Distichlis spicata)* salt hay *(Spartina patens)* big cordgrass *(Spartina cynosuroides)*
Nontidal Forests	Develop in riparian zone along rivers and backwater areas beyond direct influence of seawater in estuaries with restricted tidal signal	Bottom-land Hard-wood Forests (season-ally flooded)	Alluvial input Peat accumulation	Roanoke River, North Carolina; Albemarle Sound, North Carolina	bald cypress *(Taxodium distichum)* blackgum *(Nyssa sylvatica)* oak *(Quercus* spp.) Green ash, *Fraxinus pennsylvanica*
4. Delta	Develop on riverine sediments in shallow open water during active deposition; reworked by marine processes after abandonment		Alluvial input Peat accumulation Compaction/Sub sidence Storm sedimentation Marine Processes	Mississippi Delta, Louisiana	smooth cordgrass *(Spartina alterniflora)* black needlerush *(Juncus roemerianus)* spike grass *(Distichlis spicata)* salt hay *(Spartina patens)* glasswort *(Salicornia* spp.) saltwort *(Batis maritima)* maidencane *(Panicum haemitomon)* arrowhead *(Sagitarria spp.)*

4.3. Vertical Development and Elevation Change

A coastal marsh will survive if it builds vertically at a rate equal to the rise in sea level; that is, if it maintains its elevation relative to sea level. It is well established that marsh surface elevation changes in response to sea-level rise. Tidal wetland surfaces are frequently considered to be closely coupled with local mean sea level (*e.g.*, Pethick, 1981; Allen, 1990). If a marsh builds vertically at a slower rate than the sea rises, however, then a marsh area cannot maintain its elevation relative to sea level. In such a case, a marsh will gradually become submerged and convert to an intertidal mudflat or to open water over a period of many decades (Morris *et al.*, 2002).

The processes contributing to the capacity of a coastal wetland to maintain a stable relationship with changing sea levels are complex and often nonlinear (Cahoon *et al.*, 2006). For example, the response of tidal wetlands to future sea-level rise will be influenced not only by local site characteristics, such as slope and soil erodibility influences on sediment flux, but also by changes in drivers of vertical accretion, some of which are themselves influenced by climate change (Figure 4.1). In addition to the rate of sea-level rise, vertical accretion dynamics are sensitive to changes in a suite of human and climate-related drivers, including alterations in river and sediment discharge from changes in precipitation patterns and in discharge and runoff related to dams and increases in impervious surfaces, increased frequency and intensity of hurricanes, and increased atmospheric temperatures and carbon dioxide concentrations. Vertical accretion is also affected by local environmental drivers such as shallow (local) and deep (regional) subsidence and direct alterations by human activities (*e.g.*, dredging, diking). The relative roles of these drivers of wetland vertical development vary with geomorphic setting.

4.3.1. Wetland Vertical Development

Projecting future wetland sustainability is made more difficult by the complex interaction of processes by which wetlands build vertically (Figure 4.2) and vary across geomorphic settings (Table 4.1). Figure 4.2 shows how environmental drivers, mineral and organic soil development processes, and wetland elevation interact. Tidal wetlands build vertically through the accumulation of mineral sediments and plant organic matter (primarily plant roots). The suite of processes shown in Figure 4.2 controls the rates of mineral sediment deposition and accumulation of plant organic matter in the soil, and ultimately elevation change. Overall mineral sedimentation represents the balance between sediment import and export, which is influenced by sediment supply and the relative abundance of various particle sizes, and varies among geomorphic settings and different tidal and wave energy regimes. Sediment deposition occurs when the surface of a tidal wetland is flooded. Thus, flooding depth and duration are important controls on deposition. The source of sediment may be supplied from within the local estuary (Reed, 1989), and by transport from riverine and oceanic sources. Sediments are remobilized by storms, tides, and, in higher latitudes, ice rafting.

The formation of organic-rich wetland soils is an important contributor to elevation in both mineral sediment rich and mineral sediment poor wetlands (see review by Nyman *et al.*, 2006). Organic matter accumulation represents the balance between plant production (especially by roots and rhizomes) and decomposition and export of plant organic matter (Figure 4.2). Accumulation comes from root and rhizome growth, which contributes mass, volume, and structure to the sediments. The relative importance of mineral and organic matter accumulation can vary depending on local factors such as rates of subsidence and salinity regimes.

4.3.2 Influence of Climate Change on Wetland Vertical Development

Projections of wetland sustainability are further complicated by the fact that sea-level rise is not the only factor influencing accretionary dynamics and sustainability (Figure 4.1). The influence of sea-level rise and other human- and climate-related environmental drivers on mineral sediment delivery systems is complex. For example, the timing and amount of river flows are altered by changes in discharge related to both the effects of dams and impervious surfaces built by humans and to changes in precipitation patterns from changing climate. This results in a change in the balance of forces between river discharge and the tides that control the physical processes of water circulation and mixing, which in turn determines the fate of sediment within an estuary. Where river discharge dominates, highly stratified estuaries prevail, and where tidal motion dominates, well-mixed estuaries tend to develop (Dyer,

1995). Many mid-Atlantic estuaries are partially mixed systems because the influence of river discharge and tides are more balanced.

River discharge is affected by interannual and interseasonal variations and intensities of precipitation and evapotranspiration patterns, and by alterations in land use (*e.g.*, impervious surfaces and land cover types) and control over river flows (*e.g.*, impoundments and withdrawals). Sea-level rise can further change the balance between river discharge and tides by its effect on tidal range (Dyer, 1995). An increase in tidal range would increase tidal velocities and, consequently, tidal mixing and sediment transport, as well as extend the reach of the tide landward. In addition, sea-level rise can affect the degree of tidal asymmetry in an estuary (*i.e.*, ebb *versus* flood dominance). In flood dominant estuaries, marine sediments are more likely to be imported to the estuary. However, an increase in sea level without a change in tidal range may cause a shift toward ebb dominance, thereby reducing the input of marine sediments that might otherwise be deposited on intertidal flats and marshes (Dyer, 1995). Estuaries with relatively small intertidal areas and small tidal amplitudes would be particularly susceptible to such changes. The current hydrodynamic status of estuaries today is the result of thousands of years of interaction between rising sea level and coastal landforms.

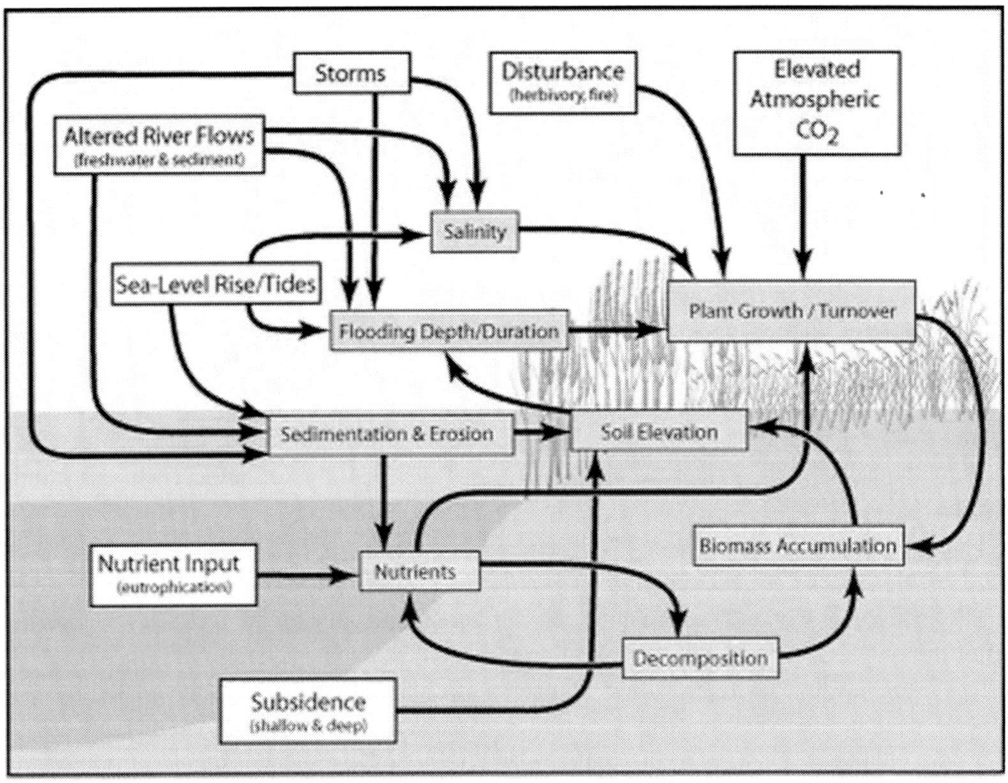

Figure 4.2. A conceptual diagram illustrating how environmental drivers (white boxes) and accretionary processes (grey boxes) influence vertical wetland developmen

The degree of influence of sea-level rise on wetland flooding, sedimentation, erosion, and salinity is directly linked with the influence of altered river flows and storm impacts (Figure 4.2). Changes in freshwater inputs to the coast can affect coastal wetland community structure and function (Sklar and Browder, 1998) through fluctuations in the salt balance up and down the estuary. Low-salinity and freshwater wetlands are particularly affected by increases in salinity. In addition, the location of the turbidity maximum zone (the region in many estuaries where suspended sediment concentrations are higher than in either the river or sea) can shift seaward with increases in river discharge, and the size of this zone will increase with increasing tidal ranges (Dyer, 1995). Heavy rains (freshwater) and tidal surges (salty water) from storms occur over shorter time periods than interannual and interseasonal variation. This can exacerbate or alleviate (at least temporarily) salinity and inundation effects of altered freshwater input and sea-level rise in all wetland types. The direction of elevation change depends on the storm characteristics, wetland type, and local conditions at the area of storm landfall (Cahoon, 2006). Predicted increases in the magnitude of coastal storms from higher sea surface temperatures (Webster *et al.,* 2005) will likely increase storm-induced wetland sedimentation in the mid-Atlantic regional wetlands. Increased storm intensity could increase the resuspension of nearshore sediments and the storm-related import of oceanic sediments into tidal marshes.

In addition to sediment supplies, accumulation of plant organic matter is a primary process controlling wetland vertical development of soil. The production of organic matter is influenced by factors associated with climate change, including increases in atmospheric carbon dioxide concentrations, rising temperatures, more frequent and extensive droughts, higher nutrient loading from floodwaters and ground waters, and increases in salinity of flood waters. Therefore, a critical question that scientists must address is: "How will these potential changes in plant growth affect wetland elevations and the capacity of the marsh to keep pace with sea-level rise?" Some sites depend primarily on plant matter accumulation to build vertically. For example, many brackish marshes dominated by salt hay (*Spartina patens*) (McCaffrey and Thomson, 1980) and mangroves on oceanic islands with low mineral sediment inputs (McKee *et al.*, 2007), changes in root production (Cahoon *et al.*, 2003, 2006) and nutrient additions (McKee *et al.*, 2007) can significantly change root growth and wetland elevation trajectories. These changes and their interactions warrant further study.

4.4. Horizontal Migration

Wetland vertical development can lead to horizontal expansion of wetland area (both landward and seaward; Redfield, 1972), depending on factors such as slope, sediment supply, shoreline erosion rate, and rate of sea-level rise. As marshes build vertically, they can migrate inland onto dry uplands, given that the slope is not too steep and there is no human-made barrier to migration (Figure 4.1). Some of the best examples of submerged upland types of wetlands in the mid-Atlantic region are found on the Eastern Shore of Chesapeake Bay, a drowned river valley estuary (Darmody and Foss, 1979). Given a setting with a low gradient slope, low wave energy, and high sediment supply (*e.g.*, Barnstable Marsh on Cape Cod, Massachusetts), a marsh can migrate both inland onto uplands and seaward onto sand flats as the shallow lagoon fills with sediment (Redfield, 1972). Most coasts, however, have enough wave energy to prevent seaward expansion of the wetlands. The more common alternative is erosion of the seaward boundary of the marsh and retreat. In these settings, as long as wetland vertical development keeps pace with sea-level rise, wetland area will expand where inland migration is greater than erosion of the seaward boundary, remain unchanged where inland migration and erosion of the seaward boundary are equal, or decline where erosion of the seaward boundary is greater than inland migration (*e.g.*, Brinson *et al.*, 1995). If wetland vertical development lags behind sea-level rise (*i.e.*, wetlands do not keep pace), the wetlands will eventually become submerged and deteriorate even as they migrate, resulting in an overall loss of wetland area, as is occurring at Blackwater National Wildlife Refuge in Dorchester County, Maryland (Stevenson *et al.*, 1985). Thus, wetland migration is dependent on vertical accretion, which is the key process for both wetland survival and expansion. If there is a physical obstruction preventing inland wetland migration, such as a road or a bulkhead, and the marsh is keeping pace with sea-level rise, then the marsh will not expand but will survive in place as long as there is no lateral erosion at its seaward edge. Otherwise, the wetland will become narrower as waves erode the shoreline. Thus, having space available with a low gradient slope for inland expansion is critical for maintaining wetland area in a setting where seaward erosion of the marsh occurs.

4.5. Vulnerability of Wetlands to Twentieth Century Sea-Level Rise

A recent evaluation of accretion and elevation trends from 49 salt marshes located around the world, including sites from the Atlantic, Gulf of Mexico, and Pacific coasts of the United States, provides insights into the mechanisms and variability of wetland responses to twentieth century trends of local sea-level rise (Cahoon *et al.,* 2006). Globally, average wetland surface accretion rates were greater than and positively related to local relative sea-level rise, suggesting that the marsh surface level was being maintained by surface accretion within the tidal range as sea level rose. In contrast, average rates of elevation rise were not significantly related to sea-level rise and were significantly lower than average surface accretion rates, indicating that shallow soil subsidence occurs at many sites. Regardless, elevation changes at many sites were greater than local sea-level rise (Cahoon *et al.,* 2006). Hence, understanding elevation change, in addition to surface accretion, is important when determining wetland sustainability. Secondly, accretionary dynamics differed strongly among geomorphic settings, with deltas and embayments exhibiting high accretion and high shallow subsidence compared to back-barrier and estuarine settings (see Cahoon *et al.,* 2006). Thirdly, strong regional differences in accretion dynamics were observed for the North American salt marshes evaluated, with northeastern U.S. marshes exhibiting high rates of both accretion and elevation change, southeastern Atlantic and Gulf of Mexico salt marshes exhibiting high rates of accretion and low rates of elevation change, and Pacific salt marshes exhibiting low rates of both accretion and elevation change (see Cahoon *et al.,* 2006). The marshes with low elevation change rates are likely vulnerable to current and future sea-level rise, with the exception of those in areas where the land surface is rising, such as on the Pacific Northwest Coast of the United States.

4.5.1. Sudden Marsh Dieback

An increasing number of reports available online (see *e.g.*, <http://wetlands.neers.org/>, <www.inlandbays.org>, <www.brownmarsh.net>, <www.lacoast.gov/watermarks/2004-04/3crms/index.htm>) of widespread "sudden marsh dieback" and "brown marsh dieback" from Maine to Louisiana, along with published studies documenting losses of marshes dominated by saltmarsh cordgrass (*Spartina alterniflora*) and other halophytes (plants that naturally grow in salty soils), suggest that a wide variety of marshes may be approaching or have actually gone beyond their tipping point where they can continue to accrete enough inorganic material to survive (Delaune *et al.,* 1983; Stevenson *et al.,* 1985; Kearney *et al.,* 1988, 1994; Mendelssohn and McKee, 1988; Hartig *et al.,* 2002; McKee *et al.,* 2004; Turner *et al.,* 2004). Sudden dieback was documented over 40 years ago by marsh ecologists (Goodman and Williams, 1961). However, it is not known whether all recently identified events are the same phenomenon and caused by the same factors. There are biotic factors, in addition to insufficient accretion, that have been suggested to contribute to sudden marsh dieback, including fungal diseases and overgrazing by animals such as waterfowl, nutria, and snails. Interacting factors may cause marshes to decline even more rapidly than scientists would predict from one driver, such as sea-level rise. There are few details about the onset of sudden dieback because most studies are done after it has already occurred (Ogburn and Alber, 2006). Thus, more research is needed to understand sudden marsh dieback. The apparent increased frequency of this phenomenon over the last several years suggests an additional risk factor for marsh survival over the next century (Stevenson and Kearney, in press).

4.6. Predicting Future Wetland Sustainability

Projections of future wetland sustainability on regional to national scales are constrained by the limitations of the two modeling approaches used to evaluate the relationship between future sea-level rise and coastal wetland elevation: landscape scale models and site-specific models. Large scale landscape models, such as the Sea Level Affecting Marshes Model (SLAMM) (Park *et al.*, 1989), simulate general trends over large areas, but typically at a very coarse resolution. These landscape models do not mechanistically simulate the processes that contribute to wetland elevation; the processes are input as forcing functions and are not simulated within the model. Thus, this modeling approach does not account for infrequent events that influence wetland vertical development, such as storms and floods, or for frequent elevation feedback mechanisms affecting processes (for example, elevation change alters flooding patterns that in turn affect sediment deposition, decomposition, and plant production). In addition, these models are not suitable for site-specific research and management problems because scaling down of results to the local level is not feasible. Therefore, although landscape models can simulate wetland sustainability on broad spatial scales, their coarse resolution limits their accuracy and usefulness to the local manager.

On the other hand, process oriented site-specific models (*e.g.*, Morris *et al.*, 2002; Rybczyk and Cahoon, 2002) are more mechanistic than landscape models and are used to simulate responses for a specific site with a narrow range of conditions and settings. These site-specific models can account for accretion events that occur infrequently, such as hurricanes and major river floods, and the feedback effects of elevation on inundation and sedimentation that influence accretionary processes over timeframes of a century. The use of site-specific conditions in a model makes it possible to predict long-term sustainability of an individual wetland in a particular geomorphic setting. However, like the landscape models, site-specific models also have a scaling problem. Using results from an individual site to make long-term projections at larger spatial scales is problematic because accretionary and process data are not available for the variety of geomorphic settings across these larger-scale landscapes for calibrating and verifying models. Thus, although site-specific models provide high resolution simulations for a local site, at the present time future coastal wetland response to sea-level rise over large areas can be predicted with only low confidence.

Recently, two different modeling approaches have been used to provide regional scale assessments of wetland response to climate change. In a hierarchical approach, detailed site-specific models were parameterized with long-term data to generalize landscape-level trends with moderate confidence for inland wetland sites in the Prairie Pothole Region of the Upper Midwest of the United States (Carroll *et al.*, 2005; Voldseth *et al.*, 2007; Johnson *et al.*, 2005). The utility of this approach for coastal wetlands has not yet been evaluated. Alternatively, an approach was used to assess coastal wetland vulnerability at regional-to-global scales from three broad environmental drivers: (1) ratio of relative sea-level rise to tidal range, (2) sediment supply, and (3) lateral accommodation space (*i.e.*, barriers to wetland migration) (McFadden *et al.*, 2007). This model suggests that, from 2000 to 2080, there will be global wetland area losses of 33 percent for a 36 centimeter (cm) rise in sea level and 44 percent for a 72 cm rise; and that regionally, losses on the Atlantic and Gulf of Mexico coasts of the United States will be among the most severe (Nicholls *et al.*, 2007). However, this model, called the Wetland Change Model, remains to be validated and faces similar challenges when downscaling, as does the previously described model when scaling up.

Taking into account the limitations of current predictive modeling approaches, the following assessments can be made about future wetland sustainability at the national scale:

- It is *virtually certain* that tidal wetlands already experiencing submergence by sea-level rise and associated high rates of loss (*e.g.*, Mississippi River Delta in Louisiana, Blackwater National Wildlife Refuge marshes in Maryland) will continue to lose area under the influence of future accelerated rates of sea-level rise and changes in other climate and environmental drivers.
- It is *very unlikely* that there will be an overall increase in tidal wetland area on a national scale over the next 100 years, given current wetland loss rates and the relatively minor accounts of new tidal wetland development (*e.g.*, Atchafalaya Delta in Louisiana).
- Current model projections of wetland vulnerability on regional and national scales are uncertain because of the coarse level of resolution of landscape scale models. In contrast, site-specific model projections are quite good where local information has been acquired on factors that control local accretionary processes in specific wetland settings. However, the authors have low confidence that site-specific model simulations, as currently portrayed, can be successfully scaled up to provide realistic projections at regional or national scales.

The following information is needed to improve the confidence in projections of future coastal wetland sustainability on regional and continental scales:

- *Models and validation data.* To scale up site-specific model outputs to regional and continental scales with high confidence, detailed data are needed on the various local drivers and processes controlling wetland elevation across all tidal geomorphic settings of the United States. Obtaining and evaluating the necessary data will be an enormous and expensive task, but not an impractical one. It will require substantial coordination with various private and government organizations in order to develop a large, searchable database. Until this type of database becomes a reality, current modeling approaches need to improve or adapt such that they can be applied across a broad spatial scale with better confidence. For example, evaluating the utility of applying the multi-tiered modeling approach used in the Prairie Pothole Region to coastal wetland systems and validating the broad scale Wetland Change Model for North American coastal wetlands will be important first steps. Scientists' ability to predict coastal wetland sustainability will improve as specific ecological and geological processes controlling accretion and their interactions on local and regional scales are better understood.
- *Expert opinion.* Although models driven by empirical data are preferable, given the modeling limitations described, an expert opinion (*i.e.*, subjective) approach can be used to develop spatially explicit landscape-scale predictions of coastal wetland responses to future sea-level rise with a low-to-moderate level of confidence. This approach requires convening a group of scientists with expert knowledge of coastal wetland geomorphic processes, with conclusions based on an understanding of the processes driving marsh survival during sea-level rise and of how the magnitude and nature of these processes might change due to the effects of climate change and other factors. Because of the enormous complexity of these issues at the continental scale, the expert opinion approach would be applied with greater confidence at the regional scale. Two case studies are presented in Sections 4.6.1 and 4.6.2; the first, using the expert opinion approach applied to the mid-Atlantic region from New York to Virginia, the second, using a description of North Carolina wetlands from the Albemarle–Pamlico Region and an evaluation of their potential response to sea-level rise, based on a review of the literature.

4.6.1. Case Study: Mid-Atlantic Regional Assessment, New York to Virginia

A panel of scientists with diverse and expert knowledge of wetland accretionary processes was convened to develop spatially explicit landscape-scale predictions of coastal wetland response to the three scenarios of sea-level rise assessed in this Product (see Chapter 1) for the mid-Atlantic region from New York to Virginia (see Text Box 4.1). The results of the panel's effort (Reed *et al.*, 2008) inform this Product assessment of coastal elevations and sea-level rise.

Box 4.1. The Wetland Assessment Process Used by a Panel of Scientists

As described in this Product, scientific consensus regarding regional-scale coastal changes in response to sea-level rise is currently lacking. To address the issue of future changes to mid-Atlantic coastal wetlands, Denise Reed, a wetlands specialist at the University of New Orleans, was contracted by the U.S.EPA to assemble a panel of coastal wetland scientists to evaluate the potential outcomes of the sea-level rise scenarios used in this Product. Denise Reed chose the 8 members of this panel on the basis of their technical expertise and experience in the coastal wetland research community, particularly with coastal wetland geomorphic processes, and also their involvement with coastal management issues in the mid- Atlantic region. The panel was charged to address the question, "To what extent can wetlands vertically accrete and thus keep pace with rising sea level, that is, will sea-level rise cause the area of wetlands to increase or decrease?".

The sea-level rise impact assessment effort was conducted as an open discussion facilitated by Denise Reed over a two-day period. Deliberations were designed to ensure that conclusions were based on an understanding of the processes driving marsh survival as sea level rises and how the magnitude and nature of these processes might change in the future in response to climate change and other factors. To ensure a systematic approach across regions within the mid-Atlantic region, the panel:

13. 1. identified a range of geomorphic settings to assist in distinguishing among the different process regimes controlling coastal wetland accretion (see Figure 4.3 and Table 4.1);
14. 2. identified a suite of processes that contribute to marsh accretion (see Table 4.1) and outlined potential future changes in current process regimes caused by climate change;
15. 3. divided the mid-Atlantic into a series of regions based on similarity of process regime and current sea-level rise rates; and
16. 4. delineated geomorphic settings within each region on 1:250,000 scale maps, and agreed upon the fate of the wetlands within these settings under the three sea-level rise scenarios, with three potential outcomes: keeping pace, marginal, and loss (see Figure 4.4).

The qualitative, consensus-based assessment of potential changes and their likelihood developed by the panel is based on their review and understanding of published coastal science literature (*e.g.*, 88 published rates of wetland accretion from the mid-Atlantic region, and sea-level rise rates based on NOAA tide gauge data), as well as field observations drawn from other studies conducted in the mid-Atlantic region. A report (Reed *et al.*, 2008) summarizing the process used, basis in the published literature, and a synthesis of the resulting assessment was produced and approved by all members of the panel.

The report was peer reviewed by external subject-matter experts in accordance with U.S. EPA peer review policies. Reviewers were asked to examine locality-specific maps for localities with which they were familiar, and the documentation for how the maps were created. They were then asked to evaluate the assumptions and accuracy of the maps, and errors or omissions in the text. The comments of all reviewers were carefully considered and incorporated, wherever possible, throughout the report. The final report was published and made available online in February 2008 as a U.S. Environmental Protection Agency report: <http://epa.gov/climatechange/effects/downloads/section21.pdf>.

4.6.1.1. Panel Assessment Methods

The general approach used by the panel is summarized in Box 4.1. The panel recognized that accretionary processes differ among settings and that these processes will change in magnitude and direction with future climate change. For example, it is expected that the magnitude of coastal storms will increase as sea-surface temperatures increase (Webster *et al.*, 2005), likely resulting in an increase in storm sedimentation and oceanic sediment inputs. Also, the importance of peat accumulation to vertical accretion in freshwater systems (Neubauer 2008) is expected to increase in response to sea-level rise up to a threshold capacity, beyond which peat accumulation can no longer increase. However, if salinities also increase in freshwater systems, elevation gains from increased peat accumulation could be offset by increased decomposition from sulfate reduction. Enhanced microbial breakdown of organic-rich soils is likely to be most important in formerly fresh and brackish environments where the availability of sulfate, and not organic matter, generally limits sulfate-reduction rates (Goldhaber and Kaplan, 1974). Increases in air and soil temperatures are expected to diminish the importance of ice effects. Changes in precipitation and human land-use patterns will alter fluvial sediment inputs.

The fate of mid-Atlantic wetlands for the three sea-level rise scenarios evaluated in this Product was determined by the panel through a consensus opinion after all information was considered (see Figure 4.4). The wetlands were classified as keeping pace, marginal, or loss (Reed *et al.*, 2008):

17. 1. *Keeping pace*: Wetlands will not be submerged by rising sea levels and will be able to maintain their relative elevation.

18. 2. *Marginal*: Wetlands will be able to maintain their elevation only under optimal conditions. Depending on the dominant accretionary processes, this could include inputs of sediments from storms or floods, or the maintenance of hydrologic conditions conducive for optimal plant growth. Given the complexity and inherent variability of climatic and other factors influencing wetland accretion, the panel cannot predict the fate of these wetlands. Under optimal conditions they are expected to survive.

19. 3. *Loss*: Wetlands will be subject to increased flooding beyond that normally tolerated by vegetative communities, leading to deterioration and conversion to open water habitat.

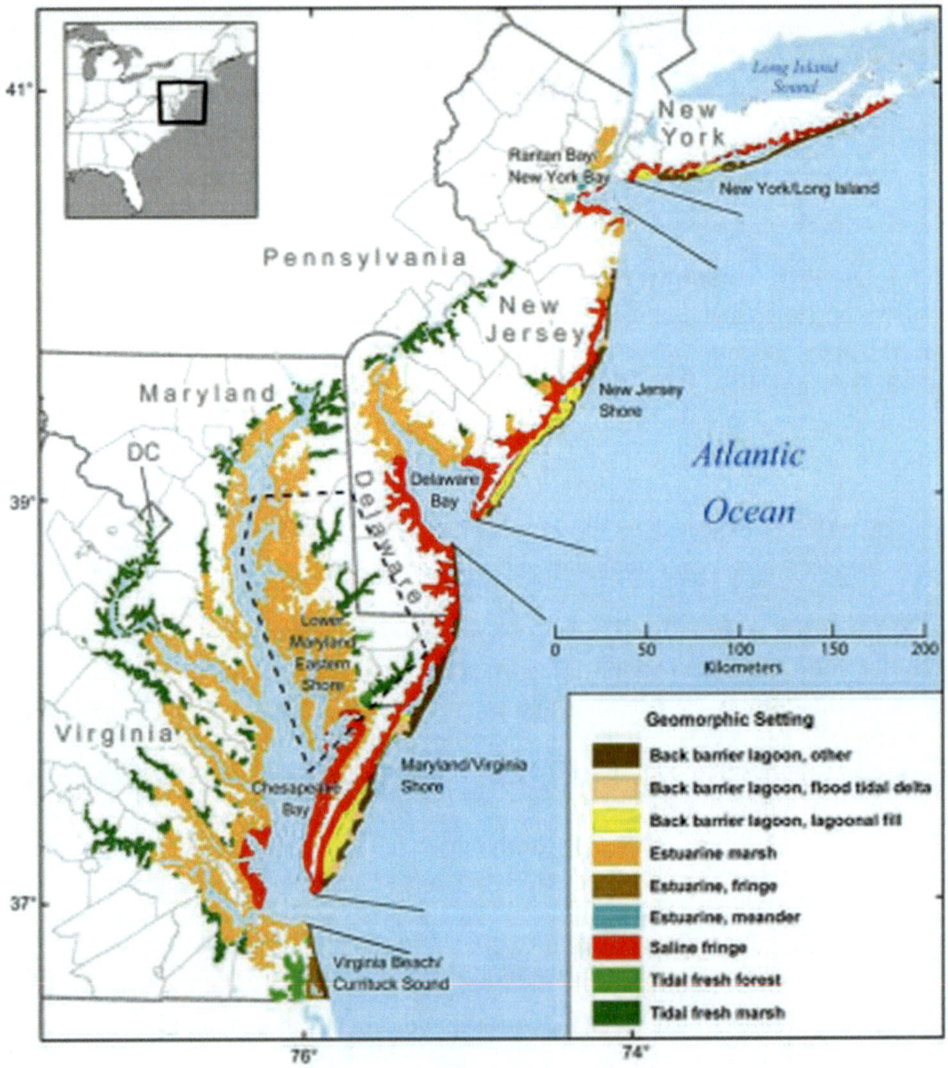

Figure 4.3. Geomorphic settings of mid-Atlantic tidal wetlands (data source: Reed *et al.,* 2008; map source: Titus *et al*., 2008).

The panel recognized that wetlands identified as marginal or loss will become so at an uneven rate and that the rate and spatial distribution of change will vary within and among similarly designated areas. The panel further recognized that wetland response to sea-level rise over the next century will depend upon the rate of sea-level rise, existing wetland condition (*e.g.*, elevation relative to sea level), and local controls of accretion processes. In addition, changes in flooding and salinity patterns may result in a change of dominant species (*i.e.*, less flood-tolerant high marsh species replaced by more flood-tolerant low marsh species), which could affect wetland sediment trapping and organic matter accumulation rates. A wetland is considered marginal when it becomes severely degraded (greater than 50 percent of vegetated area is converted to open water) but still supports ecosystem functions associated with that wetland type. A wetland is considered lost when its function shifts primarily to that of shallow open water habitat.

There are several caveats to the expert panel approach, interpretations, and application of findings. First, regional scale assessments are intended to provide a landscape-scale projection of wetland vulnerability to sea-level rise (*e.g.*, likely trends, areas of major vulnerability) and not to replace assessments based on local process data. The authors recognize that local exceptions to the panel's regional scale assessment likely exist for some specific sites where detailed accretionary data are available. Second, the panel's projections of back-barrier wetland sustainability assume that protective barrier islands retain their integrity. Should barrier islands collapse (see Section 3.7.3), the lagoonal marshes would be exposed to an increased wave energy environment and erosive processes, with massive marsh loss likely over a relatively short period of time. (In such a case, vulnerability to marsh loss would be only one of a host of environmental problems.) Third, the regional projections of wetland sustainability assume that the health of marsh vegetation is not adversely affected by local outbreaks of disease or other biotic factors (*e.g.*, sudden marsh dieback). Fourth, the panel considered the effects of a rate acceleration above current of 2 mm per year (Scenario 2) and 7 mm per year (Scenario 3), but not rates in between. Determining wetland sustainability at sea-level rise rates between Scenarios 2 and 3 requires greater understanding of the variations in the maximum accretion rate regionally and among vegetative communities (Reed *et al.*, 2008). Currently, there are few estimates of the maximum rate at which marsh vertical accretion can occur (Bricker-Urso *et al.*, 1989; Morris *et al.*, 2002) and no studies addressing the thresholds for organic matter accumulation in the marshes considered by the panel. Lastly, the panel recognized the serious limitations of scaling down their projections from the regional to local level and would place a low level of confidence on such projections in the absence of local accretionary and process data. *Thus, findings from this regional scale approach should not be used for local planning activities where local effects on accretionary dynamics may override regional controls on accretionary dynamics.*

4.6.1.2. Panel Findings

The panel developed an approach for predicting wetland response to sea-level rise that was more constrained by available studies of accretion and accretionary processes in some areas of the mid-Atlantic region (*e.g.*, Lower Maryland Eastern Shore) than in other areas (*e.g.*, Virginia Beach/Currituck Sound). Given these inherent data and knowledge constraints, the authors classified the confidence level for all findings in Reed *et al.* (2008) as *likely* (*i.e.*, greater than 0.66 but less than 0.90).

Figure 4.4 and Table 4.2 present the panel's consensus findings on wetland vulnerability of the mid-Atlantic region. The panel determined that a majority of tidal wetlands settings in the mid-Atlantic region (with some local exceptions) are likely keeping pace with Scenario 1, that is, continued sea-level rise at the twentieth century rate, 3 to 4 mm per year (Table 4.2, and areas depicted in brown, beige, yellow, and green in Figure 4.4) through either mineral sediment deposition, organic matter accumulation, or both. However, under this scenario, extensive areas of estuarine marsh in Delaware Bay and Chesapeake Bay are marginal (areas depicted in red in Figure 4.4), with some areas currently being converted to subtidal habitat (areas depicted in blue in Figure 4.4). It is virtually certain that estuarine marshes currently so converted will not be rebuilt or replaced by natural processes. Human manipulation of hydrologic and sedimentary processes and the elimination of barriers to onshore wetland migration would be required to restore and sustain these degrading marsh systems. The removal of barriers to onshore migration invariably would result in land use changes that have other societal consequences such as property loss.

Under accelerated rates of sea-level rise (Scenarios 2 and 3), the panel agreed that wetland survival would very likely depend on optimal hydrology and sediment supply conditions. Wetlands primarily dependent on mineral sediment accumulation for maintaining elevation would be very unlikely to survive Scenario 3, (*i.e.*, at least 10 mm per year rate of sea-level rise when added to the twentieth century rate). Exceptions may occur locally where sediment inputs from inlets, overwash events, or rivers are substantial (*e.g.*, back-barrier lagoon and lagoonal fill marshes depicted in green on western Long Island, Figure 4.4).

Wetland responses to sea-level rise are typically complex. A close comparison of Figure 4.3 and Figure 4.4 reveals that marshes from all geomorphic settings, except estuarine meander (which occurs in only one subregion), responded differently to sea-level rise within and/or among subregions, underscoring why local processes and drivers must be taken into account. Given the variety of marsh responses to sea-level rise among and within subregions (Table 4.2), assessing the likelihood of survival for each wetland setting is best done by subregion, and within subregion, by geomorphic setting.

The scientific panel determined that tidal fresh marshes and forests in the upper reaches of rivers are likely to be sustainable (*i.e.*, less vulnerable to future sea-level rise than most other wetland types) (Table 4.2), because they have higher accretion rates and accumulate more organic carbon than saline marshes (Craft, 2007). Tidal fresh marshes have access to reliable and often abundant sources of mineral sediments, and their sediments typically have 20 to 50 percent organic matter content, indicating that large quantities of plant organic matter are also available. Assuming that salinities do not increase, a condition that may reduce soil organic matter accumulation rates, and current mineral sediment supplies are maintained, the panel considered it likely that tidal fresh marshes and forests would survive under Scenario 3. Vertical development, response to accelerated sea-level rise, and movement into newly submerged areas are rapid for tidal fresh marshes (Orson, 1996). For several tidal fresh marshes in the high sediment-load Delaware River Estuary vertical accretion through the accumulation of both mineral and plant matter ranged from 7 mm per year to 17.4 mm per year from the 1930s to the 1980s as tidal influences became more dominant (Orson *et al.*, 1992). Exceptions to the finding that fresh marshes and forests would survive under Scenario 3 are the New Jersey shore, where tidal fresh marsh is considered marginal under Scenario 2 and lost under Scenario 3, and Virginia Beach–Currituck Sound where fresh forest is marginal under Scenario 1, marginal or lost under Scenario 2, and lost under Scenario 3.

Different marshes from the geomorphic settings back-barrier other, back-barrier lagoonal fill, estuarine marsh, and saline fringe settings responded differently to sea-level rise within at least one subregion as well as among subregions (Table 4.2). For example, back-barrier lagoonal fill marshes on Long Island, New York were classified as either keeping pace or lost at the current rate of sea-level rise. Those marshes surviving under Scenario 1 were classified as either marginal (brown) or keeping up (beige and green) under Scenario 2 (Figure 4.4). Under Scenario 3, only the lagoonal fill marshes depicted in green in Figure 4.4 are expected to survive.

Table 4.2. The range of wetland responses to three sea level rise (slr) scenarios (20th Century rate, 20th Century rate + 2 mm/yr, and 20th Century rate + 7 mm/y) within and among geomorphic settings and subregions of the Mid-Atlantic Region from New York to Virginia

	Region																										
Geomorphic Setting	**Long Island, NY**			**Raritan Bay, NY**			**New Jersey**			**Delaware Bay**			**Maryland – Virginia**			**Chesapeake Bay**			**Lower Maryland Eastern Shore**			**Virginia Beach Currituck Sound**		
	slr	**+2**	**+7**	**slr**	**+2**	**+7**	**slr**	**+2**	**+7**	**slr**	**+2**	**+7**	**slr**	**+2**	**+7**	**slr**	**+2**	**+7**	**slr**	**+2**	**+7**	**slr**	**+2**	**+7**
Back barrier lagoon, other	K	K,M	K,L				K	M	L				K	M	L							M	M-L	L
Back barrier lagoon, flood tide delta	K	K	M				K	M	L				K	M	L									
Back barrier lagoon, lagoonal fill	K,L	M,L	L				K	M	L				K	M	L									
Estuarine marsh				K	M	L	K	M	L	K,M	M,L	L				K,M, L	M-L	L	L,M	L	L	K	M	L
Estuarine fringe				K	M	L	K	M	L													M	M-L	L
Estuarine meander				K	M	L	K	M	L															
Saline fringe	K	K,L	M	K	M	L	K	M	L	K	M	L	K,L	M,L	L									
Tidal fresh forest																			K	K	K	M	M-L	
Tidal fresh marsh				K	K	K	K	M	L	K	K	K				K	K	K	K	K	K	K	K	K

K = keeping pace, M = marginal, L = loss; multiple letters under a single slr scenario (e.g., K,M or K,M,L) indicate more than one response for that geomorphic setting; M-L indicates that the wetland would be either marginal or lost.

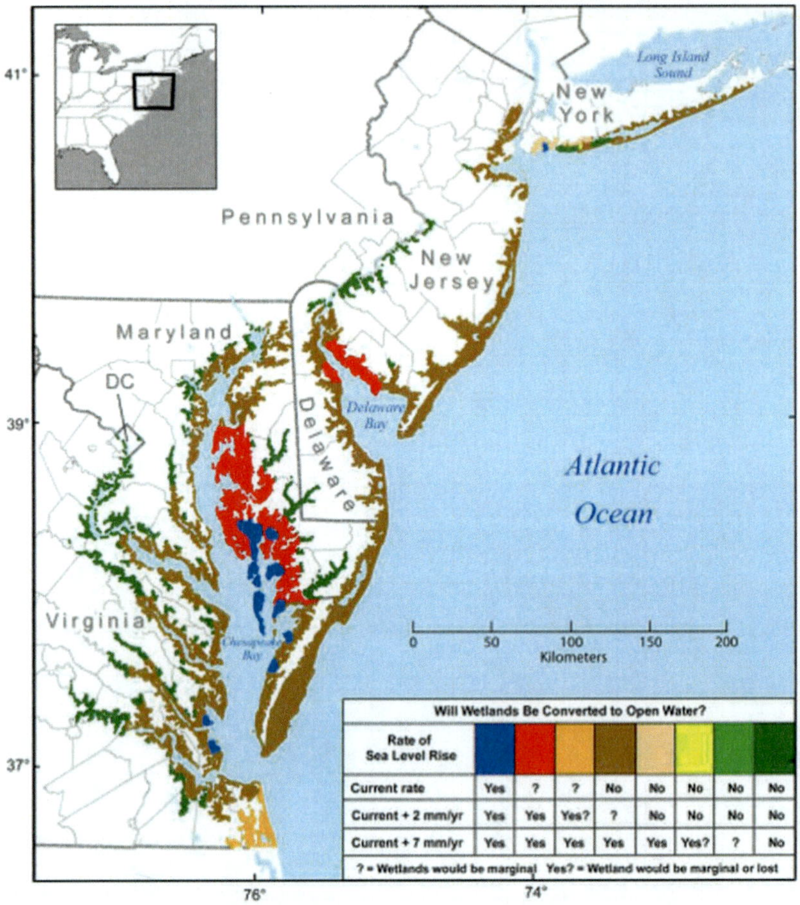

Figure 4.4. Wetland survival in response to three sea-level rise scenarios (data source: Reed *et al.*, 2008; map source: Titus *et al.*, 2008).

The management implications of these findings are important on several levels. The expert panel approach provides a regional assessment of future wetland resource conditions, defines likely trends in wetland change, and identifies areas of major vulnerability. However, the wide variability of wetland responses to sea-level rise within and among subregions for a variety of geomorphic settings underscores not only the influence of local processes on wetland elevation but also the difficulty of scaling down predictions of wetland sustainability from the regional to the local scale in the absence of local accretion data. Most importantly for managers, regional scale assessments such as this should not be used to develop local management plans because local accretionary effects may override regional controls on wetland vertical development (McFadden *et al.,* 2007). Instead, local managers are encouraged to acquire data on the factors influencing the sustainability of their local wetland site, including environmental stressors, accretionary processes, and geomorphic settings, as a basis for developing local management plans.

4.6.2. Case Study: Albemarle–Pamlico Sound Wetlands and Sea-Level Rise

The Albemarle–Pamlico (A–P) region of North Carolina is distinct in the manner and the extent to which rising sea level is expected to affect coastal wetlands. Regional wetlands influenced by sea level are among the most extensive on the U.S. East Coast because of large regions that are less than 3 meters (m) above sea level, as well as the flatness of the underlying surface. Further, the wetlands lack astronomic tides as a source of estuarine water to wetland surfaces in most of the A–P region. Instead, wind-generated water level fluctuations in the sounds and precipitation are the principal sources of water. This "irregular flooding" is the hallmark of the hydrology of these wetlands. Both forested wetlands and marshes can be found; variations in salinity of floodwater determine ecosystem type. This is in striking contrast to most other fringe wetlands on the East Coast.

4.6.2.1. Distribution of Wetland Types

Principal flows to Albemarle Sound are from the Chowan and Roanoke Rivers, and to Pamlico Sound from the Tar and Neuse Rivers. Hardwood forests occupy the floodplains of these major rivers. Only the lower reaches of these rivers are affected by rising sea level. Deposition of riverine sediments in the estuaries approximates the current rate of rising sea level (2 to 3 mm per year) (Benninger and Wells, 1993). These sediments generally do not reach coastal marshes, in part because they are deposited in subtidal areas and in part because astronomic tides are lacking to carry them to wetland surfaces. Storms, which generate high water levels (especially nor'easters and tropical cyclones), deposit sediments on shoreline storm levees and to a lesser extent onto the surfaces of marshes and wetland forests. Blackwater streams that drain pocosins (peaty, evergreen shrub and forested wetlands), as well as other tributaries that drain the coastal plain, are a minor supply of suspended sediment to the estuaries.

Most wetlands in the A–P region were formed upon Pleistocene sediments deposited during multiple high stands of sea level. Inter-stream divides, typified by the Albemarle–Pamlico Peninsula, are flat and poorly drained, resulting in extensive developments of pocosin swamp forest habitats. The original accumulation of peat was not due to rising sea level but to poor drainage and climatic controls. Basal peat ages of even the deepest deposits correspond to the last glacial period when sea level was over 100 m below its current position. Rising sea level has now intercepted some of these peatlands, particularly those at lower elevations on the extreme eastern end of the A–P Peninsula. As a result, eroding peat shorelines are extensive, with large volumes of peat occurring below sea level (Riggs and Ames, 2003).

Large areas of nontidal marshes and forested wetlands in this area are exposed to the influence of sea level. They can be classified as fringe wetlands because they occur along the periphery of estuaries that flood them irregularly. Salinity, however, is the major control that determines the dominant vegetation type. In the fresh-to-oligohaline (slightly brackish) Albemarle Sound region, forested and shrub-scrub wetlands dominate. As the shoreline erodes into the forested wetlands, bald cypress trees become stranded in the permanently flooded zone and eventually die and fall down. This creates a zone of complex habitat structure of fallen trees and relic cypress knees in shallow water. Landward, a storm levee of coarse sand borders the swamp forest in areas exposed to waves (Riggs and Ames, 2003).

Trees are killed by exposure to extended periods of salinity above approximately one-quarter to one-third sea water, and most trees and shrubs have restricted growth and reproduction at much lower salinities (Conner *et al.,* 1997). In brackish water areas, marshes consisting of halophytes replace forested wetlands. Marshes are largely absent from the shore of Albemarle Sound and mouths of the Tar and Neuse Rivers where salinities are too low to affect vegetation. In Pamlico Sound, however, large areas consist of brackish marshes with few tidal creeks. Small tributaries of the Neuse and Pamlico River estuaries grade from brackish marsh at estuary mouths to forested wetlands in oligohaline regions further upstream (Brinson *et al.,* 1985).

4.6.2.2. Future Sea-Level Rise Scenarios

Three scenarios were used to frame projections of the effects of rising sea level over the next few decades in the North Carolina non-tidal coastal wetlands. The first is a non-drowning scenario that assumes rising sea level will maintain its twentieth century, constant rate of 2 to 4 mm per year (Scenario 1). Predictions in this case can be inferred from wetland response to sea-level changes in the recent past (Spaur and Snyder, 1999; Horton *et al.*, 2006). Accelerated rates of sea-level rise (Scenarios 2 and 3), however, may lead to a drowning scenario. This is more realistic if IPCC predictions and other climate change models prove to be correct (Church and White, 2006), and the Scenario 1 rates double or triple. An additional scenario possible in North Carolina involves the collapse of barrier islands, as hypothesized by Riggs and Ames (2003). This scenario is more daunting because it anticipates a shift from the current non-tidal regime to one in which tides would be present to initiate currents capable of transporting sediments without the need of storms and frequently possibly flooding wetland surfaces now only flooded irregularly. The underlying effects of these three scenarios and effects on coastal wetlands are summarized in Table 4.3.

Under the non-drowning scenario, vertical accretion would keep pace with rising sea level as it has for millennia. Current rates (Cahoon, 2003) and those based on basal peats suggest that vertical accretion roughly matches the rate of rising sea level (Riggs *et al.,* 2000; Erlich, 1980; Whitehead and Oakes, 1979). Sources of inorganic sediment to supplement vertical marsh accretion are negligible due to both the large distance between the mouths of piedmont-draining Neuse, Tar, Roanoke and Chowan Rivers and the absence of tidal currents and tidal creeks to transport sediments to marsh surfaces.

Under the drowning scenario, the uncertainty of the effects of accelerated rates lies in the untested capacity of marshes and swamp forests to biogenically accrete organic matter at sea-level rise rates more rapid than experienced currently. It has been suggested that brackish marshes of the Mississippi Delta cannot survive when subjected to relative rates of sea-level rise of 10 mm per year (Day *et al.,* 2005), well over twice the rate currently experienced in Albemarle and Pamlico Sounds. As is the case for the Mississippi Delta (Reed *et al.,* 2006), external sources of mineral sediments would be required to supplement or replace the process of organic accumulation that now dominates wetlands of the A–P region. Where abundant supplies of sediment are available and tidal currents strong enough to transport them, as in North Inlet, South Carolina, Morris *et al.* (2002) reported that the high salt marsh (dwarf *Spartina*) could withstand a 12 mm per year rate. In contrast to fringe wetlands, swamp forests along the piedmont-draining rivers above the freshwater–seawater interface are likely to sustain themselves under drowning scenario conditions because there is a general abundance of mineral sediments during flood stage. This applies to regions within the floodplain but not at river mouths where shoreline recession occurs in response to more localized drowning.

Pocosin peatlands and swamp forest at higher elevations of the coastal plain will continue to grow vertically since they are both independent of sea-level rise. Under the drowning scenario, however, sea-level influenced wetlands of the lower coastal plain would convert to aquatic ecosystems, and the large, low, and flat pocosin areas identified by Poulter (2005) would transform to aquatic habitat. In areas of pocosin peatland, shrub and forest vegetation first would be killed by brackish water. It is unlikely that pocosins would undergo a transition to marsh for two reasons: (1) the pocosin root mat would collapse due to plant mortality and decomposition, causing a rapid subsidence of several centimeters, and resulting in a transition to ponds rather than marshes and (2) brackish water may accelerate decomposition of peat due to availability of sulfate to drive anaerobic decomposition. With the simultaneous death of woody vegetation and elimination of potential marsh plant establishment, organic-rich soils would be exposed directly to the effects of decomposition, erosion, suspension, and transport without the stabilizing properties of vegetation.

Table 4.3. Comparison of three scenarios of rising sea level and their effects on coastal processes.

Scenario	Vertical accretion of wetland surface	Shoreline erosion rate	Sediment supply
Non-drowning: historical exposure of wetlands (past hundreds to several thousand yrs) is predictive of future behavior. Vertical accretion will keep pace with rising sea level (~2-4 mm/yr)	Keeps pace with rising sea level	Recent historical patterns are maintained	Low due to a lack of sources; vertical accretion mostly biogenic
Drowning: vertical accretion rates cannot accelerate to match rates of rising sea level; barrier islands remain intact	Wetlands undergo collapse and marshes break up from within	Rapid acceleration when erosion reaches collapsed egions	Local increases of organic and inorganic suspended sediments as wetlands erode
Barrier islands breached: change to tidal regime throughout Pamlico Sound	Biogenic accretion replaced by inorganic sediment supply	Rapid erosion where high tides overtop wetland shorelines	Major increase in sediments and their redistribution; tidal creeks develop along antecedent drainages mostly in former upland regions

Under the collapsed barrier island scenario (see Section 3.7.3), the A–P regions would undergo a change from a non-tidal estuary to one dominated by astronomic tides due to the collapse of some portions of the barrier islands. A transition of this magnitude is difficult to predict in detail. However, Poulter (2005), using the ADCIRC-2DDI model of Luettich *et al.* (1992), estimated that conversion from a non-tidal to tidal estuary might flood hundreds of square kilometers. The effect is largely due to an increase in tidal amplitude that produces the flooding rather than a mean rise in sea level itself. While the mechanisms of change are speculative, it is doubtful that an intermediate stage of marsh colonization would occur on former pocosin and swamp forest areas because of the abruptness of change. Collapse of the barrier islands in this scenario would be so severe due to the sediment-poor condition of many barrier segments that attempts to maintain and/or repair them would be extremely difficult, or even futile.

The conversion of Pamlico Sound to a tidal system would likely re-establish tidal channels where ancestral streams are located, as projected by Riggs and Ames (2003). The remobilization of sediments could then supply existing marshes with inorganic sediments. It is more likely, however, that marshes would become established landward on newly inundated mineral soils of low-lying uplands. Such a state change has not been observed elsewhere, and computer models are seldom robust enough to encompass such extreme hydrodynamic transitions.

4.7. Data Needs

A few key uncertainties must be addressed in order to increase confidence in the authors' predictions of wetland vulnerability to sea-level rise. First, determining the fate of coastal wetlands over a range of accelerated sea-level rise rates requires more information on variations in the maximum accretion rate regionally, within geomorphic settings, and among vegetative communities. To date, few studies have specifically addressed the maximum rates at which marsh vertical accretion can occur, particularly the thresholds for organic accumulation. Second, although the interactions among changes in wetland elevation, sea level, and wetland flooding patterns are becoming better understood, the interaction of these feedback controls between flooding and changes in other accretion drivers, such as nutrient supply, sulfate respiration, and soil organic matter accumulation is less well understood. Third, scaling up from numerical model predictions of local wetland responses to sea-level rise to long-term projections at regional or continental scales is severely constrained by a lack of available accretionary and process data at these larger landscape scales. Newly emerging numerical models used to predict wetland response to sea-level rise need to be applied across the range of wetland settings. Fourth, scientists need to better understand the role of changing land use on tidal wetland processes, including space available for wetlands to migrate landward and alteration in the amount and timing of freshwater runoff and sediment supply. Finally, sediment supply is a critical factor influencing wetland vulnerability, but the amount and source of sediments available for wetland formation and development is often poorly understood. Coastal sediment budgets typically evaluate coarse-grain sediments needed for beach and barrier development. In contrast, fine-grain cohesive sediments needed for wetland formation and development are typically not evaluated. Improving our understanding of each of these factors is critical for predicting the fate of tidal marshes.

REFERENCES

* Indicates non-peer reviewed literature; available upon request

Allen, J. R. L., (1990). The formation of coastal peat marshes under an upward tendency of relative sea level. *Journal of the Geological Society, 147(5)*, 743-745.

Benninger, L. K. & Wells, J. T. (1993). Sources of sediment to the Neuse River estuary, North Carolina. *Marine Chemistry, 43(1-4)*, 137-156.

Bricker-Urso, S., Nixon, S. W., Cochran, J. K., Hirschberg, D. J. & Hunt, C. (1989). Accretion rates and sediment accumulation in Rhode Island salt marshes. *Estuaries*, 12(4), 300-3 17.

Brinson, M. M., Bradshaw, H. D. & Jones, M. N. (1985). Transitions in forested wetlands along gradients of salinity and hydroperiod. *Journal of the Elisha Mitchell Scientific Society, 101*, 76-94.

Brinson, M. M., Christian, R. R. & Blum, L. K. (1995). Multiple states in the sea-level induced transition from terrestrial forest to estuary. *Estuaries, 18(4)*, 648-659.

Cahoon, D. R., (2003). Storms as agents of wetland elevation change: their impact on surface and subsurface sediment processes. *Proceedings of the International Conference on Coastal Sediments 2003*, May 18-23, 2003, Clearwater Beach FL. World Scientific Publishing Corporation, Corpus Christi, TX.

Cahoon, D. R., (2006). A review of major storm impacts on coastal wetland elevation. *Estuaries and Coasts, 29(6A)*, 889-898.

Cahoon, D. R., French, J. R., Spencer, T., Reed, D. & Moller, I. (2000). Vertical accretion versus elevational adjustment in UK saltmarshes: an evaluation of alternative methodologies. In: *Coastal and Estuarine Environments: Sedimentology, Geomorphology and Geoarchaeology*. [Pye, K. & J.R.L. Allen (eds.)]. Geological Society, London, pp. 223-23 8.

Cahoon, D. R., Hensel, P. Rybczyk, J., McKee, K., Proffitt, C. E. & Perez, B. (2003). Mass tree mortality leads to mangrove peat collapse at Bay Islands, Honduras after Hurricane Mitch. *Journal of Ecology, 91*, 1093-1105.

Cahoon, D. R., Hensel, P. F., Spencer, T., Reed, D. J., McKee, K. L. & Saintilan, N. (2006). Coastal wetland vulnerability to relative sea level rise: wetland elevation trends and process controls. In: *Wetlands and Natural Resource Management* [Verhoeven, J.T.A., B. Beltman, R. Bobbink, and D. Whigham (eds.)]. Ecological studies volume 190, Springer, Berlin and New York, pp. 27 1-292.

Carroll, R., Pohll, G., Tracy, J., Winter, T. & Smith, R. (2005). Simulation of a semipermanent wetland basin in the Cottonwood Lake area, east-central North Dakota. *Journal of Hydrologic Engineering, 10(1)*, 70-84.

Church, J. A. & White, N. J. (2006). A 20th century acceleration in global sea level rise. *Geophysical Research Letters, 33(1)*, L0 1602, doi: 10.1 029/2005GL024826.

Conner, W. H., McLeod, K. W. & McCarron, J. K. (1997). Flooding and salinity effects on growth and survival of four common forested wetland species. *Wetlands Ecology and Management*, *5(2)*, 99-109.

Craft, C., (2007). Freshwater input structures soil properties, vertical accretion, and nutrient accumulation of Georgia and U.S. tidal marshes. *Limnology and Oceanography*, *52(3)*, 1220-1230.

Darmody, R. G. & Foss, J. E. (1979). Soil-landscape relationships of the tidal marshes of Maryland. *Soil Science Society of America Journal*, *43(3)*, 534-541.

Day, J. W., Jr., Barras, J., Clairain, E., J. Johnston, D. Justic, G.P. Kemp, J.-Y. Ko, Lane, R., Mitsch, W.J., Steyer, G., Templet, P. & Yañez-Arancibia, A. (2005). Implications of global climatic change and energy cost and availability for the restoration of the Mississippi delta. *Ecological Engineering*, *24(4)*, 253-265.

DeLaune, R. D., Baumann, R .H. & Gosselink, J. G. (1983). Relationships among vertical accretion, coastal submergence, and erosion in a Louisiana Gulf Coast marsh. *Journal of Sedimentary Petrology*, *53(1)*, 147-157.

Dyer, K., (1995). Response of estuaries to climate change. In: *Climate Change: Impact on Coastal Habitation* [Eisma, D. (ed.)]. Lewis Publishers, Boca Raton, FL, pp. 85- 110.

* Erlich, R. N., (1980). *Early Holocene to Recent Development and Sedimentation of the Roanoke River Area, North Carolina*. M.S. thesis, Department of Geology. University of North Carolina, Chapel Hill, 83 pp.

Goldhaber, M. B. & Kaplan, I. R. (1974). The sulfur cycle. In: *Marine Chemistry*. [Goldberg, E.D. (ed.)]. Wiley, New York, pp. 569-655.

Goodman, P. J. & Williams, W. T. (1961). Investigations into 'die-back' of *Spartina townsendii* agg.: III. Physical correlates of 'die-back'. *Journal of Ecology*, *49(2)*, 391-398.

Hartig, E. K., Gornitz, V., Kolker, A., Muschacke, F. & Fallon, D. (2002). Anthropogenic and climate-change impacts on salt marshes of Jamaica Bay, New York City. *Wetlands*, *22(1)*, 71-89.

Horton, B. P., Corbett, R., Culver, S. J., Edwards, C. Hillier, (2006). Modern salt marsh diatom distributions of the Outer Banks, North Carolina, and the development of a transfer function for high resolution reconstructions of sea level. *Estuarine, Coastal and Shelf Science, 69(3-4)*, 381-394.

Johnson, W. C., Millett, B. V., Gilmanov, T., Voldseth, R.A., Guntenspergen, G. R. & Naugle, D. E. (2005). Vulnerability of northern prairie wetlands to climate change. *Bioscience*, *55(10)*, 863-872.

Kearney, M.S., Grace, R.E. & Stevenson, J.C. (1988). Marsh loss in the Nanticoke Estuary, Chesapeake Bay. *Geographical Review*, *78*, 205-220.

Kearney, M. S., Stevenson, J. C. & Ward, L. G. (1994). Spatial and temporal changes in marsh vertical accretion rates at Monie Bay: implications for sea level rise. *Journal of Coastal Research*, *10(4)*, 1010-1020.

Luettich, R. A., Jr., Westerink, J. J. & Scheffener, N. W. (1992). ADCIRC: An Advanced Three-dimensional Circulation Model for Shelves, Coasts, and Estuaries. Report 1: Th*eory and Methodology of ADCIRC-2DDI and ADCIRC-3DL*. Technical report DRP-92-6. U.S. Army Engineers Waterways Experiment Station, Vicksburg, MS, 141 pp.

McCaffrey, R.J. & Thomson, J. (1980). A record of the accumulation of sediment and trace metals in a Connecticut salt marsh. In: *Estuarine Physics and Chemistry: Studies in Long*

Island Sound. [Saltzman, B. & R.C. Aller (eds.)]. Advances in geophysics volume 22. Academic Press, New York, 424 pp.

McFadden, L., T. Spencer, and Nicholls, R. J. (2007). Broad-scale modelling of coastal wetlands: what is required? *Hydrobiologia, 577*, 5-15.

McKee, K. L., Mendelssohn, I. A. & Materne, M. D. (2004). Acute salt marsh dieback in the Mississippi deltaic plain: a drought induced phenomenon? *Global Ecology and Biogeography, 13(1)*, 65-73.

McKee, K. L., Cahoon, D. R. & Feller, I .C. (2007). Caribbean mangroves adjust to rising sea level through biotic controls on change in soil elevation. *Global Ecology and Biogeography, 16(5)*, 545-556.

Mendelssohn, I. A. & McKee, K. L. (1988). *Spartina alterniflora* die-back in Louisiana: time-course investigation of soil waterlogging effects. *Journal of Ecology, 76(2),* 509-521.

Morris, J. T., Sundareshwar, P. V., Nietch, C. T., Kjerfve, B. & Cahoon, (2002). Responses of coastal wetlands to rising sea level. *Ecology, 83(10)*, 2869-2877.

Neubauer, S.C. 2008: Contributions of mineral and organic components to tidal freshwater marsh accretion. *Estuarine Coastal and Shelf Science, 78(1)*, 78-88.

Nicholls, R. J., Wong, P. P. Burkett, V. R. Codignotto, J. O. Hay, J. E. McLean, R. F. S. Ragoonaden, and Woodroffe, C. D. (2007). Coastal systems and low-lying areas. *Climate Change 2007: Impacts, Adaptation and Vulnerability.* Contribution of Working Group II to the Fourth Assessment Report of the Intergovernmental Panel on Climate Change [Parry, M.L., O.F. Canziani, J.P. Palutikof, P.J. vaan der Linden, and C.E. Hanson (eds.)]. Cambridge University Press, Cambridge, UK, and New York, pp. 315-356.

Nyman, J. A., Walters, R. J. Delaune, R. D. & Patrick, W. H. Jr., (2006). Marsh vertical accretion via vegetative growth. *Estuarine Coastal and Shelf Science, 69(3-4)*, 370-380.

Ogburn, M. B. & Alber, M. (2006). An investigation of salt marsh dieback in Georgia using field transplants. *Estuaries and Coasts, 29(1)*, 54-62.

Orson, R. A., (1996). Some applications of paleoecology to the management of tidal marshes. Estuaries, *19(2A)*, 238-246.

Orson, R. A., Simpson, R. L. & Good, R. E. (1992). A mechanism for the accumulation and retention of heavy metals in tidal freshwater marshes of the upper Delaware River estuary. *Estuarine, Coastal and Shelf Science, 34(2)*, 171-186.

Park, R. A., Trehan, M. S., Mausel, P.W. & Howe, P. W. (1989). The effects of sea level rise on U.S. coastal wetlands. In: *The Potential Effects of Global Climate Change on the United States. Appendix: B, Sea Level Rise.* EPA 230-05-89-052. U.S. Environmental Protection Agency, Washington, DC, 1-1 to 1-55. <http://epa.gov/climatechange/effects/coastal/appB .html>

Pethick, J., (1981). Long-term accretion rates on tidal salt marshes. *Journal of Sedimentary Petrology, 51*, 571-577.

*Poulter, B., (2005). *Interactions Between Landscape Disturbance and Gradual Environmental Change: Plant Community Migration in Response to Fire and Sea Level Rise.* Ph.D. dissertation, Nicholas School of the Environment and Earth Sciences. Duke University, Durham, NC, 216 pp.

Redfield, A. C., (1972). Development of a New England salt marsh. *Ecological Monographs, 42*, 201-237.

Reed, D. J., (1989). Patterns of sediment deposition in subsiding coastal salt marshes, Terrebonne Bay, Louisiana: the role of winter storms. *Estuaries*, *12(4)*, 222-227.

Reed, D. J., Peterson, M. S. & Lezina, B. J. (2006). Reducing the effects of dredged material levees on coastal marsh function: Sediment deposition and nekton utilization. *Environmental Management*, *37(5)*, 671-685.

Reed, D. J., Bishara, D., Cahoon, D., Donnelly, J., Kearney, M., Kolker, A., Leonard, L. Orson, R.A. & Stevenson, J.C. (2008). *Site-specific scenarios for wetlands accretion as sea level rises in the mid-Atlantic region.* Section 2.1 in: Background Documents Supporting Climate Change Science Program Synthesis and Assessment Product 4.1, J.G. Titus and E.M. Strange (eds.). EPA 430R07004. U.S. Environmental Protection Agency, Washington, DC.

Riggs, S. R. & Ames, D. V. (2003). *Drowning the North Carolina Coast: Sea-level Rise and Estuarine Dynamics*. Publication no. UNC-SG-03-04, North Carolina Sea Grant, Raleigh, NC, 152 pp.

Riggs, S. R., Rudolph, G. L. & Ames, D.V. (2000). *Erosional Scour and Geologic Evolution of Croatan Sound, Northeastern North Carolina*. Report nNo. FHWA/NC/2000-002. North Carolina Department of Transportation, Raleigh, NC, 115 pp.

Rybczyk, J.M. & D.R. Cahoon, 2002: Estimating the potential for submergence for two subsiding wetlands in the Mississippi River delta. *Estuaries*, *25(5)*, 985-998.

Sklar, F. H. & Browder, J. A. (1998). Coastal environmental impacts brought about by alterations to freshwater flow in the Gulf of Mexico. *Environmental Management*, *22(4)*, 547-562.

Spaur, C. C. & Snyder, S. W. (1999). Coastal wetlands evolution at the leading edge of the marine transgression: Jarrett Bay, North Carolina. *Journal of the Elisha Mitchell Scientific Society*, *115*, 20-46.

Stevenson, J. C. & Kearney, M. S. In press: Impacts of global change and sea level rise on tidal wetlands. In: *Human Impacts on Marshes: A Global Perspective* [Silliman. B.R., M.D. Bertness, and E. Grosholz (eds.)]. University of California Press, Berkeley, 49 pp. (expected June 2009).

Stevenson, J. C., Kearney, M. S. & Pendleton, E. C. (1985). Sedimentation and erosion in a Chesapeake Bay brackish marsh system. *Marine Geology*, *67(3-4)*, 2 13-235.

Titus, J. G., Jones, R. & Streeter, R. (2008). Maps depicting site-specific scenarios for wetlands accretion as sea level rises in the Mid-Atlantic region. Section 2.2 in: *Background Documents Supporting Climate Change Science Program Synthesis and Assessment Product 4.1: Coastal Elevations and Sensitivity to Sea Level Rise* [Titus, J. G. & Strange E. M. (eds.)]. EPA 430R07004. U.S. Environmental Protection Agency, Washington, DC, pp. 176-186. <http://epa.gov/climatechange/effects/ coastal/ background. html>

Turner, R. E., Swenson, E. M., Milan, C. S., Lee, J. M. & Oswald, T. A. (2004). Below-ground biomass in healthy and impaired salt marshes. *Ecological Research*, *19(1)*, 29-35.

Voldseth, R. A., Johnson, W. C. Gilmanov, T., Guntenspergen, G. R. & Millet, B.V. (2007). Model estimation of land use effects on water levels of northern prairie wetlands. *Ecological Applications*, *17(2)*, 527-540.

Webster, P. J., Holland, G. J., Curry, J. A. & Chang, H. R. (2005). Changes in tropical cyclone number, duration, and intensity in a warming environment. *Science*, *309(5742)*, 1844-1846.

Whitehead, D.R. & Oaks, R.Q. (1979). Developmental history of the Dismal Swamp. In: *The Great Dismal Swamp* [Kirk, P.W. (ed.)]. University Press of Virginia, Charlottesville, VA, pp. 25-43.

Woodroffe, C. I., (2002). *Coasts: Form, Process and Evolution*. Cambridge University Press, Cambridge, UK, and New York, 623 pp.

In: Coastal Sensitivity to Sea Level Rise-Focusing on the Mid-… ISBN: 978-1-60741-440-7
Editor: Melvin C. Urajner

Chapter 5

VULNERABLE SPECIES: THE EFFECTS OF SEA-LEVEL RISE ON COASTAL HABITATS

Ann Shellenbarger Jones[1], *Christina Bosch*[1] *and Elizabeth Strange*[2]
[1]Industrial Economics Inc.
[2]Stratus Consulting, Inc.

Key Findings

- The quality, quantity, and spatial distribution of coastal habitats continuously change as a result of shore erosion, salinity changes, and wetland dynamics; however, accelerated rates of sea-level rise will change some of the major controls of coastal wetland maintenance. Shore protection and development now prevents migration of coastal habitats in many areas. Vulnerable species that rely on these habitats include an array of biota ranging from endangered beetles to commercially important fish and shellfish; and from migratory birds to marsh plants and aquatic vegetation.
- Three key determinants of future tidal marsh acreage are: (1) the capacity of the marsh to raise its surface to match the rate of rising sea level, (2) the rate of erosion of the seaward boundary of the marsh, and (3) the availability of space for the marsh to migrate inland. Depending on local conditions, a tidal marsh may be lost or migrate landward in response to sea-level rise.
- Where tidal marshes become submerged or are eroded, the expected overall loss of wetlands would cause wetland-dependent species of fish and birds to have reduced population sizes. Tidal marshes and associated submerged aquatic plant beds are important spawning, nursery, and shelter areas for fish and shellfish, including commercially important species like the blue crab.
- Many estuarine beaches may also be lost in areas with vertical shore protection and insufficient sediment supply. Endangered beetles, horseshoe crabs, the red knot shorebird, and diamondback terrapins are among many species that rely on sandy beach areas.

- Loss of isolated marsh islands already undergoing submersion will reduce available nesting for bird species, especially those that rely on island habitat for protection from predators. Additional temporary islands may be formed as tidal marshes are inundated, although research on this possibility is limited.
- Many of the freshwater tidal forest systems such as those found in the Mid-Atlantic are considered globally imperiled, and are at risk from sea-level rise among other threats.
- Tidal flats, a rich source of invertebrate food for shorebirds, may be inundated, though new areas may be created as other shoreline habitats are submerged.

5.1. Introduction

Coastal ecosystems consist of a variety of environments, including tidal marshes, tidal forests, aquatic vegetation beds, tidal flats, beaches, and cliffs. For tidal marshes, Table 4.1 outlines the major marsh types, relevant accretionary processes, and the primary vegetation. These environments provide important ecological and human use services, including habitat for endangered and threatened species. The ecosystem services, described in detail within this Chapter, include not only those processes that support the ecosystem itself, such as nutrient cycling, but also the human benefits derived from those processes, including fish production, water purification, water storage and delivery, and the provision of recreational opportunities that help promote human well-being. The high value that humans place on these services has been demonstrated in a number of studies, particularly of coastal wetlands (NRC, 2005).

The services provided by coastal ecosystems could be affected in a number of ways by sea-level rise and coastal engineering projects designed to protect coastal properties from erosion and inundation. As seas rise, coastal habitats are subject to inundation, storm surges, salt water intrusion, and erosion. In many cases, the placement of hard structures along the shore will reduce sediment inputs from upland sources and increase erosion rates in front of the structures (USGS, 2003). If less sediment is available, marshes that are seaward of such structures may have difficulty maintaining appropriate elevations in the face of rising seas. Wetlands that are unable to accrete sufficient substrate as sea level rises will gradually convert to open water, even if there is space available for them to migrate inland, thereby eliminating critical habitat for many coastal species. In addition, landward migration of wetlands may replace current upland habitats that are blocked from migration (NRC, 2007; MEA, 2005). Shallow water and shore habitats are also affected by shore responses. Table 6.1 provides a preliminary overview of the expected environmental effects of human responses to sea-level rise.

Habitat changes in response to sea-level rise and related processes may include structural changes (such as shifts in vegetation zones or loss of vegetated area) and functional changes (such as altered nutrient cycling). In turn, degraded ecosystem processes and habitat fragmentation and loss may not only alter species distributions and relative abundances, but may ultimately reduce local populations of the species that depend on coastal habitats for feeding, nesting, spawning, nursery areas, protection from predators, and other activities that affect growth, survival, and reproductive success.

Habitat interactions are extremely complex. Each habitat supports adjacent systems—for example, the denitrifying effects of wetlands aid adjacent submerged vegetation beds by reducing algal growth; the presence of nearshore oyster or mussel beds reduces wave energy which decreases erosion of marsh edges; and primary productivity is exported from marsh to open waters (see Box 5.1). This Chapter presents simplifications of these interactions in order to identify primary potential effects of both increased rates of sea- level rise and likely shore protections on vulnerable species. In particular, sea-level rise is just one factor among many affecting coastal areas; sediment input, nutrient runoff, fish and shellfish management, and other factors all contribute to the ecological condition of the various habitats discussed in this Section. Sea-level rise may also exacerbate pollution through inundation of upland sources of contamination such as landfills, industrial storage areas, or agricultural waste retention ponds. Under natural conditions, habitats are also continually shifting; the focus of this Chapter is the effect that shoreline management will have on the ability for those shifts to occur (*e.g.*, for marshes or barrier islands to migrate, for marsh to convert to tidal flat or vice versa) and any interruption to the natural shift.

Box 5.1. Finfish, Tidal Salt Marshes, and Habitat Interconnectedness

Tidal salt marshes are among the most productive habitats in the world (Teal, 1986). While this productivity is used within the marshes, marsh-associated organic matter is also exported to food webs supporting marine transient fish production in open waters. Marine transients are adapted to life on a "coastal conveyor belt", often spawning far out on the continental shelf and producing estuarine-dependent young that are recruited into coastal embayments year-round (Deegan *et al.*, 2000). These fish comprise more than 80 percent of species of commercial and recreational value that occupy inshore waters.

Tidal salt marshes serve two critical functions for young finfish (Boesch and Turner, 1984). First, abundant food and the warm shallow waters of the marsh are conducive to rapid growth of both resident and temporary inhabitants. Second, large predators are generally less abundant in subtidal marsh creeks; consequently marshes and their drainage systems may serve as a shelter from predators for the young fish. Protection, rapid growth, and the ability to deposit energy reserves from the rich marsh diet prepare young fish for the rigors of migration and/or overwintering (Weinstein *et al.*, 2005; Litvin and Weinstein, in press).

Effects of Sea-Level Rise

Intertidal and shallow subtidal waters of estuarine wetlands are "epicenters" of material exchange, primary (plant) and secondary (animal) production, and are primary nurseries for the young of many fish and shellfish species (Childers *et al.*, 2000; Weinstein, 1979; Deegan *et al.*, 2000). The prospect of sea-level rise, sometimes concomitant with land subsidence, human habitation of the shore zone, and shore stabilization place these critical resources at risk. Such ecological hotspots could be lost as a result of sea-level rise because human presence in the landscape leaves tidal wetlands little or no room to migrate inland. Because of lack of a well-defined drainage system, small bands of intertidal marsh located seaward of armored shorelines have little ecological value in the production of these finfish (Weinstein *et al.*, 2005; Weinstein, 1983). Due to its interconnectedness with adjacent habitats, loss of tidal salt marshes would significantly affect fish populations, both estuarine and marine, throughout the mid-Atlantic region.

While habitat migration, loss, and gain have all occurred throughout geological history, the presence of developed shorelines introduces a new barrier. Although the potential ecological effects are understood in general terms, few studies have sought to demonstrate or quantify how the interactions of sea-level rise and different types of shore protections may affect the ecosystem services provided by coastal habitats, and in particular the abundance and distribution of animal species (see Chapter 6 for discussion of shore protections). While some studies have examined impacts of either sea-level rise (*e.g.*, Erwin *et al.*, 2006; Galbraith *et al.*, 2002) or shore protections (*e.g.*, Seitz *et al.*, 2006) on coastal fauna, minimal literature is available on the combined effects of rising seas and shore protections. Nonetheless, it is possible in some cases to identify species most likely to be affected based on knowledge of species-habitat associations. Therefore, this Chapter draws upon the ecological literature to describe the primary coastal habitats and species that are vulnerable to the interactive effects of sea-level rise and shore protection activities, and highlights those species that are of particular concern. While this Chapter provides a detailed discussion on a region-wide scale, Appendix 1 of this Product provides much more detailed discussions of specific local habitats and animal populations that may be at risk on a local scale along the mid-Atlantic coast.

5.2. Tidal Marshes

In addition to their dependence on tidal influence, tidal marshes are defined primarily in terms of their salinity: salt, brackish, and freshwater. Chapter 4 describes the structure and flora of these marshes as well as their likely responses to sea-level rise. Table 5.1 presents a general overview of the habitat types, fauna, and vulnerability discussed in this Chapter. Localized information on endangered or threatened species is available through the state natural heritage programs (see Box 5.2).

Salt marshes (back-barrier lagoon marsh or saline fringe marsh, described in Table 4.1) are among the most productive systems in the world because of the extraordinarily high amount of above- and below-ground plant matter that many of them produce, up to 25 metric tons per hectare (ha) aboveground alone (Mitsch and Gosselink, 1993). In turn, this large reservoir of primary production supports a wide variety of invertebrates, fish, birds, and other animals that make up the estuarine food web (Teal, 1986). Insects and other small invertebrates feed on this organic material of the marsh as well as detritus and algae on the marsh surface. These in turn provide food for larger organisms, including crabs, shrimp, and small fishes, which then provide food for larger consumers such as birds and estuarine fishes that move into the marsh to forage (Mitsch and Gosselink, 1993).

Box 5.2. Identifying Local Ecological Communities and Species at Risk

Every state and Washington, D.C. has Natural Heritage Programs (NHPs) that inventory and track the natural diversity of the state, including rare or endangered species. These programs provide an excellent resource for identifying local ecological communities and species at risk. Contact information for NHPs throughout the mid-Atlantic region is provided in Box Table 5.1.

A useful resource for species data outside of each state's own NHP is *NatureServe Explorer*. NatureServe (<http://www.natureserve.org/>) is a non-profit conservation organization which represents the state Natural Heritage Programs and other conservation data centers. *NatureServe Explorer* allows users to search for data on the geographic incidence of plant and animal species in the United States and Canada. The program provides an extensive array of search criteria, including species' taxonomies, classification status, ecological communities, or their national and sub-national distribution. For example, one could 2006 search for all vertebrate species federally listed as threatened that live in Delaware's section of the Chesapeake Bay. For identifying threatened and endangered species extant in vulnerable areas, the smallest geographic unit of analysis is county-level.

Box Table 5.1. State Natural Heritage Program Contact Information

Office	Website	Phone
New York State Department of Environmental Conservation, Division of Fish, Wildlife and Marine Resources	<http://www.nynhp.org/>	(518) 402- 8935
New Jersey Department of Environmental Protection, Division of Parks and Forestry, Office of Natural Lands Management	<http://www.state.nj.us/dep/parksandforests/natural/heritage/index.html>	(609) 984-1339
Pennsylvania Department of Conservation and Natural Resources, Office of Conservation Science	<http://www.naturalheritage.state.pa.us/>	(717) 783-1639
Delaware Department of Natural Resources and Environmental Control, Division of Fish and Wildlife	<http://www.dnrec.state.de.us/nhp/>	(302) 653-2880
Maryland Department of Natural Resources, Wildlife and Heritage Service	<http://www.dnr.state.md.us/wildlife/>	(410) 260-8DNR
The District of Columbia's Department of Health, Fisheries and Wildlife Division	<http://doh.dc.gov/doh/cwp/view,a,1374, Q,584468,dohNav_GID,1810,.asp>	(202) 671-5000
Virginia Department of Conservation and Recreation	<http://www.dcr.virginia.gov/natural_heritage/index.shtml>	(804) 786-7951
North Carolina Department of Environment and Natural Resources, Office of Conservation and Community Affairs	<http://www.ncnhp.org/index.html>	(919) 715-4195

Although much of the primary production in a marsh is used within the marsh itself, some is exported to adjacent estuaries and marine waters. In addition, some of the secondary production of marsh resident fishes, particularly mummichog, and of juveniles, such as blue crab, is exported out of the marsh to support both nearshore estuarine food webs as well as fisheries in coastal areas (Boesch and Turner, 1984; Kneib, 1997, 2000; Deegan *et al.*, 2000; Beck *et al.*, 2003; Dittel *et al.*, 2006; Stevens *et al.*, 2006)[2]. As studies of flood pulses have shown, the extent of the benefits provided by wetlands may be greater in regularly flooded tidal wetlands than in irregularly flooded areas (Bayley, 1991; Zedler and Calloway, 1999).

Table 5.1. Key Fauna/Habitat Associations and Degree of Dependence Habitat Type

	Tidal Marsh	Forested Wetland	Sea-Level Fens	SAV	Tidal Flats	Estuarine Beaches	Unvegetated Cliffs
Fauna							
Fish (Juvenile)	◆	-	-	◆	⬩	⬩	-
Fish (Adult)	♦	-	-	◆	⬩	⬩	-
Crustaceans/	◆			◆	◆	◆	-
Mollusks		-	-				
Other							◆
invertebrates	◆	♦	♦	♦		◆	
Turtles/Terrapins	♦	♦	⬩	♦	-	◆	-
Other reptiles/	⬩	♦	⬩				
amphibians							
Wading Birds	♦	-	-	-			-
Shorebirds	♦	-	-	-		⬩	-
Waterbirds	◆	-	-		◆	⬩	-
Songbirds	⬩	◆	-	-	-	-	⬩
Mammals	♦	◆	-	-	-	⬩	⬩

Notes: Symbols represent the degree of dependence that particular fauna have on habitat types, as described in the sections below. ◆ indicates that multiple species, or certain rare or endangered species, depend heavily on that habitat. ⬩ indicates that the habitat provides substantial benefits to the fauna. ♦ indicates that some species of that fauna type may rely on the habitat, or that portions of their lifecycle may be carried out there. – indicates that negligible activity by a type of fauna occurs in the habitat. Further details on these interactions, including relevant references, are in the sections by habitat below. SAV is submerged aquatic vegetation, discussed later in this Chapter (Section 5.5).

Figure 5.1. Marsh and tidal creek, Bethels Beach (Mathews County) Virginia. June 2002.

Tidal creeks and channels (Figure 5.1) frequently cut through low marsh areas, draining the marsh surface and serving as routes for nutrient-rich plant detritus (dead, decaying organic material) to be flushed out into deeper water as tides recede and for small fish, shrimp, and crabs to move into the marsh during high tides (Mitsch and Gosselink, 1993; Lippson and Lippson, 2006). In addition to mummichog, fish species found in tidal creeks at low tide include Atlantic silverside, striped killifish, and sheepshead minnow (Rountree and Able, 1992). Waterbirds such as great blue herons and egrets are attracted to marshes to feed on the abundant small fish, snails, shrimp, clams, and crabs found in tidal creeks and marsh ponds.

Brackish marshes support many of the same wildlife species as salt marshes, with some notable exceptions. Bald eagles forage in brackish marshes and nest in nearby wooded areas. Because there are few resident mammalian predators (such as red fox and raccoons), small herbivores such as meadow voles thrive in these marshes. Fish species common in the brackish waters of the Mid-Atlantic include striped bass and white perch, which move in and out of brackish waters year-round. Anadromous fish found in the Mid-Atlantic (those that live primarily in salt water but return to freshwater to spawn) include herring and shad, while marine transients such as Atlantic menhaden and drum species are present in summer and fall (White, 1989).

Tidal fresh marshes are characteristic of the upper reaches of estuarine tributaries. In general, the plant species composition of freshwater marshes depends on the degree of flooding, with some species germinating well when completely submerged, while others are relatively intolerant of flooding (Mitsch and Gosselink, 2000). Some tidal fresh marshes possess higher plant diversity than other tidal marsh types (Perry and Atkinson, 1997).

Tidal fresh marshes provide shelter, forage, and spawning habitat for numerous fish species, primarily cyprinids (minnows, shiners, carp), centrarchids (sunfish, crappie, bass), and ictalurids (catfish). In addition, some estuarine fish and shellfish species complete their life cycles in freshwater marshes. Tidal fresh marshes are also important for a wide range of bird species. Some ecologists suggest that freshwater tidal marshes support the greatest diversity of bird species of any marsh type (Mitsch and Gosselink, 2000). The avifauna of these marshes includes waterfowl; wading birds; rails and shorebirds; birds of prey; gulls, terns, kingfishers, and crows; arboreal birds; and ground and shrub species. Perching birds such as red-winged blackbirds are common in stands of cattail. Tidal freshwater marshes support additional species that are rare in saline and brackish environments, such as frogs, turtles, and snakes (White, 1989).

Marsh islands are a critical subdivision of the tidal marshes. These islands are found throughout the mid-Atlantic study region, and are particularly vulnerable to sea-level rise (Kearney and Stevenson, 1991). Islands are common features of salt marshes, and some estuaries and back-barrier bays have islands formed by deposits of dredge spoil. Many islands are a mixture of habitat types, with vegetated and unvegetated wetlands in combination with upland areas[3]. These isolated areas provide nesting sites for various bird species, particularly colonial nesting waterbirds, where they are protected from terrestrial predators such as red fox. Gull-billed terns, common terns, black skimmers, and American oystercatchers all nest on marsh islands (Rounds *et al.*, 2004; Eyler *et al.*, 1999; McGowan *et al.*, 2005).

As discussed in Chapter 4, tidal marshes can keep pace with sea-level rise through vertical accretion (*i.e.*, soil build up through sediment deposition and organic matter accumulation) as long as a sufficient sediment supply exists. Where inland movement is not impeded by artificial shore structures (Figure 5.2) or by geology (*e.g.*, steeply sloping areas between geologic terraces, as found around Chesapeake Bay) (Ward *et al.*, 1998; Phillips, 1986), tidal marshes can expand inland, which would increase wetland area if the rate of migration exceeds that of erosion of the marsh's seaward boundary. However, wetland area would decrease even when a marsh migrates inland if the rate of erosion of the seaward boundary exceeds the rate of migration. Further, in areas where sufficient accretion does not occur, increased tidal flooding will stress marsh plants through waterlogging and changes in soil chemistry, leading to a change in plant species composition and vegetation zones. If marsh plants become too stressed and die, the marsh will eventually convert to open water or tidal flat (Callaway *et al.*, 1996; Morris *et al.*, 2002)[4].

Sea-level rise is also increasing salinity upstream in some rivers, leading to shifts in vegetation composition and the conversion of some tidal fresh marshes into brackish marshes (MD DNR, 2005). At the same time, brackish marshes can deteriorate as a result of ponding and smothering of marsh plants by beach wrack (seaweed and other marine detritus left on the shore by the tide) as salinity increases and storms accentuate marsh fragmentation[5] (Strange *et al.*, 2008). While this process may allow colonization by lower-elevation marsh species, that outcome is not certain (Stevenson and Kearney, 1996). Low brackish marshes can change dynamically in area and composition as sea level rises. If they are lost, forage fish and invertebrates of the low marsh, such as fiddler crabs, grass shrimp, and ribbed mussels, may also be lost, which would affect fauna further up the food chain (Strange *et al.*, 2008). Though more ponding may provide some additional foraging areas as marshes deteriorate, the associated increase in salinity due to evaporative loss can also inhibit the growth of marsh plants (MD DNR, 2005). Many current marsh islands will be inundated; however, in areas with sufficient sediment, new islands may form, although research on this possibility is limited (Cleary and Hosler, 1979). New or expanded marsh islands are also formed through dredge spoil projects[6].

Effects of marsh inundation on fish and shellfish species are likely to be complex. In the short term, inundation may make the marsh surface more accessible, increasing production. However, benefits will decrease as submergence decreases total marsh habitat (Rozas and Reed, 1993). For example, increased deterioration and mobilization of marsh peat sediments increases the immediate biological oxygen demand and may deplete oxygen in marsh creeks and channels below levels needed to sustain fish. In these oxygen-deficient conditions, mummichogs and other killifish may be among the few species able to persist (Stevenson *et al.*, 2002).

Figure 5.2. Fringing marsh and bulkhead, Monmouth County, New Jersey.

In areas where marshes are reduced, remnant marshes may provide lower quality habitat, fewer nesting sites, and greater predation risk for a number of bird species that are marsh specialists and are also important components of marsh food webs, including the clapper rail, black rail, least bittern, Forster's tern, willet, and laughing gull (Figure 5.3) (Erwin *et al.*, 2006). The majority of the Atlantic Coast breeding populations of Forster's tern and laughing gull are considered to be at risk because of loss of lagoonal marsh habitat due to sea-level rise (Erwin *et al.*, 2006). In a Virginia study, scientists found that the minimum marsh size to support significant marsh bird communities was 4.1 to 6.7 hectares (ha) (10.1 to 16.6 acres [ac]) (Watts, 1993). Some species may require even larger marsh sizes; minimum marsh size for successful communities of the saltmarsh sharp-tailed sparrow and the seaside sparrow, both on the Partners in Flight Watch List, are estimated at 10 and 67 ha (25 and 166 ac), respectively (Benoit and Askins, 2002).

Figure 5.3. Marsh drowning and hummock in Blackwater Wildlife Refuge, Maryland. November, 2002.

Figure 5.4. Pocosin in Green Swamp, North Carolina

5.3. Freshwater Forested Wetlands

Forested wetlands influenced by sea level line the mid-Atlantic coast. Limited primarily by their requirements for low-salinity water in a tidal regime, tidal fresh forests occur primarily in upper regions of tidal tributaries in Virginia, Maryland, Delaware, New Jersey, and New York (NatureServe, 2006). The low-lying shorelines of North Carolina also contain large stands of forested wetlands, including cypress swamps and pocosins (Figure 5.4). Also in the mid-Atlantic coastal plains (*e.g.*, around Barnegat Bay, New Jersey) are Atlantic white cedar swamps, found in areas where a saturated layer of peat overlays a sandy substrate (NatureServe, 2006). Forested wetlands support a variety of wildlife, including the prothonotary warbler, the two-toed amphiuma salamander, and the bald eagle. Forested wetlands with thick understories provide shelter and food for an abundance of breeding songbirds (Lippson and Lippson, 2006). Various rare and greatest conservation need (GCN) species reside in mid-Atlantic tidal swamps, including the Delmarva fox squirrel (federally listed as endangered), the eastern red bat, bobcats, bog turtles, and the redbellied watersnake (MD DNR, 2005).

Tidal fresh forests, such as those found in the Mid-Atlantic, face a variety of threats, including sea-level rise, and are currently considered globally imperiled[7]. The responses of these forests to sea-level rise may include retreat at the open-water boundary, drowning in place, or expansion inland. Fleming *et al.* (2006) noted that, "Crown dieback and tree mortality are visible and nearly ubiquitous phenomena in these communities and are generally attributed to sea-level rise and an upstream shift in the salinity gradient in estuarine rivers". Figure 5.5 presents an example of inundation and tree mortality. In Virginia, tidal forest research has indicated that where tree death is present, the topography is limiting inland migration of the hardwood swamp and the understory is converting to tidal marsh (Rheinhardt, 2007).

Figure 5.5. Inundation and tree mortality in forested wetlands at Swan's Point, Lower Potomac River. These wetlands are irregularly flooded by wind-generated tides, unaffected by astronomic tides; their frequency of inundation is controlled directly by sea level.

5.4. Sea-Level Fens

Sea-level fens are a rare type of coastal wetland with a mix of freshwater tidal and northern bog vegetation, resulting in a unique assemblage that includes carnivorous plants such as sundew and bladderworts (Fleming *et al.*, 2006; VNHP, 2006). Their geographic distribution includes isolated locations on Long Island's South Shore; coastal New Jersey; Sussex County, Delaware; and Accomack County, Virginia. The eastern mud turtle and the rare elfin skimmer dragonfly are among the animal species found in sea-level fens. Fens may occur in areas where soils are acidic and a natural seep from a nearby slope provides nutrient-poor groundwater (VNHP, 2006). Little research has been conducted on the effects of sea-level rise on groundwater fens; however, the Virginia Natural Heritage Program has concluded that sea-level rise is a primary threat to the fens (VNHP, 2006).

5.5. Submerged Aquatic Vegetation

Submerged aquatic vegetation (SAV) is distributed throughout the mid-Atlantic region, dominated by eelgrass in the higher-salinity areas and a large number of brackish and freshwater species elsewhere (*e.g.*, widgeon grass, wild celery) (Hurley, 1990). SAV plays a key role in estuarine ecology, helping to regulate the oxygen content of nearshore waters, trapping sediments and nutrients, stabilizing bottom sediments, and reducing wave energy (Short and Neckles, 1999). SAV also provides food and shelter for a variety of fish and shellfish and the species that prey on them. Organisms that forage in SAV beds feed on the plants themselves, the detritus and the epiphytes on plant leaves, and the small organisms found within the SAV bed (*e.g.*, Stockhausen and Lipcius [2003] for blue crabs; Wyda *et al.* [2002] for fish). The commercially valuable blue crab hides in eelgrass during its molting periods, when it is otherwise vulnerable to predation. In Chesapeake Bay, summering sea turtles frequent eelgrass beds. The Kemp's ridley sea turtle, federally listed as endangered, forages in eelgrass beds and flats, feeding on blue crabs in particular (Chesapeake Bay Program, 2007). Various waterbirds feed on SAV, including brant, canvasback, and American black duck (Perry and Deller, 1996).

Forage for piscivorous birds and fish is also provided by residents of nearby marshes that move in and out of SAV beds with the tides, including mummichog, Atlantic silverside, naked goby, northern pipefish, fourspine stickleback, and three spine stickleback (Strange *et al.*, 2008). Juveniles of many commercially and recreationally important estuarine and marine fishes (such as menhaden, herring, shad, spot, croaker, weakfish, red drum, striped bass, and white perch) and smaller adult fish (such as bay and striped anchovies) use SAV beds as nurseries (NOAA Chesapeake Bay Office, 2007; Wyda *et al.*, 2002). Adults of estuarine and marine species such as sea trout, bluefish, perch, and drum search for prey in SAV beds (Strange *et al.*, 2008).

Effects of sea-level rise on SAV beds are uncertain because fluctuations in SAV occur on a year-to-year basis, a significantly shorter timescale than can be attributed to sea-level rise[8]. However, Short and Neckles (1999) estimate that a 50 centimeter (cm) increase in water depth as a result of sea-level rise could reduce light penetration to current seagrass beds in coastal areas by 50 percent. This would result in a 30 to 40 percent reduction in seagrass growth in those areas due to decreased photosynthesis (Short and Neckles, 1999). Increased erosion, with concomitant increased transport and delivery of sediment, would also reduce available light (MD DNR, 2000).

Although plants in some portion of an SAV bed may decline as a result of such factors, landward edges may migrate inland depending on shore slope and substrate suitability. SAV growth is significantly better in areas where erosion provides sandy substrate, rather than fine-grained or high organic matter substrates (Stevenson *et al.*, 2002).

Sea-level rise effects on the tidal range could also impact SAV, and the effect could be either detrimental or beneficial. In areas where the tidal range increases, plants at the lower edge of the bed will receive less light at high tide, increasing plant stress (Koch and Beer, 1996). In areas where the tidal range decreases, the decrease in intertidal exposure at low tide on the upper edge of the bed will reduce plant stress (Short and Neckles, 1999).

Shore construction and armoring will impede shoreward movement of SAV beds (Short and Neckles, 1999) (see Chapter 6 for additional information on shore protections). First, hard structures tend to affect the immediate geomorphology as well as any adjacent seagrass habitats (Strange *et al.*, 2008). Particularly during storm events, wave reflection off of bulkheads or seawalls can increase water depth and magnify the inland reach of waves on downcoast beaches (Plant and Griggs, 1992; USGS, 2003; Small and Carman, 2005). Second, as sea level rises in armored areas, the nearshore area deepens and light attenuation increases, restricting and finally eliminating seagrass growth (Strange *et al.*, 2008). Finally, high nutrient levels in the water limit vegetation growth. Sediment trapping behind breakwaters, which increases the organic content, may limit eelgrass success (Strange *et al.*, 2008). Low-profile armoring, including stone sills and other "living shorelines" projects, may be beneficial to SAV growth (NRC, 2007). Projects to protect wetlands and restore adjacent SAV beds are taking place and represent a potential protection against SAV loss (*e.g.*, U.S. Army Corps of Engineers restoration for Smith Island in Chesapeake Bay) (USACE, 2004).

Loss of SAV affects numerous animals that depend on the vegetation beds for protection and food. By one estimate, a 50-percent reduction in SAV results in a roughly 25-percent reduction in Maryland striped bass production (Kahn and Kemp, 1985). For diving and dabbling ducks, a decrease in SAV in their diets since the 1 960s has been noted (Perry and Deller, 1996). The decreased SAV in Chesapeake Bay is cited as a major factor in the substantial reduction in wintering waterfowl (Perry and Deller, 1996).

5.6. Tidal Flats

Tidal flats are composed of mud or sand and provide habitat for a rich abundance of invertebrates. Tidal flats are critical foraging areas for numerous birds, including wading birds, migrating shorebirds, and dabbling ducks (Strange *et al.*, 2008).

In marsh areas where accretion rates lag behind sea-level rise, marsh will eventually revert to unvegetated flats and eventually open water as seas rise (Brinson *et al.*, 1995). For example, in New York's Jamaica Bay, several hundred acres of low salt marsh have converted to open shoals (see Appendix 1.B for additional details). In a modeling study, Galbraith *et al.* (2002) predicted that under a 2°C global warming scenario, sea-level rise could inundate significant areas of intertidal flats in some regions. In some cases where tidal range increases with increased rates of sea-level rise; however, there may be an overall increase in the acreage of tidal flats (Field *et al.*, 1991).

In low energy shores with high sediment supplies, where sediments accumulate in shallow waters, flats may become vegetated as low marsh encroaches waterward, which will increase low marsh at the expense of tidal flats (Redfield, 1972). If sediment inputs are not sufficient, tidal flats will convert to subtidal habitats, which may or may not be vegetated depending on substrate composition and water transparency (Strange *et al.*, 2008).

Figure 5.6. Estuarine beach and bulkhead along Arthur Kills, Woodbridge Township, New Jersey (August 2003).

Figure 5.7. Peconic Estuary Beach, Riverhead, New York (September 2006).

Loss of tidal flats would eliminate a rich invertebrate food source for migrating birds, including insects, small crabs, and other shellfish (Strange *et al.*, 2008). As tidal flat area declines, increased crowding in remaining areas could lead to exclusion and reductions in local shorebird populations (Galbraith *et al.*, 2002). At the same time, ponds within marshes may become more important foraging sites for the birds if flats are inundated by sea-level rise (Erwin *et al.*, 2004).

5.7. Estuarine Beaches

Throughout most of the mid-Atlantic region and its tributaries, estuarine beaches front the base of low bluffs and high cliffs as well as bulkheads and revetments (see Figure 5.6)

(Jackson *et al.*, 2002). Estuarine beaches can also occur in front of marshes and on the mainland side of barrier islands (Jackson *et al.*, 2002).

The most abundant beach organisms are microscopic invertebrates that live between sand grains, feeding on bacteria and single-celled protozoa. It is estimated that there are over two billion of these organisms in a single square meter of sand (Bertness, 1999). They play a critical role in beach food webs as a link between bacteria and larger consumers such as sand diggers, fleas, crabs, and other macroinvertebrates that burrow in sediments or hide under rocks (Strange *et al.*, 2008). Various rare and endangered beetles also live on sandy shores. Diamondback terrapin and horseshoe crabs bury their eggs in beach sands. In turn, shorebirds such as the piping plover, American oystercatcher, and sandpipers feed on these resources (USFWS, 1988). The insects and crustaceans found in deposits of wrack on estuarine beaches are also an important source of forage for birds (Figure 5.7) (Dugan *et al.*, 2003).

As sea level rises, the fate of estuarine beaches depends on their ability to migrate and the availability of sediment to replenish eroded sands (Figure 5.8) (Jackson *et al.*, 2002). Estuarine beaches continually erode, but under natural conditions the landward and waterward boundaries usually retreat by about the same distance. Shoreline protection structures may prevent migration, effectively squeezing beaches between development and the water. Armoring that traps sand in one area can limit or eliminate longshore transport, and, as a result, diminish the constant replenishment of sand necessary for beach retention in nearby locations (Jackson *et al.*, 2002). Waterward of bulkheads, the foreshore habitat will likely be lost through erosion, frequently even without sea-level rise. Only in areas with sufficient sediment input relative to sea-level rise (*e.g.*, upper tributaries and upper Chesapeake Bay) are beaches likely to remain in place in front of bulkheads.

In many developed areas, estuarine beaches may be maintained with beach nourishment if there are sufficient sources and the public pressure and economic ability to do so. However, the ecological effects of beach nourishment remain uncertain. Beach nourishment will allow retention in areas with a sediment deficit, but may reduce habitat value through effects on sediment characteristics and beach slope (Peterson and Bishop, 2005).

Beach loss will cause declines in local populations of rare beetles found in Calvert County, Maryland. While the Northeastern beach tiger beetle is able to migrate in response to changing conditions, suitable beach habitat must be available nearby (USFWS, 1994).

At present, the degree to which horseshoe crab populations will decline as beaches are lost remains unclear. Early research results indicate that horseshoe crabs may lay eggs in intertidal habitats other than estuarine beaches, such as sandbars and the sandy banks of tidal creeks (Loveland and Botton, 2007). Nonetheless, these habitats may only provide a temporary refuge for horseshoe crabs if they are inundated as well (Strange *et al.*, 2008).

Where horseshoe crabs decline because of loss of suitable habitat for egg deposition, there can be significant implications for migrating shorebirds, particularly the red knot, a candidate for protection under the federal Endangered Species Act, which feeds almost exclusively on horseshoe crab eggs during stopovers in the Delaware Estuary (Karpanty *et al.*, 2006).

In addition, using high-precision elevation data from nest sites, researchers are beginning to examine the effects that sea-level rise will have on oystercatchers and other shore birds (Rounds and Erwin, 2002). To the extent that estuarine and riverine beaches, particularly on islands, survive better than barrier islands, shorebirds like oystercatchers might be able to migrate to these shores (McGowan *et al.*, 2005).

Figure 5.8. Beach with beach wrack and marsh in Bethel Beach (Mathews County), Virginia.

5.8. Cliffs

Unvegetated cliffs and the sandy beaches sometimes present at their bases are constantly reworked by wave action, providing a dynamic habitat for cliff beetles and birds. Little vegetation exists on the cliff face due to constant erosion, and the eroding sediment augments nearby beaches. Cliffs are present on Chesapeake Bay's western shore and tributaries and its northern tributaries (see Figure 5.9), as well as in Hempstead Harbor on Long Island's North Shore and other areas where high energy shorelines intersect steep slopes (Strange *et al.*, 2008).

If the cliff base is armored to protect against rising seas, erosion rates may decrease, eliminating the unvegetated cliff faces that are sustained by continuous erosion and provide habitat for species such as the Puritan tiger beetle and bank swallow. Cliff erosion also provides a sediment source to sustain the adjacent beach and littoral zone (the shore zone between high and low water marks) (Strange *et al.*, 2008). Naturally eroding cliffs are "severely threatened by shoreline erosion control practices" according to the Maryland Department of Natural Resource's Wildlife Diversity Conservation Plan (MD DNR, 2005). Shoreline protections may also subject adjacent cliff areas to wave undercutting and higher recession rates as well as reduction in beach sediment (Wilcock *et al.*, 1998). Development and shoreline stabilization structures that interfere with natural erosional processes are cited as threats to bank-nesting birds as well as two species of tiger beetles (federally listed as threatened) at Maryland's Calvert Cliffs (USFWS, 1993, 1994; CCB, 1996).

Figure 5.9. Crystal Beach, along the Elk River, Maryland (May 2005).

5.9. Summary of Impacts to Wetland-Dependent Species

Based on currently available information, it is possible to identify particular taxa and even some individual species that appear to be at greatest risk if coastal habitats are degraded or diminished in response to sea-level rise and shoreline hardening:

- Degradation and loss of tidal marshes will affect fish and shellfish production in both the marshes themselves and adjacent estuaries.
- Bird species that are marsh specialists, including the clapper rail, black rail, least bittern, Forster's tern, willet, and laughing gull, are particularly at risk. At present, the majority of the Atlantic Coast breeding populations of Forster's tern and laughing gull are considered to be at risk from loss of lagoonal marshes.
- Increased turbidity and eutrophication in nearshore areas and increased water depths may reduce light penetration to SAV beds, reducing photosynthesis, and therefore the growth and survival of the vegetation. Degradation and loss of SAV beds will affect the numerous organisms that feed, carry on reproductive activities, and seek shelter in seagrass beds.
- Diamondback terrapin are at risk of losing both marsh habitat that supports growth and adjoining beaches where eggs are buried.
- Many marsh islands along the Mid-Atlantic, and particularly in Chesapeake Bay, have already been lost or severely reduced as a result of lateral erosion and flooding

related to sea-level rise. Loss of such islands poses a serious, near-term threat for island-nesting bird species such as gull-billed terns, common terns, black skimmers, and American oystercatchers.

- Many mid-Atlantic tidal forest associations may be at risk from sea-level rise and a variety of other threats, and are now considered globally imperiled.
- Shoreline stabilization structures interfere with natural erosional processes that maintain unvegetated cliff faces that provide habitat for bank-nesting birds and tiger beetles.
- Loss of tidal flats could lead to increased crowding of foraging birds in remaining areas, resulting in exclusion of many individuals; if alternate foraging areas are unavailable, starvation of excluded individuals may result, ultimately leading to reductions in local bird populations.
- Where horseshoe crabs decline because of loss of suitable beach substrate for egg deposition, there could be significant implications for migrating shorebirds, particularly the red knot, a candidate for protection under the federal Endangered Species Act. Red knot feed almost exclusively on horseshoe crab eggs during stopovers in the Delaware Estuary.

REFERENCES

* Indicates non-peer reviewed literature; available upon request

Bayley, P. B., (1991). The flood pulse advantage and the restoration of river-floodplain systems. *Regulated Rivers: Research and Management, 6(2)*, 75-86.

Beck, M.W., K.L. Heck Jr., K.W. Able, D.L. Childers, D.B. Eggleston, B.M. Gillanders, B.S. Halpern, C.G. Hays, K. Hoshino, T.J. Minello, R.J. Orth, P.F. Sheridan, and M.P. Weinstein, 2003: The role of nearshore ecosystems as fish and shellfish nurseries. *Issues in Ecology, 11*, 1-12.

Benoit, L. K. & Askins, R. A. (2002). Relationship between habitat area and the distribution of tidal marsh birds. *The Wilson Bulletin, 114(3)*, 314-323.

Bertness, M. B., (1999). *The Ecology of Atlantic Shorelines*. Sinauer Associates, Sunderland, MA, 417 pp.

Boesch, D. F. & Turner, R. E. (1984(: Dependence of fishery species on salt marshes: the role of food and refuge. *Estuaries, 7(4A)*, 460-468.

Brinson, M. M., Christian, R. R. & Blum, L. K. (1995). Multiple states in the sea level induced transition from terrestrial forest to estuary. *Estuaries, 18(4)*, 648-659.

Callaway, J. C., Nyman, J. A. & DeLaune, R. D. (1996). Sediment accretion in coastal wetlands: A review and a simulation model of processes. *Current Topics in Wetland Biogeochemistry, 2*, 2-23.

CCB (The Center for Conservation Biology), (1996). Fieldwork concluded on bank- nesting bird study. *Cornerstone Magazine*, 2, 1. <https://www.denix.osd.mil/denix /Public/ESPrograms/Conservation/Legacy/Corn er-stone/corner.html>

* Chesapeake Bay Program [sea turtles], (2007). *Sea turtles guide.* [web site] <http://www.chesapeakebay.net/seaturtle.htm>

Childers, D. L., J. W. Day Jr., and H. N. Kellar Jr., (2000). Twenty more years of marsh and estuarine flux studies: revisiting Nixon (1980). In: *Concepts and Controversies in Tidal Marsh Ecology.* [Weinstein, M.P. and D.A. Kreeger (eds.)]. Kluwer Academic, Dordrecht, the Netherlands, pp. 391-424.

Cleary, W. J. & Hosler, P. E. (1979). Genesis and significance of marsh islands within southeastern North Carolina lagoons. *Journal of Sedimentary Research, 49(3)*, 703-709.

Deegan, L. A., Hughes, J. E. & Rountree, R. A. (2000). Salt marsh ecosystem support of marine transient species. In: *Concepts and Controversies in Tidal Marsh Ecology.* [Weinstein, M.P. and D.A. Kreeger (eds.)]. Kluwer Academic, Dordrecht, the Netherlands, pp. 333-368.

Dittel, A. I., Epifanio, C. E. & Fogel, M. L. (2006). Trophic relationships of juvenile blue crabs (*Callinectes sapidus*) in estuarine habitats. *Hydrobiologia, 568(1)*, 379-390.

Dugan, J. E., Hubbard, D. M., McCrary, M. D. & Pierson, M.O. (2003). The response of macrofauna communities and shorebird communities to macrophyte wrack subsidies on exposed sandy beaches of southern California. *Estuarine, Coastal, and Shelf Science, 58S*, 25-40.

Erwin, R. W., Sanders, G. M. & Prosser, D. J. (2004). Changes in lagoonal marsh morphology at selected northeastern Atlantic coast sites of significance to migratory waterbirds. *Wetlands, 24(4)*, 89 1-903.

Erwin, R. M., Sanders, G. M. Prosser, D. J. & Cahoon, D. R. (2006). High tides and rising seas: potential effects on estuarine waterbirds. In: *Terrestrial Vertebrates in Tidal Marshes: Evolution, Ecology, and Conservation.* [Greenberg, R. (ed.)]. Studies in avian biology no. 32. Cooper Ornithological Society, Camarillo, CA, pp. 2 14-228.

Eyler, T. B., Erwin, R. M., Stotts, D. B. & Hatfield, J. S. (1999). Aspects of hatching success and chick survival in gull-billed terns in coastal Virginia. *Waterbirds, 22(1)*, 54-59.

Field, D. W., Reyer, A. J., Genovese, P. V. & Shearer, B. D. (1991). *Coastal Wetlands of the United States.* National Oceanic and Atmospheric Administration and U.S. Fish and Wildlife Service, [Washington, DC], 58 pp.

Fleming, G. P., Coulling, P. P., Patterson, K. D. & Taverna, K. (2006). *The Natural Communities of Virginia: Classification of Ecological Community Groups.* Second approximation, version 2.2. Virginia Department of Conservation and Recreation, Division of Natural Heritage, Richmond, VA. <http://www.dcr.virginia.gov/ natural>

Galbraith, H., Jones, R., Park, R., Clough, J., Herrod-Julius, S. Harrington, B. & Page, G. (2002). Global climate change and sea level rise: potential losses of intertidal habitat for shorebirds. *Waterbirds, 25(2)*, 173-183.

Hurley, L. M., (1990). *Field Guide to the Submerged Aquatic Vegetation of Chesapeake Bay.* U.S. Fish and Wildlife Service, Chesapeake Bay Estuary Program, Annapolis, MD, 48 pp.

Jackson, N. L., Nordstrom, K. F. & Smith, D. R. (2002). Geomorphic-biotic interactions on beach foreshores in estuaries. *Journal of Coastal Research*, Special issue 36, 414-424.

Kahn, J. R. & Kemp, W. M. (1985). Economic losses associated with the degradation of an ecosystem: the case of submerged aquatic vegetation in Chesapeake Bay. *Journal of Environmental Economics and Management, 12(3)*, 246-263.

Karpanty, S. M., Fraser, J. D., Berkson, J. M., Niles, L., Dey, A. & Smith, E. P. (2006). Horseshoe crab eggs determine red knot distribution in Delaware Bay habitats. *Journal of Wildlife Management, 70(6)*, 1704-1710.

Kearney, M. S. & Stevenson, J. C. (1991). Island land loss and marsh vertical accretion rate evidence for historical sea-level changes in Chesapeake Bay. *Journal of Coastal Research*, *7(2)*, 403-416.

Kneib, R. T., (1997). The role of tidal marshes in the ecology of estuarine nekton. *Oceanography and Marine Biology, 35*, 163-220.

Kneib, R. T., (2000). Salt marsh ecoscapes and production transfers by estuarine nekton in the southeastern U.S. In: *Concepts and Controversies in Tidal Marsh Ecology*. [Weinstein, M.P. and D.A. Kreeger (eds.)]. Kluwer Academic, Dordrecht, the Netherlands, pp. 267-291.

Koch, E. W. & Beer, S. (1996). Tides, light and the distribution of *zostera marina* in Long Island Sound, USA. *Aquatic Botany*, *53(1-2)*, 97-107. Referenced in: Short, F.A. and H.A. Neckles, 1999: The effects of global climate change on seagrasses. *Aquatic Botany, 63(3-4)*, 169-196.

Lippson, A. J. & Lippson, R. L. (2006). *Life in the Chesapeake Bay*. Johns Hopkins University Press, Baltimore, MD, 3rd ed., 324 pp.

Litvin, S. Y. & Weinstein, M. P. (2009). Energy density and the biochemical condition of juvenile weakfish (*Cynoscion regalis*) in the Delaware Bay estuary, USA. *Canadian Journal of Fisheries and Aquatic Sciences* (in press).

* Loveland, R. E. & Botton, M. L. (2007). The importance of alternative habitats to spawning horseshoe crabs (*Limulus polyphemus*) in lower Delaware Bay, New Jersey. Presentation at: *Delaware Estuary Science Conference*, Cape May, NJ. Program available online at <http://www.delawareestuary.org/pdf/ScienceConferenceProgram2007.pdf>

McGowan, C. P., Simons, T. R., Golder, W. & Cordes, J. (2005). A comparison of American oystercatcher reproductive success on barrier beach and river island habitats in coastal North Carolina. *Waterbirds*, *28(2)*, 150-155.

MD DNR (Maryland Department of Natural Resources), (2000). *State of Maryland Shore Erosion Task Force*. Maryland Department of Natural Resources, Annapolis, 65 pp. <http://www.dnr. state.md.us/ccws/sec/sccreport.html>

MD DNR (Maryland Department of Natural Resources), (2005). *Maryland DNR Wildlife Conservation Diversity Plan — Final Draft.* <http://www.dnr.state.md.us/ wildlife>

MEA (Millennium Ecosystem Assessment), (2005). Climate change. In: *Ecosystems and Human Well-Being: Policy Responses*. Findings of the Responses Working Group. Millennium Ecosystem Assessment series volume 3. Island Press, Washington, DC, chapter 13.

Mitsch, W. J. & Gosselink, J. G. (1993). *Wetlands*. Van Nostrand Reinhold, New York, 2nd ed., 722 pp.

Mitsch, W.J. & Gosselink, J. G. (2000). *Wetlands*. Van Nostrand Reinhold, New York, 3rd ed., 920 pp.

Morris, J. T., Sundareshwar, P. V., Nietch, C. T., Kjerfve, B. & Cahoon, D. R. (2002). Responses of coastal wetlands to rising sea level. *Ecology*, *83(10)*, 2869-2877.

NatureServe, (2006). *NatureServe Explorer: An Online Encyclopedia of Life* [web application]. Version 5.0. NatureServe, Arlington, VA. <http://www.natureserve. org/explorer>. "Northern Atlantic Coastal Plain Tidal Swamp" CES203 .282 <http://www. natureserve. org/explorer/servlet/Nature Serve?searchSystemUid = ELEMENT GLOBAL.2.723205>

NOAA Chesapeake Bay Office, 2007: *SA V overview.* [web site]<http://noaa. chesapeakebay. net/HabitatSav.aspx>

NRC (National Research Council), (2005). *Valuing Ecosystem Services: Toward Better Environmental Decision-Making.* National Academies Press, Washington, DC, 277 pp.

NRC (National Research Council), (2007). *Mitigating Shore Erosion Along Sheltered Coasts.* National Academies Press, Washington, DC, 174 pp.

Perry, J. E. & Atkinson, R. B. (1997). Plant diversity along a salinity gradient of four marshes on the York and Pamunkey Rivers in Virginia. *Castanea, 62(2)*, 112-118.

Perry, M. C. & Deller, A. S. (1996). Review of factors affecting the distribution and abundance of waterfowl in shallow-water habitats of Chesapeake Bay. *Estuaries, 19(2A)*, 272-278.

Peterson, C. H. & Bishop, M. J. (2005). Assessing the environmental impacts of beach nourishment. *BioScience, 55(10)*, 887-896.

Phillips, J. D., (1986). Coastal submergence and marsh fringe erosion. *Journal of Coastal Research, 2(4)*, 427-43 6.

Plant, N. G. & Griggs, G. B. (1992). Interactions between nearshore processes and beach morphology near a seawall. *Journal of Coastal Research*, 8(1), 183-200.

Redfield, A. C., (1972). Development of a New England salt marsh. *Ecological Monographs,* 42, 201-237.

Rheinhardt, R., (2007). Tidal freshwater swamps of a lower Chesapeake Bay subestuary. In: *Ecology of Tidal Freshwater Forested Wetlands of the Southeastern United States.* [Conner, W. H., Doyle, T. W. & Krauss K. W. (eds.)]. Springer, Dordrecht, the Netherlands, 505 pp.

* Rounds, R. & Erwin, R. M. (2002). Flooding and sea level rise at waterbird colonies in Virginia. Presentation at: *26th Annual Waterbird Society Meeting*, November, La Crosse, WI. <http://www.vcrlter.virginia.edu/presentations/rounds0211 /rounds02 11 .pdf>

Rounds, R.A., Erwin, R. M. & Porter, J. H. (2004). Nest-site selection and hatching success of waterbirds in coastal Virginia: some results of habitat manipulation. *Journal of Field Ornithology, 75(4)*, 3 17-329.

Rountree, R. A. & Able, K. W. (1992). Fauna of polyhaline subtidal marsh creeks in southern New Jersey: composition, abundance and biomass. *Estuaries, 15(2)*, 17 1-185.

Rozas, L. P. & Reed, D. J. (1993). Nekton use of marsh-surface habitats in Louisiana (USA) deltaic salt marshes undergoing submergence. *Marine Ecology Progress Series*, *96*, 147-157.

Seitz, R. D., Lipcius, R. N., Olmstead, N. H., Seebo, M. S. & Lambert, D. M. (2006). Influence of shallow-water habitats and shoreline development on abundance, biomass, and diversity of benthic prey and predators in Chesapeake Bay. *Marine Ecology Progress Series, 326*, 11-27.

Short, F.T. & Neckles, H. A. (1999). The effects of global climate change on seagrasses. *Aquatic Botany, 63(3-4)*, 169-196.

* Small, D. & Carman, R. (2005). A history of the Washington State Hydraulic Code and marine shoreline armoring in Puget Sound. Presented at: *2005 Puget Sound Georgia Basin Research Conference*, March 29-3 1, Seattle, WA, [14 pp.] <http://www.engr.washington.edu/epp/psgb/2005psgb/2005proceedings/papers/P 5_SMALL.pdf>

Stevens, P.W., C.L. Montague, and K.J. Sulak, 2006: Fate of fish production in a seasonally flooded saltmarsh. *Marine Ecology Progress Series, 327*, 267-277.

Stevenson, J. C. & Kearney, M. S. (1996). Shoreline dynamics on the windward and leeward shores of a large temperate estuary. In: *Estuarine Shores: Evolution, Environments, and Human Alterations*. [Nordstrom, K.F. and C.T. Roman (eds.)]. Wiley, New York, pp. 233-259.

Stevenson, J. C., Kearney, M. S. & Koch, E. W. (2002). Impacts of sea level rise on tidal wetlands and shallow water habitats: a case study from Chesapeake Bay. *American Fisheries Society Symposium*, *32*, 23-36.

Stockhausen, W. T. & Lipcius, R. N. (2003). Simulated effects of seagrass loss and restoration on settlement and recruitment of blue crab postlarvae and juveniles in the York River, Chesapeake Bay. *Bulletin of Marine Science, 72(2)*, 409-422.

Strange, E. M., Shellenbarger Jones, A., Bosch, C., Jones, R., Kreeger, D. & Titus, J. G. (2008). Mid-Atlantic coastal habitats and environmental implications of sea level rise. Section 3 in: *Background Documents Supporting Climate Change Science Program Synthesis and Assessment Product 4.1: Coastal Elevations and Sensitivity to Sea Level Rise* [Titus, J.G. and E.M. Strange (eds.)]. EPA 430R07004. U.S. Environmental Protection Agency, Washington, DC, pp. 188- 342. <http://epa.gov/climatechange/ effects/coastal/background.html>

Teal, J. M., (1986). *The Ecology of Regularly Flooded Salt Marshes of New England: A Community Profile.* Biological report 85(7.4). U.S. Fish and Wildlife Service, Washington, DC, 69 pp.

USACE (U.S. Army Corps of Engineers), (2004). *Smith Island, Maryland Environmental Restoration and Protection Project.* [web site] U.S. Army Corps of Engineers, Baltimore Division.<http://www.nab.usace.army.mil/projects/Maryland/smithisland.htm>

USFWS (U.S. Fish and Wildlife Service), Undated: *Nutrient Pollution.* [web site] USFWS Chesapeake Bay Field Office. <http://www.fws.gov/chesapeakebay/nutrient>

USFWS (U.S. Fish and Wildlife Service), (1988). Endangered Species Information Booklet: *Piping Plover.* U.S. Fish and Wildlife Service, Arlington, VA.

USFWS (U.S. Fish and Wildlife Service), (1993). *Puritan Tiger Beetle (*Cicindela puritana *G. Horn) Recovery Plan.* Hadley, MA, 45 pp.

USFWS (U.S. Fish and Wildlife Service), (1994). *Recovery Plan for the Northeastern Beach Tiger Beetle (*Cicindela dorsalis dorsalis*).* Hadley, MA.

USGS (U.S. Geological Survey), (2003). *A Summary Report of Sediment Processes in Chesapeake Bay and Watershed.* [Langland, M. & T. Cronin (eds.)]. Water resources investigations report 03-4123. U.S. Geological Survey, New Cumberland, PA, 109 pp.

VNHP (Virginia Natural Heritage Program), (2006). Natural Heritage Resources Fact Sheet: *Virginia's Rare Natural Environments: Sea-level Fens.* Virginia Department of Conservation and Recreation, [Richmond, VA,] 2 pp. <http://www.dcr.virginia.gov /natural>

Ward, L. G., Kearney, M. S. & Stevenson, J. C. (1998). Variations in sedimentary environments and accretionary patterns in estuarine marshes undergoing rapid submergence, Chesapeake Bay. *Marine Geology, 151(1-4)*, 111-134.

Watts, B. D., (1993). *Effects of Marsh Size on Incidence Rates and Avian Community Organization Within the Lower Chesapeake Bay.* Center for Conservation Biology technical report CCBTR-93-03. College of William and Mary, Williamsburg, VA, 53 pp.

Weinstein, M. P., (1979). Shallow marsh habitats as primary nurseries for fishes and shellfish, Cape Fear River, North Carolina. *Fishery Bulletin, 77(2)*, 339-357.

Weinstein, M. P., (1983). Population dynamics of an estuarine-dependent fish, the spot (*Leisotomus xanthurus*) along a tidal creek-seagrass meadow coenocline. *Canadian Journal of Fisheries and Aquatic Sciences*, *40(10)*, 1633-1638.

Weinstein, M. P., Litvin, S. Y. & Guida, V. G. (2005). Considerations of habitat linkages, estuarine landscapes, and the trophic spectrum in wetland restoration design. *Journal of Coastal Research*, Special issue *40,* 51-63.

White, C. P., (1989). *Chesapeake Bay: Nature of the Estuary: A Field Guide*. Tidewater Publishers, Centreville, MD, 212 pp.

Wilcock, P. R., Miller, D. S. Shea, R. H. & Kerhin, R. T. (1998). Frequency of effective wave activity and the recession of coastal bluffs: Calvert Cliffs, Maryland. *Journal of Coastal Research, 14(1)*, 256-268.

Wyda, J. C., Deegan, L. A., Hughes, J. E. & Weaver, M. J. (2002). The response of fishes to submerged aquatic vegetation complexity in two ecoregions of the mid-Atlantic bight: Buzzards Bay and Chesapeake Bay. *Estuaries, 25(1)*, 86-100.

Zedler, J. B. & Callaway, J. C. (1999). Tracking wetland restoration: do mitigation sites follow desired trajectories? *Restoration Ecology, 7(1)*, 69-73.

End Notes

[1] Fred Anders (New York State, Dept. of State, Albany, NY), Eric Anderson (USGS, NOAA Coastal Services Center, Charleston, SC), Mark Byrnes (Applied Coastal Research and Engineering, Mashpee, MA), Donald Cahoon (USGS, Beltsville, MD), Stewart Farrell (Richard Stockton College, Pomona, NJ), Duncan FitzGerald (Boston University, Boston, MA), Paul Gayes (Coastal Carolina University, Conway, SC), Benjamin Gutierrez (USGS, Woods Hole, MA), Carl Hobbs (Virginia Institute of Marine Science, Gloucester Pt., VA), Randy McBride (George Mason University, Fairfax, VA), Jesse McNinch (Virginia Institute of Marine Science, Gloucester Pt., VA), Stan Riggs (East Carolina University, Greenville, NC), Antonio Rodriguez (University of North Carolina, Morehead City, NC), Jay Tanski (New York Sea Grant, Stony Brook, NY), E. Robert Thieler (USGS, Woods Hole, MA), Art Trembanis (University of Delaware, Newark, DE), S. Jeffress Williams (USGS, Woods Hole, MA).

[2] See Glossary for a list of correspondence between common and scientific names.

[3] Thompson's Island in Rehoboth Bay, Delaware, is a good example of a mature forested upland with substantial marsh and beach area. The island hosts a large population of migratory birds. See *Maryland and Delaware Coastal Bays* in Strange *et al.* (2008).

[4] The Plum Tree Island National Wildlife Refuge is an example of a marsh deteriorating through lack of sediment input. Extensive mudflats front the marsh (see Appendix 1 .F for additional details).

[5] Along the Patuxent River, Maryland, refuge managers have noted marsh deterioration and ponding with sea-level rise. See Appendix 1 .F for additional details.

[6] For example, see discussions of Hart-Miller and Poplar Islands in Chesapeake Bay in Appendix 1.F.

[7] As presented in NatureServe (<http://www.natureserve.org/>), the prevalent tidal forest associations such as freshwater tidal woodlands and tidal freshwater cypress swamps are considered globally imperiled.

[8] For example, nutrient enrichment and resultant eutrophication are a common problem for SAV beds (USFWS, undated)

PART II OVERVIEW. SOCIETAL IMPACTS AND IMPLICATIONS

Authors: James G. Titus, U.S. EPA; Stephen K. Gill, NOAA

The previous chapters in Part I examined some of the impacts of sea-level rise on the Mid-Atlantic, with a focus on the natural environment. Part II examines the implications of sea-level rise for developed lands. Although the direct effects of sea-level rise would be similar to those on the natural environment, people are part of this "built environment"; and people will generally respond to changes as they emerge, especially if important assets are threatened. The choices that people make could be influenced by the physical setting, the properties of the built environment, human aspirations, and the constraints of laws and economics.

The chapters in Part II examine the impacts on four human activities: shore protection/retreat, human habitation, public access, and flood hazard mitigation. This assessment does not predict the choices that people *will* make; instead it examines some of the available options and assesses actions that federal and state governments and coastal communities can take in response to sea-level rise.

As rising sea level threatens coastal lands, the most fundamental choice that people face is whether to attempt to hold back the sea or allow nature to takes its course. Both choices have important costs and uncertainties. "Shore protection" allows homes and businesses to remain in their current locations, but often damages coastal habitat and requires substantial expenditure. "Retreat" can avoid the costs and environmental impacts of shore protection, but often at the expense of lost land and—in the case of developed areas—the loss of homes and possibly entire communities. In nature reserves and major cities, the preferred option may be obvious. Yet because each choice has some unwelcome consequences, the decision may be more difficult in areas that are developing or only lightly developed. Until this choice is made, however, preparing for long-term sea-level rise in a particular location may be impossible.

Chapter 6 outlines some of the key factors likely to be a part of any dialogue on whether to protect or retreat in a given area:

- What are the technologies available for shore protection and the institutional measures that might help foster a retreat?
- What is the relationship between land use and shore protection?

- What are the environmental and social consequences of shore protection and retreat?
- Is shore protection sustainable?

Most areas lack a plan that specifically addresses whether the shore will retreat or be protected. Even in those areas where a state plans to hold the line or a park plans to allow the shore to retreat, the plan is based on existing conditions. Current plans do not consider the costs or environmental consequences of sustaining shore protection for the next century and beyond.

One of the most important decisions that people make related to sea-level rise is the decision to live or build in a low-lying area. Chapter 7 provides an uncertainty range of the population and number of households with a direct stake in possible inundation as sea level rises. The results are based on census data for the year 2000, and thus are not estimates the number of people or value of structures that *will* be affected, but rather estimate the number of people who have a stake *today* in the possible future consequences of rising sea level. Because census data estimates the total population of a given census block, but does not indicate where in that block the people live or the elevation of their homes, the estimates in Chapter 7 should not be viewed as the number of people whose homes would be lost. Rather, it estimates the number of people who inhabit a parcel of land with at least some land within a given elevation above the sea. The calculations in this Chapter build quantitatively on some of the elevation studies discussed in Chapter 2, and consider uncertainties in both the elevation data and the location of homes within a given census block. Chapter 7 also summarizes a study sponsored by the U.S Department of Transportation on the potential impacts of global sea-level rise on the transportation infrastructure.

Chapter 8 looks at the implications of sea-level rise for public access to the shore. The published literature suggests that the direct impact of sea level rise on public access would be minor because the boundary between public and private lands moves inland as the shore retreats. But responses to sea-level rise could have a substantial impact. One common response (publicly funded beach nourishment) sometimes increases public access *to* the shore; but another class of responses (privately funded shoreline armoring) can eliminate public access *along* the shore if the land seaward of the shore protection structure erodes. In parts of New Jersey, regulations governing permits for shoreline armoring avoid this impact by requiring property owners to provide access along the shore *inland* of the new shore protection structures.

Finally, Chapter 9 examines the implications of rising sea level for flood hazard mitigation, with a particular focus on the implications for the Federal Emergency Management Agency (FEMA) and other coastal floodplain managers. Rising sea level increases the vulnerability of coastal areas to flooding because higher sea level increases the frequency of floods by providing a higher base for flooding to build upon. Erosion of the shoreline could also make flooding more likely because erosion removes dunes and other natural protections against storm waves. Higher sea level also raises groundwater levels, which can increase basement flooding and increase standing water. Both the higher groundwater tables and higher surface water levels can slow the rate at which areas drain, and thereby increase the flooding from rainstorms.

Chapter 9 opens with results of studies on the relationship of coastal storm tide elevations and sea-level rise in the Mid-Atlantic. It then provides background on government agency floodplain management and on state activities related to flooding and sea-level rise under the

Coastal Zone Management Act. Federal agencies, such as FEMA, are beginning to specifically plan for future climate change in their strategic planning. Some coastal states, such as Maryland, have conducted state-wide assessments and studies of the impacts of sea-level rise and have taken steps to integrate this knowledge with local policy decisions.

The chapters in Part II incorporate the underlying sea-level rise scenarios of this Product differently, because of the differences in the underlying analytical approaches. Chapter 6 evaluates the population and property vulnerable to a 100-centimeter rise in sea level, and summarizes a study by the U.S. Department of Transportation concerning the impact of a 59-centimeter rise. Chapters 6, 8 and 9 provide qualitative analyses that are generally valid for the entire uncertainty range of future sea level rise.

In: Coastal Sensitivity to Sea Level Rise-Focusing on the Mid-… ISBN: 978-1-60741-440-7
Editor: Melvin C. Urajner

Chapter 6

SHORE PROTECTION AND RETREAT

Lead Authors: James G. Titus[1] and Michael Craghan[2]
[1]U.S. EPA
[2]Middle Atlantic Center for Geography and Environmental Studies
Contributing Authors: Stephen K. Gill[3] and S. Jeffress Williams[4]
[3]NOAA
[4]USGS

Key Findings

- Many options are available for protecting land from inundation, erosion, and flooding ("shore protection"), or for minimizing hazards and environmental impacts by removing development from the most vulnerable areas ("retreat").
- Coastal development and shore protection can be mutually reinforcing. Coastal development often encourages shore protection because shore protection costs more than the market value of undeveloped land, but less than the value of land and structures. Shore protection sometimes encourages coastal development by making a previously unsafe area safe for development. Under current policies, shore protection is common along developed shores and rare along shores managed for conservation, agriculture, and forestry. Policymakers have not decided whether the practice of protecting development should continue as sea level rises, or be modified to avoid adverse environmental consequences and increased costs of shore protection.
- Most shore protection structures are designed for the current sea level, and retreat policies that rely on setting development back from the coast are designed for the current rate of sea-level rise. Those structures and policies would not necessarily accommodate a significant acceleration in the rate of sea-level rise.
- Although shore protection and retreat both have environmental impacts, the long-term impacts of shore protection are likely to be greater.

- In the short-term, retreat is more socially disruptive than shore protection. In the longterm, however, shore protection may be more disruptive—especially if it fails or proves to be unsustainable.
- We do not know whether "business as usual" shore protection is sustainable.
- A failure to plan now could limit the flexibility of future generations to implement preferred adaptation strategies. Short-term shore protection projects can impair the flexibility to later adopt a retreat strategy. By contrast, short-term retreat does not significantly impair the ability to later erect shore protection structures inland from the present shore.

6.1. Techniques for Shore Protection and Retreat

Most of the chapters in this Product discuss some aspect of shore protection and retreat. This Section provides an overview of the key concepts and common measures for holding back the sea or facilitating a landward migration of people, property, wetlands, and beaches. Chapter 9 discusses floodproofing and other measures that accommodate rising sea level without necessarily involving choosing between shore protection and retreat.

6.1.1. Shore Protection

The term "shore protection" generally refers to a class of coastal engineering activities that reduce the risk of flooding, erosion, or inundation of land and structures (USACE, 2002). The term is somewhat of a misnomer because shore-protection measures protect land and structures immediately inland of the shore rather than the shore itself[1]. Shore- protection structures sometimes eliminate the existing shore, and shore protection does not necessarily mean environmental preservation. This Product focuses on shore-protection measures that prevent dry land from being flooded, or converted to wetlands or open water.

Shore-protection measures can be divided into two categories: shoreline armoring and elevating land surfaces. Shoreline armoring replaces the natural shoreline with an artificial surface, but areas inland of the shore are generally untouched. Elevating land surfaces, by contrast, can maintain the natural character of the shore, but requires rebuilding all vulnerable land. Some methods are hybrids of both approaches. For centuries, people have used both shoreline armoring (Box 6.1) and elevating land surfaces (Box 6.2) to reclaim dry land from the sea. This Section discusses how those approaches might be used to prevent a rising sea level from converting dry land to open water. For a comprehensive discussion, see the *Coastal Engineering Manual* (USACE, 2002).

Box 6.1. Historic use of Dikes to Reclaim Land in the Delaware Estuary

Until the twentieth century, tidal wetlands were often converted to dry land through the use of dikes and drainage systems very similar to the systems that might be used to prevent land from being inundated as sea level rises. Nowhere in the United States was more marsh converted to dry land than along the Delaware River and Delaware Bay. A Dutch governor of New Jersey diked the marsh on Burlington Island, New Jersey. In

1680, after the English governor took possession of the island, observers commented that the marsh farm had achieved greater yields of grain than nearby farms created by clearing woodland (Danckaerts, 1913). In 1675, an English governor ordered the construction of dikes to facilitate construction of a highway through the marsh in New Castle County, Delaware (Sebold, 1992).

Colonial (and later state) governments in New Jersey chartered and authorized "meadow companies" to build dikes and take ownership of the reclaimed lands. During the middle of the nineteenth century, the state agriculture department extolled the virtues of reclaimed land for growing salt hay. By 1866, 20,000 acres of New Jersey's marshes had been reclaimed from Delaware Bay, mostly in Salem and Cumberland counties (Sebold, 1992). In 1885, the U.S. Department of Agriculture cited land reclamation in Cumberland County, New Jersey, as among the most impressive in the nation (Nesbit, 1885, as quoted in Sebold, 1992). By 1885, land reclamation had converted 10,000 out of 15,000 acres of the marsh in New Castle County to agricultural lands, as well as 8,000 acres in Delaware's other two counties (Nesbit, 1885). In Pennsylvania, most of the reclaimed land was just south of the mouth of the Schuylkill along the Delaware River, near the present location of Philadelphia International Airport.

During the twentieth century, these land reclamation efforts were reversed. In many cases, lower prices for salt hay led farmers to abandon the dikes (DDFW, 2007). In some cases, where dikes remain, rising sea level has limited the ability of dikes to drain the land, and the land behind the dike has converted to marsh, such as the land along the Gibbstown Levee (See Box A1.4 in Appendix 1and Figure 11.4 c and d). Efforts are under way to restore the hydrology of many lands that were formerly diked (DDFW, 2007). In areas where dikes protect communities from flooding, however, public officials area also considering the possibility of upgrading the dikes and drainage systems.

6.1.1.1 Shoreline Armoring

Shoreline armoring involves the use of structures to keep the shoreline in a fixed position or to prevent flooding when water levels are higher than the land. Although the term is often synonymous with "shoreline hardening", some structures are comprised of relatively soft material, such as earth and sand.

Keeping the shoreline in a fixed position

Seawalls are impermeable barriers designed to withstand the strongest storm waves and to prevent overtopping during a storm. During calm periods, their seaward side may either be landward of a beach or in the water. Seawalls are often used along important transportation routes such as highways or railroads (Figure 6.1 a).

Box 6.2. Creation of the National Monument Area in Washington D.C. through Nineteenth Century Dredge and Fill

Like many coastal cities, important parts of Washington, D.C. are on land that was previously created by filling wetlands and navigable waterways. When the city of

Washington was originally planned, the Potomac River was several times as wide immediately south of Georgetown as above Georgetown (see Box Figure 6.2). L'Enfant's plan put the President's residence just northeast of the mouth of Tiber Creek. Thus, the White House grounds originally had a tidal shoreline. To improve navigation, canals connected Tiber Creek to the Anacostia River (Bryan, 1914). The White House and especially the Capitol were built on high ground immune from flooding, but much of the land between the two was quite low.

During the nineteenth century, soil eroded from upstream farming was deposited in the wide part of the river where the current slowed, which created wide mudflats below Georgetown. The success of railroads made canals less important, while the increasing population converted the canals into open sewers. During the early 1 870s, Governor Boss Shephard had the canals filled and replaced with drain pipes. A large dredge-and-fill operation excavated Washington Channel from the mudflats, and used the material to create the shores of the Tidal Basin and the dry land on which the Lincoln Memorial, Jefferson Memorial, Reflecting Pool, East Potomac Park, and Hains Point sit today (Bryan, 1914). Similarly, about half of the width of the Anacostia River was filled downstream from Poplar Point, creating what later became the U.S. Naval Air Station (now part of Bolling Air Force Base).

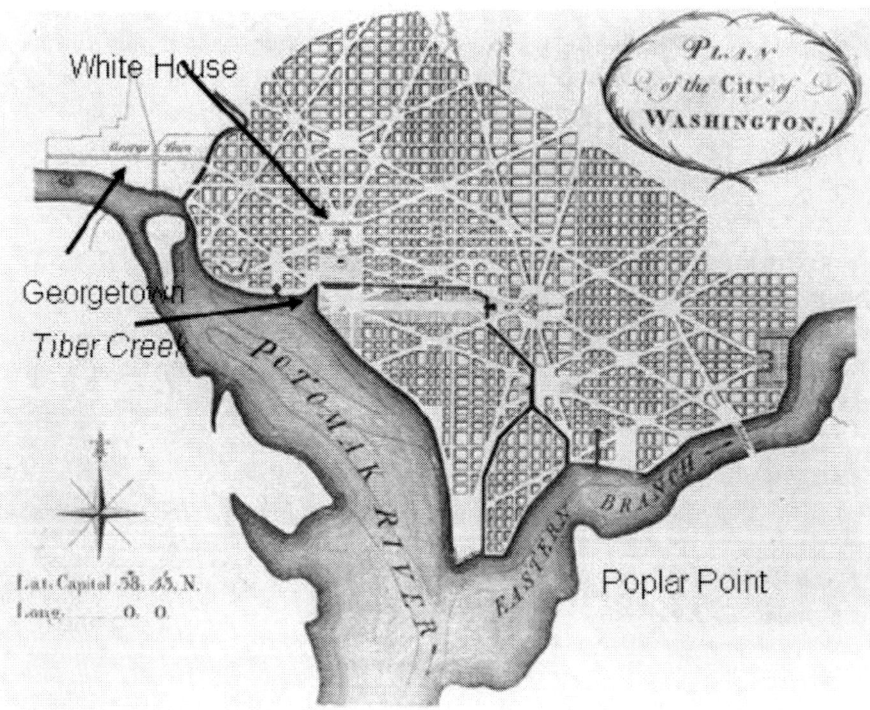

Source: Library of Congress (Labels for White House, Georgetown, and Tiber Creek added).

Figure Box 6.2. L'Enfant's Plan for the City of Washington

Figure 6.1. Seawalls and Bulkheads (a) Galveston Seawall in Texas (May 2003) and (b) Bulkheads with intervening beach along Magothy River in Anne Arundel County, Maryland (August 2005).

Figure 6.2. Geotube (a) before and (b) after being buried by beach sand at Bolivar Peninsula, Texas (May 2003).

Bulkheads are vertical walls designed to prevent the land from slumping toward the water (Figure 6.1b). They must resist waves and currents to accomplish their design intent, but unlike seawalls, they are not designed to withstand severe storms. They are usually found along estuarine shores where waves have less energy, particularly in marinas and other places where boats are docked, and residential areas where homeowners prefer a tidy shoreline. Bulkheads hold soils in place, but they do not normally extend high enough to keep out foreseeable floods. Like seawalls, their seaward sides may be inland of a beach (or marsh) or in the water.

Retaining structures include several types of structures that serve as a compromise between a seawall and a bulkhead. They are often placed at the rear of beaches and are unseen. Sometimes they are sheet piles driven downward into the sand; sometimes they are long, cylindrical, sand-filled "geo-tubes" (Figure 6.2). Retaining structures are often concealed as the buried core of an artificial sand dune. Like seawalls, they are intended to be a final line of defense against waves after a beach erodes during a storm; but they can not survive wave attack for long.

Figure 6.3. Two types of stone revetments (a) near Surfside, Texas and (b) at Jamestown, Virginia.

Figure 6.4. (a) A dike in Miami-Dade County, Florida, and (b) a newly-created dune in Surf City, New Jersey.

Revetments are walls whose sea side follows a slope. Like the beach they replace, their slope makes them more effective at dissipating the energy of storm waves than bulkheads and seawalls. As a result, revetments are less likely than bulkheads and seawalls to cause the beach immediately seaward to erode (USACE, 1995), which makes them less likely to fail during a storm (Basco, 2003; USACE, 1995). Some revetments are smooth walls, while others have a very rough appearance (Figure 6.3).

Protecting against Flooding or Permanent Inundation

Dikes are high, impermeable earthen walls designed to keep the area behind them dry. They can be set back from the shoreline if the area to be protected is a distance inland and usually require an interior drainage system. Land below mean low water requires a pumping system to remove rainwater and any water that seeps through the ground below the dike. Land whose elevation is between low and high tide can be drained at low tide, except during storms (Figure 6.4a).

Dunes are accumulations of windblown sand and other materials which function as a temporary barrier against wave runup and overwash (Figure 6.4b, see also Section 6.1.1.2).

Figure 6.5. The tide gate at the mouth of Army Creek on the Delaware side of the Delaware River. The tide gate drains flood and rain water out of the creek to prevent flooding. The five circular mechanisms on the gate open and close to control water flow (courtesy NOAA Photo Library).

Figure 6.6. Storm surge barriers. (a) Fox Point Hurricane Barrier, Providence, Rhode Island (March 1966) and (b) Moses Lake Floodgate, Texas City, Texas (March 2006).

Tide gates are barriers across small creeks or drainage ditches. By opening during low tides and closing during high tides, they enable a low-lying area above mean low water to drain without the use of pumps (Figure 6.5).

Storm surge barriers are similar to tide gates, except that they close only during storms rather than during high tides, and they are usually much larger, closing off an entire river or inlet. The barrier in Providence, Rhode Island (Figure 6.6) has gates that are lowered during a storm; the Thames River Barrier in London, by contrast, has a submerged barrier, which allows tall ships to pass. As sea level rises and storm surges become higher (see Chapter 9), these barriers must be closed more frequently. The gates in Providence, Rhode Island (Figure 6.6), for example, are currently closed an average of 19 days per year (NOAA Coastal Services Center, 2008).

6.1.1.2. Elevating Land Surfaces

A second general approach to shore protection is to elevate land and structures. Tidal marshes have long adapted to sea-level rise by elevating their land surfaces to keep pace with the rising sea (Chapter 4). Elevating land and structures by the amount of sea-level rise can keep a community's assets at the same elevation relative to the sea and thereby prevent them from becoming more vulnerable as sea level rises. These measures are sometimes collectively known as "soft" shore protection.

Beachfill, also known as *beach nourishment* or *sand replenishment,* involves the purposeful addition of the native beach material (usually sand but possibly gravel) to a beach to make it higher and wider. Sand from an offshore or inland source is added to a beach to provide a buffer against wave action and flooding (USACE, 2002; Dean and Dalrymple, 2002). Placing sand onto an eroding beach can offset the erosion that would otherwise occur over a limited time; but erosion processes continue, necessitating periodic re-nourishment.

Dunes are often part of a beach nourishment program. Although they also occur naturally, engineered dunes are designed to intercept wind-transported sand and keep it from being blown inland and off the beach. Planting dune grass and installing sand fencing increases the effectiveness and stability of dunes.

Elevating land and structures is the equivalent of a beachfill operation in the area landward of the beach. In most cases, existing structures are temporarily elevated with hydraulic jacks and a new masonry wall is built up to the desired elevation, after which the house is lowered onto the wall (See Figure 12.5). In some cases the house is moved to the side, pilings are drilled, and the house is moved onto the pilings. Finally, sand, soil, or gravel are brought to the property to elevate the land surface. After a severe hurricane in 1900, most of Galveston, Texas was elevated by more than one meter (NRC, 1987). This form of shore protection can be implemented by individual property owners as needed, or as part of a comprehensive program. Several federal and state programs exist for elevating homes, which has become commonplace in some coastal areas, especially after a major flood (see also Chapters 9 and 10).

Dredge and fill was a very common approach until the 1 970s, but it is rarely used today because of the resulting loss of tidal wetlands. Channels were dredged through the marsh, and the dredge material was used to elevate the remaining marsh to create dry land (*e.g.*, Nordstrom, 1994). The overall effect was that tidal wetlands were converted to a combination of dry land suitable for home construction and navigable waterways to provide boat access to the new homes. The legacy of previous dredge-and-fill projects includes a large number of very low-lying communities along estuaries, including the bay sides of many developed barrier islands. Recently, some wetland restoration projects have used a similar approach to create wetlands, by using material from dredged navigation channels to elevate shallow water up to an elevation that sustains wetlands. (USFWS, 2008; see Section 11.2.2 in Chapter 11).

6.1.1.3. Hybrid Approaches to Shore Protection

Several techniques are hybrids of shoreline armoring and the softer approaches to shore protection. Often, the goal of these approaches is to retain some of the storm-resistance of a hard structure, while also maintaining some of the features of natural shorelines. *Groins* are hard structures perpendicular to the shore extending from the beach into the water, usually made of large rocks, wood, or concrete (see Figure 6.7b.). Their primary effect is to diminish forces that transport sand along the shore. Their protective effect is often at the expense of increased erosion farther down along the shore; so they are most useful where an area requiring protection is updrift from an area where shore erosion is more acceptable. *Jetties* are similar structures intended to guard a harbor entrance, but they often act as a groin, causing large erosion on one side of the inlet and accretion on the other side.

Breakwaters are hard structures placed offshore, generally parallel to the shore (see Figure 6.7a). They can mitigate shore erosion by preventing large waves from striking the shore. Like groins, breakwaters often slow the transport of sand along the shore, and thereby increase erosion of shores adjacent to the area protected by the breakwaters.

Dynamic revetments (also known as *cobble beaches*) are a hybrid of beach nourishment and hard structures, in which an eroding mud or sand beach in an area with a light wave climate is converted to a cobble or pebble beach (see Figure 6.7d). The cobbles are heavy enough to resist erosion, yet small enough to create a type of beach environment (USACE, 1998; Komar, 2007; Allan *et al.*, 2005).

Recently, several state agencies, scientists, environmental organizations, and property owners have become interested in measures designed to reduce erosion along estuarine shores, while preserving more habitat than bulkheads and revetments (see Box 6.3). "*Living Shorelines*" are shoreline management options that allow for natural coastal processes to remain through the strategic placement of plants, stone, sand fill, and other structural and organic materials. They often rely on native plants, sometimes supplemented with groins, breakwaters, stone sills, or biologs[2] to reduce wave energy, trap sediment, and filter runoff, while maintaining (or increasing) beach or wetland habitat (NRC, 2007).

Box 6.3. Shore Protection Alternatives in Maryland: Living Shorelines

Shore erosion and methods for its control are a major concern in estuarine and marine ecosystems. However, awareness of the negative impacts that many traditional shoreline protection methods have, including loss of wetlands and their buffering capacities, impacts on nearshore biota, and ability to withstand storm events, has grown in recent years. Non-structural approaches, or hybrid-type projects that combine a marsh fringe with groins or breakwaters, are being considered along all shorelines except for those with large waves (from either boat traffic or a long fetch). The initial cost for these projects is often significantly less than for bulkheads or revetments; the long-run cost can be greater or less depending on how frequently the living shoreline must be rebuilt.

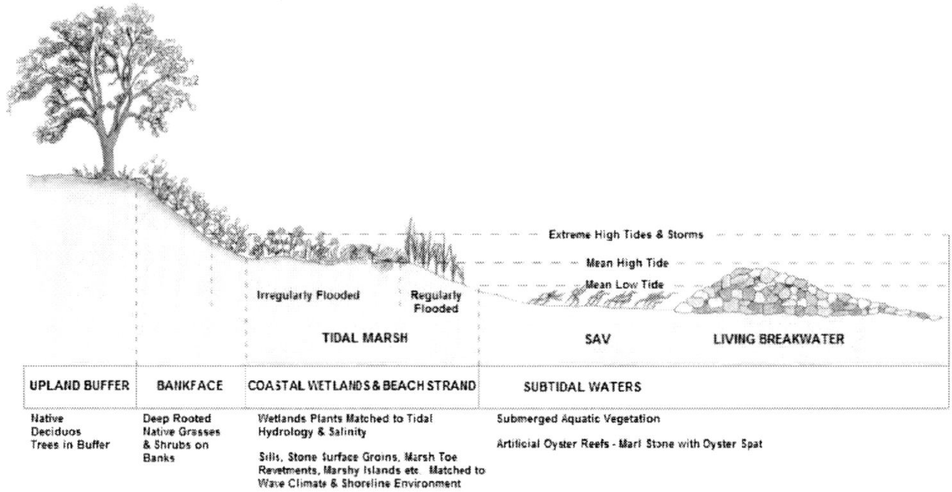

Sources: Content developed by David G. Burke for Jefferson Patterson Park and Museum, <www.jefpat.org>.

Box Figure 6.1. Depiction of Living Shoreline Treatments from the Jefferson Patterson Park and Museum, Patuxent River.

These projects typically combine marsh replanting (generally *Spartina patens* and *Spartina alterniflora*) and stabilization through sills, groins, or breakwaters. A survey of projects on the eastern and western sides of Chesapeake Bay (including Wye Island, Epping Forest near Annapolis, and the Jefferson Patterson Park and Museum on the Patuxent) found that the sill structures or breakwaters were most successful in attenuating wave energy and allowing the development of a stable marsh environment.

In addition to the hybrid techniques, communities often use a combination of shoreline armoring and elevation. Many barrier island communities apply beach nourishment on the ocean side, while armoring the bay side. Ocean shore protection projects in urban areas sometimes include both beach nourishment and a seawall to provide a final line of defense if the beach erodes during a storm. Beach nourishment projects along estuaries often include breakwaters to reduce wave erosion (Figure 6.7a), or a terminal groin to keep the sand within the area meant to be nourished (see Figure 6.7 c).

6.1.2. Retreat

The primary alternative to "shore protection" is commonly known as *retreat* (or *relocation*). Shore protection generally involves coastal engineering to manage the forces of nature and environmental engineering to manage environmental consequences. By contrast, retreat often emphasizes the management of human expectations, so that people do not make investments inconsistent with the eventual retreat.

A retreat can either occur as an unplanned response in the aftermath of a severe storm or as a planned response to avoid the costs or other adverse effects of shore protection. In Great Britain, an ongoing planned retreat is known as "managed realignment" (Rupp- Armstrong and Nicholls, 2007; Shih and Nicholls, 2007; UK Environment Agency, 2007; Midgley and McGlashan, 2004). An optimal retreat generally requires a longer lead time than shore protection (*e.g.*, Yohe and Neumann, 1997; Titus, 1998; IPCC CZMS, 1992) because the economic investments in buildings and infrastructure, and human investment in businesses and communities, can have useful lifetimes of many decades or longer. Therefore, planning, regulatory, and legal mechanisms usually play a more important role in facilitating a planned retreat than for shore protection, which for most projects can be undertaken in a matter of months or years. Some retreat measures are designed to ensure that a retreat occurs in areas where shores would otherwise be protected; other measures are designed to decrease the costs of a retreat but not necessarily change the likelihood of a retreat occurring. For a comprehensive review, see *Shoreline Management Technical Assistance Toolbox* (NOAA, 2006). The most widely assessed and implemented measures are discussed below.

Relocating structures is possibly the most engineering-related activity involved in a retreat. The most ambitious relocation in the Mid-Atlantic during the last decade has been the landward relocation of the Cape Hatteras Lighthouse (Figure 6.8a; see also Section A1 .G.4.2 in Appendix 1). More commonplace are the routine "structural moving" activities involved in relocating a house back several tens of meters within a given shorefront lot, and the removal of structures threatened by shore erosion (Figure 6. 8b).

Figure 6.7. Hybrid approaches to shore protection. (a) Breakwaters and groins along Chesapeake Bay in Bay Ridge (near Annapolis) Maryland (July 2008). The rock structures parallel to the shore in the bay are breakwaters; the structures perpendicular to the shore are groins; (b) wooden groins and bulkhead along the Peconic Estuary on Long Island, New York (September 2006). The beach is wider near the groin and narrower between groins; (c) a nourished beach with a terminal groin at North Beach (Maryland) (September 2008); (d) a dynamic revetment placed over the mud shore across Swan Creek from the Fort Washington (Maryland) unit of National Capital Parks East. Logs have washed onto the shore since the project was completed (July 2008).

Figure 6.8. Relocating structures along the Outer Banks (a) Cape Hatteras Lighthouse after relocation at the Cape Hatteras National Seashore, Buxton, North Carolina (June 2002); the original location is outlined in the foreground, and.(b) a home threatened by shore erosion in Kitty Hawk, North Carolina (June 2002). The geotextile sand bags are used to protect the septic system.

Buyout programs provide funding to compensate landowners for losses from coastal hazards by purchasing vulnerable property. In effect, these programs transfer some of the risk of sea-level rise from the property owner to the public, which pays the cost (see Chapter 12).

Conservation easements are an interest in land that allows the owner of the easement to prevent the owner of the land from developing it. Land conservation organizations have purchased non-development easements along coastal bays and Chesapeake Bay in Maryland (MALPF, 2003). In most cases, the original motivation for these purchases has been the creation of a buffer zone to protect the intertidal ecology (MDCPB, 1999; MALPF, 2003). These vacant lands also leave room for landward migration of wetlands and beaches (a concept also recognized in New Jersey Coastal Management Program 2006). Organizations can also create buffers specifically for the purpose of accommodating rising sea level. Blackwater Wildlife Refuge in Maryland and Gateway National Recreation Area in New York both own considerable amounts of land along the water onto which wetlands and beaches, respectively, could migrate inland.

Acquisition programs involve efforts by government or a conservation entity to obtain title to the land closest to the sea. Titles may be obtained by voluntary transactions, eminent domain, or dedication of flood-prone lands as part of a permitting process. In Barnegat Light, New Jersey and Virginia Beach, Virginia, for example, governments own substantial land along the shore between the Atlantic Ocean and the oceanside development.

Setbacks are the regulatory equivalent to conservation easements and purchase programs. The most common type of setback used to prepare for sea-level rise is the *erosion-based setback*, which prohibits development on land that is expected to erode within a given period of time. North Carolina requires new structures to be set back from the primary dune based on the current erosion rate times 30 years for easily moveable homes, or 60 years for large immoveable structures (see Section A1.G.4.1 in Appendix 1). Maine's setback rule assumes a 60 centimeter (cm) rise in sea level during the next 100 years[3,4]

Flood hazard regulations sometimes prohibit development based on elevation, rather than proximity to the shore. Aside from preventing flood damages, these *elevation-based setbacks* can ensure that there is room for wetlands or other intertidal habitat to migrate inland as sea level rises in areas that are vulnerable to inundation rather than wave- generated erosion. Two counties in Delaware prohibit development in the 100-year floodplain along the Delaware River and Delaware Bay (Section A1 .D.2.2 in Appendix 1).

Rolling easements are regulatory mechanisms (Burka, 1974) or interests in land (Titus, 1998) that prohibit shore protection and instead allow wetlands or beaches to migrate inland as sea level rises. Rolling easements transfer some of the risk of sea-level rise from the environment or the public to the property owner (Titus, 1998). When implemented as a regulation, they are an alternative to prohibiting all development in the area at risk, which may be politically infeasible, inequitable, or a violation of the "takings clause" of the U.S. Constitution (Titus, 1998; Caldwell and Segall, 2007). When implemented as an interest in land, they are an alternative to outright purchases or conservation easements (Titus, 1998).

The purpose of a rolling easement is to align the property owner's expectations with the dynamic nature of the shore (Titus, 1998). If retreat is the eventual objective, property owners can more efficiently prepare for that eventuality if they expect it than if it takes them by surprise (Yohe *et al.*, 1996; Yohe and Neumann, 1997). Preventing development in the area at risk through setbacks, conservations easements, and land purchases can also be effective—but such restrictions could be costly if applied to thousands of square kilometers of valuable coastal lands (Titus, 1991). Because rolling easements allow development but preclude shore protection, they are most appropriate for areas where preventing development is not feasible and shore protection is unsustainable. Conversely, rolling easements are not useful in areas where shore protection or preventing development are preferred outcomes.

Rolling easements were recognized by the common law along portions of the Texas Gulf Coast (*Feinman v. State; Matcha v. Mattox*) and reaffirmed by the Texas Open Beaches Act[5], with the key purpose being to preserve the public right to traverse the shore. Massachusetts and Rhode Island prohibit shoreline armoring along some estuarine shores so that ecosystems can migrate inland, and several states limit armoring along ocean shores (see Chapter 11). Rolling easements can also be implemented as a type of conservation easement, purchased by government agencies or conservancies from willing sellers, or dedicated as part of a planning review process (Titus, 1998); but to date, rolling easements have only been implemented by regulation.

Density restrictions allow some development but limit densities near the shore. In most cases, the primary motivation has been to reduce pollution runoff into estuaries; but they also can facilitate a retreat by decreasing the number of structures potentially lost if shores retreat. Maryland limits development to one home per 8.1 hectares (20 acres) within 305 meters (m) (1000 feet [ft]) of the shore in most coastal areas (see Section A1.F.2.1 in Appendix 1). In areas without public sewer systems, zoning regulations often restrict densities (*e.g.*, Accomack County, 2008; U.S. EPA, 1989).

Size limitations also allow development but limit the intensity of the development placed at risk. Moreover, small structures are relocated more easily than a large structure. North

Carolina limits the size of new commercial or multi-family residential buildings to 464 square meters (sq m) (5000 square feet [sq ft]) in the area that would be subject to shore erosion during the next 60 years given the current rate of shore erosion, or within 36 m (120 ft) of the shore, whichever is farther inland[6]. Maine's Sand Dune Rules prohibit structures taller than 10.7 m (35 ft) or with a "footprint" greater than 232 sq m (2500 sq ft) in all areas that are potentially vulnerable to a 60 cm rise in sea level[7].

6.1.3. Combinations of Shore Protection and Retreat

Although shore protection and retreat are fundamentally different responses to sea-level rise, strategies with elements of both approaches are possible. In most cases, a given parcel of land at a particular time is either being protected or not—but a strategy can vary with both time and place, or hedge against uncertainty about the eventual course of action.

Time. Sometimes a community switches from retreat to protection. It is common to allow shores to retreat as long as only vacant land is lost, but to erect shore protection structures once homes or other buildings are threatened. Setbacks make it more likely that an eroding shore will be allowed to retreat (Beatley *et al.*, 2002; NRC, 1987; NOAA, 2007); once the erosion reaches the setback line, the economics of shore protection are similar to what they would have been without the setback. Conversely, protection can switch to retreat. Property owners sometimes erect low-cost shore protection (*e.g.*, geotextile sandbags, shown in Figure 6.7b) that extends the lifetimes of their property, but ultimately fails in a storm. Increasing environmental implications or costs of shore protection may also motivate a switch from protection to retreat (see Section 6.5). To minimize economic and human impacts, retreat policies based on rolling easements can be designed to take effect 50 to 100 years hence, until which time protection might be allowed (Titus, 1998).

Place. Different responses operate on different scales. In general, a project to retreat or protect a given parcel will usually have effects on other parcels. For example, sand provided to an open stretch of ocean beach will be transported along the shore a significant distance by waves and currents; hence, beach nourishment along the ocean coast generally involves at least a few kilometers of shoreline or an entire island. Along estuaries, however, sands are not transported as far—especially when the shoreline has an indentation—so estuarine shore protection can operate on a smaller scale. Shoreline armoring that protects one parcel may cause adjacent shores to erode or accrete. Nevertheless, along tidal creeks and other areas with small waves, it is often feasible to protect one home with a hard structure, while allowing an adjacent vacant lot to erode. In areas with low density zoning, it may be possible to protect the land immediately surrounding a home while the rest of the lot converts to marsh, mudflat, or shallow water habitat.

Uncertainty. Some responses to sea-level rise may be appropriate in communities whose eventual status is unknown. Floodproofing homes (see Chapter 9), elevating evacuation routes, and improving drainage systems can provide cost-effective protection from flooding in the short term, whether or not a given neighborhood will eventually be protected or become subjected to tidal inundation. A setback can reduce hazards whether or not a shore protection project will eventually be implemented.

6.2. What Factors Influence the Decision Whether to Protect or Reteat?

6.2.1. Site-Specific Factors

Private landowners and government agencies who contemplate possible shore protection are usually motivated by either storm damages or the loss of land (NRC, 2007). They inquire about possible shore protection measures, investigate the costs and consequences of one or more measures, and consider whether undertaking the costs of shore protection is preferable to the consequences of not doing so. For most homeowners, the costs of shore protection include the costs of both construction and necessary government permits; the benefits include the avoided damages or loss of land and structures. Businesses might also consider avoided disruptions in business operations. Regulatory authorities that issue or deny permits for private shore protection consider possible impacts of shore protection on the environment, public access along ocean shores, and whether the design minimizes those impacts (NRC, 2007). Government agencies consider the same factors as private owners as well as public benefits of shore protection, such as greater recreational opportunities from wider beaches, increased development made possible by the shore protection (where applicable), and public safety.

Accelerated sea-level rise does not change the character of those considerations, but it would increase the magnitude of both the benefits and the consequences (monetary and otherwise) of shore protection. In some areas, accelerated sea-level rise would lead communities that are unprotected today to adopt shore protection; in other areas, the increased costs of shore protection may begin to outweigh the benefits. No published study provides a comprehensive assessment of how sea-level rise changes the costs and benefits of shore protection. However, the available evidence suggests that the environmental and social impacts could increase more than proportionately with the rate of sea-level rise (see Section 6.3 and 6.4). A case study of Long Beach Island, New Jersey (a densely developed barrier island with no high-rise buildings) concluded that shore protection is more cost-effective than retreat for the first 50 to 100 cm of sea-level rise (Titus, 1990). If the rise continues to accelerate, however, then eventually the costs of protection would rise more rapidly than the benefits, and a strategic retreat would then become the more cost-effective response, assuming that the island could be sustained by a landward migration. An economic analysis by Yohe *et al.* (1996) found that higher rates of sea-level rise make shore protection less cost-effective in marginal cases.

6.2.2. Regional Scale Factors

Potential benefits and consequences are usually the key to understanding whether a particular project will be adopted. At a broader scale, however, land use and shoreline environment are often indicators of the likelihood of shore protection. Land use provides an indicator of the resources being protected, and the shoreline environment provides an indicator of the type of shore protection that would be needed.

Most land along the mid-Atlantic ocean coast is either developed or part of a park or conservation area. This region has approximately 1,100 kilometers (almost 700 miles) of shoreline along the Atlantic Ocean. Almost half of this coastline consists of ocean beach resorts with dense development and high property values. Federal shore protection has been authorized along most of these developed shores. These lands are fairly evenly spread throughout the mid-Atlantic states, except Virginia (see Section A1.E.2.1 in Appendix 1). However, a large part of the coast is owned by landowners who are committed to allowing natural shoreline processes to operate, such as The Nature Conservancy, National Park Service (see Section 11.2.1), and U.S. Fish and Wildlife Service. These shores include most of North Carolina's Outer Banks, all of Virginia's Atlantic coast except for part of Virginia Beach and a NASA installation, more than two-thirds of the Maryland coast and New York's Fire Island. The rest of the ocean coast in this region is lightly developed, yet shore protection is possible for these coasts as well due to the presence of important coastal highways.

Development is less extensive along many estuaries than along the ocean coast. The greatest concentrations of low-lying undeveloped lands along estuaries are in North Carolina, the Eastern Shore of Chesapeake Bay, and portions of Delaware Bay. Development has come more slowly to the lands along the Albemarle and Pamlico Sounds in North Carolina than to other parts of the mid-Atlantic coast (Hartgen, 2003.) Maryland law prevents development along much of the Chesapeake Bay shore (Section A1.F.2.1 in Appendix 1), and a combination of floodplain regulations and aggressive agricultural preservation programs limit development along the Delaware Bay shore in Delaware (Section A1.D.2.2 in Appendix 1). Yet there is increasing pressure to develop land along tidal creeks, rivers, and bays (USCOP, 2004; DNREC, 2000; Titus, 1998), and barrier islands are in a continual state of redevelopment in which seasonal cottages are replaced with larger homes and high-rises (*e.g.*, Randall, 2003).

If threatened by rising sea level, these developed lands (*e.g.*, urban, residential, commercial, industrial, transportation) would require shore protection for current land uses to continue. Along estuaries, the costs of armoring, elevating, or nourishing shorelines are generally less than the value of the land to the landowner, suggesting that under existing trends shore protection would continue in most of these areas. But there are also some land uses for which the cost and effort of shore protection may be less attractive than allowing the land to convert to wetland, beach, or shallow water. Those land uses might include marginal farmland, conservations lands, portions of some recreational parks, and even portions of back yards where lot sizes are large. Along the ocean, shore protection costs are greater—but so are land values.

Shore protection is likely along much of the coastal zone, but substantial areas of undeveloped (but developable) lands remain along the mid-Atlantic estuaries, where either shore protection or wetland migration could reasonably be expected to occur (NRC, 2007; Yohe *et al.*, 1996; Titus *et al.*, 1991). Plans and designs for the development of those lands generally do not consider implications of future sea-level rise (see Chapter 11). A series of studies have been undertaken that map the likelihood of shore protection along the entirety of the U.S. Atlantic Coast as a function of land use (Nicholls *et al.*, 2007; Titus, 2004, 2005; Clark, 2001; Nuckols, 2001).

6.2.3. Mutual Reinforcement between Coastal Development and Shore Protection

Lands with substantial shore protection are more extensively developed than similar lands without shore protection, both because shore protection encourages development and development encourages shore protection. People develop floodplains, which leads to public funding for flood control structures, which in turn leads to additional development in the area protected (*e.g.*, Burby, 2006). Few studies have measured this effect, but possible mechanisms include:

- Flood insurance rates that are lower in protected areas (see Chapter 10);
- Development that may be allowed in locations that might otherwise be off limits;
- Erosion-based setbacks that require less of a setback if shore protection slows or halts erosion (see Section 6.1); and
- Fewer buildings that are destroyed by storms, so fewer post-disaster decisions to abandon previously developed land (*e.g.*, Weiss, 2006) would be expected.

The impact of coastal development on shore protection is more firmly established. Governments and private landowners generally implement a shore protection project only when the value of land and structures protected is greater than the cost of the project (see Sections 6.1 and 12.2.3).

6.3. What Are the Environmental Consequences of Retreat and Shore Protection?

In the natural setting, sea-level rise can significantly alter barrier islands and estuarine environments (Chapters 3, 4, and 5). Because a policy of retreat allows natural processes to work, the environmental impacts of retreat in a developed area can be similar to the impacts of sea-level rise in the natural setting, provided that management practices are adopted to restore lands to approximately their natural condition before they are inundated, eroded, or flooded. In the absence of management practices, possible environmental implications of retreat include:

- Contamination of estuarine waters from flooding of hazardous waste sites (Flynn *et al.*, 1984) or areas where homes and businesses store toxic chemicals;
- Increased flooding (Wilcoxen, 1986; Titus *et al.,* 1987) or infiltration into public sewer systems (Zimmerman and Cusker, 2001);
- Groundwater contamination as septic tanks and their drain fields become submerged;
- Debris from abandoned structures; and
- Interference with the ability of wetlands to keep pace or migrate inland due to features of the built landscape (*e.g.*, elevated roadbeds, drainage ditches, and impermeable surfaces).

Shore protection generally has a greater environmental impact than retreat (see Table 6.1). The impacts of beach nourishment and other soft approaches are different than the impacts of shoreline armoring.

Beach nourishment affects the environment of both the beach being filled and the nearby seafloor "borrow areas" that are dredged to provide the sand. Adding large quantities of sand to a beach is potentially disruptive to turtles and birds that nest on dunes and to the burrowing species that inhabit the beach (NRC, 1995), though less disruptive in the long term than replacing the beach and dunes with a hard structure. The impact on the borrow areas is a greater concern: The highest quality sand for nourishment is often contained in a variety of shoals which are essential habitat for shellfish and related organisms (USACE, 2002). For this reason, the U.S. Army Corps of Engineers has denied permits to dredge sand for beach nourishment in New England (*e.g.*, NOAA Fisheries Service, 2008; USACE, 2008a). As technology improves to recover smaller, thinner deposits of sand offshore, a greater area of ocean floor must be disrupted to provide a given volume of sand. Moreover, as sea level rises, the required volume is likely to increase, further expanding the disruption to the ocean floor.

As sea level rises, shoreline armoring eventually eliminates ocean beaches (IPCC, 1990); estuarine beaches (Titus, 1998), wetlands (IPCC, 1990), mudflats (Galbraith *et al.*, 2002), and very shallow open water areas by blocking their landward migration. By redirecting wave energy, these structures can increase estuarine water depths and turbidity nearby, and thereby decrease intertidal habitat and submerged aquatic vegetation. The more environmentally sensitive "living shoreline" approaches to shore protection preserve a narrow strip of habitat along the shore (NRC, 2007); however, they do not allow large- scale wetland migration. To the extent that these approaches create or preserve beach and marsh habitat, it is at the expense of the shallow water habitat that would otherwise develop at the same location.

The issue of wetland and beach migration has received considerable attention in the scientific, planning, and legal literature for the last few decades (NRC, 1987; Barth and Titus, 1984; IPCC, 1990). Wetlands and beaches provide important natural resources, wildlife habitat, and storm protection (see Chapter 5). As sea level rises, wetlands and beaches can potentially migrate inland as new areas become subjected to waves and tidal inundation—but not if human activities prevent such a migration. For example, early estimates (*e.g.*, U.S. EPA, 1989) suggested that a 70 cm rise in sea level over the course of a century would convert 65 percent of the existing mid-Atlantic wetlands to open water, and that this region would experience a 65 percent overall loss if all shores were protected so that no new wetlands could form inland. The results in Chapter 4 are broadly consistent with the 1989 study. That loss would only be 27 percent, however, if new wetlands were able to form on undeveloped lands, and 16 percent if existing developed areas converted to marsh as well.

Very little land has been set aside for the express purpose of ensuring that wetlands and other tidal habitat can migrate inland as sea level rises (see Chapter 11 of this Product; Titus, 2000), but those who own and manage estuarine conservation lands do allow wetlands to migrate onto adjacent dry land. With a few notable exceptions[8], the managers of most conservation lands along the ocean and large bays allow beaches to erode as well (see Chapter 11) The potential for landward migration of coastal wetlands is limited by the likelihood that many shorelines will be preserved for existing land uses (*e.g.*, U.S. EPA, 1989; IPCC, 1990; Nicholls *et al.*, 1999). Some preliminary studies (*e.g.*, Titus, 2004) indicate that in the mid-Atlantic region, the land potentially available for new wetland formation would be almost twice as great if future shore protection is limited to lands that are already developed, than if both developed and legally developable lands are protected.

Table 6.1. Selected Measures for Responding to Sea-Level Rise: Objective and Environmental Effects

Response Measure	Method for Protection or Retreat	Key Environmental effects
Shoreline armoring that interferes with waves and currents		
Breakwater	Reduce erosion	May attract marine life; downdrift erosion
Groin	Reduce erosion	May attract marine life; downdrift erosion

Table 6.1. (Continued)

Response Measure	Method for Protection or Retreat	Key Environmental effects
Shoreline armoring used s to define a shoreline		
Seawall	Reduce erosion, protect against flood and wave overtopping	Elimination of beach; scour and deepening in front of wall; erosion exacerbated at terminus front of wall; erosion exacerbated at terminus
Bulkhead	Reduce erosion, protect new land fill	Prevents inland migration of wetlands and beaches. Wave reflection erodes bay bottom, preventing SAY. Prevents amphibious movement from water to land.
Revetment	Reduce erosion, protect land from storm waves, protect new land fill	Prevents inland migration of wetlands and beaches. Traps horseshoe crabs and prevents amphibious movement. May create habitat for oysters and refuge for some species.
Shoreline armoring used protect against floods and/ or permanent inundation		
Dike	Prevents flooding and permanent inundation (when combined with a drainage system).	Prevents wetlands from migrating inland. Thwarts ecological benefits of floods (*e.g.*, annual sedimentation, higher water tables, habitat during migrations, productivity transfers)
Tide gate	Reduces tidal range by draining water at low tide and closing at high tide.	Restricts fish movement. Reduced tidal range reduces intertidal habitat. May convert saline habitat to freshwater habitat.
Storm surge barrier	Eliminates storm surge flooding; could protect against all floods if	Necessary storm surge flooding in salt marshes is eliminated.
Elevating land		
Dune	Protect inland areas from storm waves, provide a source of sand during storms to offset erosion.	Can provide habitat; can set up habitat for secondary dune colonization behind it
Beachfill	Reverses shore erosion, and provide some protection from storm waves.	Short-term loss of shallow marine habitat; could provide beach and dune habitat
Elevate land and structures	Avoid flooding and inundation from sea-level rise by elevating everything	Deepening of estuary unless bay bottoms are elevated as well.
Retreat		
Setback	Delay the need for shore prote-ction by keeping development out of the	Impacts of shore protection delayed until shore erodes up to the setback line. Impacts of development also reduced.
Rolling easement	Prohibit shore protection structures.	Impacts of shore protection structures avoided.
Density or size restriction	Reduce the benefits of shore protection and thereby make it less likely.	Depends on whether owners of large lots decide to protect shore. Impacts of intense development reduced.

6.4. What Are the Societal Consequences of Shore Protection and Retreat As Sea Level Rises?

6.4.1. Short-Term Consequences

Shore protection generally is designed to enable existing land uses to continue. By insulating a community from erosion, storms, and other hazards, the social consequences of sea-level rise can be minimal, at least for the short term. In the Netherlands, shore protection helped to foster a sense of community as residents battled a common enemy (Disco, 2006). In other cases, the interests of some shorefront property owners may diverge from the interests of other residents (NRC, 2007). For example, many property owners in parts of Long Beach Island, New Jersey strongly supported beach nourishment—but some shorefront owners in areas with wide beaches and dunes have been reluctant to provide the state with the necessary easements (NJDEP, 2006; see Section A1.C.2 in Appendix 1).

Allowing shores to retreat can be disruptive. If coastal erosion is gradual, one often sees a type of coastal blight in what would otherwise be a desirable community, with exposed septic tanks and abandoned homes standing on the beach, and piles of rocks or geotextile sand bags in front of homes that remain occupied (Figures 6.8b and 6.9). If the loss of homes is episodic, communities can be severely disrupted by the sudden absence of neighbors who previously contributed to the local economy and sense of community (IPCC, 1990; Perrin *et al.*, 2008; Birsch and Wachter, 2006). People forced to relocate after disasters are often at increased risk to both health problems (Yzermans *et al.*, 2005) and depression (Najarian *et al.*, 2001).

6.4.2. Long-Term Consequences

The long-term consequences of a retreat can be similar to the short-term consequences. In some areas, however, the consequences may become more severe over time. For example, a key roadway originally set far back from the shore may become threatened and have to be relocated. In the case of barrier islands, the long-term implications of retreat depend greatly on whether new land is created on the bay side to offset oceanfront erosion. If so, communities can be sustained as lost oceanfront homes are rebuilt on the bay side; if not, the entire community could be eventually lost.

Figure 6.9. The adverse impacts of retreat on safety and aesthetic appeal of recreational beaches (a) Exposed septic tank and condemned houses at Kitty Hawk, North Carolina (June 2002); (b) Beach unavailable for recreation where homes were built to withstand shore erosion and storms, at Nags Head, North Carolina (June 2007).

The long-term consequences of shore protection could be very different from the short-term consequences. As discussed below, shore protection costs could escalate. The history of shore protection in the United States suggests that some communities would respond to the increased costs by tolerating a lower level of shore protection, which could lead eventually to dike failures (Seed *et al.*, 2005; Collins, 2006) and resulting unplanned retreat. In other cases, communities would not voluntarily accept a lower level of protection, but the reliance on state or federal funding can lead to a lower level while awaiting funds (a common situation for communities awaiting beach nourishment). For communities that are able to keep up with the escalated costs, tax burdens would increase, possibly leading to divisive debates over a reconsideration of the shore protection strategy.

6.5. How Sustainable Are Shore Protection and Retreat?

Coastal communities were designed and built without recognition of rising sea level. Thus, people in areas without shore protection will have to flood-proof structures (see Chapter 9), implement shore protection, (Section 6.1.1) or plan a retreat (Section 6.1.2). Those who inhabit areas with shore protection are potentially vulnerable as well. Are the known approaches to shore protection and retreat sustainable, that is, can they be maintained for the foreseeable future?

Most shore protection structures are designed for current sea level and may not accommodate a significant rise. Seawalls (Kyper and Sorenson, 1985; NRC, 1987), bulkheads (Sorenson *et al.*, 1984.), dikes, (NRC, 1987), sewers (Wilcoxen, 1986) and drainage systems (Titus *et al.*, 1987) are designed based on the waves, water levels, and rainfall experienced in the past. If conditions exceed what the designers expect, disaster can result—especially when sea level rises above the level of the land surface. The failure of dikes protecting land below sea level resulted in the deaths of approximately 1800 people in the Netherlands in a 1953 storm (Roos and Jonkman, 2006), and more than 1000 people in the New Orleans area from Hurricane Katrina in 2005 (Knabb *et al.*, 2005). A dike along the Industrial Canal in New Orleans which failed during Katrina had been designed for sea level approximately 60 cm lower than today, because designers did not account for the land subsidence during the previous 50 years (Interagency Performance Evaluation Taskforce, 2006).

Figure 6.10. Historic homes along the Charleston Battery. Charleston, South Carolina. (April 2004).

One option is to design structures for future conditions. Depending on the incremental cost of designing for higher sea level compared with the cost of rebuilding later, it may be economically rational to build in a safety factor today to account for future conditions, such as higher and wider shore protection structures (see Chapter 10). But doing so is not always practical. Costs generally rise more than proportionately with higher water levels[9]. Project managers would generally be reluctant to overdesign a structure for today's conditions (Schmeltz, 1984). Moreover, aesthetic factors such as loss of waterfront views or preservation of historic structures (*e.g.*, Charleston Battery in South Carolina; see Figure 6.10) can also make people reluctant to build a dike or seawall higher than what is needed today.

6.5.1. Is "Business as Usual" Shore Protection Sustainable?

Public officials and property owners in densely developed recreational communities along the mid-Atlantic coast generally expect governmental actions to stabilize shores. But no one has assessed the cost and availability of sand required to keep the shorelines in their current locations through beach nourishment even if required sand is proportional to sea-level rise, which previous assessments of the cost of sea level rise have assumed (*e.g.*, U.S. EPA, 1989; Leatherman, 1989; Titus *et al.*, 1991). The prospects of barrier island disintegration and segmentation examined in Chapter 3 would require much more sand to stabilize the shore. Maintaining the shore may at first seem to require only the simple augmentation of sand along a visible beach, but over a century or so other parts of the coastal environment would capture increasing amounts of sand to maintain elevation relative to the sea. In effect, beach nourishment would indirectly elevate those areas as well (by replacing sand from the beach that is transported to raise those areas), including the ocean floor immediately offshore, tidal deltas, and eventually back-barrier bay bottoms and the bay sides of barrier islands. Similarly, along armored shores in urban areas, land that is above sea level today would become farther and farther below sea level, increasing the costs of shore protection and setting up greater potential disasters in the event of a dike failure. It is not possible to forecast whether these costs will be greater than what future generations will choose to bear. But in those few cases where previous generations have bequeathed this generation with substantial communities below sea level, a painful involuntary relocation sometimes occurs after severe storms (*e.g.*, New Orleans after Katrina).

Most retreat policies are designed for current rates of sea-level rise and would not necessarily accommodate a significant acceleration in the rate of sea-level rise. Erosion-based setbacks along ocean shores generally require homes to be set back from the primary dune by a distance equal to the annual erosion rate times a number years intended to represent the economic lifetime of the structure (*e.g.*, in North Carolina, 60 years times the erosion rate for large buildings [see Section A1.G.1 in Appendix 1). If sea-level rise accelerates and increases the erosion rate, then the buildings will not have been protected for the presumed economic lifetimes. Yet larger setback distances may not be practicable if they exceed the depth of buildable lots. Moreover, erosion-based setback policies generally do not articulate what will happen once shore erosion consumes the setback. The retreat policies followed by organizations that manage undeveloped land for conservation purposes may account for foreseeable erosion, but not for the consequences of an accelerated erosion that consumes the entire coastal unit.

6.5.2. Sustainable Shore Protection May Require Regional Coordination

Regional Sediment Management is a relatively new strategy or planning tool for managing sand as a resource (NRC, 2007). The strategy recognizes that coastal engineering projects have regional impacts on sediment transport processes and availability. This approach includes:

- Conservation and management of sediments in along the shore and immediate offshore areas, viewing sand as a resource;
- Attempt to design with nature, understanding sediment movement in a region and the interrelationships of projects and management actions;
- Conceptual and programmatic connections among all activities that involve sediment in a region (*e.g.*, navigation channel maintenance, flood and storm damage reduction, ecosystem restoration and protection, beneficial uses of dredged material);
- Connections between existing and new projects to use sediment more efficiently;
- Improved program effectiveness through collaborative partnerships between agencies; and
- Overcoming institutional barriers to efficient management (Martin, 2002).

The Philadelphia and New York Districts of the U.S. Army Corps of Engineers have a joint effort at regional sediment management for the Atlantic coast of New Jersey (USACE, 2008b). By understanding sediment sources, losses, and transport; how people have altered the natural flow; and ways to work with natural dynamics, more effective responses to rising sea level are possible.

One possible way to promote better regional sediment management would be the development of a set of "best sediment management practices". Previously, standard practices have been identified to minimize the runoff of harmful sediment into estuaries (NJDEP, 2004; City of Santa Cruz, 2007). A similar set of practices for managing sediments along shores could help reduce the environmental and economic costs of shore protection, without requiring each project to conduct a regional sediment management study.

6.5.3. Either Shore Protection or a Failure to Plan can Limit the Flexibility of Future Generations

The economic feasibility of sustained shore protection as sea level rises is unknown, as is the political and social feasibility of a planned retreat away from the shore. The absence of a comprehensive long-term shoreline plan often leaves property owners with the assumption that the existing development can and should be maintained. Property-specific shoreline armoring and small beach nourishment projects further reinforce the expectation that the existing shoreline will be maintained indefinitely, often seeming to justify additional investments by property owners in more expensive dwellings (especially if there is a through-road parallel to the shore).

Shore protection generally limits flexibility more than retreat. Once shore protection starts, retreat can be very difficult to enact because investments and expectations are based on the protection, which in turn increases the economic justification for continued shore protection. A policy of retreat can be more easily replaced with a policy of shore protection, because people do not make substantial investments on the assumption that the shore will retreat. This is not to say that all dikes and seawalls would be maintained and enlarged indefinitely if sea level continues to rise. Nevertheless, the abandonment of floodprone communities rarely (if ever) occurs because of the potential vulnerability or cost of flood protection, but rather in the aftermath of a flood disaster (*e.g.*, Missouri State Emergency Management Agency, 1995).

REFERENCES

[*] Indicates non-peer reviewed literature; available upon request

Accomack County, (2008). *Respecting the Past, Creating the Future: The Accomack County Comprehensive Plan:* Revised Draft. Accomac, Virginia: Planning Department, Accomack County. [Note: The town and county are spelled differently.]

Allan, J. C., Geitgey, R. & Hart, R. (2005). *Dynamic Revetments for Coastal Erosion in Oregon: Final Report.* Oregon Department of Transportation, Salem. <http://www.oregon.gov/ODOT/TD/TP_RES/docs/Reports/DynamicRevetments. pdf>

Basco, D. R., (2003). Shore protection projects. In: *Coastal Engineering Manual.* Engineer manual 1110-2-1100. U.S. Army Corps of Engineers, Washington, DC, Part V, chapter 3.

Barth, M. C. & Titus, J. G. (eds.), (1984). *Greenhouse Effect and Sea Level Rise: A Challenge for this Generation.* Van Nostrand Reinhold, New York, 325 pp.

Beatley, T., Brower, D. J. & Schwab, A. K. (2002). *An Introduction to Coastal Zone Management.* Island Press, Washington DC, 285 pp.

Birch, E. L. & Wachter, S. M. (eds.), (2006). *Rebuilding Urban Places after Disaster: Lessons from Katrina.* University of Pennsylvania Press, Philadephia, 375 pp.

Bryan, W. B, (1914). *A Hisory of the National Capital from its Foundation Through the Period of Adoption of the Organic Act.* The Macmillan Company, New York, 2 volumes.

Burka, P. (1974). Shoreline erosion: implications for public rights and private ownership. *Coastal Zone Management Journal, 1(2),* 175-195.

Burby, R. J. (2006). Hurricane Katrina and the paradoxes of government disaster policy: bringing about wise governmental decisions for hazardous areas. *The Annals of the American Academy of Political and Social Science, 604(1),* 171-191.

Caldwell, M. & C. Segall, (2007). No day at the beach: sea level rise, ecosystem loss, and public access along the California coast. *Ecology Law Quarterly, 34(2),* 533- 578.

City of Santa Cruz, (2007). Construction work: best management practices. In: *Best Management Practices Manual for the City's Storm Water Management Program.* City of Santa Cruz Public Works Department, Planning Department, Santa Cruz, CA, chapter 4. <http://www.ci.santacruz.ca.us/pw/Stormwater2004/Att9Update.pdf>

Clark, W. (2001). Planning for sea level rise in North Carolina. In: *Coastal Zone '01.* Proceedings of the 12th Biennial Coastal Zone Conference, Cleveland, OH, July 15-19, 2001. NOAA Coastal Services Center, Charleston, SC.

Collins, D. (2006). Challenges to reducing flood risk. In: *Participatory Planning and Working with Natural Processes On The Coast.* Dutch National Institute for Coastal and Marine Management, The Hague, the Netherlands.

Danckaerts, J. (1913). Journal of Jasper Danckaerts, 1679-1680 By Jasper Danckaerts, Peter Sluyter. "Published 1913. C. Scribner's Sons. <http://books.google> Accessed on 1/14/08. The present translation is substantially that of Mr. Henry C. Murphy, as presented in his edition of 1867, under title: "Journal of a voyage to New York and a tour in several of the American colonies in 1679-80, by Jaspar Dankers and Peter Sluyter.".

* DDFW (Delaware Division of Fish and Wildlife), (2007). *Northern Delaware Wetlands Rehabilitation Program.* [website] <http://www.dnrec. state.de.us/fw/intmrmt.htm>

Dean, R. G. & Dalrymple, R.A. (2002). *Coastal Processes with Engineering Applications.* Cambridge University Press, New York, *475 pp.*

Disco, C. (2006). Delta blues. *Technology and Culture, 47(2),* 341-348

* DNREC (Department of Natural Resource and Environmental Control), (2000). *Land Use and Population.* [Delaware Department of Natural Resource and Environmental Control, Dover], *1 p.*
<http://www.dnrec.state.de.us/dnrec2000/Admin/WholeBasin/InlandBays/land f>

Feinman v. State, 717 S.W.2d 106, 111 (Tex. App. 1986).

Flynn, T. J., Walesh, S. G., Titus, J.G. & Barth, M. C. (1984). Implications of sea level rise for hazardous waste sites in coastal floodplains. In: *Greenhouse Effect and Sea Level Rise: A Challenge for this Generation* [Barth, M.C. and J.G. Titus (eds.)].*Van Nostrand Reinhold Company, New York, pp. 27* 1-294.

Galbraith, H., Jones, R., Park, R. Clough, J., Herrod-Julius, S., Harrington, B. & Page, G. (2002). Global climate change and sea level rise: potential losses of intertidal habitat for shorebirds. *Waterbirds, 25(2),* 173-183.

* Hartgen, D. T. (2003). *Highways and Sprawl in North Carolina. University of North Carolina, Charlotte, 124 pp.* <http://www.johnlocke.org/policy%5Freports/200309254 1 .html>

Interagency Performance Evaluation Taskforce, (2006). *Performance Evaluation of the New Orleans and Southeast Louisiana Hurricane Protection System.* U.S. Army Corps of Engineers, Washington, DC.

IPCC (Intergovernmental Panel on Climate Change), (1990). *Strategies for Adaptation to Sea Level Rise.* Report of the Coastal Zone, Management Subgroup, IPCC Response Strategies Working Group. Ministry of Transport, Public Works and Water Management, *The Hague, the Netherlands, 122 pp.*

IPCC CZMS (Intergovernmental Panel on Climate Change, Coastal Zone Management Subgroup), (1992). *Global Climate Change and the Rising Challenge of the Sea.* IPCC Response Strategies Working Group. The Hague, the Netherlands.

Knabb, R. D., Rhome, J.R. & Brown, D.P. (2005). *Tropical Cyclone Report: Hurricane Katrina, 23-30 August 2005.* National Hurricanc Center, Miami, FL, 43 pp. <http://www.nhc.noaa.gov/pdf/TCR 1 22005_Katrina.pdf>

Komar, P. D., (2007). The design of stable and aesthetic beach fills: learning from nature In: *Coastal Sediments '07.* Proceedings of the Sixth International Symposium on Coastal Engineering and Science of Coastal Sediment Processes, May 13–17, 2007, New Orleans, LA. [Kraus, N.C. and J.D. Rosati (eds.]. American Society of Civil Engineers, Reston, VA, pp. 420-43 3.

Kyper, T. N. & Sorensen, R.M. (1985). The impact of selected sea level rise scenarios on the beach and coastal structures at Sea Bright, N.J. In: *Coastal Zone '85.* Proceedings of the Fourth Symposium on Coastal and Ocean Management, Omni International Hotel,

Baltimore, Maryland, July 30-August 2, 1985. American Society of Civil Engineers, New York, pp. 2645-2661.

Leatherman, S. P., (1989). Nationwide assessment of beach nourishment requirements associated with accelerated sea level rise. In: *The Climate Change on the United States. Appendix: B, Sea Level Rise*. EPA 230-05-89-052. U.S. Environmental Protection Agency, Washington, DC, pp. 2-1 to 2-30. <http://epa.gov/climatechange/effects/coastal/appB .html>

MALPF (Maryland Agricultural Land Preservation Foundation), (2003). *Maryland's Land Conservation Programs: Protecting the Chesapeake Bay Watershed*. Maryland Agricultural Land Preservation Foundation, [Annapolis]. <http://www.malpf.info/reports/GovernorReport1 21703 .pdf>

Martin, L. R., (2002). *Regional Sediment Management: Background and Overview of Initial Implementation*. IWR Report 02-PS-2. U.S. Army Corps of Engineers Institute for Water Resources, Ft. Belvoir, VA, 75 pp. <http://www.iwr.usace.army.mil/inside/products/pub/iwrreports/02ps2sed_man.p df>

Matcha v. Mattox, 711 S.W.2d 95, 100 (Tex. App. 1986).

MDCBP (Maryland Coastal Bays Program), (1999). *Today's Treasures for Tomorrow: Towards a Brighter Future*. A Comprehensive Conservation and Management Plan for Maryland's Coastal Bays. Maryland Coastal Bays Program, Berlin, 181 pp. <http://mdcoastalbays.org/archive/2003/ccmp.pdf>

Midgley, S. & McGlashan, D. J. (2004). Planning and management of a proposed managed realignment project: Bothkennar, Forth Estuary, Scotland. *Marine Policy,* 28(5), 429-435.

Missouri State Emergency Management Agency, 1995: *Out of Harm's Way: The Missouri Buyout Program*. Missouri State Emergency Management Agency, Jefferson City, 16 pp.

Najarian, L. M., Goenjian, A. K. Pelcovitz, D. Mandel, F. & Najarian, B. (2001). The effect of relocation after a natural disaster. *Journal of Traumatic Stress*, *14(3),* 5 11-526.

Nesbit, D.M., (1885). *Tide Marshes of the United States*. USDA special report 7. Government Printing Office, Washington, DC (as cited in Sebold, 1992).

Nicholls, R. J., Hoozemans, F. M. J. & Marchand, M. (1999). Increasing flood risk and wetland losses due to global sea-level rise: regional and global analyses. *Global Environmental Change,* 9(Supplement 1), S69-S87.

Nicholls, R. J., Wong, P.P., Burkett, V. R., Codignotto, J.O., Hay, J. E., McLean, R.F. Ragoonaden, S. & Woodroffe, C. D. (2007). Coastal systems and low-lying areas. In: *Climate Change 2007: Impacts, Adaptation and Vulnerability*. Contribution of Working Group II to the Fourth Assessment Report of the Intergovernmental Panel on Climate Change [Parry, M.L., O.F. Canziani, J.P. Palutikof, P.J. van der Linden, and C.E. Hanson (eds.)]. Cambridge University Press, Cambridge, UK, and New York, pp. 3 15-356.

NJDEP (New Jersey Department of Environmental Protection), (2004). Impacts on development of runoff. In: *New Jersey Stormwater Best Management Practices Manual*. New Jersey Department of Environmental Protection, Trenton, [8 pp.] <http://www.nj .gov/dep/stormwater WBMP_1 %20print.pdf>

NJDEP (New Jersey Department of Environmental Protection), (2006). *Assessment and Enhancement Strategy: FY 2006 - 2010*. Coastal Management Office, New Jersey Department of Environmental Protection, Trenton, 85 pp. <http://www.state.nj .us/dep/cmp/309_combined_strat_7_06.pdf>

NOAA (National Oceanic and Atmospheric Administration), (2006). *The Shoreline Management Technical Assistance Toolbox* [web site] <http://coastalmanagement.noaa.gov/shoreline.html>

NOAA (National Oceanic and Atmospheric Administration), (2007). *Construction Setbacks.* [web site] NOAA Ocean & Coastal Resource Management. <http://coastalmanagement.noaa.gov/initiatives/shoreline_ppr_setbacks.html>

NOAA Coastal Services Center, (2008). The rising tide: how Rhode Island is addressing sea level rise. *Coastal Services, 11(3)*, 4-6, 9.

NOAA Fisheries Service, (2008). *Northeast Region, Habitat Conservation Division, Monthly Highlights*, March-April, 3 pp. <http://www.nero.noaa.gov/hcd/08highlights/March-April08.pdf>

Nordstrom, K.F., (1994). Developed coasts. In: *Coastal Evolution: Late Quaternary Shoreline Morphodynamics* [Carter, R.W.G. and C.D. Woodroffe (eds.)]. Cambridge University Press, Cambridge, UK, pp. 477-509.

NRC (National Research Council), (1987). *Responding to Changes in Sea Level: Engineering Implications*. National Academy Press, *Washington, DC, 148 pp.*

NRC (National Research Council), (1995). *Beach Nourishment and Protection*. National Academy Press, Washington, DC, *333 pp.*

NRC (National Research Council), (2007). *Mitigating Shore Erosion along Sheltered Coasts. National Academies Press, Washington, DC, p. 51* [174 pp.]

Nuckols, W., (2001). Planning for sea level rise along the Maryland shore. In: *Coastal Zone '01*: Proceedings of the 12th Biennial Coastal Zone Conference. NOAA, Silver Spring, MD.

Perrin, P. B., Brozyna, A., Berlick, A.B., Desmond, F. F., Ye, H.J. & Boycheva, E. (2008). Voices from the post-Katrina Ninth Ward: an examination of social justice, privilege, and personal growth. *Journal for Social Action in Counseling and Psychology, 1(2),* 48-61.

Randall, M. M., (2003). Coastal development run amuck: a policy of retreat may be the only hope. *Journal of Environmental Law and Litigation, 18(1),* 145-186.

Roos, A. & Jonkman, B. (2006). Flood risk assessment in the Netherlands with focus on the expected damages and loss of life. In *Flood Risk Management: Hazards, Vulnerability and Mitigation Measures*. [Schanze, J., E. Zeman and J. Marsalek (eds.)]. Springer, Berlin, pp. 169-182.

Rupp-Armstrong, S. & Nicholls, R. J. (2007). Coastal and estuarine retreat: a comparison of the application of managed realignment in England and Germany. *Journal of Coastal Research, 23(6),* 1418-1430.

Schmeltz, E. J., (1984). Comments. In: *Greenhouse Effect and Sea Level Rise: A Challenge for This Generation*. [Barth, M.C. and J.G. Titus (eds.)]. Van Nostrand Reinhold Company, New York, pp. 300-305.

Sebold, K. R., (1992). *From Marsh to Farm: The Landscape Transformation of Coastal New Jersey*. U.S. Department of Interior, National Park Service, Historic American Buildings Survey/Historic American Engineering Record, Washington, DC. <http://www.nps.gov/history/history/online 3/index.htm>

Seed, R. B., Nicholson, P.G., Dalrymple, R.A., Battjes, J. Bea, R.G., Boutwell, G., Bray, J. D., Collins, B. D., Harder, L. F., Headland, J. R., Inamine, M., Kayen, R.E. Kuhr, R., Pestana, J. M., Sanders, R., Silva-Tulla, F., Storesund, R., Tanaka, S., Wartman, J., Wolff, T.F., Wooten, L. & Zimmie, T. (2005). *Preliminary Report on the Performance of the New Orleans Levee Systems in Hurricane Katrina on August 29, 2005*. Report no. UCB/CITRIS – 05/01. University of California at Berkeley and American Society of Civil Engineers, Berkeley.

Shih, S. C. W. & Nicholls, R. J. (2007). Urban managed realignment: application to the Thames Estuary, London. *Journal of Coastal Research, 23(6),* 1525-1534.

Sorensen, R. M., Weisman, R. N. & Lennon, G. P. (1984). Control of erosion, inundation, and salinity intrusion caused by sea level rise. In: *Greenhouse Effect and Sea Level Rise: A Challenge for This Generation* [Barth, M.C. and J.G. Titus, (eds.)]. Van Nostrand Reinhold Company, New York, pp. 179-214.

Titus, J. G., (1990). Greenhouse effect, sea-level rise, and barrier Islands: case study of Long Beach Island, New Jersey. *Coastal Management*, *18*, 65-90.

Titus, J. G., (1991). Greenhouse effect and coastal wetland policy: how Americans could abandon an area the size of Massachusetts at minimum cost. *Environmental Management, 15(1),* 39-58.

Titus, J. G., (1998). Rising seas, coastal erosion and the taking clause: how to save wetlands and beaches without hurting property owners. *Maryland Law Review*, *57(4)*, 1277-1399.

Titus, J. G., (2000). Does the US government realize that the sea is rising? How to restructure federal programs so that wetlands and beaches survive. *Golden Gate University Law Review, 30(4),* 7 17-778.

* Titus, J. G., (2004). Maps that depict the business-as-usual response to sea level rise in the decentralized United States of America. Presented at: *Global Forum on Sustainable Development*, Paris, 11-12 November 2004. Organization of Economic Cooperation and Development, Paris.

Titus, J. G., (2005). Does shoreline armoring violate the Clean Water Act? Rolling easements, shoreline planning, and other responses to sea level rise. In: *America's Changing Coasts: Private Rights and Public Trust.* [Whitelaw D.M. and G.R. Visgilio (eds.)]. *Edward Elgar Publishing, Cheltenham, UK, Northampton, MA, 248 pp.*

Titus, J. G., Kuo, C. Y., Gibbs, M. J., LaRoche, T. B., Webb, M.K. & Waddell, J. O. (1987). Greenhouse effect, sea level rise, and coastal drainage systems. *Journal of Water Resources Planning and Management*, *113(2),* 216-227.

Titus, J. G., Park, R.A., Leatherman, S.P., Weggel, J.R., Greene, M.S., Mansel, P.W., Brown, S., Gaunt, G., Treehan, M. & Yohe, G. (1991). Greenhouse effect and sea level rise: the cost of holding back the sea. *Coastal Management*, *19(2),* 171-204.

UK Environment Agency, (2007). *Managed Realignment Electronic Platform.* Version 1.0. <http://www.intertidalmanagement.co.uk/>

USACE (U.S. Army Corps of Engineers), (1995). *Engineering and Design: Design of Coastal Revetments, Seawalls, and Bulkheads.* Engineer manual no. 1110-2-1614. U.S. Army Corps of Engineers, Washington, DC.

USACE (U.S. Army Corps of Engineers) (1998). Shoreline and Channel Erosion Protection: Overview of Alternatives. WRP technical note HS-RS-4.1. U.S. Army Corps of Engineers Wetlands Research Program, [Vicksburg, MS], 8 pp. http://el.erdc.usace.army.mil/elpubs/pdf/hsrs4- 1 .pdf

USACE (U.S. Army Corps of Engineers) (2002). *Coastal Engineering Manual.* Engineer manual 1110-2-1100. U.S. Army Corps of Engineers, Washington, DC, (in 6volumes issued between 2002 and 2006).

* USACE (U.S. Army Corps of Engineers) (2008a). *Corps of Engineers Public Interest Review Results in Permit Denial for Winthrop Beach* News Release April 23, 2008. <http://www.nae.usace.army.mil/news/2008-04 1 .htm>

USACE (U. S. Army Corps of Engineers) (2008b). *Project Factsheet: New Jersey Alternative Long-Term Nourishment RSM (Regional Sediment Management) Study*. USACE Philadelphia District, 2 pp. <http://www.nap.usace.army.mil/cenap-dp/projects/factsheets/NJ/NJ%20Alt%20LT%20Nourishment.pdf>

USCOP (U.S. Commission on Ocean Policy) (2004). *An Ocean Blueprint for the 21st Century*. U.S. Commission on Ocean Policy, Washington DC. <http://www.oceancommission.gov>

U.S. EPA (Environmental Protection Agency) (1989). *The Potential Effects of Global Climate Change on the United States. Appendix: B, Sea Level Rise*. EPA 23 0-05- 89-052. U.S. Environmental Protection Agency, Washington, DC. <http://epa.gov/climatechange/effects/coastal/appB .html>

* USFWS (United States Fish and Wildlife Service) (2008). *Blackwater National Wildlife Refuge: Wetland Restoration*. [web site] <http://www.fws.gov/blackwater/restore.html>

Weggel, J. R., Brown, S., Escajadillo, J.C., Breen, P. & Doheny, E. L. (1989). The cost of defending developed shorelines along sheltered waters of the United States from a two meter rise in mean sea level. In: *The Potential Effects of Global Climate Change on the United States. Appendix: B, Sea Level Rise*. EPA 230-05-89-052. U.S. Environmental Protection Agency, Washington, DC, pp. 3-1 to 3-90. <http://epa.gov/climatechange/effects/coastal/appB .html>

Weiss, N. E., (2006). *Rebuilding Housing after Hurricane Katrina: Lessons Learned and Unresolved Issues*. Congressional Research Service, Washington, DC, 13 pp. <http://assets 121 9.pdf>

Wilcoxen, P. J. (1986). Coastal erosion and sea level rise: implications for Ocean Beach and San Francisco's Westside Transport Project. *Coastal Zone Management Journal*, 14(3), 173-192.

Yohe, G. & Neumann, J. E. (1997). Planning for sea level rise and shore protection under climate uncertainty. *Climatic Change*, *37(1),* 243-270.

Yohe, G., Neumann, J. E., Marshall, P. & Ameden, H. (1996). The economic cost of greenhouse induced sea level rise in the United States. *Climatic Change*, *32(4),* 387-410.

Yzermans, C.J., G.A. Donker, J.J. Kerssens, A.J.E Dirkzwager, R.J.H. Soeteman & P.M.H. ten Veen, 2005: Health problems of victims before and after disaster: a longitudinal study in general practice. *International Journal of Epidemiology*, *34(4),* 820-826.

Zimmerman, R. & Cusker, M. (2001). Institutional decision-making. In: *Climate Change and a Global City: The Potential Consequences of Climate Variability and Change. Metro East Coast* [Rosenzweig, C. and W.D. Solecki (eds.)]. Columbia Earth Institute and Goddard Institute of Space Studies, New York, pp. 9-1 to 9-25; A11-A17.

In: Coastal Sensitivity to Sea Level Rise-Focusing on the Mid-… ISBN: 978-1-60741-440-7
Editor: Melvin C. Urajner

Chapter 7

POPULATION, LAND USE, AND INFRASTRUCTURE

Lead Authors: Stephen K. Gill[1], Robb Wright[1] and James G. Titus[2]
[1]NOAA
[2]U.S. EPA
Contributing Authors: Robert Kafalenos[3] and Kevin Wright[4]
[3]DOT
[4]ICFI, Inc.

Key Findings

- The comprehensive high-resolution and precise analyses of the spatial distributions of population and infrastructure vulnerable to sea-level rise in the Mid-Atlantic required for planning and response do not exist at the present time. Existing studies do not have the required underlying land elevation data with the degree of confidence necessary for local and regional decision-making (see Chapter 2 of this Product).
- Existing generalized data can only support a range of estimates. For instance, in the Mid-Atlantic, between approximately 900,000 and 3,400,000 people (between 3 and 10 percent of the total population in the mid-Atlantic coastal region) live on parcels of land or city blocks with at least some land less than 1 meter above monthly highest tides. Approximately 40 percent of this population is located along the Atlantic Ocean shoreline or small adjacent inlets and coastal bays (as opposed to along the interior shorelines of the large estuaries, such as Delaware Bay and Chesapeake Bay).
- Agriculture lands, forests, wetlands, and developed lands in lower elevation areas are likely to be most impacted by a 1-meter sea-level rise for the Mid-Atlantic.
- The coupling of sea-level rise with storm surge is one of the most important considerations for assessing impacts of sea-level rise on infrastructure. Sea-level rise poses a risk to transportation in ensuring reliable and sustained transportation services.

7.1. Introduction

Coastal areas in the United States have competing interests of population growth (accompanied by building of the necessary supporting infrastructure), the preservation of natural coastal wetlands and creation of buffer zones. Increasing sea level will put increasing stress on the ability to manage these competing interests effectively and in a sustained manner. This Chapter examines the current population, infrastructure, and socioeconomic activity that may potentially be affected by sea-level rise.

7.2. Population Study Assessment

The population assessment for the Mid-Atlantic can be put into a regional perspective by first examining some recent national statistics and trends that illustrate the relative socioeconomic stress on our coasts:

- Using an analysis of coastal counties defined to have a coastline bordering the ocean or associated water bodies, or those containing special velocity zones (V Zones) defined by the Federal Emergency Management Administration (FEMA), Crowell *et al.* (2007) estimate that 37 percent of the total U.S. population is found in 364 coastal counties, including the Great Lakes. Excluding the Great Lakes counties, 30 percent of the total U.S. population is found in 281 coastal counties.
- Using an analysis with a broader definition of a coastal county to include those found in coastal watersheds in addition to those bordering the ocean and associated water bodies, the National Oceanic and Atmospheric Administration (NOAA) estimates that U.S. coastal counties, including the Great Lakes and excluding Alaska, contain 53 percent of the nation's population, yet account for only 17 percent of the total U.S. land area (Crossett *et al.*, 2004)
- Twenty-three of the 25 most densely populated U.S. counties are coastal counties. From 1980 to 2003, population density (defined as persons per unit area) increased in coastal counties by 28 percent and was expected to increase another 4 percent by 2008 (Crossett *et al.*, 2004).
- Construction permits can be used to indicate economic growth and urban sprawl. More than 1,540 single family housing units are permitted for construction every day in coastal counties across the United States. From 1999 to 2003, 2.8 million building permits were issued for single family housing units (43 percent of U.S. total) and 1.0 million building permits were issued for multi-family housing units (51 percent of the U.S. total) (Crossett *et al.*, 2004).
- In 2000, there were approximately 2.1 million seasonal or vacation homes in coastal counties (54 percent of the U.S. total) (Crossett *et al.*, 2004).

Regional trends for the Mid-Atlantic can also be summarized, based on Crossett *et al.* (2004). This Product includes the mid-Atlantic states, defined in the Product to include the area from New York to Virginia, as part of their defined Northeast region, with North Carolina included in the Southeast region. The statistics serve to illustrate the relative vulnerability of the coastal socioeconomic infrastructure, either directly or indirectly, to sea-level rise.

- Of the 10 largest metropolitan areas in the United States, three (New York, Washington, D.C., and Philadelphia) are located in the coastal zone of the mid-Atlantic region.
- The coastal population in the Northeast (Maine to Virginia) is expected to increase by 1.7 million people from 2003 to 2008, and this increase will occur mostly in counties near or in major metropolitan centers. Six of the counties near metropolitan areas with the largest expected population increases are in the New York City area and four are in the Washington, D.C. area.
- The greatest percent population changes from 2003 to 2008 in the U.S. Northeast are expected to occur in Maryland and Virginia. Eight of the 10 coastal counties with the greatest expected percent population increases are located in Virginia and two are located in Maryland.
- North Carolina coastal counties rank among the highest in the U.S. Southeast for expected percent population change from 2003 to 2008. For instance, Brunswick County is expected to have the greatest percent increase, at 17 percent.

Crossett *et al.* (2004), show the mid-Atlantic states in context with the larger Atlantic Coast region. By presenting total land area and coastal land area, as well as total and coastal county population statistics, both in absolute numbers and in population density, the NOAA report quantifies the socioeconomic stressor of population change on the coastal region. As pointed out by Crowell *et al.* (2007), the coastal counties used in the NOAA study represent counties in a broader watershed area that include more than those counties that border the land-water interface and that detailed analyses and summary statistics for populations at direct risk for inundation due to sea-level rise must use only that subset of coastal counties subject to potential inundation. The analyses and statistics discussed in subsequent sections of this Product use those subsets. Crossett *et al.* (2004) is used simply to illustrate the increasing stress on coastal areas in general. The mid-Atlantic coastal counties are among the most developed and densely populated coastal areas in the nation. It is this environment that coastal managers must plan strategies for addressing impacts of climate change, including global sea-level rise.

Several regionally focused reports on examining populations at risk to sea-level rise in the Mid-Atlantic are found in the literature. For example Gornitz *et al.* (2001) includes a general discussion of population densities and flood risk zones in the New York metropolitan region and examines impacts of sea-level rise on this area. In this report, the authors also consider that low-lying areas will be more at risk to episodic flooding from storm events because storm tide elevations for a given storm will be higher with sea-level rise than without. They suggest that the overall effect for any given location will be a reduction in the return period of the 100-year storm flooding event. A similar analysis was performed for the Hampton Roads, Virginia area by Kleinosky *et al.* (2006) that attempts to take into account increased population scenarios by 2100.

Bin *et al.* (2007) studied the socioeconomic impacts of sea-level rise in coastal North Carolina, focusing on four representative coastal counties (New Hanover, Dare, Carteret, and

Bertie) that range from high-development to rural, and from marine to estuarine shoreline. Their socioeconomic analyses studied impacts of sea-level rise on the coastal real estate market, on coastal recreation and tourism, and the impacts of tropical storms and hurricanes on business activity using a baseline year of 2004.

Comprehensive assessments of impacts of sea-level rise on transportation and infrastructure are found in the CCSP Synthesis and Assessment Product (SAP) 4.7 (CCSP, 2008), which focuses on the Gulf of Mexico, but provides a general overview of the scope of the impacts on transportation and infrastructure. In the Mid-Atlantic, focused assessments on the effects of sea-level rise to infrastructure in the New York City area are available in Jacob *et al.* (2007).

Some of the recent regional population and infrastructure assessments typically use the best available information layers (described in the following section), gridded elevation data, gridded or mapped population distributions, and transportation infrastructure maps to qualitatively depict areas at risk and vulnerability (Gornitz *et al.*, 2001). The interpretation of the results from these assessments is limited by the vertical and horizontal resolution of the various data layers, the difference in resolution and matching of the fundamental digital-layer data cells, and the lack of spatial resolution of the population density and other data layers within the fundamental area blocks used (see Chapter 2 for further discussion). As discussed in Chapter 2, the available elevation data for the entire mid-Atlantic region do not support inundation modeling for sea-level rise scenarios of 1 meter or less. Therefore, the results reported in this Chapter should not be considered as reliable quantitative findings, and they serve only as demonstrations of the types of analyses that should be done when high-accuracy elevation data become available.

7.3. Mid-Atlantic Population Analysis

In this Chapter, the methodology for addressing population and land use utilizes a Geographic Information Systems (GIS) analysis approach, creating data layer overlays and joining of data tables to provide useful summary information. GIS data are typically organized in themes as data layers. Data can then be input as separate themes and overlaid based on user requirements. Essentially, the GIS analysis is a vertical layering of the characteristics of the Earth's surface and is used to logically order and analyze data in most GIS software. Data layers can be expressed visually as map layers with underlying tabular information of the data being depicted. The analysis uses data layers of information and integrates them to obtain the desired output and estimated uncertainties in the results. The GIS layers used here are population statistics, land use information, and land elevation data.

The population and land use statistics tabulated in the regional summary tables (Tables 7.1 through 7.6) use an area-adjusted system that defines regions and subregions for analysis such that they are (1) higher than the zero reference contour (Spring High Water) used in a vertical datum-adjusted elevation model, and (2) not considered a wetland or open water, according to the state and National Wetlands Inventory wetlands data compiled by the U.S. Fish and Wildlife Service (USFWS, 2007). Uncertainties are expressed in the tables in terms of low and high statistical estimates (a range of values) in each case to account for the varying quality of topographic information and the varying spatial resolution of the other data layers.

The estimated elevation of spring high water is used as a boundary that distinguishes between normal inundation that would occur due to the normal monthly highest tides and the added inundation due to a 1-meter (m) rise in sea level (Titus and Cacela, 2008) .

Census block statistics determined for the estimated area and the percent of a block affected by sea-level rise and the estimated number of people and households affected by sea-level rise are based on two methods: (1) a uniform distribution throughout the block and (2) a best estimate based on assumptions concerning elevation and population density. For instance, there is an uncertainty regarding where the population resides within the census block, and the relationship between the portion of a block's area that is lost to sea-level rise and the portion of the population residing in the vulnerable area is also uncertain. Analysis estimates of vulnerable population are based on the percentage of a census block that is inundated. Homes are not necessarily distributed uniformly throughout a census block. In addition, the differences in grid sizes between the census blocks and the elevation layers results in various blocks straddling differing elevation grids and adds to the uncertainty of the process.

Discussion on coastal elevations and mapping limitations and uncertainties as applied for inundation purposes is provided in Chapter 2. Given these limitations and uncertainties, the population and land use analyses presented here are only demonstrations of techniques using a 1-meter (m) sea-level rise scenario. More precise quantitative estimates require high-resolution elevation data and population data with better horizontal resolution.

Figure 7.1 illustrates the three GIS data layers used in the population and land use analysis: the elevation layer (Titus and Wang, 2008), a census layer (GeoLytics, 2001), and a land-use layer (USGS, 2001).

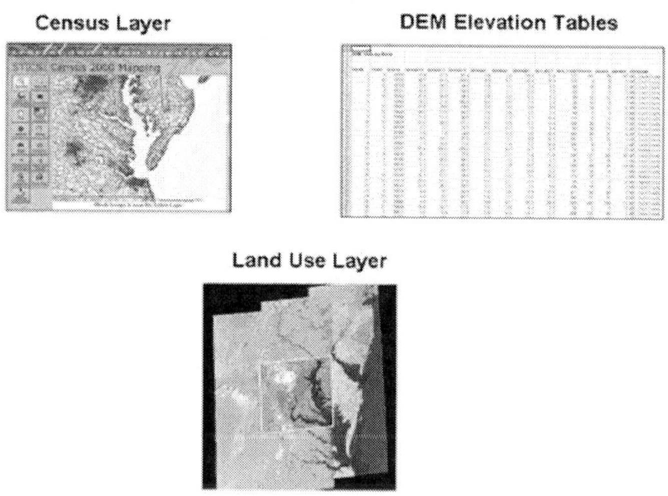

Figure 7.1. The three input data layers to the GIS analysis.

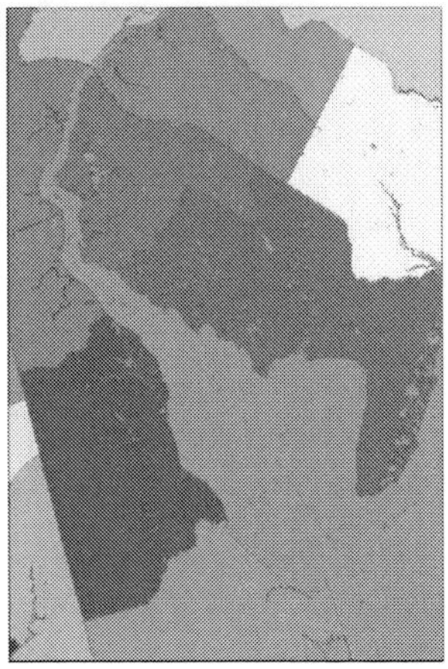

Figure 7.2. The county overlay example for Delaware Bay with each colored area depicting a county.

Figures 7.2, 7.3, and 7.4 show the fundamental underlying layers used in this study, using Delaware Bay as an example. The GIS layers used here are:

- *Elevation data*: The elevation data is the driving parameter in the population analysis. The elevation data is gridded into 30-m pixels throughout the region. All other input datasets are gridded to this system from their source format (Titus and Wang, 2008). The elevations are adjusted such that the zero-contour line is set relative to the Spring High Water vertical datum, which is interpolated from point sources derived from NOAA tide station data (Titus and Cacela, 2008).

- *Census data*: Census 2000 dataset (GeoLytics, 2001) is used in the analysis. Block boundaries are the finest-scale data available, and are the fundamental units of area of the census analysis. Tract, county, and state boundaries are derived from appropriate aggregations from their defining blocks. The census tract boundaries are the smallest census unit that contains property and tax values. Tract and county boundaries also extend fully into water bodies. For this analysis, these boundaries are cropped back to the sea-level boundary, but source census data remain intact.

- *Land use data*: The National Land Cover Data (NLCD) (USGS, 2001) dataset is used in this analysis. It consists of a 30-m pixel classification from circa 2001 satellite imagery and is consistently derived across the region. The caveat with the product is that pixels are classified as "wetland" and "open water" in places that are not classified as such by the wetland layer. Wetland layers are derived from state wetlands data (Titus and Wang, 2008). Usually, the NLCD Wetland class turns out to be forested land and the water tends to be edge effects (or uncertainty due to lack of resolution) along the shore or near farm ponds. This analysis folds the NLCD wetland pixels into forested land.

Figure 7.3. The census tract overlay example for Delaware Bay with each colored area depicting a census tract.

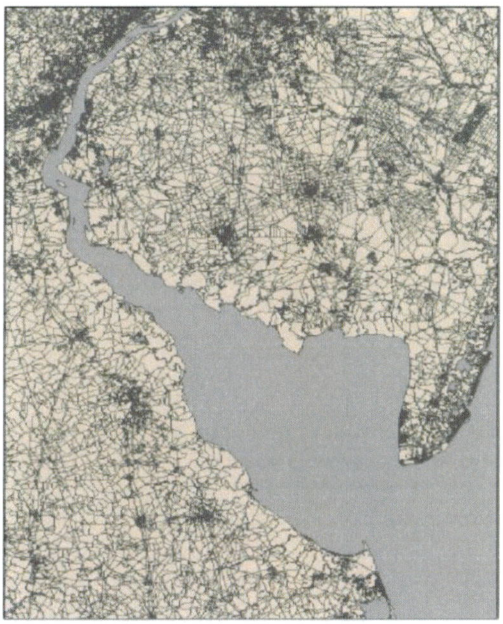

Figure 7.4. The census block overlay example for Delaware Bay with gray lines outlining individual areas of a census block.

Figure 7.2 is an example of the county overlay, and Figure 7.3 is an example of the census tract overlay. A census tract is a small, relatively permanent statistical subdivision of a county used for presenting census data. Census tract boundaries normally follow visible features such as roads and rivers, but may follow governmental unit boundaries and other non-visible features in some instances; they are always contained within counties. Census tracts are designed to be relatively homogeneous units with respect to population characteristics, economic status, and living conditions at the time of establishment, and they average about 4,000 inhabitants. The tracts may be split by any sub-county geographic entity.

Figure 7.4 provides an example of the census block overlay. A census block is a subdivision of a census tract (or, prior to 2000, a block numbering area). A block is the smallest geographic unit for which the Census Bureau tabulates data. Many blocks correspond to individual city blocks bounded by streets; however, blocks–especially in rural areas—may include many square kilometers and due to lack of roads, may have some boundaries that are other features such as rivers and streams. The Census Bureau established blocks covering the entire nation for the first time in 1990. Previous censuses back to 1940 had blocks established only for part of the United States. More than 8 million blocks were identified for Census 2000 (U.S. Census Bureau, 2007).

The Digital Elevation Model (DEM) (Titus and Wang, 2008) was the base for this analysis. The areas of various land use, counties, tracts, and blocks are rasterized (converted in a vector graphics format [shapes]) into a gridded raster image (pixels or dots) to the DEM base. This ensures a standard projection (an equal-area projection), pixel size (30 m), grid system (so pixels overlay exactly), and geographic extent. A GIS data layer intersection was completed for each of the geographic reporting units (land use, county, tract, and block) with elevation ranges to produce a table of unique combinations.

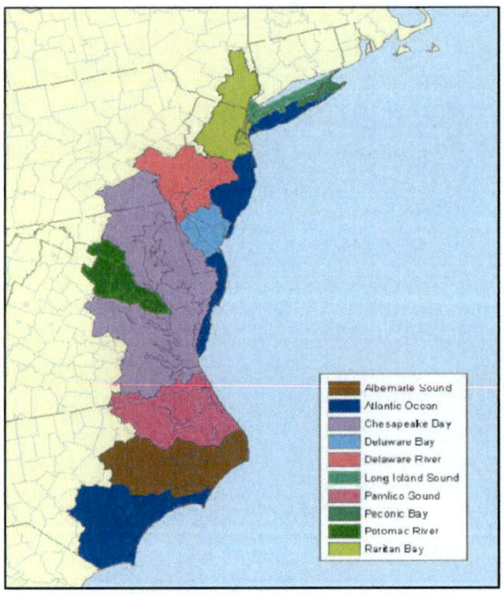

Figure 7.5. The mid-Atlantic region generalized watersheds.

This Chapter examines the mid-Atlantic region and makes some inferences on the populations that may be affected by sea-level rise. This assessment divides the mid- Atlantic region into sub-regions defined by watersheds (Crossett *et al.*, 2004), as shown in Figure 7.5. The general populations within the various watersheds, although sometimes in more than one state, have to address common problems driven by common topographies, and natural hydrological regimes. Most of the watershed boundaries are clear, for instance the Potomac River and Chesapeake Bay. The watershed boundaries used do not include the upland portions of the watershed located in upland mountains and hills; those portions are not required for the analyses of the low-lying areas. The Atlantic Ocean watershed is the most complex because it is not defined by a discrete estuarine river watershed boundary, but by exposure to the outer coastline, and it has components in several states.

7.3.1. Example Population Analysis Results

Not everyone who resides in a watershed lives in a low-lying area that may be at risk to the effects of sea-level rise. Table 7.1 provides a summary analysis of those populations in each watershed at potential risk for a 1 -m sea-level rise. The low and high estimates in Table 7.1 provide the range of uncertainty by using the low and high DEMs (Titus and Wang, 2008; Titus and Cacela, 2008). The high elevation is equal to the best estimate plus the vertical error of the elevation data; the low elevation estimate is equal to the best estimate minus the vertical error. The high vulnerability estimate uses the low elevation estimate because if elevations are lower than expected a greater population is vulnerable. Similarly, the low vulnerability estimate uses the high end of the uncertainty range of elevation estimates. These DEMs are required to express the uncertainty in the numerical results because of the varying scales and resolutions of the data in the various overlays (for instance, the census block boundaries may not line up with specific elevation contours being used and interpolation algorithms must be used to derive population statistics within certain contour intervals. As previously mentioned, this analysis is also limited by the assumption that population has uniform density within the inhabited portion of particular census block. The census data provide no information where the population resides within a particular block.

Table 7.1. Estimated mid-Atlantic low and high population estimates by watershed for a 1-meter sea-level rise (population is based on Census [2000] data). The reported numbers are subject to the caveat given at the end of Section 7.2.

Watershed	Population count	
	1m Sea level Rise	
	Low Estimate	High Estimate
Long Island Sound	1,640	191,210
Peconic Bay	7,870	29,140
NHY-Raritan Bay	35,960	678,670
Delaware Bay	22,660	62,770
Delaware River	19,380	239,480
Chesapeake Bay	326,830	807,720
Potomac River	0	124,510
Albemarle Sound	61,140	75,830
Pamlico Sound	69,720	147,290
Atlantic Ocean	362,800	1,109,280
All Watersheds	908,020	3,465,940

The uncertainty in how much of a particular census tract or block may be inundated must also be addressed by listing high and low estimates. Table 7.1 is a maximum estimate of the potential populations because it is for census blocks that could have any inundation at all and thus includes a maximum count. Similarly, it should be noted that Table 7.3 also provides maximum estimates for the Chesapeake Bay and the Atlantic Ocean.

To illustrate the nature of using the various sets of data and layers for analyses, and the uncertainty in the population distributions within a census block, a second type of analysis is useful. Because there is an uncertainty regarding where the population resides within the census block, the relationship between the portion of a block's area that is lost to sea-level rise and the portion of the population residing in the vulnerable area is also uncertain.

Analysis estimates of vulnerable population are based on the percentage of a census block that is inundated. For instance, the total 2000 population low and high estimated counts for a 1-m sea-level rise for all watersheds are 908,020 and 3,465,940 for "any inundation" of census block (see Table 7.1). However, homes are not necessarily distributed uniformly throughout a census block. If 10 percent of a block is very low, for example, that land may be part of a ravine, or below a bluff, or simply the low part of a large parcel of land. Therefore, the assumption of uniform density would often overstate the vulnerable population. Table 7.2 provides estimates that assume distributions other than uniform density regarding the percentage of a block that must be vulnerable before one assumes that homes are at risk. (This table presents the results by state rather than by subregion.) If it is assumed that 90 percent of a block must be lost before homes are at risk, and that the population is uniformly distributed across the highest 10 percent of the block, then between 26,000 and 959,000 people live less than one meter above the elevation spring high water (see NOAA, 2000 and Titus and Wang, 2008), allowing for low and high elevation estimates. The estimated elevation of spring high water is used as a boundary that distinguishes between normal inundation that would occur due to the normal monthly highest tides and the added inundation due to a 1-m rise in sea level. The spread of these estimated numbers depending upon the underlying assumptions listed at the end of Table 7.2 underscore the uncertainty inherent in making population assessments based in limited elevation data. As reported in Chapter 2, the disaggregation of population density data into a more realistic spatial distribution would be to use a Dasymetric mapping technique (Mennis, 2003) which holds promise for better analysis of population, or other socioeconomic data, and to report statistical summaries of sea-level rise impacts within vulnerable zones.

Table 7.2. Low and High estimates of population living on land within one meter above spring high water (Using assumptions other than uniform population density about how much of the land must be lost before homes are lost). The reported numbers are subject to the caveat given at the end of Section 7.2.

State	Percentage of census block within 1 m above spring high water							
	99[1]		90[2]		50[3]		0[4]	
	Low	High	Low	High	Low	High	Low	High
NY	780	421,900	780	470,900	2,610	685,500	42,320	1,126,290
NJ	12,540	302,800	15,770	352,510	41,260	498,650	177,500	834,440
DE	480	7,200	810	9,230	2,040	16,650	44,290	85,480
PA	640	7,830	640	8,940	1,530	15,090	10,360	43,450
VA	950	59,310	1,020	84,360	5,190	173,950	232,120	662,400
MD	610	4,840	1,890	8,040	4,380	17,710	46,890	137,490
DC	0	0	0	0	0	40	0	9,590
NC	1,920	14,140	5,320	25,090	17,450	60,090	283,590	345,530
Total	17,920	818,020	26,230	959,070	74,460	1,467,680	837,070	3,244,670

[1] Population estimates in this column assume that no homes are vulnerable unless 99 percent of the dry land in census block is within 1 m above spring high water.

[2] Population estimates in this column assume that no homes are vulnerable unless 90 percent of the dry land in census block is within 1 m above spring high water.

[3] Population estimates in this column assume that no homes are vulnerable unless 50 percent of the dry land in census block is within 1 m above spring high water.

[4] Assumes uniform population distribution.

Table 7.3. Low and high estimates of number of owner occupied residences in each watershed region for a 1- meter sea-level rise scenario. The reported numbers are subject to the caveat given at the end of Section 7.2.

Number of owner occupied residences	1- meter rise in sea level	
Watershed		
	Low Estimate	High Estimate
Long Island Sound	0	0
Peconic Bay	3,400	11,650
NYH-Raritan Bay	13,440	269,420
Delaware Bay	8,720	23,610
Delaware River	6,010	89,710
Chesapeake Bay	120,790	299,550
Potomac River	0	46,070
Albemarle Sound	22,760	28,720
Pamlico Sound	26,730	52,450
Atlantic Ocean	140,670	423,540
All Watersheds	342,520	1,244,720

Table 7.4. Low and high estimates of the number of renter occupied housing units by watershed for a 1-meter sea-level rise scenario. The reported numbers are subject to the caveat given at the end of Section 7.2.

Number of renter occupied residences		
	1- meter rise in sea level	
Watershed	Low Estimate	High Estimate
Long Island Sound	70	31,010
Peconic Bay	520	2,460
NYH-Raritan Bay	4,270	178,790
Delaware Bay	2,630	5,880
Delaware River	2,110	32,760
Chesapeake Bay	35,880	84,630
Potomac River	0	17,470
Albemarle Sound	5,260	6,830
Pamlico Sound	6,000	10,660
Atlantic Ocean	40,220	154,500
All Watersheds 96	960 524	990

The census information also allows further analysis of the population, broken down by owner and renter-occupied residences. This information gives a sense of the characterization of permanent home owners *versus* the more transient rental properties that could translate to infrastructure and local economy at risk as well. The estimated number of owner- and renter-occupied housing units in each watershed are shown in Tables 7.3 and 7.4. Similar to the estimates in Table 7.1, these are high estimates for which any portion of a particular census block is inundated.

The actual coastal population potentially affected by sea-level rise also includes hotel guests and those temporarily staying at vacation properties. Population census data on coastal areas are rarely able to fully reflect the population and resultant economic activity. The analysis presented in this Product does not include vacant properties used for seasonal, recreational, or occasional use nor does it characterize the "transient" population, who make up a large portion of the people found in areas close to sea level in the Mid-Atlantic during at least part of the year. These temporary residents include the owners of second homes. A significant portion of coastal homes are likely to be second homes occupied for part of the year by owners or renters who list an inland location as their permanent residence for purposes of census data. In many areas, permanent populations are expected to increase as retirees occupy their seasonal homes for longer portions of the year.

7.4. Land Use

The National Land Cover Database (USGS, 2001) is used to overlay land use onto the DEMs for a 1-m scenario of sea-level rise. Major land-use categories used for this analysis include: agriculture, barren land, developed land, forest, grassland, shrub-scrub, water, and wetland. An estimate of the area of land categorized by land use for all watersheds for the Mid-Atlantic is listed in Table 7.5. In the land-use tables, ranges of uncertainty are provided by showing the low and high estimated size of the areas for the 1-m sea-level rise scenario. The high and low estimates show significant differences in area and express the uncertainty in using this type of data layer integration.

The developed land-use acreage dominates northeast watersheds such as Long Island Sound and New York Harbor, as well as the Atlantic Coast watershed. This is in contrast to the Chesapeake Bay watershed that is dominated by agriculture and forest.

Table 7.5. Mid-Atlantic All Watersheds Summary by Land Use category, depicting low and high estimates of areas affected by a 1-meter sea-level rise (in hectares; 1 hectare is equal to 2.47 acres). The reported numbers are subject to the caveat given at the end of Section 7.2.

Area (in hectares) Land Use	1-meter rise in sea level	
Category	Low Estimate	High estimate
Agriculture	43,180	141,800
Barren Land	5,040	14,750
Developed	11,970	92,950
Forest	27,050	94,280
Grassland	7,640	14,200
Shrub-scrub	3,790	7,720
Water	1,960	4,110
Wetland	34,720	66,590

Table 7.6. Low and high area estimates by land use category for the mid-Atlantic for a 1-meter sea-level rise scenario (in hectares). The reported numbers are subject to the caveat given at the end of Section 7.2.

Area (in hectares)	For a 1-meter rise in sea level		
Watershed	**Land Use Category**	**Low Estimate**	**High Estimate**
Long Island Sound	Agriculture	0	20
	Barren Land	0	180
	Developed	90	3,280
	Forest	0	210
	Grassland	0	100
	Shrub-scrub	0	60
	Water	0	90
	Wetland	0	530
Peconic Bay	Agriculture	20	360
	Barren Land	20	340
	Developed	100	1,580
	Forest	50	760
	Grassland	0	170
	Shrub-scrub	0	70
	Water	10	150
	Wetland	70	770
NYH-Raritan Bay	Agriculture	30	870
	Barren Land	40	340
	Developed	330	21,090
	Forest	40	720
	Grassland	0	10
	Shrub-scrub	0	10
	Water	9	230
	Wetland	140	2,600
Delaware Bay	Agriculture	950	9,590
	Barren Land	280	1,040
	Developed	210	1,760
	Forest	590	4,280
	Water	80	130
	Wetland	900	2,420
Delaware River	Agriculture	310	8,190
	Barren Land	20	560
	Developed	430	10,960
	Forest	90	2,130
	Water	20	200
	Wetland	330	3,010
Chesapeake Bay	Agriculture	11,180	40,460
	Barren Land	2,070	4,650
	Developed	2,220	13,180

Table 7.6. (Continued)

Area (in hectares)	**For a 1-meter rise in sea level**		
Watershed	**Land Use Category**	**Low Estimate**	**High Estimate**
	Wetland	14,480	920
Pamlico Sound	Agriculture	1,3130	3,9670
	Barren Land	470	1,327
	Developed	1,620	4,583
	Forest	5,490	1,380
	Grassland	2,010	3,570
	Shrub-scrub	670	1,430
	Water	210	290
	Wetland	8,500	12,070
Atlantic Ocean	Agriculture	1,090	8,220
	Barren Land	1,800	5,410
	Developed	4,470	29,210
	Forest	2,980	11,540
	Grassland	820	2,010
	Shrub-scrub	380	1,360
	Water	690	1,210
	Wetland	5,260	10,870
	Forest	9,100	38,370
	Water	160	660
	Wetland	5,010	14,280
Potomac River	Agriculture	0	490
	Barren Land	0	460
	Developed	0	1,830
	Forest	0	4,630
	Water	0	130
	Wetland	0	1,120
Albemarle Sound	Agriculture	16,440	12,810
	Barren Land	320	5,900
	Developed	2,460	8,270
	Forest	8,680	4,950
	Grassland	4,790	44,720
	Shrub-scrub	2,720	10
	Water	750	8,440

7.5. Transportation Infrastructure

7.5.1. General Considerations

The coupling of sea-level rise with storm surge is one of the most important considerations for assessing impacts of sea-level rise on infrastructure. Sea-level rise poses a risk to transportation in ensuring reliable and sustained transportation services. Transportation facilities serve as the life-line to communities, and inundation of even the smallest component of an intermodal system can result in a much larger system shutdown. For instance, even though a port facility or a railway terminal may not be affected, the access roads to the port and railways could be, thus forcing the terminal to cease or curtail operation.

Sea-level rise will reduce the 100-year flood return periods and will lower the current minimum critical elevations of infrastructure such as airports, tunnels, and ship terminals (Jacob *et al.*, 2007). Some low-lying railroads, tunnels, ports, runways, and roads are already vulnerable to flooding and a rising sea level will only exacerbate the situation by causing more frequent and more serious disruption of transportation services. It will also introduce problems to infrastructure not previously affected by these factors.

The CCSP SAP 4.7 (Kafalenos *et al.,* 2008) discusses impacts of sea-level rise on transportation infrastructure by addressing the impacts generally on highways, transit systems, freight and passenger rail, marine facilities and waterways, aviation, pipelines, and implications for transportation emergency management and also specifically for the U.S. Gulf Coast region. Each of these transportation modes also apply to the mid-Atlantic region.

One impact of sea-level rise not generally mentioned is the decreased clearance under bridges. Even with precise timing of the stage of tide and passage under fixed bridges, sea-level rise will affect the number of low water windows available for the large vessels now being built. Bridge clearance has already become an operational issue for major ports, as evidenced by the installation of real-time reporting air gap/bridge clearance sensors in the NOAA Physical Oceanographic Real-Time System (PORTS) (NOAA, 2005). Clearance under bridges has become important because the largest vessels need to synchronize passage with the stage of tide and with high waters due to weather effects and high river flows. To provide pilots with this critical information, air gap sensors in the Mid-Atlantic have been deployed at the Verrazano Narrows Bridge at the entrance to New York Harbor, the Chesapeake Bay Bridge located in mid-Chesapeake Bay, and on bridges at both ends of the Chesapeake and Delaware Canal connecting the upper Chesapeake Bay with mid-Delaware Bay (NOAA, 2008).

There are other potential navigation system effects as well because of sea-level rise. Estuarine navigation channels may need to be extended landward from where they terminate now to provide access to a retreating shoreline. The corollary benefit is that less dredging will be required in deeper water because a rising water elevation will provide extra clearance.

This discussion is limited in scope to transportation infrastructure. Complete infrastructure assessments need to include other at-risk engineering and water control structures such as spillways, dams, levees and locks, with assessments of their locations and design capacities.

7.5.2. Recent U.S. Department of Transportation Studies

The U.S. Department of Transportation (US DOT) studied the impacts of sea-level rise on transportation, as discussed in US DOT (2002). The study addresses the impacts of sea-level rise on navigation, aviation, railways and tunnels, and roads, and describes various options to address those impacts, such as elevating land and structures, protecting low-lying infrastructure with dikes, and applying retreat and accommodation strategies.

The US DOT has recently completed an update of the first phase of a study, "The Potential Impacts of Global Sea Level Rise on Transportation Infrastructure" (US DOT,

2008). The study covers the mid-Atlantic region and is being implemented in two phases: Phase 1 focuses on North Carolina, Virginia, Washington, D.C., and Maryland. Phase 2 focuses on New York, New Jersey, Pennsylvania, Delaware, South Carolina, Georgia, and the Atlantic Coast of Florida. This second phase is expected to be completed by the end of 2008. This study was designed to produce rough quantitative estimates of how future climate change, specifically sea-level rise and storm surge, might affect transportation infrastructure on a portion of the East Coast of the United States. The major purpose of the study is to aid policy makers responsible for transportation infrastructure including roads, rails, airports, and ports in incorporating potential impacts of sea-level rise in planning and design of new infrastructure and in maintenance and upgrade of existing infrastructure.

The report considers that the rising sea level, combined with the possibility of an increase in the number of hurricanes and other severe weather related incidents, could cause increased inundation and more frequent flooding of roads, railroads, and airports, and could have major consequences for port facilities and coastal shipping.

The GIS approach (US DOT, 2008) produces maps and statistics that demonstrate the location and quantity of transportation infrastructure that could be regularly inundated by sea-level rise and at risk to storm surge under a range of potential sea-level rise scenarios. The elevation data for the transportation facilities is the estimated elevation of the land upon which the highway or rail line is built.)

The three basic steps involved in the US DOT analysis help identify areas expected to be regularly inundated or that are at-risk of periodic flooding due to storm surge:

- Digital Elevation Models were used to evaluate the elevation in the coastal areas and to create tidal surfaces in order to describe the current and future predicted sea water levels.
- Land was identified that, without protection, will regularly be inundated by the ocean or is at risk of inundation due to storm surge under each sea-level rise scenario.
- Transportation infrastructure was identified that, without protection, will regularly be inundated by the ocean or be at risk of inundation due to storm surge under the given sea-level rise scenario.

The US DOT study compares current conditions (for 2000) to estimates of future conditions resulting from increases in sea level. The study examines the effects of a range of potential increases in sea level up to 59 centimeters (cm). The estimates of increases in sea level are based upon two sources: (1) the range of averages of the Atmosphere-Ocean General Circulation Models for all 35 SRES (Special Report on Emission Scenarios), as reported in Figure 11.12[10] from the IPCC Third Assessment Report and (2) the highest scenario (59 cm) that corresponds with the highest emission scenario modeled by the IPCC Fourth Assessment Report (Meehl *et al.*, 2007).

As noted above, the US DOT study was not intended to create a new estimate of future sea levels or to provide a detailed view of a particular area under a given scenario; similarly, the results should not be viewed as predicting the specific timing of any changes in sea levels. The inherent value of this study is the broad view of the subject and the overall estimates identified. Due to the overview aspect of the US DOT study, and systematic and value uncertainties in the involved models, this US DOT analysis appropriately considered sea-level rise estimates from the IPCC reports as uniform sea-level rise estimates, rather than estimates for a particular geographic location. The confidence stated by IPCC in the regional distribution of sea-level change is *low*, due to significant variations in the included models; thus, it would be inappropriate to use the IPCC model series to estimate local changes. Local variations, whether caused by erosion, subsidence (sinking of land) or uplift, local steric (volumetric increase in water due to thermal expansion) factors or even coastline protection, were not considered in this study[11]. Given the analysis and cautionary statements presented in Chapter 2 regarding using the USGS National Elevation Data (NED) with small increments of sea-level rise as used in this US DOT study, only representative statistical estimations are presented here for just the largest 59-cm scenario. Because the 59-cm sea-level rise scenario is within the statistical uncertainty of the elevation data, the statistics are representative of the types of analyses that could be done if accurate elevation data were available.

The study first estimates the areas that would be regularly inundated or at risk during storm conditions, given nine potential scenarios of sea-level rise. It defines regularly inundated areas or base sea level as NOAA's mean higher high water (MHHW) for 2000. The regularly inundated areas examined are the regions of the coast that fall between MHHW in 2000 and the adjusted MHHW levels (MHHW in 2000 plus for several scenarios up to 59 cm). For at-risk areas or areas that could be affected by storm conditions, the study uses a base level of NOAA's highest observed water levels (HOWL) for 2000, and adjusts this upwards based on the nine sea-level rise scenarios. The at-risk areas examined are those areas falling between the adjusted MHHW levels and the adjusted HOWL levels.

Table 7.7. A representative output table for Virginia showing estimates of regularly inundated and at-risk areas and lengths under the 59 centimeter (cm) scenario, the highest level examined in the U.S. Department of Transportation (US DOT) study. The percent affected represent the proportion for the entire state, not only coastal areas (From US DOT, 2008). The reported numbers are subject to the caveat given at the end of Section 7.2.

State of Virginia Statistics	**For a 59-cm rise in sea level**					
	Regularly Inundated		**At-Risk to Storm Surge**		**Total**	
By Length in Kilometers (km)	Length (km)	Percent Affected	Length (km)	Percent Affected	Length (km)	Percent Affected
Interstates	7	0%	16	1%	23	1%
Non-Interstate Principal Arterials	12	0%	62	1%	74	2%
NHS Minor Arterials	2	0%	9	0%	11	0%
National Highway System (NHS)	22	0%	64	1%	86	2%
Rails	19	0%	64	1%	83	1%
By Area in Hectares	Area (Hectares)	Percent Affected	Area (Hectares)	Percent Affected	Area (Hectares)	Percent Affected
Ports	60	11%	132	24%	192	35%
Airport Property	277	2%	365	3%	642	4%
Airport Runways	29	2%	37	3%	66	5%
Total Land Area Affected	68,632	1%	120,996	1%	189,628	2%

Table 7.8. Summary of estimated areas and lengths for the total of regularly inundated and at risk infrastructure combined for a 59 centimeters (cm) increase in sea-level rise (based on US DOT, 2008). The reported numbers are subject to the caveat given at the end of Section 7.2.

Total, Regularly Inundated and At Risk								
For a 59-cm increase in sea level	Washington, D.C.		Virginia		Maryland		North Carolina	
By Length in Kilometers(km)	Length (km)	% Affected	Length (km)	% Affected	Length (km)	% Affected	Length (km)	% Affected
Interstates	1	5%	25	1%	2	0%	1	0%
Non-Interstate Principal Arterials	7	4%	75	2%	21	1%	130	2%
Minor Arterials	0	0%	11	0%	66	4%	209	4%
National Highway System (NHS)	7	5%	87	2%	19	1%	305	4%
Rails	3	5%	84	1%	44	2%	105	1%
Ports	n/a	n/a	192	35%	120	32%	88	47%
Airport Property	n/a	n/a	642	4%	59	1%	434	3%
Airport Runways	n/a	n/a	66	5%	1	0%	27	2%
Total Land Area Affected	968	6%	189,628	2%	192,044	8%	743,029	6%

A sample of output tables from the US DOT study are shown in Table 7.7, which covers the state of Virginia. The numerical values for length and area in Tables 7.7 and 7.8 have been rounded down to the nearest whole number to be conservative in the estimates for lengths and areas at risk. This was done to avoid overstating the estimates as there are no estimates of uncertainty or error in the numbers presented.

Table 7.7 indicates there is some transportation infrastructure at risk under the 59-cm sea level rise scenario. Less than 1 percent (7 kilometers [km] of interstates, 12 km of non-interstate principal arterials) of the Virginia highways examined in the US DOT study would be regularly inundated, while an additional 1 percent (16 km of interstates, 62 km of non-interstate principal arterials) could be affected by storm conditions. It should be noted that these percentages are given as a percentage of the total for each state, not only for coastal counties.

Table 7.8 provides the areas and percent of total areas affected of the various regularly inundated and at-risk transportation categories for the US DOT (2008) 59-cm sea-level rise scenario for Washington, D.C., Virginia, Maryland, and North Carolina.

Based on the small percentage (1 to 5 percent) statistics in Table 6.8, the combination of rising sea level and storm surge appears to have the potential to affect only a small portion of highways and roads across the region. However, because these transportation systems are basically networks, just a small disruption in one portion could often be sufficient to have far-reaching effects, analogous to when a storm causes local closure of a major airport, producing ripple effects nation-wide due to scheduling and flight connections and delays. Local flooding could have similar ripple effects in a specific transportation sector.

North Carolina appears slightly more vulnerable to regular inundation due to sea-level rise, both in absolute terms and as a percentage of the state highways: less than 1 percent of interstates (0.3 km), 1 percent of non-interstate principal arterials (59 km) and 2 percent of National Highway System (NHS) minor arterials (93 km) in the state would be regularly inundated given a sea-level rise of 59 cm. This US DOT study focuses on larger roads but there are many miles of local roads and collectors that could also be affected. In general, areas at risk to storm surge are limited. Washington, D.C. shows the greatest vulnerability on a percentage basis for both interstates and NHS roads for all sea-level rise scenarios examined.

Please refer to the US DOT study for complete results, at:

<http://climate.dot.gov/publications/potential_impacts_of_global_sea_level_rise/index.html>

REFERENCES

Bin, P., Dumas, C., Poulter, B. & Whitehead, J. (2007). *Measuring the Impacts of Climate Change on North Carolina Coastal Resources.* Department of Economics, Appalachian State University, Boone, NC, 91 pp. <http://econ.appstate.edu/climate/ >

CCSP (Climate Change Science Program), (2008). *Impacts of Climate Change andVariability on Transportation Systems and Infrastructure: Gulf Coast Study, Phase I.* [Savonis, M.J., V.R. Burkett and J.R. Potter (eds.)]. Climate Change Science Program Synthesis and Assessment Product 4.7. U.S. Department of Transportation, Washington DC, 445 pp.

Crossett, K.M., Culliton, T. J., Wiley, P.C. & Goodspeed, T. R. (2004). *Population Trends along the Coastal United States 1980-2008.* NOAA National Ocean Service, Special Projects Office, [Silver Spring, MD], 47 pp.

Crowell, M., Edelman, S. Coulton, K. & McAfee, S. (2007). How many people live in coastal areas? *Journal of Coastal Research, 23(5),* iii-vi.

GeoLytics, (2001). *CensusCD 2000.* Version 1.1. GeoLytics, Inc., East Brunswick, NJ.

Gornitz, V., Couch, S. & Hartig, E. K. (2001). Impacts of sea level rise in the New York City metropolitan area. *Global and Planetary Change, 32(1),* 61-88.

Jacob, K., Gornitz, V. & Rosenzweig, C. (2007). Vulnerability of the New York City metropolitan area to coastal hazards, including sea-level rise: Inferences for urban coastal risk management and adaptation policies. In: *Managing Coastal Vulnerability.* [McFadden, L., R.J. Nicholls, and E.C. Penning-Rowsell (eds.)]. Elsevier, Amsterdam and Oxford, pp. 139-156.

Kafalenos, R. S., Leonard, K.J., Beagan, D.M., Burkett, V.R., Keim, B.D., Meyers, A., Hunt, D.T., Hyman, R.C., Maynard, M. K. Fritsche, B., Henk, R. H., Seymour, E. J., Olson, L. E., Potter, J. R. & M.J. Savonis, (2008). What are the implications of climate change and variability for Gulf Coast transportation? In: *Impacts of Climate Change and Variability on Transportation Systems and Infrastructure: Gulf Coast Study, Phase I.* [Savonis, M.J., V.R. Burkett, and J.R. Potter (eds.)]. Climate Change Science Program Synthesis and Assessment Product 4.7. U.S. Department of Transportation, Washington DC, [104 pp.]

Kleinosky, L. R., Yarnal, B. & Fisher, A. (2006). Vulnerability of Hampton Roads, Virginia to storm-surge flooding and sea level rise. *Natural Hazards,40(1),* 43-70.

Meehl, G. A., Stocker, T. F., Collins, W.D., Friedlingstein, P., Gaye, A.T., Gregory, J. M., Kitoh, A., Knutti, R., Murphy, J. M., Noda, A., Raper, S. C. B., Watterson, I. G., Weaver A. J. & Z.-C. Zhao, (2007). Global climate projections. In: *Climate Change 2007: The Physical Science Basis.* Contribution of Working Group I to the Fourth Assessment Report of the Intergovernmental Panel on Climate Change [Solomon, S., D. Qin, M. Manning, Z. Chen, M. Marquis, K.B. Averyt, M. Tignor, and H.L. Miller (eds.)]. Cambridge University Press, Cambridge, UK, and New York, pp. 747-845.

Mennis, J. (2003). Generating surface models of population using dasymetric mapping. *Professional Geographer, 55(1),* 31-42.

NOAA (National Oceanic and Atmospheric Administration), (2000). *Tide and Current Glossary.* NOAA National Ocean Service, Silver Spring, MD, 29 pp. <http://tidesand currents.noaa.gov/publications/glossary2.pdf>

NOAA (National Oceanic and Atmospheric Administration), (2005). *Microwave Air Gap-Bridge Clearance Sensor Test, Evaluation, and Implementation Report.* NOAA technical report NOS CO-OPS 042. NOAA National Ocean Service Ocean Systems Test and Evaluation Program, Silver Spring, MD, 111 pp.

NOAA (National Oceanic and Atmospheric Administration), (2008). *The Physical Oceanographic Real-Time System (PORTS).* [Website] <http://tidesandcurrents. noaa.gov/ports>

Titus, J. G. & Cacela, P. (2008). Uncertainty ranges associated with EPA's estimates of the area of land close to sea level. Section 1 .3b in: *Background Documents Supporting Climate Change Science Program Synthesis and Assessment Product 4.1: Coastal Elevations and Sensitivity to Sea Level Rise* [Titus, J.G. and E.M. Strange (eds.)]. EPA 430R07004. U.S. Environmental Protection Agency, Washington, DC, pp. 69-133. <http://epa.gov/climatechange/effects/coastal/background.html>

Titus, J.G. & J. Wang, (2008). Maps of lands close to sea level along the middle Atlantic coast of the United States: an elevation data set to use while waiting for LIDAR. Section 1.1 in: *Background Documents Supporting Climate Change Science Program Synthesis and Assessment Product 4.1: Coastal Elevations and Sensitivity to Sea Level Rise* [Titus, J.G. and E.M. Strange (eds.)]. EPA 430R07004. U.S. Environmental Protection Agency, Washington, DC, pp. 2-44.

U.S. Census Bureau, (2000). *United States Census 2000.* [Website] U.S. Census Bureau, Washington, DC. <http://www.census.gov/main/www/cen2000.html>

U.S. Census Bureau, (2007). *American FactFinder Glossary.* [U.S. Census Bureau, Washington, DC.] <http://factfinder.census.gov/home/en/epss/glossary_a.html>

US DOT (Department of Transportation), (2002): *The Potential Impacts of Climate Change on Transportation*, Workshop Proceedings; J. Titus: Does Sea Level Rise matter to Transportation along the Atlantic Coast? October 1-2, 2002, Summary and Discussion Papers. <http://www.epa.gov/climatechange/effects/downloads/Transportation_Paper.pdf >

US DOT, (2008). *The Potential Impacts of Global Sea Level Rise on Transportation Infrastructure, Phase 1 - Final Report: the District of Columbia, Maryland, North Carolina and Virginia.* Center for Climate Change and Environmental Forecasting, U.S. Department of Transportation, Washington DC.

USGS (U.S. Geological Survey), (2001). *National Land Cover Database 2001.* U.S. Geological Survey, Sioux Falls, SD. <http://www.mrlc.gov/mrlc2k_nlcd.asp>

In: Coastal Sensitivity to Sea Level Rise-Focusing on the Mid-... ISBN: 978-1-60741-440-7
Editor: Melvin C. Urajner

Chapter 8

PUBLIC ACCESS

James G. Titus
U.S. EPA

Key Findings

- The Public Trust Doctrine provides access along the shore below mean high water, but it does not include the right to cross private property to reach the shore. Therefore, access *to* the shore varies greatly, depending on the availability of roads and public paths to the shore.
- Rising sea level alone does not have a significant impact on either access to the shore or access along the shore; however, responses to sea-level rise can decrease or increase access.
- Shoreline armoring generally eliminates access along estuarine shores, by eliminating the intertidal zone along which the public has access. New Jersey has regulatory provisions requiring shorefront property owners in some urban areas to provide alternative access inland of new shore protection structures. Other mid-Atlantic states lack similar provisions to preserve public access.
- Beach nourishment has minimal impact in areas with ample access; however, it can increase access in areas where public access is restricted. Federal and state policies generally require public access to and along a shore before providing subsidized beach nourishment. In several communities, property owners have assigned public access easements in return for beach nourishment.

Responses based on allowing shores to retreat generally have minimal impact on public access to and along the shore.

8.1. Introduction

Rising sea level does not inherently increase or decrease public access to the shore, but the response to sea-level rise can. Beach nourishment tends to increase public access along the shore because federal (and some state) laws preclude beach nourishment funding unless the public has access to the beach that is being restored. Shoreline armoring, by contrast, can decrease public access along the shore, because the intertidal zone along which the public has access is eliminated.

This Chapter examines the impacts of sea-level rise on public access to the shore and describes existing public access to the shore (Section 8.2), the likely impacts of shoreline changes (Section 8.3), and how responses to sea-level rise might change public access (Section 8.4) The focus of this Chapter is on the public's legal right to access the shore, not on the transportation and other infrastructure that facilitates such access[12].

8.2. Existing Public Access and the Public Trust Doctrine

The right to access tidal waters and shores is well established. Both access to and ownership of tidal wetlands and beaches is defined by the "Public Trust Doctrine", which is part of the common law of all the mid-Atlantic states. According to the Public Trust Doctrine, navigable waters and the underlying lands were publicly owned at the time of statehood and remain so today.

The Public Trust Doctrine is so well established that it often overrides specific governmental actions that seem to transfer ownership to private parties (Lazarus, 1986; Rose, 1986). Many courts have invalidated state actions that extinguished public ownership or access to the shore (*Illinois Central R.R. v. Illinois; Arnold v. Mundy*; see also Slade, 1990). Even if a land deed states that someone's property extends into the water, the Public Trust Doctrine usually overrides that language and the public still owns the shore[13]. In those cases when government agencies do transfer ownership of coastal land to private owners, the public still has the right to access along the shore for fishing, hunting, and navigation, unless the state explicitly indicates an intent to extinguish the public trust (Lazarus, 1986; Slade, 1990).

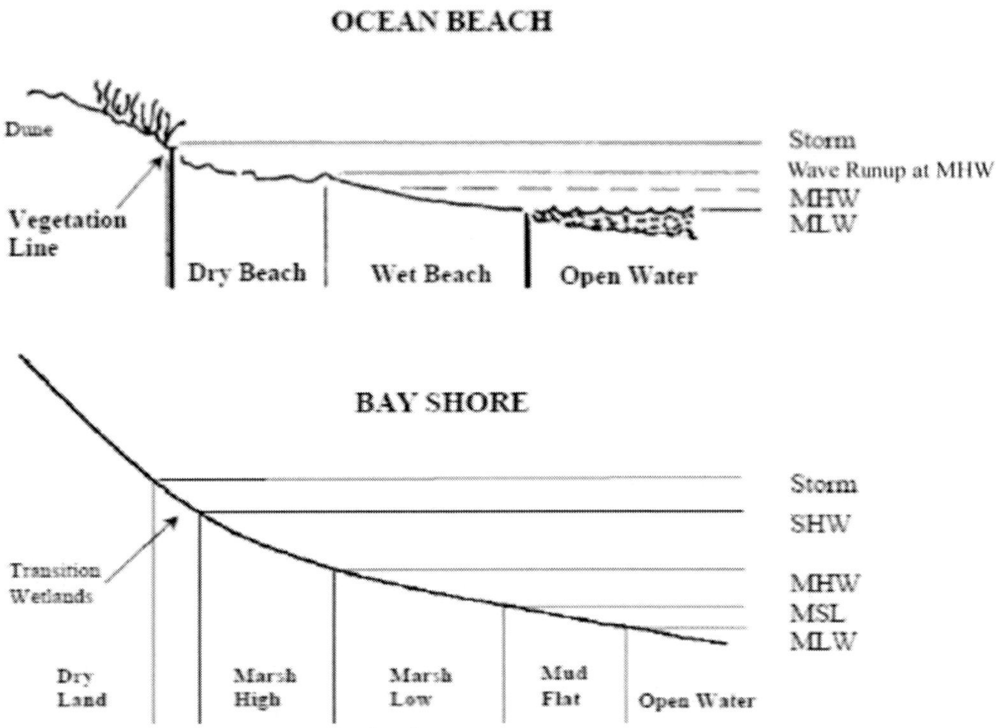

MSL	=	Mean Sea level
MLW	=	Mean Low Level
MHW	=	Mean High Water
SHW	=	Spring High Water
Storm	=	Average Annual Storm Tide

Figure 8.1. Legal and geological tideland zonation. The area below mean high water is usually publicly owned, and in all cases is subject to public access for fishing and navigation. Along the ocean, the dry beach above mean high water may be privately owned; however, in several states the public has an easement. Along the bay, the high marsh above mean high water is also privately owned, but wetland protection laws generally prohibit or discourage development.

Figure 8.2. Traditional purposes of the Public Trust Doctrine include fishing and transportation along the shore. (a) New Jersey side of Delaware River, below Delaware Memorial Bridge (March 2003). (b) Beach provided primary access to homes along the beach at Surfside, Texas (May 2003).

Figure 8.3. Privately owned dunes adjacent to publicly owned intertidal beach. Southold, New York. (September 2006).

Figure 8.1 illustrates some key terminology used in this Chapter. Along sandy shores with few waves, the wet beach lies between *mean high water* and *mean low water*. (Along shores with substantial waves, the beach at high tide is wet inland from the mean high water mark, as waves run up the beach.) The *dry beach* extends from approximately mean high water inland to the seaward edge of the dune grass or other terrestrial plant life, sometimes called the *vegetation line* (Slade, 1990). The dune grass generally extends inland from the point where a storm in the previous year struck with sufficient force to erode the vegetation (Pilkey, 1984), which is well above mean high water. Along marshy shores, mudflats are found between mean low water and mean sea level, *low marsh* is found between mean sea level and mean high water, and *high marsh* extends from mean high water to *spring high water*. Collectively, the lands between mean high water and mean low water (mudflats, low marsh, and wet beaches) are commonly known as *tidelands*.

The Public Trust Doctrine includes these wetlands and beaches because of the needs associated with hunting, fishing, transportation along the shore, and landing boats for rest or repairs (Figure 8.2). In most states, the public owns all land below the high water mark (Slade, 1990) which is generally construed as mean high water. The precise boundary varies in subtle ways from state to state. The portion of the wet beach inland of mean high water resulting from wave runup has also been part of the public trust lands in some cases (see *e.g.*, *State v. Ibbison* and Freedman and Higgins [undated]). Thus, in general, the public trust includes mudflats, low marsh, and wet beach, while private parties own the high marsh and dry beach (Figure 8.3). Nevertheless, Figure 8.4.shows that there are some exceptions. In Pennsylvania, Delaware, and Virginia, the publicly owned land extends only up to the low water mark (Slade, 1990). In New York, by contrast, the inland extent of the public trust varies; in some areas the public owns the dry beach as well[14]. The public has also obtained ownership to some beaches through government purchase, land dedication by a developer, or other means (see Slade 1990; Figure 8.5).

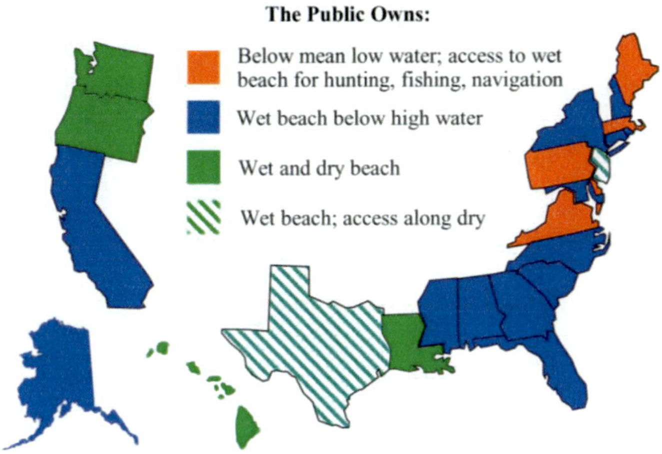

Figure 8.4. The public's common law interest in the shores of various coastal states. Source: Titus (1998)

Figure 8.5. Public beach owned by local government. Beaches that are owned by local governments sometimes have access restrictions for nonresidents. Atlantic Beach, New York (September, 2006).

Ownership, however, is only part of the picture. In Pennsylvania, Delaware, and Virginia, the Public Trust Doctrine provides an easement along the tidelands for hunting, fishing, and navigation. In New Jersey, the Public Trust Doctrine includes access along the *dry* part of the beach for recreation, as well as the traditional public trust purposes (*Matthews v. Bay Head*). Other states have gradually obtained easements for access along some dry beaches either through purchases or voluntary assignment by the property owners in return for proposed beach nourishment. The federal policy precludes funding for beach nourishment unless the public has access (USACE, 1996). Some state laws specify that any land created with beach nourishment belong to the state (*e.g.*, MD. CODE ANN., NAT. RES. II 8-1103 [1990]).

The right to access *along* the shore does not mean that the public has a right to cross private land to get *to* the shore. Unless there is a public road or path to the shore, access along the shore is thus only useful to those who either reach the shore from the water or have permission to cross private land. Although the public has easy access to most ocean beaches and large embayments like Long Island Sound and Delaware Bay, the access points to the shores along most small estuaries are widely dispersed (*e.g.*, Titus, 1998). However, New Jersey is an exception: its Public Trust Doctrine recognizes access *to* the shore in some cases

(*Matthews v. Bay Head*); and state regulations require new developments with more than three units along all tidal waters to include public access to the shore (NJAC 7:7E-8. 11 [d-f]). Given the federal policy promoting access, the lack of access to the shore has delayed several beach nourishment projects. To secure the funding, many communities have improved public access to the shore, not only with more access ways to the beach, but also by upgrading availability of parking, restrooms, and other amenities (*e.g.*, New Jersey, 2006).

8.3. Impact of Shore Erosion on Public Access

The rule that property lines retreat whenever shores erode gradually has been part of the common law for over one thousand years (*County of St. Clair v. Lovingston*; *DNR v. Ocean City*), assuming that the shoreline change is natural. Therefore, as beaches migrate landward, the public's access rights to tidal wetlands and beaches do not change, they simply migrate landward along with the wetlands and beaches. Nevertheless, the area to which the public has access may increase or decrease, if sea-level rise changes the area of wetlands or beaches.

When riparian landowners caused the shorelines to advance seaward, the common law did not vest owners with title to land reclaimed from the sea, although legislatures sometimes have (ALR, 1941). If beach nourishment or a federal navigation jetty artificially creates new land, a majority of states (*e.g.*, MD. CODE ANN., ENVIR. 16-201) award the new land to the riparian owner if he or she is not responsible for creating the land (Slade, 1990); a minority of states (*e.g.*, *Garrett v. State of New Jersey;* N.C. Gen Stat §146-6[f]) vest the state public trust with the new land. Although these two approaches were established before sea-level rise was widely recognized, legal scholars have evaluated the existing rules in the analogous context of shore erosion (*e.g.*, Slade, 1990). Awarding artificially created land to the riparian owner has two practical advantages over awarding it to the state. First, determining what portion of a shoreline change resulted from some artificial causes, (*e.g.*, sedimentation from a jetty or a river diversion) is much more difficult than determining how much the shoreline changed when the owner filled some wetlands. Second, this approach prevents the state from depriving shorefront owners of their riparian access by pumping sand onto the beach and creating new land (*e.g.*, *Board of Public Works v. Larmar Corp*). A key disadvantage is that federal and state laws generally prevent the use of public funds to create land that accrues to private parties. Therefore, part of the administrative requirements of a beach nourishment project is to obtain easements or title to the newly created land. Obtaining those rights can take time, and significantly delayed a beach nourishment project at Ocean City, Maryland (Titus, 1998).

Sea-level rise causes shores to retreat both through inundation and erosion. Although the case law generally assumes that the shore is moving as a result of sediment being transported, inundation and shore erosion are legally indistinguishable. Among the causes of natural shoreline change, the major legal distinction has been between gradual and imperceptible shifts, and sudden shifts that leave land intact but on the other side of a body of water, often known as "avulsion". Shoreline erosion changes ownership; avulsion does not. If an inlet formed 200 meters (m) west of one's home during a storm after which an existing inlet 200 m east of the home closed, an owner would still own her home because this shoreline change is considered to be avulsion. But if the inlet gradually migrated 400 m west, entirely eroding the property but later creating land in the same location, all of the newly created land will belong to the owner to the east (see Figure 8.6). The public trust has the same rights of access to beaches created through avulsion as to beaches migrating by gradual erosion in New York (*People v. Steeplechase Park Co.*) and North Carolina (Kalo, 2005). In other states, the law is less clear (Slade, 1990).

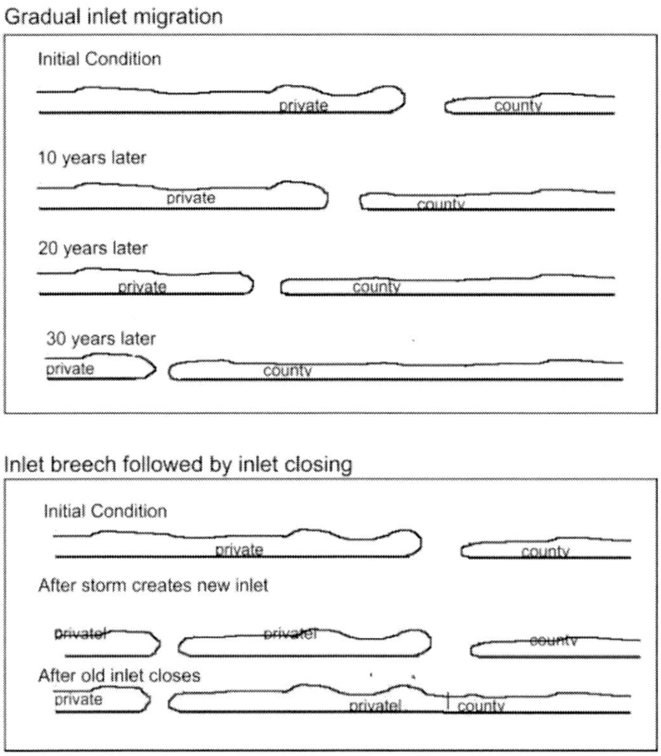

Figure 8.6. Impact of inlet migration and inlet breech on land ownership. In this example, the island to the west is privately owned while the island to the east is a county park.

Figure 8.7. Public access along a bulkheaded shore. In North Beach, Maryland, one block of Atlantic Avenue is a walkway along Chesapeake Bay (May,2006).

Because the public has access to the intertidal zone as long as it exists, the direct effect of sea-level rise on public access depends on how the intertidal zone changes. Along an undeveloped or lightly developed ocean beach, public access is essentially unchanged as the beach migrates inland (except perhaps where a beach is in front of a rocky cliff, which is rare in the Mid-Atlantic). If privately owned high marsh becomes low marsh, then the public will have additional lands on which they may be allowed to walk (provided that environmental regulations to protect the marsh do not prohibit it). Conversely, if sea-level rise reduces the area of low marsh, then pedestrian access may be less, although areas that convert to open water remain in the public trust.

8.4. Impact of Responses to Sea-Level Rise on Public Access

Although sea-level rise appears to have a small direct effect on public access to the shore, responses to sea-level rise can have a significant impact, especially in developed areas. Along developed bay beaches, by contrast, public access along the shore can be eliminated if the shorefront property owner erects a bulkhead, because the beach is eventually eliminated. A number of options are available for state governments that wish to preserve public access along armored shores, such as public purchases of the shorefront (Figure 8.7) and protecting public access in permits for shore protection structures. New Jersey requires pathways to be at least 5 m (16 feet [ft]) wide between the shore and new developments with more than three units along urban tidal rivers (NJAC 7.7E-8. 11 [e]; see also Section A1 .D.2 in Appendix 1) and some other areas, and has a more general requirement to preserve public access elsewhere. (NJAC 7.7E-8. 11 [d] [1]). However, single-family homes are generally exempt (NJAC 7.7E-8. 11 [f] [7])—and other mid-Atlantic states have no such requirements. Therefore, sea-level rise has reduced public access along many estuarine shores and is likely to do so in the future as well.

Government policies related to beach nourishment, by contrast, set a minimum standard for public access (USACE, 1996), which often increases public access along the shore. Along the ocean shore from New York to North Carolina, the public does not have access along the dry beach under the Public Trust Doctrine (except in New Jersey)[15]. However, once a federal beach nourishment project takes place, the public gains access. Beach nourishment projects have increased public access *along* the shore in Ocean City, Maryland and Sandbridge (Virginia Beach), Virginia, where property owners had to provide easements to the newly created beach before the projects began (Titus, 1998; Virginia Marine Resources Commission, 1988).

Areas where public access *to* the beach is currently limited by a small number of access points include the area along the Outer Banks from Southern Shores to Corolla, North Carolina (NC DENR, 2008); northern Long Beach Township, New Jersey (USACE, 1999); and portions of East Hampton, South Hampton, Brookhaven, and Islip along the South Shore of Long Island, New York (Section A1.A.2 in Appendix 1). In West Hampton, landowners had to provide six easements for perpendicular access from the street to the beach in order to meet the New York state requirement of public access every one-half mile (see Section A1.A.2 in Appendix 1). A planned $71 million beach restoration project for Long Beach Island has been stalled (Urgo, 2006), pending compliance with the New Jersey state requirement of perpendicular access every one- quarter mile (USACE, 1999). An additional 200 parking spaces for beachgoers must also be created in Northern Long Beach Township (USACE, 1999). Private communities along Delaware Bay have granted public access to the beaches in return for state assistance for beach protection (Beaches 2000 Planning Group, 1988).

If other communities with limited access seek federal beach nourishment in the future, public access would similarly increase. Improved access to the beach for the disabled may also become a requirement for future beach nourishment activities (*e.g.*, Rhode Island CRMC, 2007). This is not to say that all coastal communities would provide public access in return for federal funds. But aside from the portion of North Carolina southwest of Cape Lookout, the Mid-Atlantic has no privately owned gated barrier islands, unlike the Southeast, where several communities have chosen to expend their own funds on beach nourishment rather than give up their exclusivity.

Ultimately, the impact of sea-level rise on public access will depend on the policies and preferences that prevail over the coming decades. Sometimes the desire to protect property as shores erode will come at the expense of public access. Sometimes it will promote an entire re-engineering of the coast, which under today's policies generally favors public access. It is possible that rising sea level is already starting to cause people to rethink the best way to protect property along estuarine shores (NRC, 2007) to protect the environmental benefits of natural shores. If access along estuarine shores becomes a policy goal, techniques are available for preserving public access as sea level rises.

REFERENCES

* Indicates non-peer reviewed literature; available upon request

ALR, (1941). Annotation: Waters: rights in respect of changes by accretion or reliction due to artificial conditions. *American Law Review, 134,* 467-472.

Arnold v. Mundy, 6 N.J.L. 1 (1821)

Beaches 2000 Planning Group, (1988). Beaches 2000: Report to the Governor [Delaware], June 21, 1988.

Board of Pub. Works v. Larmar Corp., 277 A.2d 427, 436 (Md. 1971)

County of St. Clair v. *Lovingston, 90 U.S. (23 Wall.) 46,* 66-69 (1874) (quoting the Institutes of Justinian, Code Napoleon, and Blackstone for the universal rule that a boundary shifts with the shore)

DNR (Department of Natural Resources) v. Ocean City, 332 A.2d 630-63 8 (Md. 1975)

Freedman, J. & Higgins, M. Undated: *What Do You Mean by High Tide? The Public Trust Doctrine in Rhode Island.* Rhode Island Coastal Resources Management Council, Wakefield, 5 pp. <http://www.crmc. state.ri.us/presentations/presentations/wdymbht.pdf>

Garrett v. State [of New Jersey]. *118 N.J. Super. 594 (Ch. Div. 1972),* 289 A.2d 542 (N.J. Super 1972).

Illinois Central R.R. v. Illinois, 146 U.S. 387 (1982).

Kalo, J.J. (2005):. North Carolina oceanfront property and public waters and beaches: the rights of littoral owners in the twenty-first century. *North Carolina Law Review, 83*, 1427-1506.

Lazarus, R. J. (1986). Changing conceptions of property and sovereignty in natural resources: questioning the Public Trust Doctrine. *Iowa Law Review, 71*, 631.

Matthews v. Bay Head Improvement Association, 471 A.2d 355-358 (N.J. 1984)

New Jersey, (2006). *Highlights of the Public Access Proposal.* New Jersey Department of Environmental Protection Coastal Management Program, 4 pp. <http://www.nj .gov/dep/cmp/access/pa_rule_highlights.pdf>

NC DENR (North Carolina Department of Environment and Natural Resources), (2008).: *Public Beach & Waterfront Access Interactive Mapping.* <http://dcm2.enr. state.nc.us/Access/sites>

NRC (National Research Council) (2007). *Mitigating Shore Erosion Along Sheltered Coasts.* National Academies Press, Washington DC, 188 pp.

People v. Steeplechase Park Co., 82 Misc 247, 255-256; 143, N.Y.S. 503, 509

Pilkey, O.H., Jr. (ed.) (1984). *Living with the East Florida Shore.* Duke University Press, Durham, NC, 259 pp.

Rhode Island CRMC (Coastal Resources Management Council), (2007). Coastal Resources Management Program, as Amended (a.k.a. the "Red Book"). State of Rhode Island Coastal Resources Management Council, Providence, RI. <http://www.crmc.state.ri.us/regulations>

Rose, C. (1986). The comedy of the commons: custom, commerce, and inherently public property. *University of Chicago Law Review, 53,* 711, 715-723.

Slade, D.C. (1990). Lands, waters and living resources subject to the Public Trust Doctrine. In: *Putting the Public Trust Doctrine to Work.* Coastal States Organization, Washington, DC, pp. 13, 59.

State v. Ibbison. 448 A.2d 728 (Rhode Island, 1982). Titus, J.G. (1998). Rising seas, coastal erosion, and the takings clause: how to save wetlands and beaches without hurting property owners. *Maryland Law Review, 57(4),* 1277-1399.

Titus, J. G. (1998). Rising seas, coastal erosion and the taking clause: how to save wetlands and beaches without hurting property owners. *Maryland Law Review, 57(4),* 1277-1399.

* Urgo, J. L. (2006). A standoff over sand: A state beach project requires more public access, but in Loveladies and North Beach, it's meeting some resistance. *Philadelphia Inquirer*, June 11, 2006.

USACE (U.S. Army Corps of Engineers) (1996). *Digest of Water Resources Policies and Authorities 14-1: Shore Protection.* EP 1165-2-1, U.S. Army Corps of Engineers, Washington, DC.

USACE (U.S. Army Corps of Engineers), (1999). *Barnegat Inlet to Little Egg Inlet.* Final Feasibility Report and Integrated Final Environmental Impact Statement.

USACE (U.S. Army Corps of Engineers) & New Jersey Department of Environmental Protection, (1999). *Barnegat Inlet to Little Egg Inlet, Final Feasibility and Integrated Final Environmental Impact Statement.*

Virginia Marine Resources Commission (1988). *Criteria for the Placement of Sandy Dredged Material along Beaches in the Commonwealth.* Regulation VAC 20- 400-10 et seq. <http://www.mrc.virginia.gov/regulations. shtm>

In: Coastal Sensitivity to Sea Level Rise-Focusing on the Mid-… ISBN: 978-1-60741-440-7
Editor: Melvin C. Urajner

Chapter 9

MANAGEMENT ISSUES

Lead Authors: Stephen K. Gill and Doug Marcy
NOAA
Contributing Author: Zoe Johnson
Maryland Department of Natural Resources

Key Findings

- Rising sea level increases the vulnerability of coastal areas to flooding. The higher sea level provides a higher base for storm surges to build upon. It also diminishes the rate at which low-lying areas drain, thereby increasing the risk of flooding from rainstorms. Increased shore erosion can further increase flood damages by removing protective dunes, beaches, and wetlands, thus leaving previously protected properties closer to the water's edge. In addition to flood damages, many other effects, responses, and decisions are likely to occur during or in the immediate aftermath of severe storms. Beach erosion and wetlands loss often occur during storms, and the rebuilding phase after a severe storm often presents the best opportunity for developed areas to adapt to future sea-level rise.
- Coastal storms could have higher flooding potential in the future due to higher sea levels relative to the land.
- The most recent Federal Emergency Management Agency (FEMA) study on the potential effects of sea-level rise on the Nation's flood insurance program was published in 1991. Because of the uncertainties in the projections of potential changes in sea level at the time and the ability of the rating system to respond easily to a 0.3 meter rise in sea level, the 1991 FEMA study (FEMA, 1991) concluded that no immediate program changes were needed.
- The mid-Atlantic coastal zone management community is increasingly recognizing that sea-level rise is a high-risk coastal hazard as evidenced by the recent comprehensive analyses and studies needed to make recommendations for state policy formulation performed by Maryland.

9.1. Introduction

This Chapter examines the effects of sea-level rise on coastal floodplains and on coastal flooding management issues confronting the U.S. Federal Emergency Management Agency (FEMA), the floodplain management community, the coastal zone management community, coastal resource managers, and the public, including private industry. Sea- level rise is just one of numerous complex scientific and societal issues these groups face. There is also uncertainty in the local rate of sea-level change, which needs to be taken into account along with the interplay with extreme storm events (see Chapter 1). In addition, impacts of increased flooding frequency and extent on coastal areas can be significant for marine ecosystem health and human health in those areas (Boesch *et al.*, 2000). This Chapter provides a discussion of the current state of knowledge and provides assessments for a range of actions being taken by many state and federal agencies and other groups related to coastal flooding.

9.2. Physical Characteristics

9.2.1. Floodplain

In general, a floodplain is any normally dry land surrounding a natural water body that holds the overflow of water during a flood. Because they border water bodies, floodplains have been popular sites to establish settlements, which subsequently become susceptible to flood-related disasters. Most management and regulatory definitions of floodplains apply to rivers; however, open-coast floodplains characterized by beach, dunes, and shrub-forest are also important since much of the problematic development and infrastructure is concentrated in these areas (see Chapter 3 for a detailed description of this environment).

The federal regulations governing FEMA (2008) via Title 44 of the Code of Federal Regulations defines floodplains as "any land area susceptible to being inundated by flood waters from any source". The FEMA (2002) *Guidelines and Specifications for Flood Hazard Mapping Partners Glossary of Terms* defines floodplains as:

20. A flat tract of land bordering a river, mainly in its lower reaches, and consisting of alluvium deposited by the river. It is formed by the sweeping of the meander belts downstream, thus widening the valley, the sides of which may become some kilometers apart. In time of flood, when the river overflows its banks, sediment is deposited along the valley banks and plains.
21. Synonymous with the 100-year floodplain, which is defined as the land area susceptible to being inundated by stream derived waters with a 1-percent-annual-chance of being equaled or exceeded in a given year.

The National Oceanic and Atmospheric Administration (NOAA) National Weather Service (NWS) defines a floodplain as the portion of a river valley that has been inundated by the river during historic floods. None of these formal definitions of floodplains include the word "coastal". However, as river systems approach coastal regions, river base levels approach sea level, and the rivers become influenced not only by stream flow, but also by coastal processes such as tides, waves, and storm surges. In the United States, this complex interaction takes place near the governing water body, either open ocean, estuaries, or the Great Lakes.

The slope and width of the coastal plain determines the size and inland extent of coastal influences on river systems. Coastal regions are periodically inundated by tides, and frequently inundated by high waves and storm surges. Therefore, a good working definition of a coastal floodplain, borrowing from the general river floodplain definition, is any normally dry land area in coastal regions that is susceptible to being inundated by water from any natural source, including oceans (*e.g.*, tsunami runup, coastal storm surge, relative sea-level rise), rivers, streams, and lakes.

Floodplains generally contain unconsolidated sediments, often extending below the bed of the stream or river. These accumulations of sand, gravel, loam, silt, or clay are often important aquifers; the water drawn from them is prefiltered compared to the water in the river or stream. Geologically ancient floodplains are often revealed in the landscape by terrace deposits, which are old floodplain deposits that remain relatively high above the current floodplain and often indicate former courses of rivers and streams.

Floodplains can support particularly rich ecosystems, both in quantity and diversity. These regions are called riparian zones or systems. Wetting of the floodplain soil releases an immediate surge of nutrients, both those left over from the last flood and those from the rapid decomposition of organic matter that accumulated since the last flood. Microscopic organisms thrive and larger species enter a rapid breeding cycle. Opportunistic feeders (particularly birds) move in to take advantage of these abundant populations. The production of nutrients peaks and then declines quickly; however, the surge of new growth endures for some time, thus making floodplains particularly valuable for agriculture. Markedly different species grow within floodplains compared to surrounding regions. For instance, certain riparian trees species (that grow in floodplains near river banks) tend to be very tolerant of root disturbance and thus tend to grow quickly, compared to different tree species growing in a floodplain some distance from a river.

9.3. Potential Impacts of Sea-Level Rise on Coastal Floodplains

Assessing the impacts of sea-level rise on coastal floodplains is a complicated task, because those impacts are coupled with impacts of climate change on other coastal and riverine processes and can be offset by human actions to protect life and property. Impacts may range from extended periods of drought and lack of sediments to extended periods of above-normal freshwater runoff and associated sediment loading. Some seasons may have higher than normal frequency and intensity of coastal storms and flooding events. Impacts will also depend on construction and maintenance of dikes, levees, waterways, and diversions for flood management.

With no human intervention, the hydrologic and hydraulic characteristics of coastal and river floodplain interactions will change with sea-level rise. Fundamentally, the floodplains will become increasingly vulnerable to inundation. In tidal areas, the tidal inundation characteristics of the floodplain may change with the range of tide and associated tidal currents increasing with sea-level rise. With this inundation, floodplains will be vulnerable to increased coastal erosion from waves, river and tidal currents, storm-induced flooding, and tidal flooding. Upland floodplain boundaries will be vulnerable to horizontal movement. Coastal marshes could be vulnerable to vertical buildup or inundation (see Chapter 4 for further discussion).

In a study for the state of Maine (Slovinsky and Dickson, 2006), the impacts of sea-level rise on coastal floodplains were characterized by marsh habitat changes and flooding implications. The coast of Maine has a significant spring tidal range of 2.6 to 6.7 meters (m) (8.6 to 22.0 feet [ft]), such that impacts of flooding are coupled with the timing of storms and the highest astronomical tides on top of sea-level rise. The study found that there was increasing susceptibility to inlet and barrier island breaches where existing breach areas were historically found, increased stress on existing flood-prevention infrastructure (levees, dikes, roads), and a gradual incursion of low marsh into high marsh with development of a steeper bank topography. On the outer coast, impacts included increased overwash and erosion.

In addition, the effects of significant local or regional subsidence of the land will add to the effects of sea-level rise on coastal floodplains. Regional areas with significant subsidence include the Mississippi River Delta region (AGU, 2006), the area around the entrance to the Chesapeake Bay (Poag, 1997), and local areas such as the Blackwater National Wildlife Refuge on the Eastern Shore of Maryland (Larsen *et al.*, 2004).

9.4. Potential Effects of Sea-Level Rise on the Impacts of Coastal Storms

The potential interaction among increased sea levels, storm surges, and upstream rivers is complex. The storm surge of any individual storm is a function of storm intensity defined by storm strength and structure, forward speed, landfall location, angle of approach, and local bathymetry and topography. However, the absolute elevation of the maximum water levels observed relative to the land during a storm (operationally defined as storm tides) are a combination of the storm surge defined above, plus the non-storm-related background water level elevations due to the stage of tide, the time of year (sea level varies seasonally), river flow, local shelf circulation patterns (such as the Gulf Loop Current/eddies and the El Niño-Southern Oscillation [especially on the west coast]). Storm surge "rides" on top of these other variations, including sea level rise (NOAA, 2008). Storm surge can travel several hundred kilometers up rivers at more than 40 kilometers (km) (25 miles [mi]) per hour, as on the Mississippi River, where storm surge generated by land-falling hurricanes in the Gulf of Mexico can be detected on stream gauges upstream of Baton Rouge, Louisiana, more than 480 km (300 mi) from the mouth of the river (Reed and Stucky, 2005).

Both NWS (for flood forecasting) and FEMA (for insurance purposes and land use planning) recognize the complexity of the interactions among sea-level rise, storm surge, and river flooding. For instance, NWS uses both a hurricane storm surge model (the Sea, Lakes, and Overland Surge from Hurricanes [SLOSH] model, Jelesnianski *et al.*, 1992) and a riverine hydraulic model (the Operational Dynamic Wave Model) to forecast effects of storm surge on river stages on the Mississippi River. The two models are coupled such that the output of the storm surge model is used as the downstream boundary of the river model. This type of model coupling is needed to determine the effects of sea-level rise and storm surge on riverine systems. Other modeling efforts are starting to take into account river and coastal physical process interactions, such as use of the two-dimensional hydrodynamic model (the Advanced Circulation Model or ADCIRC; Luettich *et al.*, 1992) on the Wacammaw River in South Carolina to predict effects of storm surge on river stages as far inland as Conway, 80 km (50 mi) from the Atlantic Ocean (Hagen *et al.*, 2004). These model coupling routines are becoming increasingly more common and have been identified as future research needs by such agencies as NOAA and the U.S. Geological Survey (USGS), as scientists strive to model the complex interactions between coastal and riverine processes. As sea level rises, these interactions will become ever more important to the way the coastal and riverine floodplains respond (Pietrafesa *et al.*, 2006).

9.4.1. Historical Comparison at Tide Stations

There is the potential for higher elevations of coastal flooding from coastal storms over time as sea level rises relative to the land. Looking at storms in historical context and accounting for sea level change is one way to estimate maximum potential storm water levels. For example, this assessment can be made by analyzing the historical record of flooding elevations observed at NOAA tide stations in the Chesapeake Bay. The following analysis compares the elevation of the storm tides for a particular storm at a particular tide station; that is from when it occurred historically to as if the same exact storm occurred today under the exact same conditions, but adjusted for relative sea level rise at that station. These comparisons are enabled because NOAA carefully tabulates water level elevations over time relative to a common reference datum that is connected to the local land elevations at each tide station. From this, relative sea level trends can be determined and maximum water level elevations recorded during coastal storms can be directly compared over the time period of record (Zervas, 2001). The relative sea level trend provides the numerical adjustment needed depending on the date of each storm.

Table 9.1. Five highest water levels for Baltimore, Maryland in meters above mean higher high water. Ranked first by absolute elevation and then ranked again after adjustment for sea level rise.

Absolute water level			Corrected for sea-level rise to 2003		
Event	**Date**	**Elevation (m)**	**Event**	**Date**	**Elevation (m)**
Hurricane Isabel	Sep 2003	1.98	Hurricane	Aug 1933	2.06
Hurricane	Aug 1933	1.84	Hurricane Isabel	Sep 2003	1.98
Hurricane Connie	Aug 1955	1.44	Hurricane Connie	Aug 1955	1.59
Hurricane Hazel	Oct 1954	1.17	Hurricane	Aug 1915	1.38
Hurricane	Aug 1915	1.11	Hur. Hazel	Oct 1954	1.32

Table 9.2. Five highest water levels for Annapolis, Maryland in meters above mean higher high water. Ranked first by absolute elevation and then ranked again after adjustment for sea level rise.

Absolute water level			Corrected for sea-level rise to 2003		
Event	Date	Elevation (m)	Event	Date	Elevation (m)
Hurricane Isabel	Sep 2003	1.76	Hurricane Isabel	Sep 2003	1.76
Hurricane	Aug 1933	1.45	Hurricane	Aug 1933	1.69
Hurricane Connie	Aug 1955	1.08	Hurricane Connie	Aug 1955	1.25
Hurricane Fran	Sep 1996	1.04	Hurricane Hazel	Oct 1954	1.19
Hurricane Hazel	Oct 1954	1.02	Hurricane Fran	Sep 1996	1.06

The NOAA post-hurricane report (Hovis, 2004) on the observed storm tides of Hurricane Isabel assessed the potential effects of sea-level rise on maximum observed storm tides for four long-term tide stations in the Chesapeake Bay. Prior to Hurricane Isabel, the highest water levels reached at the NOAA tide stations at Baltimore, Maryland; Annapolis, Maryland; Washington, D.C.; and Sewells Point, Virginia occurred during the passage of an unnamed hurricane in August, 1933. At the Washington, D.C. station, the 1933 hurricane caused the third highest recorded water level, surpassed only by river floods in October 1942 and March 1936. Hurricane Isabel caused water levels to exceed the August 1933 levels at Baltimore, Annapolis and Washington, D.C. by 0.14, 0.31, and 0.06 meters (m), respectively. At Sewells Point, the highest water level from Hurricane Isabel was only 0.04 m below the level reached in August 1933. Zervas (2001) calculated sea-level rise trends for Baltimore, Annapolis, Washington, and Sewells Point of 3.12, 3.53, 3.13, and 4.42 millimeters (mm) per year, respectively. Using these rates, the time series of monthly highest water level were adjusted for the subsequent sea-level rise up to the year 2003. The resulting time series, summarized in Tables 9.1, 9.2, 9.3, and 9.4, indicate the highest level reached by each storm as if it had taken place in 2003 under the same conditions, thus allowing an unbiased comparison of storms. The purpose of Tables 9.1 through 9.4 is to show that the relative ranking of the flooding elevations from particular storm events changes at any given station once the adjustment for sea level trend is taken into account. The 1933 hurricane, especially, moves up in ranking at Baltimore and Washington, DC once adjusted for the local sea level trend. Hurricane Hazel moved up in ranking at Annapolis. If the 1933 hurricane occurred today under the same conditions, it would have had the highest water level of record at Baltimore, not Hurricane Isabel. Elevations are relative to the tidal datum of mean higher high water (MHHW). Noting the earlier discussion in this section on the operational difference between storm surge and the actual observed storm tide elevation, the tables suggest that, while not affecting intensity of storms and the resulting amplitude of storm surges, sea-level rise could increasingly add to the potential maximum water level elevations observed relative to the land during coastal storms.

9.4.2. Typical 100-Year Storm Surge Elevations Relative to Mean Higher High Water within the Mid-Atlantic Region

A useful application of long-term tide gauge data is a return frequency analysis of the monthly and annual highest and lowest observed water levels. This type of analysis provides information on how often extreme water levels can be expected to occur (*e.g.*, once every 100 years, once every 50 years, once every 10 years?) On the East Coast and in the Gulf of Mexico, hurricanes and winter storms interact with the wide, shallow, continental shelf to produce large extreme storm tides. A generalized extreme value distribution can be derived for each station after correcting the values for the long-term sea-level trend (Zervas, 2005). Theoretical exceedance probability statistics give the 99-percent, 50-percent, 10-percent, and 1-percent annual exceedance probability levels. These levels correspond to average storm tide return periods of 1, 2, 10, and 100 years. The generalized extreme value analyses are run on the historical data from each tide station. Interpolating exceedance probability results away from the tide station location is not recommended as elevations of tidal datums and the extremes are highly localized. Figures 9.1 and 9.2 show the variations in these statistics along the mid-Atlantic coast. Figure 9.1 shows exceedance elevations above local mean sea level (LMSL) at mid-Atlantic stations relative to the 1983 to 2001 National Tidal Datum Epoch (NTDE).

Figure 9.2 shows the same exceedance elevations, except the elevations are relative to mean higher high water (MHHW) computed for the same 1983 to 2001 NTDE.

In Figure 9.1, the elevations relative to LMSL are highly correlated with the range of tide at each station (Willets Point, New York has a very high range of tide, 2.2 m), except for the 1-percent level at Washington D.C., which is susceptible to high flows of the Potomac River. Due to their varying locations, the 1-percent elevation level varies the most among the stations. Figure 9.2 shows a slightly geographically decreasing trend in the elevations from north to south.

Examining the effects of sea-level rise on the highest water level during a hurricane or coastal storm does not provide a complete picture because the impacts of sea-level rise on the duration of the inundation can be as important as the maximum height. Sea-level rise, coupled with any increased frequency of extra-tropical storms (nor'easters), may also increase the durations of inundation from extra-tropical storms (NOAA, 1992). For instance, some of the most severe impacts of nor'easters are generally felt in bays where water can get in but not out for several days as the storms slowly transit parallel to the coast.

Table 9.3. Five highest water levels for Washington, D.C. in meters above mean higher high water. Ranked first by absolute elevation and then ranked again after adjustment for sea level rise.

Absolute water level			Corrected for sea-level rise to 2003		
Event	**Date**	**Elevation (m)**	**Event**	**Date**	**Elevation (m)**
Flood	Oct 1942	2.40	Flood	Oct 1942	2.59
Flood	Mar 1936	2.25	Flood	Mar 1936	2.46
Hurricane Isabel	Sep 2003	2.19	Hurricane	Aug 1933	2.35
Hurricane	Aug 1933	2.13	Hurricane Isabel	Sep 2003	2.19
Flood	Apr 1937	1.70	Flood	Apr 1937	1.91

Table 9.4. Five highest water levels for Sewells Point, Virginia in meters above mean higher high water. Ranked first by absolute elevation and then ranked again after adjustment for sea level rise.

Absolute water level			Corrected for sea-level rise to 2003		
Event	Date	Elevation (m)	Event	Date	Elevation (m)
Hurricane	Aug 1933	1.60	Hurricane	Aug 1933	1.91
Hurricane Isabel	Sep 2003	1.56	Hurricane Isabel	Sep 2003	1.56
Winter Storm	Mar 1962	1.36	Winter Storm	Mar 1962	1.54
Hurricane	Sep 1936	1.21	Hurricane	Sep 1936	1.50
Winter Storm	Feb 1998	1.16	Hurricane	Sep 1933	1.33

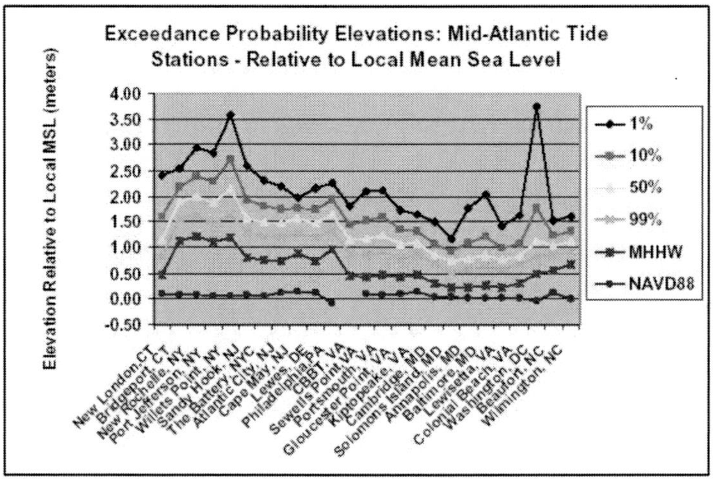

Figure 9.1. Exceedance probabilities for mid-Atlantic tide stations relative to local mean sea level.

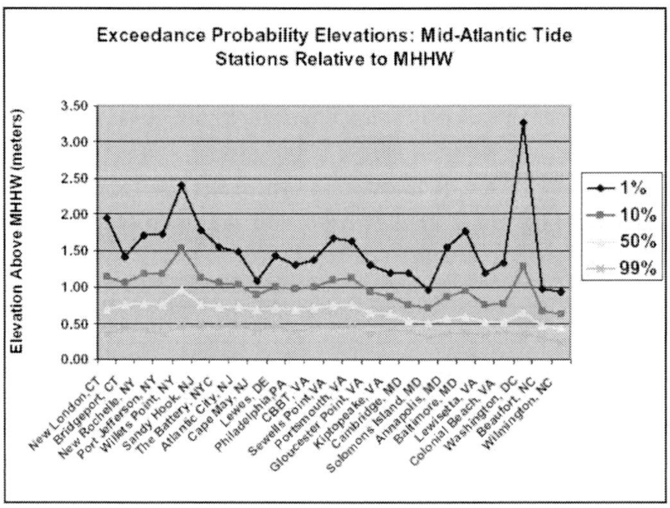

Figure 9.2. Exceedance probabilities at mid-Atlantic tide stations relative to mean higher high water.

Other federal agencies, such as NOAA, have been sponsoring applied research programs to bring an integrated approach to understanding the effects of sea-level rise into operations. One such study on the ecological effects of sea-level rise is discussed in Box 9.1 (NOAA, 2007), which is due to come out with a final report in 2009.

9.5. Floodplain Mapping and Sea-Level Rise

A nationwide study was performed by FEMA (1991) (see Box 9.2) in which costs for remapping floodplains were estimated at $150,000 per county (in 1991 dollars) or $1,500 per map panel (the standard map presentation used by FEMA). With an estimated 283 counties (5,050 map panels) potentially in need of remapping, the total cost of restudies and remapping was estimated at $30 million (in 1991). Based on this study and assuming that the maps are revised on a regular basis, such an undertaking today would cost about $46.5 million. The 1991 study concluded that "there are no immediate program changes needed" (FEMA, 1991).

At present, FEMA periodically revises Flood Insurance Rate Maps (FIRMs) to reflect new engineering, scientific, and imagery data. In addition, under their Map Modernization and post-Map Modernization Programs, FEMA intends to assess the integrity of the flood hazard data by reviewing the flood map inventory every five years. Where the review indicates the flood data integrity has degraded the flood maps (due to outdated data and known changes in hydrology and floodplain elevation since the last maps were issued), updates will be provided or new studies will be performed. Whenever an update or remap of coastal areas is made, changes that had occurred in the interim due to sea-level rise will be accounted for. An upcoming Impact of Climate Change on the National Flood Insurance Program study (scheduled to begin at the end of fiscal year 2008 and last 1.5 years) may come up with different conclusions than the 1991 study and cause FEMA to rethink the issue.

The primary floodplain management adjustment for sea-level rise is the local increase in required base flood elevation (BFE) for new construction. Elevating a building's lowest floor above predicted flood elevations by a small additional height, generally 0.3 to 0.9 meters above National Flood Insurance Program (NFIP) minimum height requirements, is termed a freeboard addition. Freeboard additions are generally justified for other more immediate purposes including the lack of safety factor in the 1-percent flood and uncertainties in prediction and modeling. FEMA encourages freeboard adoptions through the Community Rating System, which offers community-wide flood insurance premium discounts for higher local standards and for individuals through premium discounts for higher than minimum elevation on higher risk buildings. Velocity flood zones, known as V Zones or coastal high hazard areas, have been identified by FEMA as areas "where wave action and/or high velocity water can cause structural damage in the 100-year flood", a flood with a 1 percent chance of occurring or being exceeded in a given year. FEMA also defines A Zones as areas inundated in a 100-year storm event that experience conditions of less severity, for example, wave heights less than 1 m, than conditions experienced in V Zones. Accurate determination of the spatial extent of these zones is vital to understanding the level of risk for a particular property or activity.

A recent historical overview of FEMA's Coastal Risk Assessment process is found in Crowell *et al.* (2007), and includes overviews of the FEMA Map Modernization Program, revised coastal guidelines, and FEMA's response to recommendations of a Heinz Center report, *Evaluation of Erosion Hazards* (Heinz Center, 2000).

9.6. Studies of Future Coastal Conditions and Floodplain Mapping

9.6.1. FEMA Coastal Studies

Currently, communities can opt to use future conditions (projected) hydrology for mapping according to FEMA rules established in December 2001[16]. Showing future conditions flood boundaries has been provided at the request of some communities in Flood Map Modernization, but it is not a routine product. As outlined in those rules, showing a future condition boundary in addition to the other boundaries normally shown on a FIRM is acceptable. FEMA shows future condition boundaries for informational purposes only and carries with it no additional requirements for floodplain management. Insurance would not be rated using a future condition boundary. The benefits showing future condition flood boundaries relate to the fact that future increases in flood risk can lead to significant increases in both calculated and experienced flood heights, resulting in serious flood losses (structural damage and economic) as well as loss of levee certification and loss of flood protection for compliant post-FIRM structures. Providing this information to communities may lead to coordinated watershed-wide actions to manage for, or otherwise mitigate, these future risks.

A recent increase in losses from coastal storms has been recognized by FEMA (Crowell, 2008). In 2005, Hurricane Katrina clearly illustrated this, reporting the most losses of any U.S. natural disaster to date. This fact, coupled with the facts that new developments in modeling and mapping technology have allowed for more accurate flood hazard assessment over the past few years and that populations at risk are growing in coastal areas, has caused FEMA to develop a new national coastal strategy. This strategy consists of assessing coastal Flood Insurance Studies on a national scale and developing a nationwide plan for improved coastal flood hazard identification. The assessment will prioritize regional studies, look at funding allocations, and develop timelines for coastal study updates.

River models that are affected by tides and storm surge require the downstream boundary starting water surface elevation to be the "1-percent-annual-chance" base flood elevation (BFE) from an adjacent coastal study. If the coastal study BFE is raised by 0.3 m or even 0.9 m because of sea-level rise, the river study flood profile will be changed as well and this will ultimately affect the resulting FIRMs that are published. This is a complicated issue and points out the fact that simply raising the coastal BFEs to estimate a new 1-percent-annual-chance floodplain is not taking into account the more complex hydraulics that will have undetermined effects on the upstream 1-percent-annual-chance floodplains as well. The 1991 study does not factor in the complexity of different tidal regimes that would be occurring because of an increased sea level and how those regimes would affect the geomorphology of the floodplains. This is because FEMA is restricted in what it can and cannot do in the regulated NFIP process (Crowell, 2008).

Maryland has completed a comprehensive state strategy document in response to sea-level rise (MD DNR, 2000). The Maryland Department of Natural Resources (MD DNR, 2000) requires all communities to adopt standards that call for all structures in the non-tidal floodplain to be elevated 0.3 m (1 ft) above the 100-year floodplain elevation, and all coastal counties except Worcester, Somerset, and Dorchester (the three most vulnerable to exacerbated flooding due to sea-level rise) have adopted the 1-ft freeboard standard. Although 1 foot of freeboard provides an added cushion of protection to guard against uncertainty in floodplain projections, it may not be enough in the event of 0.6 to 0.9 m (2 to 3 ft) of sea-level rise, as MD DNR (2000) points out.

Crowell *et al.* (2007) identified a need for a tide-gauge analysis for FEMA Region III, which encompasses the mid-Atlantic states, similar to new studies being done currently on Chesapeake Bay by the state of Maryland. Each coastal FEMA region has been evaluated and new guidelines and specifications have been developed by FEMA for future coastal restudies, the first of which was for the Pacific Coast region. These guidelines outline new coastal storm surge modeling and mapping procedures and allow for new flooding and wave models to be used for generating coastal BFEs.

To aid in ongoing recovery and rebuilding efforts, FEMA initiated short-term projects in 2004 and 2005 to produce coastal flood recovery maps for areas that were most severely affected by Hurricanes Ivan, Katrina, and Rita. The Katrina maps, for example, show high water marks surveyed after the storm, an inundation limit developed from these surveyed points, and FEMA's Advisory Base Flood Elevations (ABFEs) and estimated zone of wave impacts.

These maps and associated ABFEs (generated for Katrina and Rita only) were based on new flood risk assessments that were done immediately following the storms to assist communities with rebuilding. The recovery maps provide a graphical depiction of ABFEs and coastal inundation associated with the observed storm surge high water mark values, in effect documenting the flood imprint of the event to be used in future studies and policy decisions. Adherence to the ABFEs following Katrina affected eligibility for certain FEMA-funded mitigation and recovery projects. They were used until the Flood Insurance Studies (FIS) were updated for the Gulf region and are available as advisory information to assist communities in rebuilding efforts.

FEMA cannot require the use of future conditions data based on planned land-use changes or proposed development for floodplain management or insurance rating purposes unless statutory and regulatory changes to the NFIP are made. In addition, using projected coastal erosion information for land-use management and insurance rating purposes through the NFIP would require a legislative mandate and regulatory changes.

9.6.2. Mapping Potential Impacts of Sea-Level Rise on Coastal Floodplains

Floodplain management regulations are intended to minimize damage as a result of flooding disasters, in conjunction with other local land-use requirements and building codes. Meeting only these minimum requirements will not guarantee protection from storm damages. Management activities that focus on mitigating a single, short-term hazard can result in structures that are built only to withstand the hazards as they are identified today, with no easy way to accommodate an increased risk of damage in the coming decades (Honeycutt and Mauriello, 2005). The concept of going above and beyond current regulations to provide additional hazards information other than BFEs and the 1-percent-annual-chance flood (coastal erosion and storm surge inundation potential) has been advocated in some quarters with a No Adverse Impact (NAI) program (Larson and Plasencia, 2002). A NAI toolkit was developed that outlines a strategy for communities to implement a NAI approach to floodplain management (ASFPM, 2003, 2008).

The International Codes (FEMA, 2005) include freeboard (elevations above the BFE) and standards for coastal A Zones that are more stringent than the NFIP criteria. The International Codes also incorporate criteria from the national consensus document ASCE 24-05 *Flood Resistant Design and Construction Standard* (ASCE, 2006).

9.7. How Coastal Resource Managers Cope With Sea-Level Rise and Issues They Face

9.7.1. Studies by the Association of State Floodplain Managers

The Association of State Floodplain Mangers (ASFPM) recently completed a study that contains a broad spectrum of recommendations for improving the management of U.S. floodplains (ASFPM, 2007). In their study, ASFPM noted that changing climate was one of the major challenges for the significant changes in social, environmental, and political realities and their impact on floodplain management, and highlights the wide spread implications for flood protection.

9.7.2 The Response through Floodproofing

The U.S. Army Corps of Engineers heads the national floodproofing committee, established through the USACE's floodplain management services program, to promote the development and use of proper floodproofing techniques throughout the United States (USACE, 1996). The USACE publication on floodproofing techniques, programs, and references gives an excellent overview of currently accepted flood mitigation practices from an individual structure perspective.

Mitigating flooding or "floodproofing" is a process for preventing or reducing flood damages to structures and/or to the contents of buildings located in flood hazard areas. It mainly involves altering or changing existing properties; however, it can also be incorporated into the design and construction of new buildings. There are three general approaches to floodproofing:

1. *Raising or moving the structure.* Raising or moving the structure such that floodwaters cannot reach damageable portions of it is an effective floodproofing approach.

2. *Constructing barriers to stop floodwater from entering the building.* Constructing barriers can be an effective approach used to stop floodwaters from reaching the damageable portions of structures. There are two techniques employed in constructing barriers. The first technique involves constructing free-standing barriers that are not attached to the structure. The three primary types of freestanding barriers used to reduce flood damages are berms, levees, or floodwalls. The second technique that can be used to construct a barrier against floodwaters is known as "dry floodproofing". With this technique, a building is sealed such that floodwaters cannot get inside.

3. *Wet Floodproofing*. This approach to floodproofing involves modifying a structure to allow floodwaters inside, but ensuring that there is minimal damage to the building's structure and to its contents. Wet floodproofing is often used when dry floodproofing is not possible or is too costly. Wet floodproofing is generally appropriate in cases where an area is available above flood levels to which damageable items can be relocated or temporarily stored.

The recommended techniques of levees, berms, floodwalls and wet floodproofing are not allowed under the NFIP to protect new individual structures. These techniques may also have limited use in protecting older existing structures in coastal areas. Although dry floodproofing is allowed in A Zones (not V Zones), FEMA does not generally recommend its use for new non-residential structures in the coastal A Zones due to the potential flood forces. Under the NFIP, all new construction and substantial improvements of residential buildings in A Zones must have the lowest floor elevated to or above the BFE. All new construction and substantial improvement of non-residential buildings in A Zones must have either the lowest floor elevated to or above the BFE or the building must be dry floodproofed to the BFE. In V Zones, all new construction and substantial improvements must have the bottom of the lowest horizontal structural member of the lowest floor elevated to or above the BFE on a pile or column foundation. Although the NFIP allows dry floodproofing in coastal A Zone areas, FEMA does not recommend its use in the coastal A Zone because of the potential for severe flood hazards. While Base Flood Elevations in coastal A Zones contain a wave height of less than 3 feet, the severity of the hazard in coastal A Zones is often much greater than in non-coastal A Zones due to the combination of water velocity, wave action, and debris impacts that can occur in these areas. For existing, older structures in the coastal area, the best way to protect the structure is elevating or relocating the structure.

9.7.3. Coastal Zone Management Act

Dramatic population growth along the coast brings new challenges to managing national coastal resources. Coastal and floodplain managers are challenged to strike the right balance between a naturally changing shoreline and the growing population's desire to use and develop coastal areas. Challenges include protecting life and property from coastal hazards; protecting coastal wetlands and habitats while accommodating needed economic growth; and settling conflicts between competing needs such as dredged material disposal, commercial development, recreational use, national defense, and port development. Coastal land loss caused by chronic erosion has been an ongoing management issue in many coastal states that have Coastal Zone Management (CZM) programs and legislation to mitigate erosion using a basic retreat policy. With the potential impacts of sea-level rise, managers and lawmakers must now decide how or whether to adapt their current suite of tools and regulations to face the prospect of an even greater amount of land loss in the decades to come.

The U.S. Congress recognized the importance of meeting the challenge of continued growth in the coastal zone and responded by passing the Coastal Zone Management Act in 1972. The amended act (CZMA, 1996), administered by NOAA, provides for management of U.S. coastal resources, including the Great Lakes, and balances economic development with environmental conservation.

As a voluntary federal–state partnership, the CZMA is designed to encourage state-tailored coastal management programs. It outlines two national programs, the National Coastal Zone Management Program and the National Estuarine Research Reserve System, and aims to balance competing land and water issues in the coastal zone, while estuarine reserves serve as field laboratories to provide a greater understanding of estuaries and how humans impact them. The overall program objectives of CZMA remain balanced to "preserve, protect, develop, and where possible, to restore or enhance the resources of the nation's coastal zone" (CZMA, 1996).

9.7.4. The Coastal Zone Management Act and Sea-Level Rise Issues

The CZMA language (CZMA, 1996) refers specifically to sea-level rise issues (16 U.S.C. § 1451). Congressional findings (§ 302) calls for coastal states to anticipate and plan for sea-level rise and climate change impacts.

In 16 U.S.C. § 1452, Congressional declaration of policy (§ 303), the Congress finds and declares that it is the national policy to manage coastal development to minimize the loss of life and property caused by improper development in flood-prone, storm surge, geological hazard, and erosion-prone areas, and in areas likely to be affected by or vulnerable to sea-level rise, land subsidence, and saltwater intrusion, and by the destruction of natural protective features such as beaches, dunes, wetlands, and barrier islands; to study and develop plans for addressing the adverse effects upon the coastal zone of land subsidence and of sea-level rise; and to encourage the preparation of special area management plans which provide increased specificity in protecting significant natural resources, reasonable coastal-dependent economic growth, improved protection of life and property in hazardous areas, including those areas likely to be affected by land subsidence, sea-level rise, or fluctuating water levels of the Great Lakes, and improved predictability in governmental decision-making.

9.7.5 The Coastal Zone Enhancement Program

The reauthorization of CZMA in 1996 by the U.S. Congress led to the establishment of the Coastal Zone Enhancement Program (CZMA §309), which allows states to request additional funding to amend their coastal programs in order to support attainment of one or more coastal zone enhancement objectives. The program is designed to encourage states and territories to develop program changes in one or more of the following nine coastal zone enhancement areas of national significance: wetlands, coastal hazards, public access, marine debris, cumulative and secondary impacts, special area management plans, ocean/Great Lakes resources, energy and government facility citing, and aquaculture. The Coastal Zone Enhancement Grants (§ 309) defines a "Coastal zone enhancement objective" as "preventing or significantly reducing threats to life and destruction of property by eliminating development and redevelopment in high-hazard areas, managing development in other hazard areas, and anticipating and managing the effects of potential sea-level rise and Great Lakes level rise".

Through a self-assessment process, state coastal programs identify high-priority enhancement areas. In consultation with NOAA, state coastal programs then develop five-year strategies to achieve changes (enhancements) to their coastal management programs within these high-priority areas. Program changes often include developing or revising a law, regulation or administrative guideline, developing or revising a special area management plan, or creating a new program such as a coastal land acquisition or restoration program.

For coastal hazards, states base their evaluation on the following criteria:

1. What is the general level or risk from specific coastal hazards (*i.e.*, hurricanes, storm surge, flooding, shoreline erosion, sea-level rise, Great Lakes level fluctuations, subsidence, and geological hazards) and risk to life and property due to inappropriate development in the state?
2. Have there been significant changes to the state's hazards protection programs (*e.g.*, changes to building setbacks/restrictions, methodologies for determining building setbacks, restriction of hard shoreline protection structures, beach/dune protection, inlet management plans, local hazard mitigation planning, or local post-disaster redevelopment plans, mapping/GIS/tracking of hazard areas)?
3. Does the state need to direct future public and private development and redevelopment away from hazardous areas, including the high hazard areas delineated as FEMA V Zones and areas vulnerable to inundation from sea- and Great Lakes level rise?
4. Does the state need to preserve and restore the protective functions of natural shoreline features such as beaches, dunes, and wetlands?
5. Does the state need to prevent or minimize threats to existing populations and property from both episodic and chronic coastal hazards?

Section 309 grants have benefited states such as Virginia in developing local conservation corridors that identify and prioritize habitat areas for conservation and restoration; and New Jersey for supporting new requirements for permittees to submit easements for land dedicated to public access, when such access is required as a development permit condition and is supporting a series of workshops on the Public Trust Doctrine and ways to enhance public access (see <http://coastalmanagement.noaa.gov/nationalsummary.html>).

9.7.6. Coastal States Strategies

Organizations such as the Coastal States Organization have recently become more proactive in how coastal zone management programs consider adaptation to climate change, including sea-level rise (Coastal States Organization, 2007) and are actively leveraging each other's experiences and approaches as to how best obtain baseline elevation information and inundation maps, how to assess impacts of sea-level rise on social and economic resources and coastal habitats, and how to develop public policy. There have also been several individual state-wide studies on the impact of sea-level rise on local state coastal zones (*e.g.*, Johnson [2000] for Maryland; Cooper *et al.* [2005] for New Jersey). Many state coastal management websites show an active public education program with regards to providing information on impacts of sea-level rise:

New Jersey: <http://www.nj .gov/dep/njgs/enviroed/infocirc/sealevel.pdf>Delaware: <http://www.dnrec.delaware.gov/Climate+change+shoreline+erosion>

Maryland: <http://www.dnr. state.md.us/Bay/czm/sea_level_rise.html>

9.7.6.1. Maryland's Strategy

The evaluation of sea-level rise response planning in Maryland and the resulting strategy document constituted the bulk of the state's CZMA §309 *Coastal Hazard Assessment and Strategy for 2000–2005* and in the 2006-20 10 Assessment and Strategy (MD DNR, 2006). Other mid-Atlantic states mention sea-level rise as a concern in their assessments, but have not yet developed a comprehensive strategy.

The sea-level rise strategy is designed to achieve the desired outcome within a five-year time horizon. Implementation of the strategy is evolving over time and is crucial to Maryland's ability to achieve sustainable management of its coastal zone. The strategy states that planners and legislators should realize that the implementation of measures to mitigate impacts associated with erosion, flooding, and wetland inundation will also enhance Maryland's ability to protect coastal resources and communities whether sea level rises significantly or not.

Maryland has taken a proactive step towards addressing a growing problem by committing to implementation of this strategy and increasing awareness and consideration of sea-level rise issues in both public and governmental arenas. The strategy suggests that Maryland will achieve success in planning for sea-level rise by establishing effective response mechanisms at both the state and local levels. Sea-level rise response planning is crucial in order to ensure future survival of Maryland's diverse and invaluable coastal resources.

Since the release of Maryland's sea-level rise response strategy (Johnson, 2000), the state has continued to progressively plan for sea-level rise. The strategy is being used to guide Maryland's current sea-level rise research, data acquisition, and planning and policy development efforts at both the state and local level. Maryland set forth a design vision for "resilient coastal communities" in its *CZMA §309 Coastal Hazard Strategy for 2006– 2010* (MD DNR, 2006). The focus of the approach is to integrate the use of recently acquired sea-level rise data- and technology-based products into both state and local decision-making and planning processes. Maryland's coastal program is currently working with local governments and other state agencies to: (1) build the capacity to integrate data and mapping efforts into land-use and comprehensive planning efforts; (2) identify specific opportunities (*i.e.*, statutory changes, code changes, comprehensive plan amendments) for advancing sea-level rise at the local level; and (3) improve state and local agency coordination of sea-level rise planning and response activities (MD DNR, 2006).

In April 2007, Maryland's Governor, Martin O'Malley, signed an Executive Order establishing a Commission on Climate Change (Maryland, 2007) that is charged with advising both the Governor and Maryland's General Assembly on matters related to climate change and is charged with developing a Plan of Action that will address climate change on all fronts, including both the drivers and the consequences. The Maryland Commission on Climate Change released its Climate Action Plan in August 2008 (Maryland, 2008). A key component of the Action Plan is The Comprehensive Strategy to Reduce Maryland's Vulnerability to Climate Change. The Strategy, which builds upon Maryland's sea-level rise response strategy (Johnson, 2000), sets forth specific actions necessary to protect Maryland's people, property, natural resources, and public investments from the impacts of climate change, sea-level rise, and coastal storms. A comprehensive strategy and plan of action were presented to the Maryland's Governor and General Assembly in April 2008.

The Maryland Department of Natural Resources has been active in developing an online mapping tool for general information and educational purposes that provides user-driven maps for shoreline erosion and for various sea-level rise scenarios (see <http://shorelines.dnr.state.md.us/coastal_hazards>) and has completed case studies with other agencies (see Box 9.3) for studying implication of sea-level rise for county level planning. Although this particular case study did not base results on a numerical storm surge model, it represents the type of initial analyses that local planners need to undertake.

REFERENCES

AGU (American Geophysical Union), (2006). *Hurricanes and the U.S. Gulf Coast: Science and Sustainable Rebuilding*. American Geophysical Union, [Washington, DC], 29 pp. <http://www.agu.org/report/hurricanes>

ASCE (American Society of Civil Engineers), (2006). *Flood Resistant Design and Construction.* ASCE/SEI 24-05. American Society of Civil Engineers, Reston, VA, 61 pp.

ASFPM (Association of State Floodplain Managers), (2003). *No Adverse Impact. A Toolkit for Common Sense Floodplain Management.* [Larson, L.A., M.J. Klitzke, and D.A Brown eds.]. Association of State Floodplain Managers, Madison, WI, 108 pp.

ASFPM (Association of State Floodplain Managers), (2007). *National Flood Programs and Policies in Review – 2007.* Association of State Floodplain Managers, Madison, WI, 92 pp.

ASFPM (Association of State Floodplain Managers), (2008). *Coastal No Adverse Impact Handbook.* Association of State Floodplain Managers, Madison, WI, 165 pp. <http://www.floods.org/CNAI/CNAI_Handbook.asp>

Boesch, D. F., Field, J.C. & Scavia, D. (eds.) (2000). *The Potential Consequences of Climate Variability and Change on Coastal Areas and Marine Resources: Report of the Coastal Areas and Marine Resources Sector Team, U.S. National Assessment of the Potential Consequences of Climate Variability and Change, U.S. Global Change Research Program.* NOAA Coastal Ocean Program decision analysis series no. 21. NOAA Coastal Ocean Program, Silver Spring, MD, 163 pp.

Coastal States Organization, (2007). *The Role of Coastal Zone Management Programs in Adaptation to Climate Change.* CSO Climate Change Work Group, Washington, DC, 27 pp. <http://www.coastalstates.org/uploads/PDFs/CSO%20Climate%20Change%20Report%20-%20Final%20 10-1 -07.pdf>

Cooper, M. J. P., Beevers, M. D. & M. Oppenheimer, (2005). *Future Sea Level Rise And The New Jersey Coast: Assessing Potential Impacts and Opportunities.* Science, Technology and Environmental Policy Program, Woodrow Wilson School of Public and International Affairs, Princeton University, Princeton, NJ, 36 pp. <http://www.princeton.edu/~cmi/news/Future%20of%20Sea%20Level%20Rise% 20and%20the%20New%20Jersey%20Coast.pdf>

Crowell, M., Hirsch, E. & Hayes, T. L. (2007). Improving FEMA's coastal risk assessment through the National Flood Insurance Program: an historical overview. *Marine Technology Science, 41(1),* 18-27.

CZMA (Coastal Zone Management Act) (1996). The Coastal Zone Management Act of 1972, as amended through P.L. 104-150, The Coastal Zone Protection Act of 1996, 16 U.S.C. § 1451 through 16 U.S.C. § 1465.

FEMA (Federal Emergency Management Agency), (1991). *Projected Impact of Relative Sea-level Rise on the National Flood Insurance Program*. Federal Emergency Management Agency, Washington, DC, 61 pp.

FEMA (Federal Emergency Management Agency) (2002). *Guidelines and Specifications for Flood Hazard Mapping Partners Archive*. [web site] Federal Emergency Management Agency. <http://www.fema.gov/plan/prevent/fhm/gs_arch. shtm>

FEMA (Federal Emergency Management Agency) (2005). *Reducing Flood Losses Through the International Codes, Meeting the Requirements of the National Flood Insurance Program*. Federal Emergency Management Agency and International Code Council, [Washington, DC], 2nd edition, 156 pp.

FEMA (Federal Emergency Management Agency) (2008). *National Flood Insurance Program Definitions* [web page] <http://www.fema.gov/business/nfip/1 9def2. shtm>

Hagen, S.C., W. Quillian & R. Garza (2004). *A Demonstration of Real-time Tide and Hurricane Storm Surge Predictions for the National Weather Service River Forecast System*. Technical report, UCAR Contract No. S01-32794. CHAMPS Laboratory, University of Central Florida, Orlando.

Heinz Center (2000). *Evaluation of Erosion Hazards*. The H. John Heinz III Center for Science, Economics and the Environment, Washington, DC, 203 pp.

Honeycutt, M.G. & Mauriello, M.N. (2005). Multi-hazard mitigation in the coastal zone: when meeting the minimum regulatory requirements isn't enough: In: *Solutions to Coastal Disasters 2005*, Proceedings of the conference, May 8-11, Charleston, SC. American Society of Civil Engineers, Reston VA, pp. 7 13-722.

Hovis, J., Popovich, W., Zervas, C., Hubbard, J., Shih, H. H.& Stone, P. (2004). *Effect of Hurricane Isabel on Water Levels: Data Report*. NOAA technical Report NOS CO-OPS 040. NOAA, Silver Spring, MD, 120 pp.

Jelesnianski, C. P., Chen, J. & Shaffer, W.A. (1992). *SLOSH: Sea, Lake, and Overland Surges from Hurricanes*. NOAA technical report NWS 48. National Weather Service, Silver Spring, MD, 71 pp.

Johnson, Z. P. (2000). *A Sea Level Response Strategy for the State of Maryland*. Maryland Department of Natural Resources, Coastal Zone Management Division, Annapolis, 49 pp.

Larsen, C., Clark, I., Guntenspergen, G. R., Cahoon, D. R., Caruso, V., Hupp, C. & Yanosky, T. (2004). *The Blackwater NWR Inundation Model. Rising Sea Level on a Low-lying Coast: Land Use Planning for Wetlands*. U.S. Geological Survey open file report 04-1302. U.S. Geological Survey, Reston, VA. <http://pubs.usgs.gov/of/2004/1 302/>

Luettich, R. A., Jr., Westerink, J.J. & Scheffner, N.W. (1992). *ADCIRC: An Advanced Three-dimensional Circulation Model of Shelves, Coasts and Estuaries, Report 1: Theory and Methodology of ADCIRC-2DD1 and ADCIRC-3DL*. Technical report DRP-92-6. U.S. Army Engineer Waterways Experiment Station, Vicksburg, MS, 141 pp.

Maryland, (2007). Maryland Executive Order 01.01.2007.07, Commission on Climate Change. State of Maryland, Executive Department, [Annapolis].

Maryland, (2008). *Climate Action Plan*. Maryland Commission on Climate Change and Maryland Department of Environment, Baltimore, 356 pp. <http://www.mdclimatechange.us/>

MD DNR (Maryland Department of Natural Resources), Coastal Zone Management Program, (2006): Coastal Zone Management Assessment, Section 309, Assessment and Strategy.

NOAA (National Oceanic and Atmospheric Administration), (1992). *Effects of the Late October 1991 North Atlantic Extra-Tropical Storm on Water Levels: Data Report*. NOAA National Ocean Service, Rockville, MD, 46 pp.

NOAA (National Oceanic and Atmospheric Administration), (2007). *NOAA 's Sea level Rise Research Program: North Carolina Managers Meetings Fact Sheet*. NOAA National Centers for Coastal Ocean Science, Silver Spring, MD, 2 pp. http://www.cop.noaa.gov/stressors out.pdf

NOAA (National Oceanic and Atmospheric Administration), (2008). *Glossary of NHC Terms*. NOAA National Weather Service, National Hurricane Center, Miami, FL. See defintions for Storm Surge and for Storm Tide. <http://www.nhc.noaa.gov/aboutgloss.shtml>

Pietrafesa, L. J., Buckley, E.B. Peng, M. Bao, S. Liu, H. Peng, S. Xie, L. & Dickey, D.A. (2007). On coastal ocean systems, coupled model architectures, products and services: morphing from observations to operations and applications. *Marine Technology Societyl, 41(1),* 44-52.

Poag, C. W. (1997). Chesapeake Bay bolide impact: a convulsive event in Atlantic Coastal Plain evolution. *Sedimentary Geology, 108(1-4),* 45-90.

Reed, D. B. & Stucky, B. E. (2005). Forecasting storm surge on the Mississippi River. In: *Solutions to Coastal Disasters*, Proceedings of the conference, May 8-11, Charleston, SC. American Society of Civil Engineers, Reston, VA, pp. 52-60.

Slovinsky, P. A. & Dickson, S. M. (2006). *Impacts of Future Sea Level Rise on the Coastal Floodplain*. MGS open-file 06-14. Maine Geological Survey, Augusta, ME, [26 pp.] <http://www.maine.gov/doc/nrimc/mgs/explore/marine/sealevel/contents .htm>

USACE (U.S. Army Corps of Engineers), (1996). *Flood Proofing Techniques, Programs and References*. U.S. Army Corps of Engineers, Washington, DC, 25 pp.

Worcester County Planning Commission, (2006). *Comprehensive Plan, Worcester County Maryland*. Worcester County Commissioners, Snow Hill, MD, 96 pp. <http://www.co.worcester.md.us/cp/finalcomp3 1 406.pdf>

Zervas, C. E. (2001). *Sea Level Variations for the Unites States 1854-1999*. NOAA technical report NOS CO-OPS 36. NOAA National Ocean Service, Silver Spring, MD, 186 pp. <http://tidesandcurrents.noaa.gov/publications/techrpt3 6doc.pdf>

Zervas, C. E. (2005). *Extreme Storm Tide Levels of the United States 1897-2004*. Poster paper presented at NOAA Climate Program Office, Office of Climate Observations Annual Meeting, Silver Spring, MD.

End Notes

[1] The shore is the land immediately in contact with the water.

[2] Biologs are assemblages of woody, organic, and biodegradable material in a log-shaped form.

[3] 06-096 Code of Maine Rules §355.5(C), (2007).

[4] 06-096 Code of Maine Rules §355.5(C), (2007).

[5] TEX. NAT. RES. CODE ANN. §§ 61.001-.178 (West 1978 & Supp. 1998).

[6] 15A NCAC 07H. 0305-0306. The required setback for single-family homes and smaller commercial structures is half as great (see Section A1 .G.4 in Appendix 1 for details).

[7] 06-096 Code of Maine Rules §355 (5) (D). (2007).

[8] Exceptions include Cape May Meadows in New Jersey (protecting freshwater wetlands near the ocean), beaches along both sides of Delaware Bay (horseshoe crab habitat) and Assateague Island, Maryland (to prevent the northern part of the island from disintegrating).

[9] Weggel *et al.*, (1989) estimate that costs are proportional to the height of the design water level raised to the 1.5 power.

[10] IPCC3, WG1, c.1 1, page 671. <http://www.grida.no/climate/ipcc_tar 1.PDF>

[11] It is recognized that protection such as bulkheads, seawalls or other protective measures may exist or be built that could protect specific land areas but, due to the overview nature of this study, they were not included in the analysis.

[12] Chapter 7 discusses impacts on transportation infrastructure.

[13] The "mean low water states" (i.e., Virginia, Delaware, and Pennsylvania), are an exception. See Figure 8.2.

[14] e.g. Dolphin Lane Assocs. v. Town of Southampton, 333 N.E.2d 358, 360 (N.Y. 1975)

[15] In some places, the public has obtained access through government purchase, land dedication by a developer, or other means. See Slade (1990).

[16] Input to author team during CCSP SAP 4.1 Federal Advisory Committee review, Mark Crowell, FEMA.

PART III OVERVIEW. PREPARING FOR SEA-LEVEL RISE

Author: James G. Titus, U.S. EPA

For at least the last four centuries, people have been erecting permanent settlements in the coastal zone of the Mid-Atlantic without regard to the fact that the sea is rising. Because the sea has been rising slowly and only a small part of the coast was developed, the consequences have been relatively isolated and manageable. Part I of this Product suggests, however, that a 2 millimeter per year acceleration of sea-level rise *could* transform the character of the mid-Atlantic coast, with a large scale loss of tidal wetlands and possible disintegration of barrier islands. A 7 millimeter per year acceleration is likely to cause such a transformation, although shore protection may prevent some developed barrier islands from disintegrating and low-lying communities from being taken over by wetlands.

For the last quarter century, scientific assessments have concluded that regardless of possible policies to reduce emissions of greenhouse gases, people will have to adapt to a changing climate and rising sea level. Adaptation assessments differentiate "reactive adaptation" from "anticipatory adaptation".

Part III focuses on what might be done to prepare for sea-level rise. Chapter 10 starts by asking whether preparing for sea-level rise is even necessary. In many cases, reacting later is more justifiable than preparing now, both because the rate and timing of future sea-level rise is uncertain and the additional cost of acting now can be high when the impacts are at least several decades in the future. Nevertheless, for several types of impacts, the cost of preparing now is very small compared to the cost of reacting later. Examples where preparing can be justified include:

- *Coastal wetland protection.* It may be possible to reserve undeveloped lands for wetland migration, but once developed, it is very difficult to make land available for wetland migration. Therefore, it is far more feasible to aid wetland migration by setting aside land before it is developed, than to require development to be removed as sea level rises.

- *Some long-lived infrastructure.* Whether it is beneficial to design coastal infrastructure to anticipate rising sea level depends on economic analysis of the incremental cost of designing for a higher sea level now, and the retrofit cost of modifying the structure at some point in the future. Most long-lived infrastructure in the threatened areas is sufficiently sensitive to rising sea level to warrant at least an assessment of the costs and benefits of preparing for rising sea level.

- *Floodplain management.* Rising sea level increases the potential disparity between rates and risk. Even without considering the possibility of accelerated sea-level rise, the National Academy of Sciences and a Federal Emergency Management Agency (FEMA)-supported study by the Heinz Center recommended to Congress that insurance rates should reflect the changing risks resulting from coastal erosion.

Chapter 11 discusses organizations that are preparing for a possible acceleration of sea-level rise. Few organizations responsible for managing coastal resources vulnerable to sea-level rise have modified their activities. Most of the best examples of preparing for the environmental impacts of sea-level rise are in New England, where several states have enacted policies to enable wetlands to migrate inland as sea-level rise. Ocean City, Maryland is an example of a town considering future sea-level rise in its infrastructure planning.

Chapter 12 examines the institutional barriers that make it difficult to take the potential impacts of future sea-level rise into account for coastal planning. Although few studies have discussed the challenge of institutional barriers and biases in coastal decision making, their implications for sea-level rise are relatively straightforward:

- *Inertia and short-term thinking.* Most institutions are slow to take on new challenges, especially those that require preparing for the future rather than fixing a current problem.

- *The interdependence of decisions* reinforces institutional inertia. In many cases, preparing for sea-level rise requires a decision as to whether a given area will ultimately be given up to the sea, protected with structures and drainage systems, or elevated as the sea rises. Until communities decide which of those three pathways they will follow in a given area, it is difficult to determine which anticipatory or initial response measures should be taken.

- *Policies favoring protection of what is currently there.* In some cases, longstanding preferences for shore protection (as discussed in Chapter 6) discourage planning measures that foster retreat. Because retreat may require a greater lead time than shore protection, the presumption that an area will be protected may imply that planning in unnecessary. On the other hand, these preferences may help accelerate the response to sea-level rise in areas where shore protection is needed.
- *Policies Favoring Coastal Development.* One possible response to sea-level rise is to invest less in the lands likely to be threatened. However, longstanding policies that encourage coastal development can discourage such a response. On the other hand, increasingly dense coastal development improves the ability to raise funds required for shore protection. Therefore, policies that encourage coastal development may be

part of an institutional bias favoring shore protection, but they are not necessarily a barrier to responding to sea-level rise.

Although most institutions have not been preparing for a rising sea, (Chapter 11) , that may be changing. As these chapters were drafted, several states have started to seriously examine possible responses. For example, Maryland enacted a statute to limit the adverse environmental impact of shore protection structures as sea level rises; and FEMA is beginning to assess possible changes to the National Flood Insurance Program. It is too soon to tell whether the increased interest in the consequences of climate change will overtake—or be thwarted by—the institutional barriers that have discouraged actionuntil now.

In: Coastal Sensitivity to Sea Level Rise-Focusing on the Mid-… ISBN: 978-1-60741-440-7
Editor: Melvin C. Urajner

Chapter 10

IMPLICATIONS FOR DECISIONS

Lead Author: James G. Titus
U.S. EPA
Contributing Author: James E. Neumann
Industrial Economics, Inc.

Key Findings

- In many cases, it is difficult to determine whether taking a specific action to prepare for sea-level rise is justified, due to uncertainty in the timing and magnitude of impacts, and difficulties in quantifying projected benefits and costs. Nevertheless, published literature has identified some cases where acting now can be justified.
- Key opportunities for preparing for sea-level rise concern coastal wetland protection, flood insurance rates, and the location and elevation of coastal homes, buildings, and infrastructure.
- Incorporating sea-level rise into coastal wetlands programs can be justified because the Mid-Atlantic still has substantial vacant land onto which coastal wetlands could migrate as sea level rises. Policies to ensure that wetlands are able to migrate inland are likely to be less expensive and more likely to succeed if the planning takes place before people develop these dry lands than after the land becomes developed. Possible tools include rolling easements, density restrictions, coastal setbacks, and vegetative buffers.
- Sea-level rise does not threaten the financial integrity of the National Flood Insurance Program. Incorporating sea-level rise into the program, however, could allow flood insurance rates to more closely reflect changing risk and enable participating local governments to more effectively manage coastal floodplains.
- Long-term shoreline planning is likely to yield benefits greater than the costs; the more sea level rises, the greater the value of that planning.

10.1. Introduction

Most decisions of everyday life in the coastal zone have little to do with the fact that the sea is rising. Some day-to-day decisions depend on today's water levels. For example, sailors, surfers, and fishermen all consult tide tables before deciding when to go out. People deciding whether to evacuate during a storm consider how high the water is expected to rise above the normal level of the sea. Yet the fact that the normal sea level is rising about 0.01 millimeters (mm) per day does not affect such decisions.

Sea-level rise can have greater impacts on the outcomes of decisions with long-term consequences. Those impacts do not all warrant doing things differently today. In some cases, the expected impacts are far enough in the future that people will have ample time to respond. For example, there is little need to anticipate sea-level rise in the construction of docks, which are generally rebuilt every few decades, because the rise can be considered when they are rebuilt (NRC, 1987). In other cases, the adverse impacts of sea- level rise can be more effectively addressed by preparing now than by reacting later. If a dike will eventually be required to protect a community, for example, it can be more cost-effective to leave a vacant right-of-way when an area is developed or redeveloped, rather than tear buildings down later.

People will have to adapt to a changing climate and rising sea level (NRC, 1983; Hoffman *et al.*, 1983; IPCC 1990, 1996, 2001, 2007). The previous chapters (as well as Appendix 1) discuss vulnerable private property and public resources, including ecosystems, real estate, infrastructure (*e.g.*, roads, bridges, parks, playgrounds, government buildings), and commercial buildings (*e.g.*, hotels, office buildings, industrial facilities). Those responsible for managing those assets will have to adapt to changing climate and rising sea level regardless of possible efforts to reduce greenhouse gases, because society has already changed the atmosphere and will continue to do so for at least the next few decades (NRC, 1983; Hoffman *et al.*, 1983; IPCC 1990, 1996, 2001, 2007). Some of these assets will be protected or preserved in their current locations, while others must be moved inland or be lost. Chapters 6, 8, and 9 examine government policies that are, in effect, the current response to sea-level rise. Previous assessments have emphasized the need to distinguish the problems that can be solved by future generations reacting to changing climate from problems that could be more effectively solved by preparing today (Titus, 1990; Scheraga and Grambsch, 1998; Klein *et al.*, 1999; Frankhauser *et al.*, 1999; OTA 1993). Part III (*i.e.*, this Chapter and the next two chapters) makes that distinction.

This Chapter addresses the question: "Which decisions and activities (if any) have outcomes sufficiently sensitive to sea-level rise so as to justify doing things differently, depending on how much the sea is expected to rise?" (CCSP, 2006). Doing things differently does not always require novel technologies or land-use mechanisms; most measures for responding to erosion or flooding from sea-level rise have already been used to address erosion or flooding caused by other factors (see Section 6.1 in Chapter 6). Section 10.2 describes some categories of decisions that may be sensitive to sea-level rise, focusing on the idea that preparing now is not worthwhile unless the expected present value of the benefits of preparing is greater than the cost. Sections 10.3 to 10.7 examine five issues related to rising sea level: wetland protection, shore protection, long- lived structures, elevating homes, and floodplain management.

The examples discussed in this Chapter focus on activities by governments and homeowners, not by corporations. Most published studies about responses to sea-level rise have been funded by governments, with a goal to improve government programs, communicate risk, or provide technical support to homeowners and small businesses. Corporations also engage in many of the activities discussed in this Chapter. It is possible that privately funded (and unpublished) strategic assessments have identified other near- term decisions that are sensitive to sea-level rise.

A central premise of this Chapter is that the principles of economics and risk management provide a useful paradigm for thinking about the implications of sea-level rise for decision making. In this paradigm, decision makers have a well-defined objective concerning potentially vulnerable coastal resources, such as maximizing return on an investment (for a homeowner or investor) or maximizing overall social welfare (for a government). Box 10.1 elaborates on this analytical framework. Although economic analysis is not the only method for evaluating a decision, emotions, perceptions, ideology, cultural values, family ties, and other non-economic factors are beyond the scope of this Chapter.

This Chapter is not directly tied to specific sea-level rise scenarios. Instead, it considers a wide range of plausible sea-level rise over periods of time ranging from decades to centuries, depending on the decision being examined. The Chapter does not quantify the extent to which decisions might be affected by sea-level rise. All discussions of costs assume constant (inflation-adjusted) dollars.

Box 10.1. Conceptual Framework for Decision Making with Sea-Level Rise

This Chapter's conceptual framework for decision making starts with the basic assumption that homeowners or governments with an interest in coastal resources seek to maximize the value of those resources to themselves (homeowners) or to the public as a whole (governments), over a period of time (planning horizon). Each year, coastal resources provide some value to its owner. In the case of the homeowner, a coastal property might provide rental income, or it might provide "imputed rent" that the owner derives from owning the home rather than renting a similar home. The market value of a property reflects an expectation that property will generate similar income over many years. Because a dollar of income today is worth more than a dollar in the future, however, the timing of the income stream associated with a property also affects the value (see explanation of "discounting" in Section 10.2).

Natural hazards and other risks can also affect the income a property provides over time. Erosion, hurricane winds, episodic flooding, and other natural hazards can cause damages that reduce the income from the property or increase the costs of maintaining it, even without sea-level rise,. These risks are taken into account by owners, buyers, and sellers of property to the extent that they are known and understood.

Sea-level rise changes the risks to coastal resources, generally by increasing existing risks. This Chapter focuses on investments to mitigate those additional risks.

In an economic framework, investing to mitigate coastal hazards will only be worthwhile if the cost of the investment (incurred in the short term) is less than net expected returns (which accrue over the long-term). Therefore, these investments are more likely to be judged worthwhile when: (1) there is a large risk of near- term damage (and it can be effectively reduced); (2) there is a small cost to effectively reduce the risk; or (3) the investment shifts the risk to future years.

10.2. Decisions Where Preparing for Sea-Level Rise Is Worthwhile

Sea-level rise justifies changing what people do today if the outcome from considering sea-level rise has an expected net benefit, that is, the benefit is greater than the cost. Thus, when considering decisions where sea-level rise justifies doing things differently, one can exclude from further consideration those decisions where either (1) the administrative costs of preparing are large compared to the impacts, or (2) the net benefits are likely to be small or negative. Few, if any, studies have analyzed the administrative costs of preparing for sea-level rise. Nevertheless, one can infer that administrative costs exceed any benefits from preparing for a very small rise in sea level.[1] Most published studies that investigate which decisions are sensitive to sea-level rise (IPCC, 1990; NRC 1987; Titus and Narayanan, 1996) concern decisions whose consequences last decades or longer, during which time a significant rise in sea level might occur. Those decisions mostly involve long-lived structures, land-use planning, or infrastructure, which can influence the location of development for centuries, even if the structures themselves do not remain that long.

For what type of decision is a net benefit likely from considering sea-level rise? Most analyses of this question have focused on cases where (1) the more sea level rises, the greater the impact; (2) the impacts will mostly occur in the future and are uncertain because the precise impact of sea-level rise is uncertain; and (3) preparing now will reduce the eventual adverse consequences (see Figure 10.1).

In evaluating a specific activity, the first question is whether preparing now would be better than never preparing. If so, a second question is whether preparing now is also better than preparing during some future year. Preparing now to avoid possible effects in the future involves two key economic principles: uncertainty and discounting.

Uncertainty. Because projections of sea-level rise and its precise effects are uncertain, preparing now involves spending today for the sake of uncertain benefits. If sea level rises less than expected, then preparing now may prove, in retrospect, to have been unnecessary. Yet if sea level rises more than expected, whatever one does today may prove to be insufficient. That possibility tends to justify waiting to prepare later, if people expect that a few years later (1) they will know more about the threat and (2) the opportunity to prepare will still be available[2]. Given these reasons to delay, responding now may be difficult to justify, unless preparing now is either fairly inexpensive, or part of a "robust" strategy (*i.e.*, it works for a wide range of possible outcomes). For example, if protecting existing development is important, beach nourishment is a robust way to prepare, because the sand will offset some shore erosion no matter how fast or slow the sea rises.

Discounting. Discounting is a procedure by which economists determine the "present value" of something given or received at a future date (U.S. EPA, 2000). A dollar today is preferred over a dollar in the future, even without inflation (Samuelson and Nordhaus, 1989); therefore, a future dollar must be discounted to make costs and benefits received in different years comparable. Economists generally agree that the appropriate way to discount is to choose an assumed annual interest rate and compound it year-by-year (just as interest compounds) and use the result to discount future dollars (U.S. EPA, 2000; Congressional Research Service, 2003; OMB, 1992; Nordhaus, 2007a b; Dasgupta, 2007).

Most of the decisions where preparing now has a positive net benefit fall into at least one of three categories: (1) the near-term impact may be large; (2) preparing now costs little compared to the cost of the possible impact; or (3) preparing now involves options that reallocate (or clarify) risk.

10.2.1. Decisions that Address Large Near-Term Impacts

If the near-term impact of sea-level rise is large, preparing now may be worthwhile. Such decisions might include:

- *Beach nourishment* to protect homes that are in imminent danger of being lost. The cost of beach nourishment is often less than the value of the threatened structures (USACE, 2000a).

- *Enhancing vertical accretion* (build-up) of wetlands that are otherwise in danger of being lost in the near term (Kentula, 1999; Kussler, 2006). Once wetlands are lost, it can be costly (or infeasible) to bring them back.

- *Elevating homes* that are clearly below the expected flood level due to historic sea-level rise (see Sections 10.6 and 10.7). If elevating the home is infeasible (*e.g.*, historic row houses), flood-proofing walls, doors, and windows may provide a temporary solution (see Chapter 9).

- *Fortifying dikes* to the elevation necessary to protect from current floods. Because sea level is rising, dikes that once protected against a 100-year storm would be overtopped by a similar flood on top of today's higher sea level (see *e.g.*, IPET, 2006).

10.2.2. Decisions Where Preparing Now Costs Little

These response options can be referred to as "low regrets" and "no regrets", depending on whether the cost is little or nothing. The measures are justifiable, in spite of the uncertainty about future sea-level rise, because little or nothing is invested today, in return for possibly averting or delaying a serious impact. Examples include:

- *Setting a new home back from the sea within a given lot.* Setting a home back from the water can push the eventual damages from sea-level rise farther into the future, lowering their expected present value[3]. Unlike the option of not building, this approach retains almost the entire value of using the property—especially if nearby homes are also set back so that all properties retain the complete panorama view of the waterfront—provided that the lot is large enough to build the same house as would have been built without the setback requirement.

- *Building a new house with a higher floor elevation.* While elevating an existing house can be costly, building a new house on pilings one meter (a few feet) higher only increases the construction cost by about 1 percent (Jones *et al.*, 2006).

- *Designing new coastal drainage systems with larger pipes to incorporate future sea-level rise.* Retrofitting or rebuilding a drainage system can cost 10 to 20 times as much as including larger pipes in the initial construction (Titus *et al.*, 1987).

- *Rebuilding roads to a higher elevation during routine reconstruction.* If a road will eventually be elevated, it is least expensive to do so when it is rebuilt for other purposes.

- *Designing bridges and other major facilities.* As sea level rises, clearance under bridges declines, impairing navigation (TRB, 2008). Building the bridge higher in the first place can be less expensive than rebuilding it later.

10.2.3. Options That Reallocate or Clarify Risks from Sea-Level Rise

Instead of imposing an immediate cost to avoid problems that may or may not occur, these approaches impose a future cost, but only if and when the problem emerges. The premise for these measures is that current rules or expectations can encourage people to behave in a fashion that increases costs more than necessary. People make better decisions when all of the costs of a decision are internalized (Samuelson and Nordhaus, 1989). Changing rules and expectations can avoid some costs, for example, by establishing today that the eventual costs of sea-level rise will be borne by a property owner making a decision sensitive to sea-level rise, rather than by third parties (*e.g.*, governments) not involved in the decision. Long-term shoreline planning and rolling easements are two example approaches.

Figure 10.1. Homes set back from the shore. Myrtle Beach, South Carolina. (April, 2004)

Long-term shoreline planning can reduce economic or environmental costs by concentrating development in areas that will not eventually have to be abandoned to the rising sea. People logically invest more along eroding shores if they assume that the government will provide subsidized shore protection (see Box 10.2) than in areas where owners must pay for the shore protection or where government rules require an eventual abandonment. The value to a buyer of that government subsidy is capitalized into higher land prices, which can further encourage increased construction. Identifying areas that will not be protected can avoid misallocation of both financial and human resources. If residents wrongly assume that they can expect shore protection and the government does not provide it, then real estate prices can decline; in extreme cases, people can lose their homes unexpectedly. People's lives and economic investments can be disrupted if dunes or dikes fail and a community is destroyed. A policy that clearly warns that such an area will *not* be protected (see Section 12.3 in Chapter 12) could lead owners to strategically depreciate the physical property[4] and avoid some of the noneconomic impacts that can occur after an unexpected relocation (see Section 6.4.1). (see Section 12.3 for further discussion).

Box 10.2. Erosion, Coastal Programs, and Property Values

Do government shore protection and flood insurance programs increase property values and encourage coastal development? Economic theory would lead one to expect that in areas with high land values, the benefits of coastal development are already high compared to the cost of development, and thus most of these areas will become developed unless the land is acquired for other purposes. In these areas, government programs that reduce the cost of maintaining a home should generally be reflected in higher land values; yet they would not significantly increase development because development would occur without the programs. By contrast, in marginal areas with low land prices, coastal programs have the potential to reduce costs enough to make a marginal investment profitable.

Several studies have investigated the impact of flood insurance on development, with mixed results. Leatherman (1997) examined North Bethany Beach, Delaware, a community with a checkerboard pattern of lands that were eligible and ineligible for federal flood insurance due to the Coastal Barrier Resources Act. He found that ocean-front lots generally sold for $750,000, with homes worth about $250,000. Development was indistinguishable between areas eligible and ineligible for flood insurance. In the less affluent areas along the back bays, however, the absence of federal flood insurance was a deterrent to developing some of the lower priced lots. Most other studies have not explicitly attempted to distinguish the impact of flood insurance on low- and high-value lands. Some studies (*e.g.*, Cordes and Yezer, 1998; Shilling *et al.*, 1989) have concluded that the highly subsidized flood insurance policies during the 1970s increased development, but the actuarial policies since the early 1980s have had no detectable impact on development. Others have concluded that flood insurance has a minimal impact on development (*e.g.*, GAO 1982; Miller, 1981). The Heinz Center (2000) examined the impacts of the National Flood Insurance Program (NFIP) and estimated that "the density of structures built within the V Zone after 1981 may be 15 percent higher than it would have been if the NFIP had not been adopted. However, the expected average annual flood and erosion damage to these structures dropped close to 35 percent. Thus, overall, the damage to V Zone structures built after 1981 is between 25 and 30 percent lower than it would have been if development had occurred at the lower densities, but higher expected damage that would have occurred absent the NFIP". A report to the Federal Emergency Management Agency (FEMA) reviewed 36 published studies and commentaries concerning the impacts of flood insurance on development and concluded that none of the studies offer irrefutable evidence that the availability, or the lack of availability, of flood insurance is a primary factor in floodplain development today (Evatt, 1999, 2000).

Considering shore protection and flood insurance together, The Heinz Center (2000) estimated that "in the absence of insurance and other programs to reduce flood risk, development density would be about 25 percent lower in areas vulnerable to storm wavers (*i.e.*, V Zones) than in areas less susceptible to damage from coastal flooding". Cordes and Yezer (1998) modeled the impact on new building permit activity in coastal areas of shore protection activity in 42 coastal counties, including all of the counties with developed ocean coasts in New York, New Jersey, Maryland, and Virginia. They did not find a statistically significant relationship between shore protection and building permits.

The impact of federal programs on property values has not been assessed to the same extent. The Heinz Center (2000) reported that along the Atlantic coast, a house with a remaining lifetime of 10 to 20 years before succumbing to erosion is worth 20 percent less than a home expected to survive 200 years. Landry *et al.* (2003) found that property values tend to be higher with wide beaches and low erosion risk. It would therefore follow that shore protection programs that widen beaches, decrease erosion risk, and lengthen a home's expected lifetime would increase property values. Nevertheless, estimates of the impact on property values are complicated by the fact that proximity to the shore increases the risk of erosion but also improves access to the beach and views of the water (Bin *et al.*, 2008).

Rolling easements can also reallocate or clarify the risks of sea-level rise, depending on the pre-existing property rights of a given jurisdiction (Titus, 1998). A rolling easement is an arrangement under which property owners have no right or expectation of holding back the sea if their property is threatened. Rolling easements have been implemented by regulation along ocean and sheltered shores in three New England states (see Section 11.2 in Chapter 11 and along ocean shores in Texas and South Carolina. Rolling easements can also be implemented as a type of conservation easement, with the easement donated, purchased at fair market value, or exacted as a permit condition for some type of coastal development (Titus, 1998). In either case, they prevent property owners from holding back the sea but otherwise do not alter what an owner can do with the property. As the sea advances, the easement automatically moves or "rolls" landward. Without shoreline armoring, sediment transport remains undisturbed and wetlands and other tidal habitat can migrate naturally. Because the dry beach and intertidal land continues to exist, the rolling easement also preserves the public's lateral access right to walk along the shore[5] (*Matcha versus Mattox*, 1986).

Under a rolling easement, the property owner bears all of the risk of sea-level rise. Without a rolling easement, property owners along most shores invest as if their real estate is sustainable, and then expend resources—or persuade governments to expend resources—to sustain the property. The overall effect of the rolling easement is that a community clearly decides to pursue retreat instead of shore protection in the future. The same result could also be accomplished by purchasing (or prohibiting development on) the land that would potentially be eroded or submerged as sea level rises. That approach, however, would have a large near-term social cost because the coastal land would then be unavailable for valuable uses. By contrast, rolling easements do not prevent the property from being used for the next several decades while the land remains dry. (Even if the government purchases the rolling easement, the purchase price is a transfer of wealth, not a cost to society[6].) The landward migration from the rolling easement should also have lower eventual costs than having the government purchase property at fair market value as it becomes threatened (Titus, 1991). Property owners can strategically depreciate their property and make other decisions that are consistent with the eventual abandonment of the property (Yohe *et al.*, 1996; Titus, 1998), efficiently responding to information on sea-level rise as it becomes available. Figure 9.1 shows how a rolling easement might work over time in an area already developed when rolling easements are obtained.

10.3. Protecting Coastal Wetlands

The nation's wetland programs generally protect wetlands in their current locations, but they do not explicitly consider retreating shorelines. As sea level rises, wetlands can adapt by accreting vertically (Chapter 4) and migrating inland. Most tidal wetlands are likely to keep pace with the current rate of sea-level rise but could become marginal with an acceleration of 2 millimeters (mm) per year, and are likely to be lost if sea-level rise accelerates by 7 mm per year (see Chapter 4). Although the dry land available for potential wetland migration or formation is estimated to be less than 20 percent of the current area of wetlands (see Titus and Wang 2008), these lands could potentially become important wetland areas in the future. However, given current policies and land-use trends, they may not be available for wetland migration and formation (Titus 1998, 2001). Much of the coast is developed or being developed, and those who own developed dry land adjacent to the wetlands increasingly take measures to prevent the wetlands from migrating onto their property (see Figure 10.4 and Chapter 6).

Continuing the current practice of protecting almost all developed estuarine shores could reverse the accomplishments of important environmental programs (*e.g.*, Titus 1991, 2001, 2005). Until the mid-twentieth century, tidal wetlands were often converted to dredge-and-fill developments (see Section 6.1.1.2 in Chapter 6 for an explanation of these developments and their vulnerability to sea-level rise). By the 1970s, the aggregate result of the combination of federal and state regulations had, for all practical purposes, halted that practice. Today, most tidal wetlands in the Mid-Atlantic are off-limits to development. Coastal states generally prohibit the filling of low marsh, which is publicly owned in most states under the Public Trust Doctrine (see Section 8.2).

A landowner who wants to fill tidal wetlands on private property must usually obtain a permit from the U.S. Army Corps of Engineers (USACE)[7]. These permits are generally not issued unless the facility is inherently water-related, such as a marina[8]. Even then, the owners usually must mitigate the loss of wetlands by creating or enhancing wetlands elsewhere (U.S. EPA and USACE, 1990). (Activities with small impacts on wetlands, however, are often covered by a nationwide permit, which exempts the owner from having to obtain a permit [see Section 12.2]). The overall effect of wetland programs has been to sharply reduce the rate of coastal wetland loss (*e.g.*, Stockton and Richardson, 1987; Hardisky and Klemas,

1983) and to preserve an almost continuous strip of marshes, beaches, swamps, and mudflats along the U.S. coast. If sea-level rise accelerates, these coastal habitats could be lost by submergence and—in developed areas where shores are protected—by prevention of their natural inland migration (Reed *et al.*, 2008), unless future generations use technology to ensure that wetland surfaces rise as rapidly as the sea (NRC, 2007).

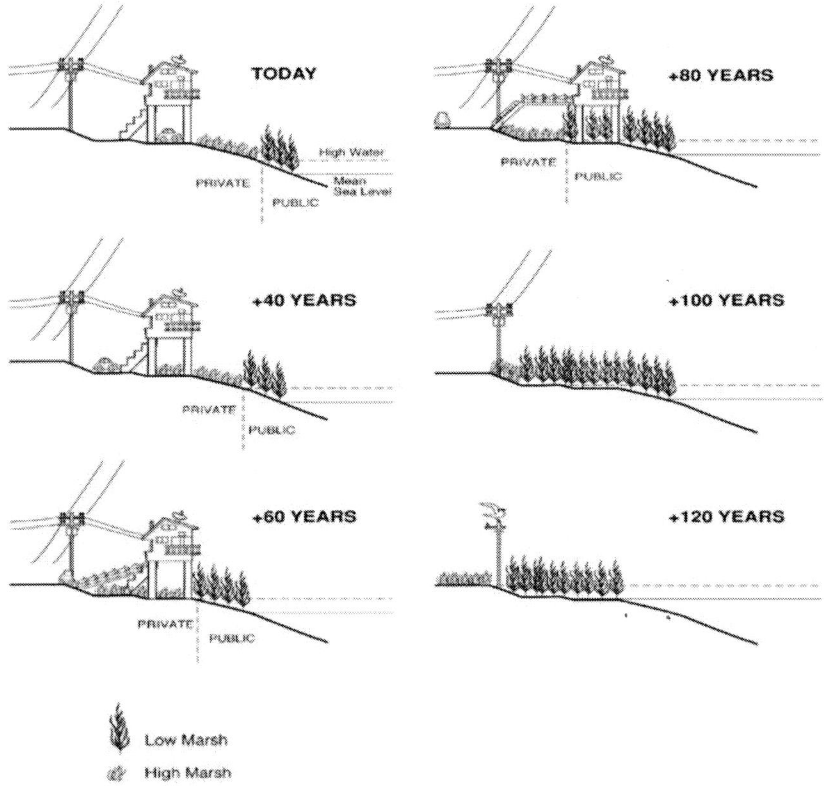

Figure 10.2. The landward migration of wetlands onto property subject to a rolling easement. A rolling easement allows construction near the shore, but requires the property owner to recognize nature's right-of- way to advance inland as sea level rises. In the case depicted, the high marsh reaches the footprint of the house 40 years later. Because the house is on pilings, it can still be occupied (assuming that it is hooked to a sewerage treatment plant. A flooded septic system would probably fail, because the drainfield must be a minimum distance above the water table). After 60 years, the marsh has advanced enough to require the owner to park their car along the street and construct a catwalk across the front yard. After 80 years, the marsh has taken over the entire yard; moreover, the footprint of the house is now seaward of mean high water and hence, on public property. At this point, additional reinvestment in the property is unlikely. Twenty years later, the particular house has been removed, although other houses on the same street may still be occupied. Eventually, the entire area returns to nature. A home with a rolling easement would depreciate in value rather than appreciate like other coastal real estate. But if the loss is expected to occur 100 years from today, it would only offset the current property value by 1 to 5 percent, which could be compensated or offset by other permit considerations (Titus, 1998).

Figure 10.3. Coastal Wetlands migrating onto previously dry lowland. Webbs Island, just east of Machipongo, in Northampton County, Virginia (June, 2007).

Figure 10.4. Wetland Migration thwarted by development and shore protection. Elevating the land surface with fill prevents wetlands from migrating into the back yard with a small or modest rise in sea level. The bulkhead prevents waves from eroding the land, which would otherwise provide sand and other soil materials to help enable the wetlands to accrete with rising sea level (Monmouth New Jersey, August, 2003).

Current approaches would *not* protect wetlands for future generations if sea level rises beyond the ability of wetlands to accrete, which is likely for most of Chesapeake Bay's wetlands if sea level rises 50 centimeters (cm) in the next century, and for most of the Mid-Atlantic if sea level rises 100 cm (see Figure 4.4).

Current federal statutes are designed to protect existing wetlands, but the totality of the nation's wetland protection program is the end result of decisions made by many actors. Federal programs discourage destruction of most *existing* coastal wetlands, but the federal government does little to allow tidal wetlands to migrate inland (Titus, 2000). North Carolina, Maryland, New Jersey, and New York own the tidal wetlands below Mean High Water; and Virginia, Delaware, and Pennsylvania have enough ownership interest under the Public Trust Doctrine to preserve them (Titus, 1998). However, most states give property owners a near-universal permit to protect property by preventing wetlands from migrating onto dry land. Farmers rarely erect shore protection structures, but homeowners usually do (Titus, 1998;

NRC, 2006). Only a few coastal counties and states have decided to keep shorefront farms and forests undeveloped, (see Sections A1.D, A1.E, and A1.F in Appendix 1). Government agencies that hold land for conservation purposes are not purchasing the land or easements necessary to enable wetlands to migrate inland (Section 11.2.1 discusses private conservancies). In effect, the nation has decided to *save* its existing wetlands. Yet the overall impact of the decisions made by many different agencies is very likely to *eliminate* wetlands by blocking their landward migration as a rising sea erodes their outer boundaries.

Not only is the long-term success of wetland protection sensitive to sea-level rise, it is also sensitive to when people decide to prepare. The political and economic feasibility of allowing wetlands to take over a given parcel as sea level rises is much greater if appropriate policies are in place before that property is intensely developed. Many coastal lands are undeveloped today, but development continues. Deciding now that wetlands will have land available to migrate inland could protect more wetlands at a lower cost than deciding later (Titus, 1991). In some places, such policies might discourage development in areas onto which wetlands may be able to migrate. In other areas, development could occur with the understanding that eventually land will revert to nature if sea level rises enough to submerge it. As with beach nourishment, artificially elevating the surfaces of tidal wetlands would not always require a lead-time of several decades; but developing technologies to elevate the wetlands, and determining whether and where they are appropriate, could take decades. Finally, in some areas, the natural vertical accretion (build-up) of tidal wetlands is impaired by human activities, such as water flow management, development that alters drainage patterns, and beach nourishment and inlet modification, which thwarts barrier island overwash. In those areas, restoring natural processes before the wetlands are lost is more effective than artificially re-creating them (U:S. EPA, 1995; U.S. EPA and USACE, 1990; Kruczynski, 1990).

Although the long-term success of the nation's efforts to protect wetlands is sensitive to sea-level rise, most of the individual decisions that ultimately determine whether wetlands can migrate inland depend on factors that are not sensitive to sea-level rise. The desire of bay-front homeowners to keep their homes is strong, and unlikely to diminish even with a significant acceleration of sea-level rise[9]. State governments must balance the public interest in tidal wetlands against the well-founded expectations of coastal property owners that they will not have to yield their property. Only a few states (none in the Mid-Atlantic) have decided in favor of the wetlands (see Section 11.2.1). Local government decisions regarding land use reflect many interests. Objectives such as near- term tax revenues (often by seasonal residents who make relatively few demands for services) and a reluctance to undermine the economic interests of landowners and commercial establishments are not especially sensitive to rising sea level.

Today's decentralized decision-making process seems to protect existing coastal wetlands reasonably well at the current rate of sea-level rise; however, it will not enable wetlands to migrate inland as sea-level rise continues or accelerates. A large-scale landward migration of coastal wetlands is very unlikely to occur in most of the Mid- Atlantic unless a conscious decision is made for such a migration by a level of government with authority to do so. Tools for facilitating a landward migration include coastal setbacks, density restrictions, rolling easements, vegetation buffers, and building design standards (see Sections 6.1.2 and A1 .D, and A1 .F in Appendix 1 for further details).

10.4. Shore Protection

The case for anticipating sea-level rise as part of efforts to prevent erosion and flooding has not been as strong as the case for wetland protection. Less lead time is required for shore protection than for a planned retreat and wetland migration (NRC, 1987). Dikes, seawalls, bulkheads, and revetments can each be built within a few years. Beach nourishment is an incremental periodic activity; if the sea rises more than expected, communities can add more sand.

The U.S. Army Corps of Engineers (USACE) has not evaluated whether sea-level rise will ultimately require fundamental changes in shore protection; such changes do not appear to be urgent. Since the early 1990s, USACE has recommended robust strategies: "Feasibility studies should consider which designs are most appropriate for a range of possible future rates of rise. Strategies that would be appropriate for the entire range of uncertainty should receive preference over those that would be optimal for a particular rate of rise but unsuccessful for other possible outcomes" (USACE, 2000a). To date, this guidance has not significantly altered USACE's approach to shore protection. Nevertheless, there is some question as to whether continued beach nourishment would be sustainable in the future if the rate of sea-level rise accelerates. It may be possible to double or triple the rate at which USACE nourishes beaches and to elevate the land surfaces of barrier islands 50 to 100 cm, and thereby enable land surfaces to keep pace with rising sea level in the next century. Yet continuing such a practice indefinitely would eventually leave back-barrier bays much deeper than today (see Chapter 5), with unknown consequences for the environment and the barrier islands themselves. Similarly, it may be possible to build a low bulkhead along mainland shores as sea level rises 50 to 100 cm; however, it could be more challenging to build a tall dike along the same shore because it would block waterfront views, require continual pumping, and expose people behind the dike to the risk of flooding should that dike fail (Titus, 1990).

10.5. Long-Lived Structures: Should We Plan Now or Later?

The fact that eventually a landowner will either hold back the sea or allow it to inundate a particular parcel of land does not, by itself, imply that the owner must respond today. A community that will not need a dike until the sea rises 50 to 100 cm has little reason to build that dike today. Nevertheless, if the land where the dike would eventually be constructed is vacant now, the prospect of future sea-level rise might be a good reason to leave that land vacant. A homeowner whose house will be inundated (or eroded) in 30 to 50 years has little reason to move the house back today, but if it is damaged by fire or storms, it might be advisable to rebuild the house on a higher (or more inland) part of the lot to provide the rebuilt structure a longer lifetime.

Whether one must be concerned about long-term sea-level rise ultimately depends on the lead time of the response options and on the costs and benefits of acting now *versus* acting later. A fundamental premise of cost-benefit analysis is that resources not yet deployed can be invested profitably in another activity and yield a return on investment. Delaying the response is economically efficient if the most effective response can be delayed with little or no additional cost, which is the case with most engineering responses to sea-level rise. For a given level of protection, dikes, seawalls, beach nourishment, and elevating structures and roadways are unlikely to cost more in the future than they cost today (USACE, 2000b, 2007). Moreover, these approaches can be implemented within the course of a few years. If shore protection is the primary approach to sea-level rise, responding now may not be necessary, with two exceptions.

The first exception could be called the "retrofit penalty" for failure to think long-term. It may be far cheaper to design for rising sea level in the initial design of a new (or rebuilt) road or drainage system than to modify it later because modifying it later requires the facility, in effect, to be built twice. For example, in a particular watershed in Charleston, South Carolina, if sea level rises 30 cm (1 ft), the planned drainage system would fail and need to be rebuilt, but it would only cost an extra 5 percent to initially design the system for a 30-cm rise (Titus *et al.*, 1987). Similarly, bridges are often designed to last for 100 years, and although roads are paved every 10 to 20 years, the location of a road may stay the same for centuries. Thus, choices made today about the location and design of transportation infrastructures can have a large impact on the feasibility and cost of accommodating rising sea level in the future (TRB, 2008). The design and location of a house is yet another example. If a house is designed to be movable, it can be relocated away from the shore; but non-moveable houses, such as a brick house on a slab foundation, could be more problematic. Similarly, the cost of building a house 10 meters (m) farther from the shore may be minor if the lot is large enough, whereas the cost of moving it back 10 m could be substantial (U.S. EPA, 1989).

The second exception concerns the incidental benefits of acting sooner. If a dike is not needed until the sea rises 0.5 m, because at that point a 100-year storm would flood the streets with 1 m of water, the decision to not build the dike today implicitly accepts the 0.5 m of water that such a storm would provide today. If a dike is built now, it would stop this smaller flood as well as protect from the larger flood that will eventually occur. This reasoning was instrumental in leading the British to build the Thames River Barrier, which protects London. Some people argued that this expensive structure was too costly given the small risk of London flooding, but rising sea level implied that such a structure would eventually have to be built. Hence, the Greater London Council decided to build it during the 1 970s (Gilbert and Horner, 1984). As expected, the barrier closed 88 times to prevent flooding between 1983 and 2005 (Lavery and Donovan 2005).

While most engineering responses can be delayed with little penalty, failure to consider sea-level rise when making land-use decisions could be costly. Once an area is developed, the cost of vacating it as the sea rises is much greater than that cost would have been if the area was not developed. This does not mean that eventual inundation should automatically result in placing land off-limits to development. Even if a home has to be torn down 30 to 50 years hence, it might still be worth building. In some coastal areas where demand for beach access is great and land values are higher than the value of the structures, rentals may recover the cost of home construction in less than a decade. However, once an area is developed, it is unlikely to be abandoned unless either the eventual abandonment was part of the original construction plan, or the owners can not afford to hold back the sea. Therefore, the most effective way to preserve natural shores is to make such a decision before an area is developed. Because the coast is being developed today, a failure to deal with this issue now is, in effect, a decision to allow the loss of wetlands and bay beaches along most areas where development takes place.

Many options can be delayed, because the benefits of preparing for sea-level rise would still accrue later. Delaying action decreases the present value of the cost of acting and may make it easier to tailor the response to what is actually necessary. Yet delay can also increase the likelihood that people do not prepare until it is too late. One way to address this dilemma is to consider the lead times associated with particular types of adaptation (IPCC CZMG, 1992; O'Callahan, 1994). Emergency beach nourishment and bulkheads along estuarine shores can be implemented in less than a year. Large-scale beach nourishment generally takes a few years. Major engineering projects to protect London and the Netherlands took a few decades to plan, gain consensus, and construct (*e.g.*, Gilbert and Horner, 1984). To minimize the cost of abandoning an area, land use planning requires a lead time of 50 to 100 years (Titus, 1991, 1998).

10.6. Decisions by Coastal Property Owners on Elevating Homes

People are increasingly elevating homes to reduce the risk of flooding during severe storms and, in very low-lying areas, people are also elevating their yards. The cost of elevating even a small wood-frame cottage on a block foundation is likely to be $15,000 to $20,000; larger houses cost proportionately more (Jones *et al.*, 2006; FEMA, 1998). If it is necessary to drill pilings, the cost is higher because the house must be moved to the side and then moved back onto the pilings. If elevating the home prevents its subsequent destruction within a few decades, it will have been worthwhile. At a 5 percent discount rate, for example, it is worth investing 25 percent of the value of a structure to avoid a guaranteed loss 28 years later[10]. In areas where complete destruction is unlikely, people sometimes elevate homes to obtain lower insurance rates and to avoid the risk of water damages to walls and furniture. The decision to elevate involves other factors, both positive and negative, including better views of the water, increased storage and/or parking spaces, and greater difficulty for the elderly or disabled to enter their homes. Rising sea level can also be a motivating factor when an owner is uncertain about whether the current risks justify elevating the house, because rising water levels would eventually make it necessary to elevate it (unless there is a good chance that the home will be rebuilt or replaced before it is flooded).

In cases where a new home is being constructed, or an existing home is elevated for reasons unrelated to sea-level rise (such as a realization of the risk of flooding), rising sea level would justify a higher floor elevation that would otherwise be the case. For example, elevating a $200,000 home on pilings to 30cm above the base flood elevation when the home is built would increase the construction cost by approximately $500-1000 more than building the home at the base flood elevation (Jones *et al.*, 2006). Yet a 30 cm rise in sea level would increase the actuarial annual flood insurance premium by more than $2000 if the home was not elevated the extra 30 cm (NFIP, 2008).

10.7. Floodplain Management

The Federal Emergency Management Agency (FEMA) works with state and local governments on a wide array of activities that are potentially sensitive to rising sea level, including floodplain mapping, floodplain regulations, flood insurance rates, and the various hazard mitigation activities that often take place in the aftermath of a serious storm. Although the outcomes of these activities are clearly sensitive to sea-level rise, previous assessments have focused on coastal erosion rather than on sea-level rise. Because implications of sea-level rise and long-term erosion overlap in many cases, previous efforts provide insights on cases where the risks of future sea-level rise may warrant changing the way things are done today.

10.7.1. Floodplain Regulations

The flood insurance program requires new or substantially rebuilt structures in the coastal floodplain to have the first floor above the base flood elevation, *i.e.*, 100-year flood level. (see Chapter 9). The program vests considerable discretion in local officials to tailor specific requirements to local conditions, or to enact regulations that are more stringent than FEMA's minimum requirements. Several communities have decided to require floor levels to be 30 cm (or more) above the base flood elevation (*e.g.*, Township of Long Beach, 2008; Town of Ocean City, 1999; see also Box A1.5 in Appendix 1). In some cases, past or future sea-level rise has been cited as one of the justifications for doing so (Cape Cod Commission, 2002). There is considerable variation in both the costs and benefits of designing buildings to accommodate future sea-level rise. If local governments believe that property owners need an incentive to optimally address sea- level rise, they can require more stringent (*i.e.*, higher) floor elevations. A possible reason for requiring higher floor elevations in anticipation of sea-level rise (rather than allowing the owner to decide) is that, under the current structure of the program, the increased risk from sea-level rise does not lead to proportionately higher insurance rates (see Section 10.7.3.1) (although rates can rise for other reasons).

10.7.2. Floodplain Mapping

Local jurisdictions have pointed out (see Box A1.6 in Appendix 1) that requiring floor elevations above the base flood elevation to prepare for sea level rise can create a disparity between property inside and outside the existing 100-year floodplain.

Unless floodplain mapping also takes sea-level rise into account, a building in the current floodplain would have to be higher than adjacent buildings on higher ground just outside the floodplain (see Figure 10.5). Thus, the ability of local officials to voluntarily prepare for rising sea level is somewhat constrained by the lack of floodplain mapping that takes sea-level rise into account. Incorporating sea-level rise into floodplain maps would be a low-regrets activity, because it is relatively inexpensive and would enable local officials to modify requirements where appropriate.

10.7.3. Federal Flood Insurance Rates

The available reports on the impacts of rising sea level or shoreline retreat on federal flood insurance have generally examined one of two questions:

- What is the risk to the financial integrity of the flood insurance program?
- Does the program discourage policyholders from preparing for sea-level rise by shielding them from the consequences of increased risk?

Figure 10.5. Rationale for incorporating sea-level rise into floodplain mapping. In this figure, the (left) three houses in the existing floodplain have first floor elevations about 80 centimeters (cm) above the level of the 100-year storm, to account for a projected 50-cm rise in sea level and the standard requirement for floors to be 30 cm above the base flood elevation. The (right) three homes outside of the regulated floodplain are exempt from the requirement. Actual floods, however, do not comply with floodplain regulations. A 100-year storm on top of the higher sea level would thus flood the buildings to the right which are outside of today's floodplain, while the regulated buildings would escape the flooding. This potential disparity led the city of Baltimore to suggest that floodplain mapping should account for sea level rise as part of any process to increase the freeboard requirement (see Box A1 .7, Section A1 .F in Appendix 1).

No assessment has found that sea-level rise threatens the federal program's financial integrity. A 1991 report to Congress by FEMA, for example, concluded that there was little need to change the Flood Insurance Program because rates would be adjusted as sea level rises and flood maps are revised (FEMA, 1991). Nevertheless, the current rate structure can discourage some policyholders from preparing for increases in flood risks caused by sea-level rise, shore erosion, and other environmental changes. For new and rebuilt homes, the greater risks from sea-level rise cause a roughly proportionate increase in flood insurance premiums. For existing homes, however, the greater risks from sea-level rise cause premiums to rise much less than proportionately, and measures taken to reduce vulnerability to sea-level rise do not necessarily cause rates to decline.

Flood insurance policies can be broadly divided into actuarial and subsidized. "Actuarial" means that the rates are designed to cover the expected costs; "subsidized" means that the rates are designed to be less than the cost, with the government making up the difference. Most of the subsidized policies apply to "pre-FIRM" construction, that is, homes that were built before the Flood Insurance Rate Map (FIRM) was adopted for a given locality[11]; and most actuarial policies are for post-FIRM construction. Nevertheless, there are also a few small classes of subsidized policies for post-FIRM construction; and some owners of pre-FIRM homes pay actuarial rates. The following subsections discuss these two broad categories in turn.

10.7.3.1. Actuarial (Post-FIRM) Policies

Flood Insurance Rate Maps show various hazard zones, such as V (wave velocity) Zone, A (stillwater flooding during a 100-year storm) Zone and the "shaded X Zone"[12] (stillwater flooding during a 500-year storm) (see Chapter 9). These zones are used as classes for setting rates. The post-FIRM classes pay actuarial rates. For example, the total premiums by all post-FIRM policyholders in the A Zone equals FEMA's estimate of the claims and administrative costs for the A Zone[13]. Hypothetically, if sea-level rise were to double flood damage claims in the A Zone, then flood insurance premiums would double (ignoring administrative costs)[14].

Therefore, the impact of sea-level rise on post-FIRM policy holders would not threaten the program's financial integrity under the current rate structure.

The rate structure can, however, insulate property owners from the effects of sea-level rise, removing the market signal[15] that might otherwise induce a homeowner to prepare or respond to sea-level rise. Although shoreline erosion and rising sea level increase the expected flood damages of a given home, the increased risk to a specific property does not cause the rate on that specific property to rise. Unless a home is substantially changed, its assumed risk is grandfathered[16], that is, FEMA assumes that the risk has not increased when calculating the flood insurance rate (*e.g.*, NFIP, 2007; Heinz Center, 2000)[17]. Because the entire class pays an actuarial rate, the grandfathering causes a "cross-subsidy" between new or rebuilt homes and the older grandfathered homes.

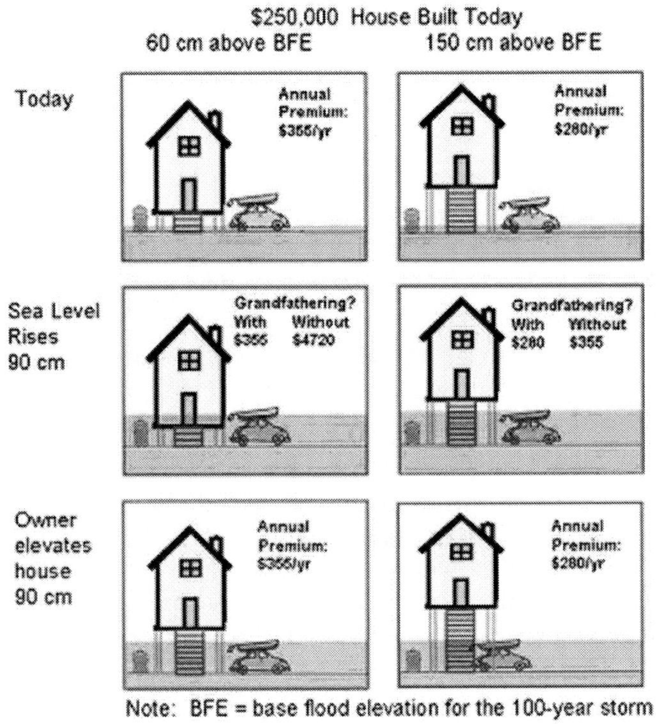

Figure 10.6. Impact of grandfathering and floor elevation on flood insurance rates in the A Zone as sea level rises. Without grandfathering, a 90-centimeter (cm) rise in sea level would increase the flood insurance rate from $355 to $4720, for a home built 60 cm above today's 100-year flood elevation (left column); if the home is built 150 cm above the 100-year flood, sea level rise increases the rate from $280 to $355. Elevating the house 90 cm after sea level rise lowers the rate to what it had been originally, Thus, if the 90 cm rise is expected during the owner's planning horizon, there would be a significant incentive to either build the house higher or elevate it later. With grandfathering, however, sea-level rise does not increase the rate and elevating the home later does not reduce the rate. Thus, grandfathering reduces the incentive to anticipate sea level rise or react to it after the fact.
Caveat: The numerical example is based on rates published in NFIP (2008), Table 3B, and does not include the impact of the annual changes in the rate structure. Such rate changes would complicate the numerical illustration, but would not fundamentally alter the incentives illustrated, because the annual rate changes are across-the-board within a given class. For example, if rates increased by 50 percent by the time sea level rises 90 cm, then all of the premiums shown in the bottom four boxes would rise 50 percent.

Grandfathering can discourage property owners from either anticipating or responding to sea-level rise. If anticipated risk is likely to increase, for example, by about a factor of 10 and a total loss would occur eventually (*e.g.*, a home on an eroding shore), grandfathering the assumed risk may allow the policy holder to secure compensation for a total loss at a small fraction of the cost of that loss. For instance, a $250,000 home built to base flood elevation in the A Zone would typically pay about $900 per year (NFIP, 2008); but if shore erosion left the property in the V Zone, the annual rate would rise to more than $10,000 (NFIP, 2008)[18], if the property was not grandfathered. Under such circumstances, the $9,000 difference in eventual insurance premiums might be enough of a subsidy to encourage owners to build in locations more hazardous than where they might have otherwise built had they anticipated that they would bear the entire risk (*cf.* Heinz Center, 2000). For homes built in the A Zone, the effect of grandfathering is less, but still potentially significant (see Figure 10.6).

Grandfathering can also remove the incentive to respond as sea level rises. Consider a home in the A Zone that is originally 30 cm (about 1 ft) above the base flood elevation. If sea level rises 30 to 90 cm (almost 1 to 3 ft), then the actuarial rates would typically rise by approximately two to ten times the original amount (NFIP, 2008), but because of grandfathering, the owners would continue to pay the same premium. Therefore, if the owner were to elevate the home 30 to 90 cm, the insurance premium would not decline because the rate already assumes that the home is 30 cm above the flood level (see the bottom four panels of Figure 10.6).

The importance of grandfathering is sensitive to the rate of sea-level rise. At the current rate of sea-level rise (3 mm per year), most homes would be rebuilt (and thus lose the grandfathering benefit) before the 100 to 300 years it takes for the sea to rise 30 to 90 cm. By contrast, if sea level rises 1 cm per year, this effect would only take 30 to 90 years—and many coastal homes survive that long.

Previous assessments have examined this issue (although they were focused on shoreline erosion from all causes, rather than from sea-level rise). The National Academy of Sciences (NAS) has recommended that the Flood Insurance Program create mechanisms to ensure that insurance rates reflect the increased risks caused by long-term coastal erosion (NAS, 1990). NAS pointed out that Congress has explicitly included storm- related erosion as part of the damages covered by flood insurance (42 U.S.C. §4121), and that FEMA's regulations (44 CFR Part 65.1) have already defined special "erosion zones", which consider storm-related erosion (NAS, 1990)[19]. A FEMA-supported report to Congress by The Heinz Center (2000) and a theme issue in the *Journal of Coastal Research* (Crowell and Leatherman, 1999) also concluded that, because of existing longterm shore erosion, there can be a substantial disparity between actual risk and insurance rates.

Would sea-level rise justify changing the current approach? Two possible alternatives would be to (1) shorten the period during which the assumed risk is kept fixed so that rates can respond to risk and property owners can respond, or (2) lengthen the duration of the insurance policy to the period of time between risk calculations, that is, instead of basing rates on the risk when the house is built, which tends to increasingly underestimate the risk, base the rate on an estimate of the average risk over the lifetime of the structure, using "erosion-hazard mapping" with assumed rates of sea-level rise, shore erosion, and structure lifetime. Both of these alternatives address changing risk by estimating risk over a time horizon equal to the period of time between risk recalculation. The erosion-hazard mapping approach has received considerable attention; the Heinz Center study also recommended that Congress authorize erosion-hazard mapping. Although Congress has not provided FEMA with authority to base rates on erosion hazard mapping, FEMA has raised rates in the V Zone by 10 percent per year (during most years) as a way of anticipating the increased flood damages resulting from the longterm erosion that The Heinz Center evaluated (Crowell *et al.*, 2007).

The Heinz Center study and recent FEMA efforts have assumed current rates of sea-level rise. FEMA has not investigated whether accelerated sea-level rise would increase the disparity between risks and insurance rates enough to institute additional changes in rates; nor has it investigated the option of relaxing the grandfathering policy so that premiums on existing homes rise in proportion to the increasing risk. Nevertheless, the Government Accountability Office (2007) recently recommended that FEMA analyze the potential long-term implications of climate change for the National Flood Insurance Program (NFIP). FEMA agreed to undertake such a study (Buckley, 2007) and initiated it in September 2008 (Department of Homeland Security, 2008).

10.7.3.2. Pre-FIRM and other Subsidized Policies

Since the 1970s, the flood insurance program has provided a subsidized rate for homes built before the program was implemented, that is, before the release of the first flood insurance rate map for a given location (Hayes *et al.*, 2006). The premium on a $100,000 home, for example, is generally $650 and $1170 for the A and V Zones, respectively—regardless of how far above or below the base flood elevation the structure may be (NFIP, 2008). Not all pre-FIRM homes obtain the subsidized policy. The subsidized rate is currently greater than the actuarial rate in the A and V Zones for homes that are at least 30 cm and 60 cm, respectively, above the base flood elevation (NFIP, 2008). But the subsidy is substantial for homes that are below the base flood elevation. Homes built in the V Zone between 1975 and 1981 also receive a subsidized rate; which is about $1500 for a $100,000 home built at the base flood elevation (NFIP, 2008).

Does sea-level rise justify changing the rate structure for subsidized policies? Economics alone can not answer that question because the subsidies are part of the program for reasons other than risk management and economic efficiency, such as the original objective of providing communities with an incentive to join the NFIP and the policy goal of not pricing people out of their homes (Hayes *et al.*, 2006). Moreover, the implications depend in large measure on whether the NFIP responds to increased damages from sea-level rise by increasing premiums or the subsidy, a question that rests on decisions that have not yet been made. Sea-level rise elevates the base flood elevation; and the subsidized rate is the same regardless of how far below the base flood elevation a home is built. Considering those factors alone, sea-level rise increases expected damages, but not the subsidized rate. However, the NFIP sets the subsidized rates to ensure that the entire program covers its costs during the average non-catastrophic year[20]. Therefore, if total damages (which include inland flooding) rise by the same proportion as damages to subsidized policies, the subsidized portion would stay the same as sea level rises.

FEMA has not yet quantified whether climate change is likely to increase total damages by a greater or smaller proportion than the increase due to sea-level rise. Without an assessment of whether the subsidy would increase or decrease, it would be premature to conclude that sea-level rise warrants a change in FEMA's rate structure. Nevertheless, sea-level rise is unlikely to threaten the financial integrity of the flood insurance program as long as subsidized rates are set high enough to cover claims during all but the catastrophic loss years, and Congress continues to provide the program with the necessary funds during the catastrophic years. Because the pre-FIRM subsidies only apply to homes that are several decades old, they do not encourage hazardous construction. As with grandfathering, the subsidized rate discourages owners of homes below the base flood elevation from elevating or otherwise reducing the risk to their homes as sea level rises, because the premium is already as low as it would be from elevating the home to the base flood elevation[21].

The practical importance of the pre-FIRM subsidy is sensitive to the future rate of sea-level rise. Today, pre-FIRM policies account for 24 percent of all policies (Hayes *et al.*, 2006). However, that fraction is declining (Crowell *et al.*, 2007) because development continues in coastal floodplains, and because the total number of homes eligible for pre-FIRM rates is declining, as homes built before the 1970s are lost to fire and storms, enlarged, or replaced with larger homes. A substantial rise in sea level over the next few decades would affect a large class of subsidized policy holders by the year 2100. Nevertheless, the portion of pre-FIRM houses is likely to be very small, unless there is a shift in the factors that have caused people to replace small cottages with larger houses and higher-density development (see Section 12.2.3).

Two other classes, which together account for 2 percent of policies, also provide subsidized rates. The A99 Zone consists of areas that are currently in the A Zone, but for which structural flood protection such as dikes are at least 50 percent complete. Policyholders in such areas pay a rate as if the structural protection was already complete (and successful). The AR Zone presents the opposite situation: locations where structural protection has been decertified. Provided that the structures are on a schedule for being rebuilt, the rates are set to the rate that applies to the X Zone or the pre-FIRM subsidized rate, whichever is less. As sea level rises, the magnitude of these subsidies may increase, both because the base flood elevations (without the protection) will be higher, and because more coastal lands may be protected with dikes and other structural measures. Unlike the pre-FIRM subsidies, the A99 and AR Zone subsidies may encourage construction in hazardous areas; but unlike other subsidies, the A99 and AR Zone subsidies encourage protection measures that reduce hazards.

10.7.4. Post-Disaster Hazard Mitigation

If a coastal community is ultimately going to be abandoned to the rising sea, a major rebuilding effort in the current location may be less useful than expending the same resources to rebuild the community on higher ground. On the other hand, if the community plans to remain in its current location despite the increasing costs of shore protection, then it is important for people to understand that commitment. Unless property owners know which path the community is following, they do not know whether to reinvest. Moreover, if the community is going to stay in its current location, owners need to know whether their land will be protected with a dike or if land surfaces are likely to be elevated over time (see Section 12.3).

10.8. Conclusions

The need to prepare for rising sea level depends on the length of time over which the decision will continue to have consequences; how sensitive those consequences are to sea level; how rapidly the sea is expected to rise and the magnitude of uncertainty over that expectation; the decision maker's risk tolerance; and the implications of deferring a decision to prepare. Considering sea-level rise may be important if the decision has outcomes over a long period of time and concerns an activity that is sensitive to sea level, especially if what can be done to prepare today would not be feasible later. Those making decisions with outcomes over a short period of time concerning activities that are not sensitive to sea level probably need not consider sea-level rise, especially if preparing later is as effective as preparing today.

Instances where the existing literature provides an economic rationale for preparing for accelerated sea-level rise include:

- *Coastal wetland protection.* Wetlands and the success of wetland-protection efforts are almost certainly sensitive enough to sea-level rise to warrant examination of some changes in coastal wetland protection efforts, assuming that the objective is to ensure that most estuaries that have extensive wetlands today will continue to have tidal wetlands in the future. Coastal wetlands are sensitive to rising sea level, and many of the possible measures needed to ensure their survival as sea level rises are least disruptive with a lead time of several decades. Changes in management approaches would likely involve consideration of options at various levels of authority.

- *Coastal infrastructure.* Whether it is beneficial to design coastal infrastructure to anticipate rising sea level depends on the ratio of the incremental cost of designing for a higher sea level now, compared with the retrofit cost of modifying the structure later. No general statement is possible because this ratio varies and relatively few engineering assessments of the question have been published. However, because the cost of analyzing this question is very small compared with the retrofit cost, it is likely that most long-lived infrastructure in the coastal zone is sufficiently sensitive to rising sea level to warrant an analysis of the comparative cost of designing for higher water levels now and retrofitting later.

- *Building along the coast.* In general, the economics of coastal development alone does not currently appear to be sufficiently sensitive to sea-level rise to avoid construction in coastal areas. Land values are so high that development is often economic even if a home is certain to be lost within a few decades. The optimal location and elevation of new homes may be sensitive to how rapidly sea level is expected to rise.

- *Shoreline planning.* A wide array of measures for adapting to rising sea level depend on whether a given area will be elevated, protected with structures, or abandoned to the rising sea. Several studies have shown that in those cases where the shores will retreat and structures will be removed, the economic cost will be much less if people plan for that retreat. The human toll of an unplanned abandonment may be much greater than if people gradually relocate when it is convenient to do so. Conversely, people may be reluctant to invest in an area without some assurance that lands will not be lost to the sea. Therefore, long-term shoreline planning is generally justified and will save more than it costs; the more the sea ultimately rises, the greater the value of that planning.

- *Rolling easements, density restrictions, and coastal setbacks.* Several studies have shown that, in those cases where the shores will retreat and structures will be removed, the economic cost will be much less if people plan for that retreat. Along estuaries, a retreat in developed areas rarely occurs and thus is likely to only occur if land remains lightly developed. It is very likely that options such as rolling easements, density restrictions, coastal setbacks, and vegetative buffers, would increase the ability of wetlands and beaches to migrate inland.

- *Floodplain management: Consideration of reflecting actual risk in flood insurance rates.* Economists and other commentators generally agree that insurance works best when the premiums reflect the actual risk. Even without considering the possibility of accelerated sea-level rise, the National Academy of Sciences (NAS, 1990) and a FEMA-supported study by The Heinz Center (2000) concluded and recommended to Congress that insurance rates should reflect the changing risks resulting from coastal erosion. Rising sea level increases the potential disparity between rates and risks of storm-related flooding.

REFERENCES

* Indicates non-peer reviewed literature; available upon request

Arrow, K. J. & Fisher, A. C. (1974). Environmental preservation, uncertainty, and irreversibility. *Quarterly Journal of Economics, 88(1),* 312-319.

Bin, O., Crawford, T. Kruse, J. B. & Landry, C. E. (2008). Viewscapes and flood hazard: coastal housing market response to amenities and risk. *Land Economics, 84(3),* 434-448.

Buckley, M., (2007). Testimony of Michael Buckley, U.S. Senate Committee on Homeland Security and Government Affairs, April 19, 2007.

Cape Cod Commission, 2002: *Model Floodplain District Bylaw.* Cape Cod Commission, Barnstable, MA, section 06.1. <http://www.capecodcommission.org/ bylaws/floodplain. html>

CCSP (Climate Change Science Program), (2006). *Coastal Elevations and Sensitivity to Sea level Rise Final Prospectus for Synthesis and Assessment Product 4.1.* United States Climate Change Science Program, Washington, DC, 19 pp.

Congressional Research Service, (2003). *Benefit-Cost Analysis and the Discount Rate for the Corps of Engineers' Water Resource Projects: Theory and Practice.* [Power, K. (analyst)]. RL3 1976. Congressional Research Service, Washington, DC, 26 pp.

Cordes, J. J. & Yezer, A. M. J. (1998). In harm's way: does federal spending on beach enhancement and protection induce excessive development in coastal areas? *Land Economics*, *74(1)*, 128-145.

Crowell, M. & Leatherman, S.P. (eds.), (1999). Coastal erosion mapping and management. *Journal of Coastal Research*, Special Issue 28, 196 pp.

Crowell, M., Hirsch, E. & Hayes, T.L. (2007). Improving FEMA's coastal risk assessment through the National Flood Insurance Program: an historical overview. *Marine Technology Society Journal*, *41(1)*, 18-27.

Dasgupta, P., (2007). Commentary: The Stern Review's economics of climate change. *National Institute Economic Review*, *199*, 4-7.

Department of Homeland Security, (2008). *Impact of Climate Change on the National Flood Insurance Program*. Solicitation number HSFEHQ-08-R-0082. <http://www.Fed BizOpps.gov>

Evatt, D. S. (1999). *National Flood Insurance Program: Issues Assessment.* Federal Emergency Management Agency, Washington, DC, 123 pp.

Evatt, D.S., (2000). Does the national flood insurance program drive floodplain development? *Journal of Insurance Regulation*, *18(4)*, 497-523.

FEMA (Federal Emergency Management Agency), 1991: *Projected Impact of Relative Sea Level Rise on the National Flood Insurance Program.* FEMA Flood Insurance Administration, Washington DC, 72 pp.

Fisher, A.C. and W.M. Hanemann, (1987). Quasi-option value: some misconceptions dispelled. *Journal of Environmental Economics and Management*, *14(2)*, 183- 190.

Frankhauser, S. Smith, J.B. & Tol, R.S.J. (1999). Weathering climate change: Some simple rules to guide adaptation decisions. *Ecological Economics*, *30(1)*, 67-78.

Freeman, A.M. (2003). *The Measurement of Environmental and Resource Values: Theory and Methods*. Resources for the Future, Washington, DC, 2nd ed., 491 pp.

Gilbert, S. & Horner, R. (1984). *The Thames Barrier*. Thomas Telford, London, 182 pp.

GAO (General Accounting Office), (1982). *National Flood Insurance: Marginal Impact on Floodplain Development, Administrative Improvements Needed.* Report to the Subcommittee on Consumer Affairs, Committee on Banking, Housing, and Urban Affairs, U.S. Senate. General Accounting Office, Washington, DC, 59 pp.

GAO (Government Accountability Office), (2007). *Climate Change: Financial Risks to Federal and Private Insurers in Coming Decades are Potentially Significant.*

GAO-07-285. Government Accountability Office, Washington DC, 68 pp. <http://www.gao.gov/new.items/d07285 .pdf>

Ha-Duong, M., (1998). Quasi-option value and climate policy choices. *Energy Economics*, *20(5/6)*, 599-620.

Hardisky, M.A. & Klemas, V. (1983). Tidal wetlands natural and human-made change from 1973 to 1979 in Delaware: mapping techniques and results. *Environmental Management*, *7(4)*, 339-344.

Hayes, T.L., Spafford, D.R. & Boone, J.P. (2006). *Actuarial Rate Review*. National Flood Insurance Program, Washington DC, 34 pp. < http://www.fema.gov/library/ viewRecord.do?id 63>

Heinz Center, (2000). *Evaluation of Erosion Hazards*. The H. John Heinz III Center for Science, Economics, and the Environment, Washington, DC, 252 pp. <http://www.heinzctr.org/publications.shtml#erosionhazards>

Hoffman, J.S., Keyes, D. & Titus, J.G. (1983). *Projecting Future Sea Level Rise; Methodology, Estimates to the Year 2100, and Research Needs*. U.S. Environmental Protection Agency, Washington DC, 121 pp.

IPCC (Intergovernmental Panel on Climate Change), (1990). *Strategies for Adaptation to Sea Level Rise*. Report of the Coastal Zone Management Subgroup, IPCC Response Strategies Working Group. Ministry of Transport and Public Works, The Hague, the Netherlands, 131 pp. < http://yosemite.epa.gov/OAR%5Cglobalwarming. nsf/content/ ResourceCenterPu blicationsSLRAdaption.html>

IPCC (Intergovernmental Panel on Climate Change), (1996). *Climate Change 1995: The Science of Climate Change.* Contribution of Working Group I to the Second Assessment Report of the Intergovernmental Panel on Climate Change. [Houghton, J.J., L.G. Meiro Filho, B.A. Callander, N. Harris, A. Kattenberg, and K. Maskell (eds.)]. Cambridge University Press, Cambridge, UK, and New York, 572 pp.

IPCC (Intergovernmental Panel on Climate Change), (2001). *Climate Change 2001: The Scientific Basis.* Contribution of Working Group I to the Third Assessment Report of the Intergovernmental Panel on Climate Change [Houghton, J.T., Y. Ding, D.J. Griggs, M. Noguer, P.J. van der Linden, X. Dai, K. Maskell, and C.A. Johnston (eds.)]. Cambridge University Press, Cambridge, UK, and New York, 881 pp.

IPCC (Intergovernmental Panel on Climate Change), (2007). *Climate Change 2007: The Physical Science Basis.* Contribution of Working Group I to the Fourth Assessment Report of the Intergovernmental Panel on Climate Change [Solomon, S., D. Qin, M. Manning, Z. Chen, M. Marquis, K.B. Averyt, M. Tignor, and H.L. Miller (eds.)]. Cambridge University Press, Cambridge, UK, and New York, 996 pp.

IPCC CZMS (Intergovernmental Panel on Climate Change Coastal Zone Management Subgroup), 1992: *Global Climate Change and the Rising Challenge of the Sea.*IPCC Response Strategies Working Group, Rijkswaterstaat, The Hague, the Netherlands.

IPET (Interagency Performance Evaluation Task Force), (2006). *Performance Evaluation of the New Orleans and Southeast Louisiana Hurricane Protection System.* U.S. Army Corps of Engineers, Washington, DC. <https://ipet.wes.army.mil/>

Jones, C.P., Coulbourne, W.L. Marshall, J. & Roger Jr, S.M. (2006). *Evaluation of the National Flood Insurance Program's Building Standards.* American Institutes for Research, Washington, DC.

Kentula, M. (1999). *Restoration, creation, and recovery of wetlands: wetland restoration and creation.* In: *National Water Summary on Wetland* Resources. [Fretwell, J.D., R.J. Redman, and J.S. Williams (eds.)]. United States Geological Survey water supply paper 2425. U.S. Geological Survey, [Reston, VA]. <http://water.usgs.gov/nwsum/WSP2425/ restoration.html>

Klein, R. J. T., Nicholls, R. J. & Mimura, N. (1999). Coastal adaptation to climate change: Can the IPCC technical guidelines be applied? *Mitigation and Adaptation Strategies for Global Change, 4(3-4),* 239-252.

Kruczynski, W.L., (1990). Options to be considered in preparation and evaluation of mitigation plans. In: *Wetland Creation and Restoration: The Status of the Science* [Kusler, J.A. and M.E. Kentula (eds.)]. Island Press, Washington, DC, 594 pp.

Kussler, J., (1999). *Common Questions: Wetland Restoration, Creation, and Enhancement.* Association of State Wetland Managerss, Berne, NY, 15 pp. <http://www.aswm. org/propub/20_restoration_6_26_06.pdf>

Lavery, S. & Donovan, B. (2005). Flood risk management in the Thames Estuary looking ahead 100 years. *Philosophical Transactions of the Royal Society A, 363,* 1455-1474.

Landry, C.E., Keeler, A.G. & Kriesel, W. (2003). An economic evaluation of beach erosion management alternatives. *Marine Resource Economics, 18(2),* 105-127.

Leatherman, S. (1997). *Flood Insurance Availability in Coastal Areas: The Role It Plays in Encouraging Development Decisions.* Federal Emergency Management Agency, Washington, DC.

Matcha vs Maddox, 711 S.W.2d 95, 100 (Tex. App. 1986).

Miller, H.C. (1981). *Coastal Flood Hazards and the National Flood Insurance Program.* Federal Emergency Management Agency, Washington, DC, 50 pp.

NAS (National Academy of Sciences) (1990). *Managing Coastal Erosion.* National Academy Press, Washington, DC, 182 pp.

NFIP (National Flood Insurance Program), (2007). *Fact Sheet: Saving on Flood Insurance Information about the NFIP 's Grandfathering Rule.* FEMA, [Washington DC], 2 pp. <http://www.fema.gov/library/viewRecord.do?id>

NFIP (National Flood Insurance Program), (2008). *Flood Insurance Manual.* Federal Emergency Management Agency, Washington DC.

Nordhaus, W. D., (2007a). Critical assumptions in the Stern Review on Climate Change. *Science, 317(5835),* 201-202.

Nordhaus, W.D., (2007b). A review of the Stern Review on *The Economics of Climate Change. Journal of Economic Literature, 45,* 686-702.

NRC (National Research Council), (1983). *Changing Climate.* National Academy Press, Washington, DC, 496 pp.

NRC (National Research Council), (1987). *Responding to Changes in Sea Level: Engineering Implications.* National Academy Press, Washington, DC, 148 pp.

NRC (National Research Council), (2007). *Mitigating Shore Erosion Along Sheltered Coasts.* National Academies Press, Washington, DC, 174 pp.

O'Callahan, J. (ed.), (1994). *Global Climate Change and the Rising Challenge of the Sea.* Proceedings of the third IPCC CZMS workshop, Isla de Margarita, Venezuela, 9–13 March 1992. National Oceanic and Atmospheric Administration, Silver Spring, MD, 691 pp.

OMB (Office of Management and Budget), (1992). *Guidelines and Discount Rates for Benefit-Cost Analysis of Federal Programs.* OMB Circular A-94, Office of Management and Budget, Washington, DC. <http://www.whitehouse.gov/omb/ circulars/a094/ a094.html>

OTA (Office of Technology Assessment), (1993). *Preparing for an Uncertain Climate – Volume I.* OTA-O-567. U.S. Government Printing Office, Washington, DC, 359 pp.

Reed, D. J., Bishara, D. Cahoon, D. Donnelly, J. Kearney, M. Kolker, A. Leonard, L. Orson, R. A. & Stevenson, J.C. (2008). Site-specific scenarios for wetlands accretion as sea level rises in the mid-Atlantic region. Section 2.1 in: *Background Documents Supporting Climate Change Science Program Synthesis and Assessment Product 4.1: Coastal Elevations and Sensitivity to Sea Level Rise* [Titus, J.G. and E.M. Strange (eds.)]. EPA 430R07004. U.S. Environmental Protection Agency, Washington, DC, pp. 134-174. <http://epa.gov/climatechange/effects/coastal/background.html>

Samuelson, P. A. & Nordhaus, W. D. (1989). *Economics.* McGraw-Hill, New York, 13th ed., 1013 pp.

Scheraga, J. D. & Grambsch, A.E. (1998). Risks, opportunities, and adaptation to climate change. *Climate Research, 11(1),* 85-95.

Shilling, J.D., Sirmans, C.E. & Benjamin, J.D. (1989). Flood insurance, wealth redistribution, and urban property values. *Journal of Urban Economics*, *26(1),* 43- 53.

Stockton, M. B. & Richardson, C. J. (1987). Wetland development trends in coastal North Carolina, USA, from 1970 to 1984. *Environmental Management, 11(5),* 649-657.

Titus, J. G. (1990). Greenhouse effect, sea level rise, and barrier islands: case study of Long Beach Island, New Jersey. *Coastal Management, 18(1),* 65-90.

Titus, J. G. (1991). Greenhouse effect and coastal wetland policy: how Americans could abandon an area the size of Massachusetts at minimum cost. *Environmental Management, 15(1),* 39-58.

Titus, J. G. (1998). Rising seas, coastal erosion and the takings clause: how to save wetlands and beaches without hurting property owners. *Maryland Law Review*, *57(4)*, 1279-1299.

Titus, J. G. (2000). Does the U.S. government realize that the sea is rising? How to restructure federal programs so that wetlands and beaches survive. *Golden Gate Law Review*, *30(4),* 717–786.

Titus, J.G. (2005). Does shoreline armoring violate the Clean Water Act? Rolling easements, shoreline planning, and other responses to sea level rise. In: *America's Changing Coasts: Private Rights and Public Trust* [Whitelaw, D.M. and G.R. Visgilio (eds.)]. Edward Elgar Publishing, Cheltenham, UK, and Northampton, MA, 248 pp.

Titus, J. G. & Narayanan, V. (1996). The risk of sea level rise. *Climatic Change, 33(2),* 151-212.

Titus, J. G. & Wang, J. (2008). Maps of lands close to sea level along the middle Atlantic coast of the United States: an elevation data set to use while waiting for LIDAR. Section 1.1 in: *Background Documents Supporting Climate Change Science Program Synthesis and Assessment Product 4.1: Coastal Elevations and Sensitivity to Sea Level Rise* [Titus, J.G. and E.M. Strange (eds.)]. EPA 430R07004. U.S. Environmental Protection Agency, Washington, DC, pp. 2-44.

Titus, J. G., Kuo, C. Y., Gibbs, M.J., LaRoche, T.B., Webb, M.K.& Waddell, J.O. (1987). Greenhouse effect, sea level rise, and coastal drainage systems. *Journal of Water Resources Planning and Management, 113(2)*, 216-227.

Titus, J. G., Park, R. A., Leatherman, S. P., Weggel, J. R., Greene, M. S., Mansel, P. W., Brown, G., Gaunt, S., Treehan, M. & Yohe, G. (1991). Greenhouse effect and sea level rise: the cost of holding back the sea. *Coastal Management*, *19(2),* 171-204.

Township of Long Beach (New Jersey), (2008). *Important Flood Information.* <http://longbeachtownship.com/importantfloofing information.html>

Town of Ocean City, Maryland, (1999). City Code § 38-71.

TRB (Transportation Research Board), (2008). *Potential Impacts of Climate Change on U.S. Transportation.* National Academies Press, Washington, DC, 234 pp.

USACE (U.S. Army Corps of Engineers), (2000a). *Planning Guidance Notebook: Appendix E: Civil Works Missions and Evaluation Procedures*. ER 1105-2-100. U. S. Army Corps of Engineers, Washington, DC, 310 pp. <http://www.usace.army.mil/publications/eng-regs/er1 105-2-1 00/toc.htm>

USACE (U.S. Army Corps of Engineers), (2000b). *Civil Works Construction Cost Index System (CWCCIS).* EM 1110-2-1304. U.S. Army Corps of Engineers, Washington, DC.

For updated CWCCIS data tables, see: <http://www.usace.army.mil/publications/eng-manuals/em1 110-2-1 304/toc.htm>

USACE (U.S. Army Corps of Engineers), (2007). *Civil Works Construction Cost Index System (CWCCIS). Revised Tables.* EM 1110-2-1304. U.S. Army Corps of Engineers, Washington, DC. Tables revised 30 September 2007. <http://www.usace.army.mil/publications/eng-manuals/em1 110-2-1 304/toc.htm>

U.S. EPA (Environmental Protection Agency), (1989). *Potential Effects of Global Climate Change on the United States: Report to Congress.* EPA 230-05-89-050. U.S. Environmental Protection Agency, Washington, DC.

U.S. EPA (Environmental Protection Agency), (1995). Federal guidance for the establishment, use and operation of mitigation banks. *Federal Register, 60(228)*, 58605-58614.

U.S. EPA (Environmental Protection Agency), (2000). *Guidelines for Preparing Economic Analyses.* EPA 240-R-00-003. EPA Office of the Administrator, [Washington DC]. <http://yosemite.epa.gov/ee/epa/eed.nsf/webpages>

U.S. EPA (Environmental Protection Agency) & USACE (U.S. Army Corps of Engineers), (1990). Mitigation Memorandum of Agreement 4 (Feb. 6, 1990).

Weggel, J. R., Brown, S., Escajadillo, J.C., Breen, P.& Doheyn, E. (1989). The cost of defending developed shorelines along sheltered waters of the United States from a two meter rise in mean sea level. In: *Potential Effects of Global Climate Change on the United States. Appendix: B, Sea Level Rise.* U.S. Environmental Protection Agency, Washington, DC, pp. 3-1 to 3-90. <http://epa.gov/climatechange/effects/coastal/appB .html>

Yohe, G., Neumann, J. E. Marshall, P. & Ameden, H. (1996). The economic cost of greenhouse-induced sea-level rise for developed property in the United States. *Climatic Change, 32(4),* 387-410.

In: Coastal Sensitivity to Sea Level Rise-Focusing on the Mid-... ISBN: 978-1-60741-440-7
Editor: Melvin C. Urajner

Chapter 11

ONGOING ADAPTATION

James G. Titus
EPA

Key Findings

- Most organizations are not yet taking specific measures to prepare for rising sea level. Recently, however, many public and private organizations have begun to assess possible response options.
- Most of the specific measures that have been taken to prepare for accelerated sea-level rise have had the purpose of reducing the long-term adverse environmental impacts.

11.1. Introduction

Preparing for the consequences of rising sea level has been the exception rather than the rule in the Mid-Atlantic. Nevertheless, many coastal decision makers are now starting to consider how to prepare.

This Chapter examines those cases in which organizations are taking specific measures to consciously anticipate the effects of sea-level rise. It does not include most cases in which an organization has authorized a study but not yet acted upon the study. Nor does it catalogue the activities undertaken for other reasons that might also help to prepare for accelerated sea-level rise[22], or cases where people responded to sea level rise after the fact (see Box 11.1). Finally, it only considers measures that had been taken by March 2008. Important measures may have been adopted between the time this Product was drafted and its final publication.

11.2. Adaptation for Environmental Purposes

Many organizations that manage land for environmental purposes are starting to anticipate the effects of sea-level rise. Outside the Mid-Atlantic, some environmental regulators have also begun to address this issue.

11.2.1. Environmental Regulators

Organizations that regulate land use for environmental purposes generally have not implemented adaptation options to address the prospects of accelerated sea-level rise. Congress has given neither the U.S. Army Corps of Engineers (USACE) nor the U.S. Environmental Protection Agency (EPA) a mandate to modify existing wetland regulations to address rising sea level; nor have those agencies developed approaches for moving ahead without such a mandate (see Chapter 12). For more than a decade, Maine[23], Massachusetts[24], and Rhode Island[25] have had statutes or regulations that restrict shoreline armoring to enable dunes or wetlands to migrate inland with an explicit recognition of rising sea level (Titus, 1998).

None of the eight mid-Atlantic states require landowners to allow wetlands to migrate inland as sea level rises (NOAA, 2006). During 2008, however, the prospect of losing ecosystems to a rising sea prompted Maryland to enact the "Living Shoreline Protection Act"[26]. Under the Act, the Department of Environment will designate certain areas as appropriate for structural shoreline measures (*e.g.*, bulkheads and revetments). Outside of those areas, only nonstructural measures (*e.g.*, marsh creation, beach nourishment) will be allowed unless the property owner can demonstrate that nonstructural measures are infeasible[27]. The new statute does not ensure that wetlands are able to migrate inland; but Maryland's coastal land use statute limits development to one home per 8.09 hectares (ha) (20 acres [ac]) in most rural areas within 305 meters (m) (1000 feet [ft]) of the shore (see Section A1.F.2.1 in Appendix 1). Although that statute was enacted in the 1980s to prevent deterioration of water quality, the state now considers it to be part of its sea-level rise adaptation strategy.[28]

11.2.2. Environmental Land Managers

Those who manage land for environmental purposes have taken some initial steps to address rising sea level.

Federal Land Managers

The Department of Interior (Secretarial Order 3226, 2001) requires climate change impacts be taken into account in planning and decision making (Scarlett, 2007). The National Park Service has worked with the United States Geological Survey (USGS) to examine coastal vulnerability on 25 of its coastal parks (Pendleton *et al.*, 2004). The U.S. Fish and Wildlife Service is incorporating studies of climate change impacts, including sea-level rise, in their Comprehensive Conservation Plans where relevant.

Figure 11.1. Allowing beaches and wetlands to migrate inland in the national parks (a) Cape Hatteras National Seashore. (June 2002) Until it was relocated inland in 1999, the lighthouse was just to the right of the stone groin in the foreground. (b) Jamestown Island ,Virginia (September 2004). As sea level rises, marshes have taken over land that was cultivated during colonial times.

Figure 11.2. Responding to sea-level rise at Blackwater National Wildlife Refuge, Maryland (October 2002). (a) Marsh Deterioration. (b) Marsh Creation. The dredge fills the area between the stakes to create land at an elevation flooded by the tides, after which marsh grasses are planted

The National Park Service and the U.S. Fish and Wildlife Service each have large coastal landholdings that could erode or become submerged as sea level rises (Thieler *et al.*, 2002; Pendleton *et al.*, 2004). Neither organization has an explicit policy concerning sea- level rise, but both are starting to consider their options. The National Park Service generally favors allowing natural shoreline processes to continue (NPS Management Policies §4.8.1), which allows ecosystems to migrate inland as sea level rises (see Figure 11.1). In 1999, this policy led the Park Service to move the Cape Hatteras Lighthouse inland 900 m (2900 ft) at a cost of $12 million. The U.S. Fish and Wildlife Service generally allows dry land to convert to wetlands, but it is not necessarily passive as rising sea level erodes the seaward boundary of tidal wetlands. Blackwater National Wildlife Refuge, for example, has used dredge material to rebuild wetlands on a pilot basis, and is exploring options to recreate about 3000 ha (7000 ac) of marsh (see Figure 11.2). Neither agency has made land purchases or easements to enable parks and refuges to migrate inland.

Figure 11.3. The Albemarle Sound environment that the Nature Conservancy seeks to preserve as sea level rises (June 2002). (a) Nature Conservancy lands on Roanoke Island depict effects of rising sea level. Tidal wetlands (juncas and spartina patens) have taken over most of the area depicted as sea level rises, but a stand of trees remains in a small area of higher ground. (b) Mouth of the Roanoke River, North Carolina. Cypress trees germinate on dry land; but continue to grow in the water after the land is eroded or submerged by rising sea level.

The Nature Conservancy

The Nature Conservancy (TNC) is the largest private holder of conservation lands in the Mid-Atlantic. It has declared as a matter of policy that it is trying to anticipate rising sea level and climate change. Its initial focus has been to preserve ecosystems on the Pamlico-Albemarle Peninsula, such as those shown in Figure 11.3 (Pearsall and Poulter, 2005; TNC, 2007). Options under consideration include: plugging canals to prevent subsidence-inducing saltwater intrusion, planting cypress trees where pocosins have been converted to dry land, and planting brackish marsh grasses in areas likely to be inundated. As part of that project, TNC undertook the first attempt by a private conservancy to purchase rolling easements (although none were purchased). TNC owns the majority of barrier islands along the Delmarva Peninsula, but none of the mainland shore. TNC is starting to examine whether preserving the ecosystems as sea level rises would be best facilitated by purchasing land on the mainland side as well, to ensure sediment sources for the extensive mudflats so that they might keep pace with rising sea level.

State conservation managers have not yet started to prepare for rising sea level (NOAA, 2006). But at least one state (Maryland) is starting to refine a plan for conservation that would consider the impact of rising sea level.

11.3. Other Adaptation Options being Considered by Federal, State, and Local Governments

11.3.1. Federal Government

Federal researchers have been examining how best to adapt to sea-level rise for the last few decades, and those charged with implementing programs are also now beginning to consider implications and options. The longstanding assessment programs will enable federal agencies to respond more rapidly and reasonably if and when policy decisions are made to begin preparing for the consequences of rising sea level.

The Coastal Zone Management Act is a typical example. The Act encourages states to protect wetlands, minimize vulnerability to flood and erosion hazards, and improve public access to the coast. Since 1990, the Act has included sea-level rise in the list of hazards that states should address. This congressional mandate has induced NOAA to fund state-specific studies of the implications of sea-level rise, and encouraged states to periodically designate specific staff to keep track of the issue. But it has not yet altered what people actually do along the coast (New York, 2006; New Jersey, 2006; Pennsylvania, 2006; Delaware, 2005; Maryland, 2006; Virginia, 2006; North Carolina, 2006). Titus (2000) and CSO (2007) have examined ways to facilitate implementation of this statutory provision, such as federal guidance and/or additional interagency coordination. Similarly, the U.S. Army Corps of Engineers (USACE) has formally included the prospect of rising sea level for at least a decade in its planning guidance for the last decade (USACE, 2000), and staff have sometimes evaluated the implications for specific decisions (*e.g.*, Knuuti, 2002). But the prospect of accelerated sea-level rise has not caused a major change in the agency's overall approach to wetland permits and shore protection (see Chapter 12).

11.3.2.State Government

Maryland has considered the implications of sea-level rise in some decisions over the last few decades. Rising sea level was one reason that the state gave for changing its shore protection strategy at Ocean City from groins to beach nourishment (see Section A1.F in Appendix 1). Using NOAA funds, the state later developed a preliminary strategy for dealing with sea-level rise. As part of that strategy, the state also recently obtained a complete lidar dataset of coastal elevations.

Delaware officials have long considered how best to modify infrastructure as sea level rises along Delaware Bay, although they have not put together a comprehensive strategy (CCSP, 2007).

Because of the vulnerability of the New Jersey coast to flooding, shoreline erosion, and wetland loss (see Figure 11.4), the coastal management staff of the New Jersey Department of Environmental Protection have been guided by a long-term perspective on coastal processes, including the impacts of sea-level rise. So far, neither Delaware nor New Jersey has specifically altered their activities because of projected sea-level rise. Nevertheless, New Jersey is currently undertaking an assessment that may enable it to factor rising sea level into its strategy for preserving the Delaware Estuary (CCSP, 2007).

Figure 11.4. Vulnerability of New Jersey's coastal zone (a) Wetland fringe lacks room for wetland migration (Monmouth August 2003). (b) Low bay sides of barrier islands are vulnerable to even a modest storm surge. (Ship Bottom, September 2, 2006). (c) Gibbstown Levee and (d) associated tide gate protect lowlying areas of Greenwich Township (March 2003).

In the last two years, states have become increasingly interested in addressing the implications of rising sea level. A bill in the New York General Assembly would create a sea-level rise task force (Bill AO9002 2007-2008 Regular Session). Maryland and Virginia have climate change task forces that have focused on adapting to rising sea level. (For a comprehensive survey of what state governments are doing in response to rising sea level, see Coastal States Organization, 2007.)

11.3.3.Local Government

A few local governments have considered the implications of rising sea level for roads, infrastructure, and floodplain management (see Boxes A1.4 and A1.6 in Appendix 1). New York City's plan for the year 2030 includes adapting to climate change (City of New York, 2008). The New York City Department of Environmental Protection is looking at ways to decrease the impacts of storm surge by building flood walls to protect critical infrastructure such as waste plants, and is also examining ways to prevent the sewer system from backing up more frequently as sea level rises (Rosenzweig *et al.*, 2006). The city has also been investigating the possible construction of a major tidal flood gate across the Verizano Narrows to protect Manhattan (Velasquez-Manoff, 2006).

Outside of the Mid-Atlantic, Miami-Dade County in Florida has been studying its vulnerability to sea-level rise, including developing maps to indicate which areas are at greatest risk of inundation. The county is hardening facilities to better withstand hurricanes, monitoring the salt front, examining membrane technology for desalinating seawater, and creating a climate advisory task force to advise the county commission (Yoder, 2007).

Box 11.1. Jamestown: An Historic Example of Retreat in Response to Sea Level Rise

Established in 1607 along the James River, Jamestown was the capital of Virginia until 1699, when a fire destroyed the statehouse. Nevertheless, rising sea level was probably a contributing factor in the decision to move the capital to Williamsburg, because it was making the Jamestown peninsula less habitable than it had been during the previous century. Fresh water was scarce, especially during droughts (Blanton, 2000). The James River was brackish, so groundwater was the only reliable source of freshwater. But the low elevations on Jamestown limited the thickness of the freshwater table—especially during droughts. As Box Figure 11.1 shows, a 10 centimeter (cm) rise in sea level can reduce the thickness of the freshwater table by four meters on a low-lying island where the freshwater lens floats atop the salt water.

Rising sea level has continued to alter Jamestown. Two hundred years ago, the isthmus that connected the peninsula to the mainland eroded, creating Jamestown Island (Johnson and Hobbs, 1994). Shore erosion also threatened the location of the historic town itself, until a stone revetment was constructed (Johnson and Hobbs, 1994). As the sea rose, the shallow valleys between the ridges on the island became freshwater marsh, and then tidal marsh (Johnson and Hobbs, 1994). Maps from the seventeenth century show agriculture on lands that today are salt marsh. Having converted mainland to island, the rising sea will eventually convert the island to open water, unless the National Park Service continues to protect it from the rising water.

Other shorelines along Chesapeake Bay have also been retreating over the last four centuries. Several bay island fishing villages have had to relocate to the mainland as the islands on which they were located eroded away (Leatherman *et al.*, 1995). Today, low-lying farms on the Eastern Shore are converting to marsh, while the marshes in wildlife refuges convert to open water.

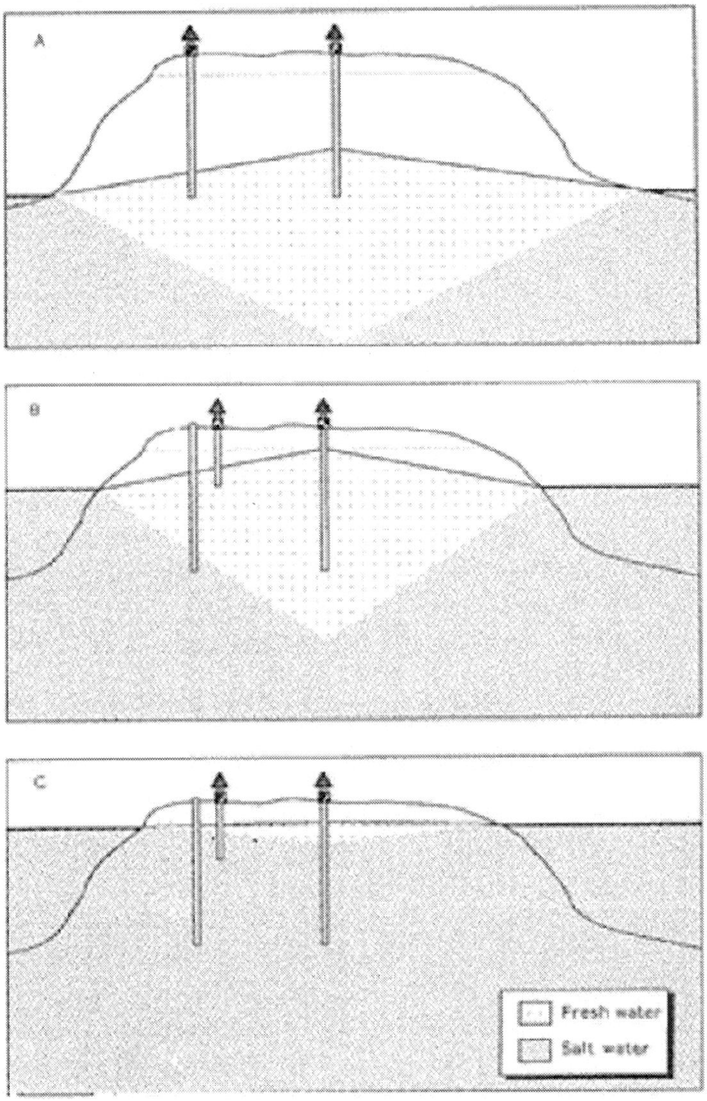

Box Figure 11.1. Impact of sea-level rise on an island freshwater table. (a) According to the Ghyben- Herzberg relation, the freshwater table extends below sea level 40 cm for every 1 cm by which it extends above sea level (Ghyve [1889] and Herzberg [1901], as cited by Freeze and Cherry [1979]). (b) For islands with substantial elevation, a 1 -m rise in sea level simply shifts the entire water table up 1 meter, and the only problem is that a few wells will have to be replaced with shallower wells. (c) However, for very low islands the water table cannot rise because of runoff, evaporation, and transpiration. A rise in sea level would thus narrow the water table by 40 cm for every 1 cm that the sea level rises, effectively eliminating groundwater supplies for the lowest islands.

REFERENCES

* Indicates non-peer reviewed literature; available upon request

Blanton, D.B. (2000). Drought as a factor in the Jamestown colony, 1607-1612. *Historical Archaeology*, *34(4),* 74-81.

CCSP (Climate Change Science Program), ((2007). *Stakeholder Meetings Final Report: Climate Change Science Program Synthesis and Assessment Product 4.1.* [National Oceanic and Atmospheric Administration, Silver Spring, MD], 81 pp. <http://www.climatescience.gov/Library/sap/sap4-1 /stakeholdermeetingfinalreport.pdf>

City of New York, (2008). *PlaNYC: A Greater, Greener New York.* City of New York, New York, 155 pp. <http://home2.nyc.gov/html/planyc2030/downloads/pdf/full_report.pdf>

(CSO) Coastal States Organization, (2007). *The Role of Coastal Zone Management Programs in Adaptation to Climate Change.* Coastal States Organization, Washington, DC, 27 pp. <http://www.coastalstates.org/documents/CSO%20Climate%20Change%20Final %20Report.pdf>

Delaware (Delaware Coastal Programs) (2005). *Delaware Coastal Programs: Section 309 Enhancement Assessment.* Department of Natural Resources and Environmental Control, Dover, DE, 46 pp.

Freeze, R. A. & Cherry, J. A. (1979). *Groundwater.* Prentice-Hall, Englewood Cliffs, NJ, 604 pp.

Johnson, G. H. & Hobbs, C. H. (1994). The geological history of Jamestown Island. *Jamestown Archaeological Assessment Newsletter*, *1(2/3),* 9-11.

Knuuti, K. (2002). Planning for sea level rise: U.S. Army Corps of Engineers policy. In: *Solutions to Coastal Disasters'02* [Ewing, L. and L. Wallendorf (eds.)]. American Society of Civil Engineers, Reston, VA, pp. 549-560.

* Leatherman, S. P., Chalfont, R. Pendleton, E. C. & McCandless, T. L. (1995). *Vanishing Lands: Sea Level, Society, and Chesapeake Bay.* Laboratory of Coastal Research, University of Maryland, College Park, and Department of the Interior, [Washington, DC], 47 pp.

Maryland, (2006). *CZMA, Secton 309 Assessment and Strategy.* Maryland Department of Natural Resources, Coastal Zone Management Division, Watershed Services Center, Annapolis, 70 pp. <http://www.dnr.state.md.us/bay/czm/assessment>

New Jersey, (2006). *New Jersey Coastal Management Program: Assessment and Enhancement Strategy, FY2006-201 0.* New Jersey Department of Environmental Protection Coastal Management Office, Trenton, 84 pp. <http://www. state.nj .us/dep/cmp/czm_309.html>

New York, (2006). *New York State Coastal Management Program: 309 Assessment and Strategies, July 1, 2006 through June 30, 2010.* Department of State, New York State, Albany, 98 pp.

North Carolina, (2006). *Assessment and Strategy of the North Carolina Coastal Management Program.* North Carolina Department of Environment and Natural Resources, Raleigh, 90pp.

NOAA (National Oceanic and Atmospheric Administration) (2006). Responses to Section 309 of the Coastal Zone Management Act. Coastal Zone Enhancement Program compiled by NOAA Office of Ocean and Coastal Resource Management, Silver Spring, MD. <http://coastalmanagement.noaa.gov/enhanc.html>

Pearsall, S. H., III & Poulter, B. (2005). Adapting coastal lowlands to rising seas: a case study. In: *Principles of Conservation Biology* [Groom, M.J., G.K. Meffe, and C.R. Carroll (eds.)]. Sinauer Associates, Sunderland, MA, 3rd edition, pp. 366- 370.

Pendleton, E. A., Williams, S. J. & Thieler, E.R. (2004). *Coastal Vulnerability Assessment of Assateague Island National Seashore (ASIS) to Sea-level Rise*. U. S. Geological Survey open-file report 2004-1020. U.S. Geological Survey, Reston, VA, 20 pp. <http://pubs.usgs.gov/of/2004/1 020/>

Pennsylvania, (2006). Section 309 Assessment and Strategy: Pennsylvania's Coastal Resources Management Program. Pennsylvania Department of Environmental ProtectionWater Planning Office, Harrisburg, 69 pp. <http://www.dep.state.pa.us/ river/docs/309_FINAL_June30_06.pdf>

Rosenzweig, C. Majors, D. Tults, M. & Demong, K. (2006). New York City Climate Change Task Force. In: *Adapting to Climate Change: Lessons for London.* Greater London Authority, London, pp. 150-153. <http://www.london.gov.uk/climatechangepartnership/>

* Scarlett, L., 2007: Testimony of P. Lynn Scarlett, Deputy Secretary Department of the Interior, before the House Appropriations Subcommittee on Interior, Environment and Related Agencies Regarding Climate Change. April 26, 2007.

Thieler, E.R., S.J. Williams, and R. Beavers, 2002: *Vulnerability of U.S. National Parks to Sea-Level Rise and Coastal Change*. U.S. Geological Survey fact sheet FS 095- 02. [U.S. Geological Survey, Reston, VA], 2 pp. <http://pubs.usgs.gov/fs/fs095- 02/>

Titus, J. G. (1998). Rising seas, coastal erosion, and the takings clause: how to save wetlands and beaches without hurting property owners. *Maryland Law Review, 57(4)*, 1376-1378.

Titus, J. G. (2000). Does the US government realize that the sea is rising? How to restructure federal programs so that wetlands and beaches survive. *Golden Gate University Law Review, 30(4),* 7 17-778.

* TNC (The Nature Conservancy) (2007). *Save of the Week: Climate Change Action on North Carolina's Albemarle Peninsula.* The Nature Conservancy, Arlington, VA. <http://www.nature.org/success/art14181.html>

USACE (U.S. Army Corps of Engineers) (2000). *Planning Guidance Notebook Appendix E: Civil Works Missions and Evaluation Procedures*. Army Corps of Engineers document # ER 1105-2-100. Department of the Army, U. S. Army Corps of Engineers, Washington, DC, 310 pp. <http://www.usace.army.mil/publications/eng-regs/erl 105-2-1 00/toc.htm>

*Velasquez-Manoff, M. (2006). How to keep New York afloat. *The Christian Science Monitor*, November 9, 2006, p. 13.

Virginia, (2006). *Virginia Coastal Zone Management Program: Section 309 Needs Assessment and Strategy*. Virginia Coastal Zone Management Program, Richmond, 104 pp. <http://www.deq.state.va.us/coastal/assess.html>

*Yoder, D. (2007). Miami Dade Water and Sewer Department presentation to American Water Works Association webcast, *Global Climate Impacts*, March 14, 2006.

In: Coastal Sensitivity to Sea Level Rise-Focusing on the Mid-... ISBN: 978-1-60741-440-7
Editor: Melvin C. Urajner

Chapter 12

INSTITUTIONAL BARRIERS

James G. Titus
U.S. EPA

Key Findings

- Most coastal institutions were designed without considering sea-level rise.
- Some regulatory programs were created in order to respond to a demand for hard shoreline structures (*e.g.*, bulkheads) to hold the coast in a fixed location, and have generally not shifted to retreat or soft shore protection (*e.g.*, beach nourishment).
- The interdependence of decisions made by property owners and federal, state, and local governments creates an institutional inertia that currently impedes preparing for sea-level rise, as long as no decision has been made regarding whether particular locations will be protected or yielded to the rising sea.

12.1. Introduction

Chapter 10 described several categories of decisions where the risk of sea-level rise can justify doing things differently today. Chapter 11, however, suggested that only a few organizations have started to prepare for rising sea level since the 1 980s when projections of accelerated sea-level rise first became widely available.

It takes time to respond to new problems. Most coastal institutions were designed before the 1980s. Therefore, land-use planning, infrastructure, home building, property lines, wetland protection, and flood insurance all were designed without considering the dynamic nature of the coast (see Chapters 6, 8, 9, 10). A common mindset is that sea level and shores are stable, or that if they are not then shores should be stabilized (NRC, 2007). Even when a particular institution has been designed to account for shifting shores, people are reluctant to give up real estate to the sea. Although scientific information can quickly change what people expect, it takes longer to change what people want.

Short-term thinking often prevails. The costs of planning for hazards like sea-level rise are apparent today, while the benefits may not occur during the tenure of current elected officials (Mileti, 1999). Local officials tend to be responsive to citizen concerns, and the public is generally less concerned about hazards and other long-term or low-probability events than about crime, housing, education, traffic, and other issues of day-to-day life (Mileti, 1999; Depoorter, 2006). Land-use and transportation planners generally have horizons of 20 to 25 years (TRB, 2008), while the effects of sea-level rise may emerge over a period of several decades. Although federal law requires transportation plans to have a time horizon of *at least* 20 years[29], some officials view that time horizon as the maximum (TRB, 2008). Uncertainty about future climate change is a logical reason to prepare for the range of uncertainty (see Chapter 10) but cognitive dissonance[30] can lead people to disregard the new information instead (Kunreuther *et al.,* 2004; Bradshaw and Borchers, 2000; Akerlof and Dickens, 1982). Some officials resist changing procedures unless they are provided guidance (TRB, 2008).

Finally, a phenomenon known as "moral hazard" can discourage people from preparing for long-term consequences. Moral hazard refers to a situation in which insurance or the expectation of a government bailout reduces someone's incentive to prevent or decrease the risk of a disaster (Pauly, 1974). The political process tends to sympathize with those whose property is threatened, rather than allowing them to suffer the consequences of the risk they assumed when they bought the property (Burby, 2006). It can be hard to say "no" to someone whose home is threatened (Viscusi and Zeckhauser, 2006).

This Chapter explores some of the institutional barriers that discourage people and organizations from preparing for the consequences of rising sea level. "Institution" refers to governmental and nongovernmental organizations and the programs that they administer. "Institutional barriers" refer to characteristics of an institution that prevent actions from being taken. This discussion has two general themes. First, institutional *biases* are more common than actual *barriers*. For example, policies that encourage higher densities in the coastal zone may be barriers to wetland migration, but they improve the economics of shore protection. Such a policy might be viewed as creating a bias in favor of shore protection over wetland migration, but it is not really a barrier to adaptation from the perspective of a community that prefers protection anyway. A bias encourages one path over another; a barrier can block a particular path entirely.

Second, interrelationships between various decisions tend to reinforce institutional inertia For instance, omission of sea-level rise from a land-use plan may discourage infrastructure designers from preparing for the rise, and a federal regulatory preference for hard structures may prevent state officials from encouraging soft structures. Although inertia has slowed current acts to respond to the risk of sea-level rise, it could just as easily help to sustain momentum toward a response once key decision makers decide which path to follow.

The barriers and biases examined in this Chapter mostly concern governmental rather than private sector institutions. Private institutions do not always exhibit foresight. In fact, their limitations have helped motivate the creation of government flood insurance (Kunreuther, 1978), wetland protection (Scodari, 1997), shore protection, and other government programs (Bator, 1958; Arrow, 1970). This Chapter omits an analysis of private institutions for two reasons. First, there is little literature available on private institutional barriers to preparing for sea-level rise. It is unclear whether this absence implies that the private barriers are less important, or simply that private organizations keep their affairs private. Second, the published literature provides no reason to expect that private institutions have important barriers different from those of public institutions. The duty of for-profit corporations to maximize shareholder wealth, for example, may prevent a business from giving up property to facilitate future environmental preservation as sea level rises. At first glance, this duty might appear to be a barrier to responding to sea-level rise, or at least a bias in favor of shore protection over retreat. Yet that same duty would lead a corporation to sell the property to an organization willing to offer a profitable price, or invest money for shore protection. Thus, the duty to maximize shareholder wealth is a bias in favor of profitable responses over money-losing responses, but not a barrier to preparing for sea level rise.

12.2. Some Specific Institutional Barriers and Biases

Productive institutions are designed to accomplish a mission, and rules and procedures are designed to help accomplish those objectives. These rules and procedures are inherently biased toward achieving the mission, and against anything that thwarts the mission. By coincidence more than design, the rules and procedures may facilitate or thwart the ability of others to achieve other missions.

No catalogue of institutional biases in the coastal zone is available; but three biases have been the subject of substantial commentary: (1) shore protection *versus* retreat; (2) hard structures *versus* soft engineering solutions; and (3) coastal development *versus* preservation.

12.2.1. Shore Protection versus Retreat

Federal, state, local, and private institutions generally have a strong bias *favoring* shore protection over retreat in developed areas. Many institutions also have a bias *against* shore protection in undeveloped areas.

Figure 12.1. Recently nourished beach and artificially created dune in Surf City, New Jersey, with recent plantings of dune grass. (June 2007).

Figure 12.2. Beach nourishment along estuaries. (a) The Department of Natural Resources provided an interest-free loan to private landowners for a combined breakwater and beach nourishment project to preserve the recreational beach and protect homes in Bay Ridge, Maryland (July 2008). (b) The Virginia Beach Board and Town of Colonial Beach nourished the public beach along the Potomac River for recreation and to protect the road and homes to the left (October 2002).

U.S. Army Corps of Engineers (USACE) Civil Works. Congressional appropriations for shore protection in coastal communities generally provide funds for various engineering projects to limit erosion and flooding (see Figure 12.1). The planning guidance documents for USACE appear to provide the discretion to relocate or purchase homes if a policy of retreat is the locally preferred approach and is more cost-effective than shore protection (USACE, 2000). In part because the federal government generally pays for 65 percent of the initial cost[31], retreat is rarely the locally preferred option (Lead and Meiners, 2002; NRC, 2004). USACE's environmental policies discourage its Civil Works program from seriously considering projects to foster the landward migration of developed barrier islands (see *Wetland Protection* discussed further below). Finally, the general mission of this agency, its history (Lockhart and Morang, 2002), staff expertise, and funding preferences combine to make shore protection far more common than a retreat from the shore.

State Shore Protection. North Carolina, Virginia, Maryland, Delaware, and New Jersey all have significant state programs to support beach nourishment along the Atlantic Ocean (see Figure 12.1 and Sections A1.C.2, A1.E.2, and A1.G.4 in Appendix 1). Virginia, Maryland, Delaware, and New Jersey have also supported beach nourishment in residential areas along estuaries (see Figure 12.2). Some agencies in Maryland encourage private shore protection to avoid the environmental effects of shore erosion (see Section A1.F.2 in Appendix 1), and the state provides interest-free loans for up to 75 percent of the cost of nonstructural erosion control projects on private property (MD DNR, 2008). Although a Maryland guidance document for property owners favors retreat over shore protection structures (MD DNR, 2006), none of these states has a program to support a retreat in developed areas.

FEMA Programs. Some aspects of the National Flood Insurance Program (NFIP) encourage shore protection, while others encourage retreat. The Federal Emergency Management Agency (FEMA) requires local governments to ensure that new homes along the ocean are built on pilings sunk far enough into the ground so that the homes will remain standing even if the dunes and beach are largely washed out from under the house during a storm[32]. The requirement for construction on pilings can encourage larger homes; after a significant expense for pilings, people rarely build a small, inexpensive cottage. These larger homes provide a better economic justification for government- funded shore protection than the smaller homes.

Beaches recover to some extent after storms, but they frequently do not entirely recover. In the past, before homes were regularly built to withstand the 100-year storm, retreat from the shore often occurred after major storms (*i.e.,* people did not rebuild as far seaward as homes had been before the storm). Now, many homes can withstand storms, and the tendency is for emergency beach nourishment operations to protect oceanfront homes. A FEMA emergency assistance program often funds such nourishment in areas where the beach was nourished before the storm[33] (FEMA, 2007a). For example, Topsail Beach, North Carolina received over $1 million for emergency beach nourishment after Hurricane Ophelia in 2005, even though it is ineligible for USACE shore protection projects and flood insurance under the Coastal Barrier Resources Act (GAO, 2007a). In portions of Florida that receive frequent hurricanes, these projects are a significant portion of total beach nourishment (see Table 12.1). They have not yet been a major source of funding for beach nourishment in the Mid-Atlantic.

Several FEMA programs are either neutral or promote retreat. In the wake of Hurricane Floyd in 1999, one county in North Carolina used FEMA disaster funds to elevate structures, while an adjacent county used those funds to help people relocate rather than rebuild (see Section A1 .G in Appendix 1.). Repetitively flooded homes have been eligible for relocation assistance under a number of programs. Because of FEMA's rate map grandfathering policy (see Section 10.7.3.1 in Chapter 10), a statutory cap on annual flood insurance rate increases, and limitations of the hazard mapping used to set rates, some properties have rates that are substantially less than the actuarial rate justified by the risk. As a result, relocation programs assist property owners and save the flood insurance program money by decreasing claims. From 1985 to 1995, the Upton-Jones Amendment to the National Flood Insurance Act helped fund the relocation of homes in imminent danger from erosion (Crowell *et al.*, 2007). FEMA's Severe Repetitive Loss Program is authorized to spend $80 million to purchase or elevate homes that have made either four separate claims or at least two claims totaling more than the value of the structure (FEMA, 2008a). Several other FEMA programs provide grants for reducing flood damages, which states and communities can use for relocating residents out of the flood plain, erecting flood protection structures, or flood-proofing homes (FEMA, 2008b, c, d, e).

Flood insurance rates are adjusted downward to reflect the reduced risk of flood damages if a dike or seawall decreases flood risks during a 100-year storm. Because rates are based on risk, this adjustment is not a bias toward shore protection, but rather a neutral reflection of actual risk.

Wetland Protection. The combination of federal and state regulatory programs to protect wetlands in the Mid-Atlantic strongly discourages development from advancing into the sea, by prohibiting or strongly discouraging the filling or diking of tidal wetlands for most purposes (see Chapter 9). Within the Mid-Atlantic, New York promotes the landward migration of tidal wetlands in some cases (see Section A1.A.2 in Appendix 1), and Maryland favors shore protection in some cases. The federal wetlands regulatory program has no policy on the question of retreat *versus* shore protection. Because the most compelling argument against estuarine shore protection is often the preservation of tidal ecosystems (*e.g.*, NRC, 2007), a neutral regulatory approach has left the strong demand for shore protection from property owners without an effective countervailing force for allowing wetlands to migrate (Titus 1998, 2000). Wetlands continue to migrate inland in many undeveloped areas (see Figure 12.3) but not in developed areas, which account for an increasing portion of the coast.

Neither federal nor most state regulations encourage developers to create buffers that might enable wetlands to migrate inland, nor do they encourage landward migration in developed areas (Titus, 2000). In fact, USACE has issued a nationwide permit for bulkheads and other erosion-control structures[34]. Titus (2000) concluded that this permit often ensures that wetlands will not be able to migrate inland unless the property owner does not want to control the erosion. For this and other reasons, the State of New York has decided that bulkheads and erosion structures otherwise authorized under the nationwide permit will not be allowed without state concurrence (NYDOS 2006; see Section A1.A.2 in Appendix 1).

Table 12.1. Selected Beach Nourishment Projects in Florida Authorized by FEMA's Public Assistance Grant Program

Year	Location	Hurricane	Authorized Volume of Sand (cubic meters[d])	Obligated Funds[a] (dollars)
1987	Jupiter Island	Floyd	90,000	637,670
1999	Jupiter Island	Irene	48,500	343,101
			0	
2001	Longboat Key	Gabrielle	48,253	596,150
2001	Collier County	Gabrielle	37,800	452,881
2001	Vanderbilt Beach	Gabrielle	61,534	1,592,582
2001	Vanderbilt Beach	Gabrielle	b	738,821
2004	Manasota Key/Knights	Charley *et al.*[c]		2,272,521
	Island		115,700	
2004	Bonita Beach	Charley *et al.*[c]	21,652	1,678,221
2004	Lovers Key	Charley *et al*.c	13,300	102,709
2004	Lido Key	Charley *et al.*[c]	67,600	2,319,322
2004	Boca Raton	Frances	297,572	3,313,688
2004	Sabastian Inlet Recreation Area	Frances	184,755	10,097,507
2004	Hillsboro Beach	Frances	83,444	1,947,228
2004	Jupiter Island	Frances	871,187	8,317,345
2004	Pensacola Beach	Ivan	2,500,000	11,069,943
2004	Bay County	Ivan	56,520	1,883,850
2005	Pensacola Beach	Dennis	400,000	2,338,248
2005	Naples Beach	Katrina	34,988	1,221,038

Table 12.1. (Continued)

Year	Location	Hurricane	Authorized Volume of Sand (cubic meters[d])	Obligated Funds[a] (dollars)
2005	Pensacola Beach	Katrina	482,000	4,141,019
2005	Naples Beach	Wilma	44,834	3,415,844
2005	Longboat Key	Wilma	66,272	1,093,011

Source: Federal Emergency Management Agency. 2008. "Project Worksheets Involving 'Beach Nourishment' Obligated Under FEMA's Public Assistance Grant Program: As of June 19, 2008."

[a] For some projects, the figure may include costs other than placing sand into the beach system, such as reconstructing dunes and planting dune vegetation, as well as associated planning and engineering costs.

[b] Supplemental grant. Applicant lost original sand source and had to go 50 kilometers offshore to collect the sand being used. This increased the cost to $30.82 per cubic meter ($23.57 per cubic yard), compared with originally assumed cost of $10.80 per cubic meter ($8.25 per cubic yard).

[c] Cumulative impact of the 2004 hurricanes Charley Frances, Ivan Jeanne.

[d] Converted from cubic yards, preserving significant digits from the original source, which varies by project.

Federal statutes appear to discourage regulatory efforts to promote landward migration of wetlands. Section 10 of the Rivers and Harbors Act of 1899 and Section 404 of the Clean Water Act require a permit to dredge or fill any portion of the navigable waters of the United States[35]. Courts have long construed this jurisdiction to include lands within the "ebb and flow of the tides", (*e.g., Gibbons v. Ogden; Zabel v. Tabb*; 40 C.F.R. § 230.3[s][1], 2004), but it does not extend inland to lands that are dry today but would become wet if the sea were to rise one meter (Titus, 2000). The absence of federal jurisdiction over the dry land immediately inland of the wetlands can limit the ability of federal wetlands programs to anticipate sea-level rise.

Although the federal wetlands regulatory program generally has a neutral effect on the ability of wetlands to migrate as sea level rises, along the bay sides of barrier islands, regulatory programs, discourage or prevent wetland migration. Under natural conditions, barrier islands often migrate inland as sea level rises (see Chapter 3). Winds and waves tend to fill the shallow water immediately inland of the islands, allowing bayside beaches and marshes to slowly advance into the bay toward the mainland (Dean and Dalrymple, 2002; Wolf 1989). Human activities on developed islands, however, limit or prevent wetland migration (Wolf, 1989). Artificial dunes limit the overwash (see Section 6.2 in Chapter 6). Moreover, when a storm does wash sand from the beach onto other parts of the island, local governments bulldoze the sand back onto the beach; wetland rules against filling tidal waters prevent people from artificially imitating the overwash process by transporting sand directly to the bay side (see Section 10.3). Although leaving the sand in place would enable some of it to wash or blow into the bay and thereby accrete (build land) toward the mainland, doing so is generally impractical. If regulatory agencies decided to make wetland migration a priority, they would have more authority to encourage migration along the bay sides of barrier islands than elsewhere, because the federal government has jurisidiction over the waters onto which those wetlands would migrate.

In addition to the regulatory programs, the federal government preserves wetlands directly through acquisition and land management. Existing statutes give the U.S. Fish and Wildlife Service and other coastal land management agencies the authority to foster the landward migration of wetlands (Titus, 2000). A 2001 Department of Interior (DOI) order directed the Fish and Wildlife Service and the National Park Service to address climate change[36]. However, resource managers have been unable to implement the order because (1) they have been given no guidance on how to address climate change and (2) preparing for climate change has not been a priority within their agencies (GAO, 2007b).

Relationship to Coastal Development. Many policies encourage or discourage coastal development, as discussed in Section 12.2.3. Even policies that subsidize relocation may have the effect of encouraging development, by reducing the risk of an uncompensated loss of one's investment.

Figure 12.3. Tidal Wetland Migration. (a) Marshes taking over land on Hooper Island (Maryland) that had been pine forest until recently, with some dead trees standing in the foreground and a stand of trees on slightly higher ground visible in the rear [October 2004]. (b) Marshes on the mainland opposite Chintoteague Island, Virginia (June 2007).

12.2.2. Shoreline Armoring versus Living Shorelines

The combined effect of federal and state wetland protection programs is a general preference for hard shoreline structures over soft engineering approaches to stop erosion along estuarine shores (see Box 12.1). USACE has issued nationwide permits to expedite the ability of property owners to erect bulkheads and revetments[37], but there are no such permits for soft solutions such as rebuilding an eroded marsh or bay beach[38]. The bias in favor of shoreline armoring results indirectly because the statute focuses on filling navigable waterways, not on the environmental impact of the shore protection. Rebuilding a beach or marsh requires more of the land below high water to be filled than building a bulkhead.

Until recently, state regulatory programs shared the preference for hard structures, but Maryland now favors "living shorelines" (see Chapter 11), a soft engineering approach that mitigates coastal erosion while preserving at least some of the features of a natural shoreline (compare Figure 12.4a with 12.4b). Nevertheless, federal rules can be a barrier to these state efforts (see *e.g.*, Section A1.F.2.2 in Appendix 1), because the living shoreline approaches generally include some filling of tidal waters or wetlands, which requires a federal permit (see Section 10.3).

The regulatory barrier to soft solutions appears to result more from institutional inertia than from a conscious bias in favor of hard structures. The nationwide permit program is designed to avoid the administrative burden of issuing a large number of specific but nearly-identical permits (Copeland, 2007). For decades, many people have bulkheaded their shores, so in the 1 970s USACE issued Nationwide Permit 13 to cover bulkheads and similar structures. Because few people were rebuilding their eroding tidal wetlands, no nationwide permit was issued for this activity. Today, as people become increasingly interested in more environmentally sensitive shore protection, they must obtain permits from institutions that were created to respond to requests for hard shoreline structures. During the last few years however, those institutions have started to investigate policies for soft shore protection measures along estuarine shores.

Figure 12.4. Hard and Soft Shore Protection. (a) Stone Revetment along Elk River at Port Herman, Maryland, May 2005 (b) Dynamic Revetment along Swan Creek, at Fort Washington, Maryland, September 2008.

BOX 12.1 THE EXISTING DECISION-MAKING PROCESS FOR SHORELINE PROTECTION ON SHELTERED COASTS

- There is an incentive to install seawalls, bulkheads, and revetments on sheltered coastlines because these structures can be built landward of the federal jurisdiction and thus avoid the need for federal permits.
- Existing biases of many decision makers in favor of bulkheads and revetments with limited footprints limit options that may provide more ecological benefits.
- The regulatory framework affects choices and outcomes. Regulatory factors include the length of time required for permit approval, incentives that the regulatory system creates, [and] general knowledge of available options and their consequences.
- Traditional structural erosion control techniques may appear to be the most cost-effective. However, they do not account for the cumulative impacts that result in

environmental costs nor the undervaluation of the environmental benefits of the nonstructural approaches.

- There is a general lack of knowledge and experience among decision makers regarding options for shoreline erosion mitigation on sheltered coasts, especially options that retain more of the shorelines' natural features.
- The regulatory response to shoreline erosion on sheltered coasts is generally reactive rather than proactive. Most states have not developed plans for responding to erosion on sheltered shores.

Source: NRC (2007)

12.2.3. Coastal Development

Federal, state, local, and private institutions all have a modest bias favoring increased coastal development in developed areas. The federal government usually discourages development in undeveloped areas, while state and local governments have a more neutral effect.

Coastal counties often favor coastal development because expensive homes with seasonal residents can substantially increase property taxes without much demand for government services (GAO, 2007a). Thus, local governments provide services (*e.g.*, police, fire, trash removal) to areas in Delaware and North Carolina that are ineligible for federal funding under the Coastal Barrier Resources Act[39]. The property tax system often encourages coastal development. A small cottage on a lot that has appreciated to $1 million can have an annual property tax bill greater than the annual rental value of the cottage.

Governments at all levels facilitate the continued human occupation of low-lying lands by providing roads, bridges, and other infrastructure. As coastal farms are replaced with development, sewer service is often extended to the new communities—helping to protect water quality but also making it possible to develop these lands at higher densities than would be permitted by septic tank regulations.

Congressional appropriations for shore protection can encourage coastal development along shores that are protected by reducing the risk that the sea will reclaim the land and structures (NRC, 1995; Wiegel, 1992). This reduced risk increases land values and property taxes, which may encourage further development In some cases, the induced development has been a key justification for the shore protection (GAO, 1976; Burby, 2006). Shore protection policies may also encourage increased densities in lightly developed areas. The benefit-cost formulas used to determine eligibility (USACE, 2000) find greater benefits in the most densely developed areas, making increased density a possible path toward federal funding for shore protection. Keeping hazardous areas lightly developed, by contrast, is not a path for federal funding (USACE, 1998; *cf.* Cooper and McKenna, 2008).

Several authors have argued that the National Flood Insurance Program (NFIP) encourages coastal development (*e.g.*, Tibbetts, 2006; Suffin, 1981; Simmons, 1988; USFWS, 1997). Insurance converts a large risk into a modest annual payment that people are willing to pay. Without insurance, some people would be reluctant to risk $250,000[40] on a home that could be destroyed in a storm. However, empirical studies suggest that the NFIP no longer has a substantial impact on the intensity of coastal development (Evatt, 2000; see Chapter 10). The program provided a significant incentive for construction in undeveloped areas during the

1970s, when rates received a substantial subsidy (Cordes and Yezer, 1998; Shilling *et al.*, 1989; Evatt, 1999). During the last few decades, however, premiums on new construction have not been subsidized and hence, the program has had a marginal impact on construction in undeveloped areas (Evatt, 2000; Leatherman, 1997; Cordes and Yezer, 1998; see Chapter 10). Nevertheless, in the aftermath of severe storms, the program provides a source of funds for reconstruction— and subsidized insurance while shore protection structures are being repaired (see Chapter 10). Thus, in developed areas the program helps rebuild communities that might be slower to rebuild (or be abandoned) if flood insurance and federal disaster assistance were unavailable. More broadly, the combination of flood insurance and the various post-disaster and emergency programs providing relocation assistance, mitigation (*e.g.*, home elevation), reconstruction of infrastructure, and emergency beach nourishment provide coastal construction with a federal safety net that makes coastal construction a safe investment.

Flood ordinances have also played a role in the creation of three-story homes where local ordinances once limited homes to two stories. Flood regulations have induced some people to build their first floor more than 2.5 meters (8 feet) above the ground (FEMA, 1984, 1994, 2000, 2007b). Local governments have continued to allow a second floor no matter the elevation of the first floor. Property owners often enclose the area below the first floor (*e.g.*, FEMA, 2002), creating ground-level (albeit illegal[41] and uninsurable[42]) living space.

The totality of federal programs, in conjunction with sea-level rise, creates moral hazard. Coastal investment is profitable but risky. If government assumes much of this risk, then the investment can be profitable without being risky—an ideal situation for investors (Loucks *et al.*, 2006). The "moral hazard" concern is that when investors make risky decisions whose risk is partly borne by someone else, there is a chance that they will create a dangerous situation by taking on too much risk (Pauly, 1974). The government may then be called upon to take on even the risks that the private investors had supposedly assumed because the risk of cascading losses could harm the larger economy (Kunreuther and Michel-Kerjant, 2007). Investors assume that shore protection is cost-effective and governments assume that flood insurance rates reflect the risk in most cases; however, if sea-level rise accelerates, will taxpayers, coastal property owners, or inland flood insurance policyholders have to pay the increased costs?

The Coastal Barrier Resources Act (16 U.S.C. U.S.C. §3501 *et seq.*) discourages the development of designated undeveloped barrier islands and spits, by denying them shore protection, federal highway funding, mortgage funding, flood insurance on new construction, some forms of federal disaster assistance[43], and most other forms of federal spending. Within the Mid-Atlantic, this statute applies to approximately 90 square kilometers of land, most of which is in New York or North Carolina (USFWS, 2002)[44]. The increased demand for coastal property has led the most developable of these areas to become developed anyway (GAO, 1992; 2007a). "Where the economic incentive for development is extremely high, the Act's funding limitations can become irrelevant" (USFWS, 2002).

12.3. Interdependence: a Barrier or a Support Network?

Uncertainty can be a hurdle to preparing for sea-level rise. Uncertainty about sea-level rise and its precise effects is one problem, but uncertainty about how others will react can also be a barrier. For environmental stresses such as air pollution, a single federal agency (U.S. EPA) is charged with developing and coordinating the nation's response. By contrast, the response to sea-level rise would require coordination among several agencies, including U.S. EPA (protecting the environment), USACE (shore protection), Department of Interior (managing conservation lands), FEMA (flood hazard management), and NOAA (coastal zone management). State and local governments generally have comparable agencies that work with their federal counterparts. No single agency is in charge of developing a response to sea-level rise, which affects the missions of many agencies.

The decisions that these agencies and the private sector make regarding how to respond to sea-level rise are interdependent. From the perspective of one decision maker, the fact that others have not decided on their response can be a barrier to preparing his or her own response. One of the barriers of this type is the uncertainty whether the response to sea-level rise in a particular area will involve shoreline armoring, elevating the land, or retreat (see Chapter 6 for a discussion of specific mechanisms for each of these pathways).

12.3.1. Three Fundamental Pathways: Armor, Elevate, or Retreat

Long-term approaches for managing low coastal lands as the sea rises can be broadly divided into three pathways:

- *Protect* the dry land with seawalls, dikes, and other structures, eliminating wetlands and beaches (also known as "*shoreline armoring*") (see Figure 12.4a and Section 6.1.1).

- *Elevate the land*, and perhaps the wetlands and beaches as well, enabling them to survive (see Figures 12.1 and 12.5)

- *Retreat* by allowing the wetlands and beaches to take over land that is dry today (see Figure 12.6).

Combinations of these three approaches are also possible. Each approach will be appropriate in some locations and inappropriate in others. Shore protection costs, property values, the environmental importance of habitat, and the feasibility of protecting shores without harming the habitat all vary by location. Deciding how much of the coast should be protected may require people to consider social priorities not easily included in a cost-benefit analysis of shore protection.

Figure 12.5. Elevating land and house. (a) Initial elevation of house in Brant Beach (New Jersey). (b) Structural beams placed under house, which is lifted approximately 1.5 meters by hydraulic jack in blue truck. (c) Three course of cinder blocks added then house set down onto the blocks. (d) Soil and gravel brought in to elevate land surface. (January through June 2005)

Figure 12.6. Retreat. (a) Houses along the shore in Kitty Hawk, North Carolina (June 2002). Geotextile sand bags protect the septic tank buried in the dunes. (b) October 2002. (c) June 2003

Like land use planning, the purpose of selecting a pathway would be to foster a coordinated response to sea-level rise, not to lock future generations into a particular approach. Shoreline armoring may be appropriate over the next few decades to halt shoreline erosion along neighborhoods that are about one meter above high water; but as sea level continues to rise, the strategy may switch to elevating land surfaces and homes rather than erecting dikes, which eventually leads to land becoming below sea level. Some towns may be protected by dikes at first, but eventually have to retreat as shore protection costs increase beyond the value of the assets protected. In other cases, retreat may be viable up to a point, past which the need to protect critical infrastructure and higher density development may justify shore protection.

12.3.2. Decisions That Cannot Be Made Until the Pathway Is Chosen

Rising sea level has numerous implications for current activities. In most cases, the appropriate response depends on which of the three pathways a particular community intends to follow. This subsection examines the relationship between the three pathways and six example activities, summarized in Table 12.2.

Coastal Drainage Systems in Urban Areas. Sea-level rise slows natural drainage and the flow of water through drain pipes that rely on gravity. If an area will not be protected from increased inundation, then larger pipes or wider ditches (see Figure 12.7) may be necessary to increase the speed at which gravity drains the area. If an area will be protected with a dike, then it will be more important to pump the water out and to ensure that seawater does not back up into the streets through the drainage system; so then larger pipes will be less important than underground storage, check valves, and ensuring that the system can be retrofitted to allow for pumping (Titus *et al.,* 1987). If land surfaces will be elevated, then sea-level rise will not impair drainage.

Table 12.2. The best way to prepare for sea-level rise Depends on whether (and how) a community intends to hold back the sea.

Activity	Pathways for responding to sea-level rise		
	Shoreline armoring (*e.g.*, dike or seawall)	**Elevate land**	**Retreat/wetland migration**
Rebuild drainage systems	Check valves, holding tanks; room for pumps	No change needed	Install larger pipes, larger rights of way for ditches
Replace septics with public sewer	Extending sewer helps improve drainage	Mounds systems; elevate septic system; extending sewer also acceptable	Extending sewer undermines policy; mounds system acceptable
Rebuild roads	Keep roads at same elevation; owners will not have to elevate lots	Rebuild road higher; motivates property owners to elevate lots	Elevate roads to facilitate evacuation
Location of roads	Shore-parallel road needed for dike maintenance	No change needed	Shore parallel road will be lost; all must have access to shore-perpendicular road
Setbacks/subdiv-isions	Setback from shore to leave room for dike	No change needed	Erosion-based setbacks
Easements	Easement or option to purchase land for dike	No change needed	Rolling easements to ensure that wetlands and beaches migrate

Figure 12.7. Tidal Ditches in the Mid-Atlantic. (a) Hoopers Island, Maryland (October 2004). (b) Poquoson, Virginia (June 2002). (c) Swan Quarter, North Carolina (October 2002). (d) Sea Level, North Carolina. (October 2002). The water rises and falls with the tides in all of these ditches, although the astronomic tide is negligible in (c) Swan Quarter. Wetland vegetation is often found in these ditches. Bulkheads are necessary to prevent the ditch from caving in and blocking the flow of water in (b).

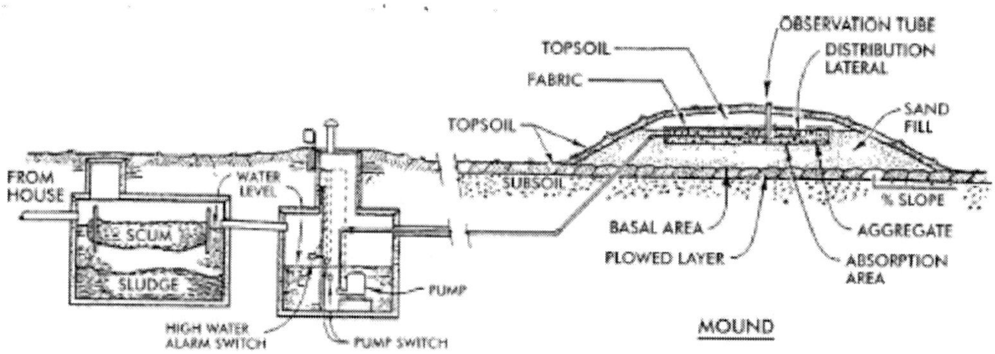

Figure 12.8 Mounds-based septic system for areas with high water tables. For areas with high water tables, where traditional septic/drainfield systems do not work, sand mounds are often used. In this system, a sand mound is contructed on the order of 50 to 100 cm above the ground level, with perforated drainage pipes in the mound above the level of adjacent ground, on top of a bed of gravel to ensure proper drainage. Effluent is pumped from the septic tank up to the perforated pipe drainage pipe. Source: Converse and Tyler (1998).

In many newly developed areas, low-impact development attempts to minimize runoff into the drainage system in favor of on-site recharge. In areas where land surfaces will be elevated over time, the potential for recharge would remain roughly constant as land surfaces generally rise as much as the water table (*i.e.*, groundwater level). In areas that will ultimately be protected with dikes, by contrast, centralized drainage would eventually be required because land below sea level can not drain unless artificial measures keep the water table even father below sea level.

Septics and Sewer. Rising sea level can elevate the water table (ground water) to the point where septic systems no longer function properly (U.S. EPA, 2002)[45]. If areas will be protected with a dike, then all of the land protected must eventually be artificially drained and sewer lines further extended to facilitate drainage. On the other hand, extending sewer lines would be entirely incompatible with allowing wetlands to migrate inland, because the high capital investment tends to encourage coastal protection; a mounds- based septic system (see Figure 12.8) is more compatible. If a community's long-term plan is to elevate the area, then either a mounds-based system or extended public sewage will be compatible.

Road Maintenance. As the sea rises, roads flood more frequently. If a community expects to elevate the land with the sea, then routine repaving projects would be a cost-effective time to elevate the streets. If a dike is expected, then repaving projects would consciously avoid elevating the street above people's yards, lest the projects cause those yards to flood or prompt people to spend excess resources on elevating them, when doing so is not necessary in the long run.

The Town of Ocean City, Maryland, currently has policies in place that could be appropriate if the long-term plan was to build a dike and pumping system, but not necessarily cost-effective if land surfaces are elevated as currently expected.the town expects to elevate instead. Currently, the town has an ordinance that requires property owners to maintain a 2 percent grade so that rainwater drains into the street. The town has interpreted this rule as imposing a reciprocal responsibility on the town itself to not elevate roadways above the level where yards can drain, even if the road is low enough to flood during minor tidal surges. Thus, the lowest lot in a given area dictates how high the street can be. As sea level rises, the town will be unable to elevate its streets, unless it changes this rule. Yet public health reasons require drainage to prevent standing water in which mosquitoes breed. Therefore, Ocean City has an interest in ensuring that all property owners gradually elevate their yards so that the streets can be elevated as the sea rises without causing public health problems. The town has developed draft rules that would require that, during any significant construction, yards be elevated enough to drain during a 10-year storm surge for the life of the project, considering projections of future sea-level rise. The draft rules also state that Ocean City's policy is for all lands to gradually be elevated as the sea rises (see Box A1.5 in Appendix 1).

Locations of Roads. As the shore erodes, any home that is accessed only by a road seaward of the house could lose access before the home itself is threatened. Homes seaward of the road might also lose access if that road were washed out elsewhere. Therefore, if the shore is expected to erode, it is important to ensure that all homes are accessible by shore-perpendicular roads, a fact that was recognized in the layout of early beach resorts along the New Jersey and other shores. If a dike is expected, then a road along the shore would be useful for dike construction and maintenance. Finally, if all land is likely to be elevated, then sea-level rise may not have a significant impact on the best location for new roads.

Subdivision and Setbacks. If a dike is expected, then houses need to be set back enough from the shore to allow room for the dike and associated drainage systems. Setbacks and larger coastal lot sizes are also desirable in areas where a retreat policy is preferred for two reasons. First, the setback provides open lands onto which wetlands and beaches can migrate inland without immediately threatening property. Second, larger lots mean lower density and hence fewer structures that would need to be moved, and less justification for investments in central water and sewer. By contrast, in areas where the plan is to elevate the land, sea-level rise does not alter the property available to the homeowner, and hence would have minor implication for setbacks and lot sizes.

Covenants and Easements Accompanying Subdivision. Although setbacks are the most common way to anticipate eventual dike construction and the landward migration of wetlands and beaches, a less expensive method would often be the purchase of (or regulatory conditions requiring) rolling easements, which allow development but prohibit hard structures that stop the landward migration of ecosystems. The primary advantage of a rolling easement is that society makes the decision to allow wetlands to migrate inland long before the property is threatened, so owners can plan around the assumption of migrating wetlands, whether that means leaving an area undeveloped or building structures that can be moved.

Local governments can also obtain easements for future dike construction. This type of easement, as well as rolling easements, would each have very low market prices in most areas, because the fair market value is equal to today's land value discounted by the rate of interest compounded over the many decades that will pass before the easement would have any effect (Titus, 1998). As with setbacks, a large area would have to be covered by the easements if wetlands are going to migrate inland; a narrow area would be required along the shore for a dike; and no easements are needed if the land will be elevated in place.

12.3.3. Opportunities for Deciding on the Pathway

At the local level, officials make assumptions about which land will be protected in order to understand which lands will truly become inundated (see Chapter 2) and how shorelines will actually change (see Chapter 3), which existing wetlands will be lost (see Chapter 4), whether wetlands will be able to migrate inland (see Chapter 6), and the potential environmental consequences (see Chapter 5); the population whose homes would be threatened (see Chapter 7) and the implications of sea-level rise for public access (see Chapter 8) and floodplain management (see Chapter 9). Assumptions about which shores will be protected are also necessary in order to estimate the level of resources that would be needed to fulfill property owners' current expectations for shore protection (*e.g.*, Titus, 2004).

Improving the ability to project the impacts of sea-level rise is not the only for such analyses utility of data regarding shore protection. Another use of such studies has been to initiate a dialogue about what *should* be protected, so that state and local governments can decide upon a plan of what will actually be protected. Just as the lack of a plan can be a barrier to preparing for sea-level rise, the adoption of a plan could remove an important barrier and signal to decision makers that it may be possible for them to plan for sea-level rise as well.

REFERENCES

* Indicates non-peer reviewed literature; available upon request

Akerlof, G. A. & Dickens, W. T. (1982). The economic consequences of cognitive dissonance. *American Economic Review*, *72*, 307, 309.

Arrow, K. J. (1970). The organization of economic activity: issues pertinent to the choice of market versus nonmarket allocation. In: *Public Expenditures and Policy Analysis* [Margolis, J. (ed.)]. Markham Publishing Company, Chicago, pp. 59-73.

Bator, F. M. (1958). The anatomy of market failure. *The Quarterly Journal of Economics,72(3),* 351-79.

Bradshaw, G.A. and Borchers, J.G. (2000). Uncertainty as information: narrowing the science-policy gap. *Conservation Ecology*, *4(1),* 7.

Burby, R. J. (2006). Hurricane Katrina and the Paradoxes of Government Disaster Policy: Bringing About Wise Governmental Decisions for Hazardous Areas. *The Annals of the American Academy of Political and Social Science*, *604(1),* 171-191.

Converse, J. C. & Tyler, E.J. (1998). Soil treatment of aerobically treated domestic wastewater with emphasis on modified mounds. In: *On-Site Wastewater Treatment: Proceedings of the Eighth National Symposium on Individual and Small Community Sewage Systems* [Sievers, D.M. (ed.)]. American Society of Agricultural Engineers, St. Joseph, MI.

Cooper, J. A. G. & McKenna, J. (2008). Social justice in coastal erosion management: the temporal and spatial dimensions. *Geoforum*, *39(1)*, 294-306.

Copeland, C. (2007). *The Army Corps of Engineers' Nationwide Permits Program: Issues and Regulatory Developments*. Congressional Research Service, Washington, DC, [30 pp.]

Cordes, J. J. & Yezer, A.M.J. (1998). In harm's way: does federal spending on beach enhancement and protection induce excessive development in coastal areas? *Land Economics, 74(1),* 128-145.

Crowell, M., Hirsch, E. & Hayes, T. L. (2007). *Marine Technology Society Journal*, *41(1)*, 18-27.

Dean, R. G. & Dalrymple, R. A. (2002). *Coastal Processes with Engineering Applications*. Cambridge University Press, Cambridge, UK, and New York, 475 pp.

Depoorter, B. (2006). Horizontal political externalities: the supply and demand of disaster management. *Duke Law* Journal, *56(1),* 101-125.

Evatt, D. S. (1999). *National Flood Insurance Program: Issues Assessment.* Federal Emergency Management Agency, Washington, DC, 123 pp.

Evatt, D. S. (2000). Does the national flood insurance program drive floodplain development? *Journal of Insurance Regulation*, *18(4),* 497-523.

FEMA (Federal Emergency Management Agency) (1984). *Elevated Residential Structures*. FEMA 54. Federal Emergency Management Agency, Washington, DC, 144 pp.

FEMA (Federal Emergency Management Agency) (1994). *Mitigation of Flood and Erosion Damage to Residential Buildings in Coastal Areas*. FEMA 257. Federal Emergency Management Agency, Washington, DC, 40 pp.

FEMA (Federal Emergency Management Agency) (2000). *Above the Flood: Elevating Your Floodprone House*. FEMA 347. Federal Emergency Management Agency, Washington, DC, 69 pp.

FEMA (Federal Emergency Management Agency) (2002). *FEMA Notifies Monroe County, Florida, of Impending Flood Insurance Probation*. Region IV News Release Number: R4-02-15. <http://www.fema.gov/news/newsrelease.fema?id>

FEMA (Federal Emergency Management Agency) (2007a). *Public Assistance Guide*. FEMA 322. Federal Emergency Management Agency, Washington, DC. <http://www.fema.gov/government. shtm>

FEMA (Federal Emergency Management Agency) (2007b). *Coastal Construction Manual*. Federal Emergency Management Agency, Washington, DC. <http://www.fema.gov/rebuild/mat/fema55. shtm>

FEMA (Federal Emergency Management Agency) (2008a). *Severe Repetitive Loss Program*. [Web site] Federal Emergency Management Agency, Washington, DC. <http://www.fema.gov/government. shtm>

FEMA (Federal Emergency Management Agency) (2008b). *Repetitive Flood Claims Program: Program Overview*. [Web site] Federal Emergency Management Agency, Washington, DC. <http://www.fema.gov/government>

FEMA (Federal Emergency Management Agency) (2008c). *Hazard Mitigation Grant Program*. [Web site] Federal Emergency Management Agency, Washington, DC. <http://www.fema.gov/government>

FEMA (Federal Emergency Management Agency) (2008d). *Flood Mitigation Assistance Program* [Web site] Federal Emergency Management Agency, Washington, DC. <http://www.fema.gov/government>

FEMA (Federal Emergency Management Agency) (2008e). *Pre-Disaster Mitigation Program*. [Web site] Federal Emergency Management Agency, Washington, DC. <http://www.fema.gov/government. shtm>

Festinger, L. (1957). *A Theory of Cognitive Dissonance*. Stanford University Press, Stanford, CA, 291 pp.

GAO (General Accounting Office) (1976). *Cost, Schedule, and Performance Problems of the Lake Pontchartrain and Vicinity, Louisiana, Hurricane Protection Project*. PSAD-76-161. General Accounting Office, Washington, DC, 25 pp.

GAO (General Accounting Office) (1992). *Coastal Barriers: Development Occurring Despite Prohibitions Against Federal Assistance*. GAO/RCED-92-115. General Accounting Office, Washington, DC, 71 pp.

GAO (Government Accountability Office) (2007a). *Coastal Barriers Resources System: Status of Development that Has Occurred and Financial Assistance Provided by Federal Agencies: Development Occurring Despite Prohibitions Against Federal Assistance*. GAO/07-356. Government Accountability Office, Washington, DC, 66 pp.

GAO (Government Accountability Office) (2007b). *Climate Change: Agencies Should Develop Guidance for Addressing the Effects on Federal Land and Water Resources*. GAO-07-863. Government Accountability Office, Washington, DC, 179 pp.

Gibbons v. Ogden, 22 U.S. 1, 217-18 (9 Wheat. 1824)

Harmon-Jones, E. & Mills, J. (1999). *Cognitive Dissonance: Progress on a Pivotal Theory in Social Psychology*. American Psychological Association, Washington, DC, 411 pp.

Kunreuther, H., Ginsberg, R., Miller, L., Sagi, P., Slovic, P., Borkan, B. & Katz, N. (1978). *Disaster Insurance Protection*. John Wiley and Sons, New York, 400 pp.

Kunreuther, H. C. & Michel-Kerjant, E.O. (2007). Climate change, insurability of large- scale disasters, and the emerging liability challenge. *University of Pennsylvania Law Review, 155(6),* 1795-1842.

Kunreuther, H., Meyer, R. & Van den Bulte, C. (2004). *Risk Analysis for Extreme Events: Economic Incentives for Reducing Future Losses*. National Institute of Standards and Technology, Gaithersburg, MD, 93 pp.

Lead, D. & Meiners, R. E. (2002). *Government vs. Environment.* Rowan and Littlefield, Lanham, MD, 207 pp.

Leatherman, S. P. (1997). *Flood Insurance Availability in Coastal Areas: The Role It Plays in Encouraging Development Decisions*. Federal Emergency Management Agency,Washington, DC.

Lockhart, J. & Morang, A. (2002). History of coastal engineering. In: *Coastal Engineering Manual, Part I* [Morang, A. (ed.)]. Engineer manual 1110-2-1100. U.S. Army Corps of Engineers, Washington, DC, [39 pp.] <http://chl.erdc.usace.army.mil/cemtoc>

Loucks, D. P., Stedinger, J. R. & Stakhiv, E. Z. (2006). Individual and societal responses to natural hazards. *Journal of Water Resources Planning & Management, 132(5),* 3 15-3 19.

MD DNR (Maryland Department of Natural Resources), 2006: *Shore Erosion Control Guidelines for Waterfront Property Owners*. Maryland Department of Natural Resources, Annapolis, 30 pp.<http://www.dnr.state.md.us/ccws/sec/download. pdf>

MD DNR (Maryland Department of Natural Resources) (2008). *Grants and Loans: Shore Erosion Control*. [web site] Maryland Department of Natural Resources, Annapolis. <http://www.dnr.state.md.us/land>

Mileti, D. S. (1999). *Disasters by Design: A Reassessment of Natural Hazards in the United States*. Joseph Henry Press, Washington, DC, 351 pp.

NRC (National Research Council) (1995). *Beach Nourishment and Protection.* National Academy Press, Washington, DC, 334 pp.

NRC (National Research Council) (2004). *River Basins and Coastal Systems Planning Within the U.S. Army Corps of Engineers*. National Academies Press, Washington, DC, 167 pp.

NRC (National Research Council) (2007). *Mitigating Shore Erosion along Sheltered Coasts.* National Academies Press, Washington, DC, 174 pp.

NYDOS (New York Department of State) (2006). Department of State review of US Army Corps of Engineers consistency determination for reissuance, modification, and issuance of new nationwide permits and conditions. Letter to USACE Buffalo and New New York districts, December 8, 2006. New York Division of Coastal Resources, Albany, 5 pp.

Pauly, M.V. (1974). Overinsurance and public provision of insurance: the roles of moral hazard and adverse selection. *Quarterly Journal of Economics*, *88(1)*, 44-62.

Scodary, P. (1997). *Measuring the Benefits of Federal Wetland Protection Programs*. Environmental Law Institute, Washington, DC, 103 pp.

Simmons, M. (1988). *The Evolving National Flood Insurance Program*. 86-641 ENR. Congressional Research Service Washington, DC.

Shilling, J. D., Sirmans, C. E. & Benjamin, J. D. (1989). Flood insurance, wealth redistribution, and urban property values. *Journal of Urban Economics*, *26(1),* 43- 53.

Suffin, W. J. (1981). Bureaucracy, entrepreneurship, and natural resources: witless policy and barrier islands. *Cato Journal, 1(1),* 293-311.

Tibbetts, J. H. (2006). After the storm. *Coastal Heritage, 20(4),* 3-11.

Titus, J. G. (1998). Rising seas, coastal erosion and the taking clause: how to save wetlands and beaches without hurting property owners. *Maryland Law Review, 57(4),* 1277-1399.

Titus, J. G. (2000). Does the U.S. government realize that the sea is rising? How to restructure federal programs so that wetlands and beaches survive. *Golden Gate University Law Review, 30(4),* 7 17-778.

* Titus, J. G. (2004). Maps that depict the business-as-usual response to sea level rise in the decentralized United States of America. Presented at: *Global Forum on Sustainable Development*, Paris, 11-12, November 2004. Organization of Economic Cooperation and Development, Paris.

Titus, J. G., Kuo, C. Y., Gibbs, M. J., LaRoche, T. B. Webb, M. K. & Waddell, J. O. (1987). Greenhouse effect, sea level rise, and coastal drainage systems. *Journal of Water Resources Planning and Management, 113(2)*, 216-227.

TRB (Transportation Research Board) (2008). *Potential Impacts of Climate Change on U.S. Transportation*. National Academies Press, Washington, DC, 234 pp.

USACE (U.S. Army Corps of Engineers) (1998). *Atlantic Coast of Long Island, Fire Island Inlet to Montauk Point: Alternative Screening Report*. Draft report. U.S. Army Corps of Engineers New York District, New York, 118 pp. <http://www.nan.usace.army.mil/fimp/pdf/montauk/screening.pdf>

USACE (U.S. Army Corps of Engineers) (2000). *Planning Guidance Notebook*. Document ER 1105-2-100. U.S. Army Corps of Engineers, Washington, DC. <http://www.usace.army.mil/publications/eng-regs/er1 105-2-100/>

U.S. EPA (Environmental Protection Agency) (2002). *Onsite Wastewater Treatment Systems Manual*. EPA/625/R-00/008. EPA Office of Water and Office of Research and Development, Washington, DC, [367 pp.] <http://purl.access.gpo.gov/GPO 1380>

USFWS (U.S. Fish and Wildlife Service) (1997). *Biological Opinion: Administration of the National Flood Insurance Program in Monroe County, Florida, by the Federal Emergency Management Agency*. U.S. Fish and Wildlife Service, Atlanta, GA.

USFWS (U.S. Fish and Wildlife Service) (2002). *The Coastal Barrier Resources Act: Harnessing the Power of Market Forces to Conserve America's Coasts and Save Taxpayers' Money*. U.S. Fish and Wildlife Service, Arlington, VA, 34 pp. <http://www.fws.gov/habitatconservation/TaxpayerSavingsfromCBRA.pdf>

Viscusi, W. K. & Zeckhauser, R. J. (2006). National survey evidence on disasters and relief: risk beliefs, self-interest, and compassion. *Journal of Risk & Uncertainty, 33(1/2),* 13-36.

Wolff, F. (1989). Environmental assessment of human interference on the natural processes affecting the barrier beaches of Long Island, New York. *Northeastern Environmental Science, 8(2),* 119-134.

Wiegel, R.L. (1992). Dade County, Florida beach nourishment and hurricane surge protection. *Shore and Beach, 60(4),* 2-28.

Zabel v. Tabb, 430 F.2d 199, 215 (5th Cir. 1970)

End Notes

[1] Administrative costs (*e.g.*, studies, regulations, compliance, training) of addressing a new issue are roughly fixed regardless of how small the impact may be, while the benefits of addressing the issue depend on the magnitude

of sea-level rise. Therefore, there would be a point below which the administrative costs would be greater than any benefits from addressing the issue.

[2] There is an extensive economic literature on decision-making and planning under uncertainty, particularly where some effects are irreversible. A review of this literature on the topic of "quasi-option value" can be found in Freeman (2003),. Quasi-option value arises from the value of information gained by delaying an irreversible decision (*e.g.*, to rebuild a structure to withstand higher water levels). In the sea- level rise context, it applies because the costs and benefits of choosing to retreat or protect are uncertain, and it is reasonable to expect that uncertainty will narrow over time concerning rates of sea level rise, the effects, how best to respond, and the costs of each response option. Two influential works in this area include Arrow and Fisher (1974) and Fisher and Hanemann (1987); an application to climate policy decisions can be found in Ha-Duong (1998).

[3] The present value of a dollar T years in the future is $1/(1+i)^T$, where i is the interest rate (discount rate) used for the calculations (see Samuelson and Nordhaus, 1989).

[4] Yohe *et al.* (1996) estimated that the nationwide value of "foresight" regarding response to sea-level rise is $20 billion, based largely on the strategic depreciation that foresight makes possible.

[5]Another mechanism for allowing wetlands and beaches to migrate inland are setbacks, which prohibit development near the shore. Setbacks can often result in successful "takings" claims if a property is deemed undevelopable due to the setback line. By contrast, rolling easements place no restrictions on development and hence are not constitutional takings (see, *e.g.*, Titus [1998]).

[6] A social cost involves someone losing something of value (*e.g*,. the right to develop coastal property) without a corresponding gain by someone else. A wealth transfer involves one party losing something of value with another party gaining something of equal value (*e.g.*, the cost of a rolling easement being transferred from the government to a land owner). For additional details, see Samuelson and Nordhaus (1989).

[7] 33 U.S.C. §§ 403, 409, 1344(a)

[8] 40 C.F.R. § 230.10(a)(3)

[9] See Weggel *et al.* (1989), Titus *et al.* (1991), and NRC (2007) for an examination of costs and options for estuarine shore protection.

[10] *i.e.*, \$25 invested today would be worth $\$25 \times (1.05)^{28} = \$98$ twenty eight years hence. Therefore, it is better to invest \$25 today than to face a certain loss of \$100 28 years hence (see glossary for definition of discount rate).

[11] Flood Insurance Rate Maps display the flood hazards of particular locations for purposes of setting flood insurance rates. The maps do not show flood insurance rates (see Chapter 9 for additional details).

[12] The shaded X Zone was formerly known as the B Zone.

[13] Owners of pre-FIRM homes can also pay the actuarial rate, if it is less than the subsidized rate.

[14] The National Flood Insurance Program (NFIP) modifies flood insurance rates every year based on the annual "Actuarial Rate Review". Rates can either be increased, decreased, or stay the same, for any given flood insurance class. The rates for post-FIRM policies are adjusted based on the risk involved and accepted actuarial principals. As part of this rate adjustment, hydrologic models are used to estimate loss exposure in flood-prone areas. These models are rerun every year using the latest hydrologic data available. As such, the models incorporate the retrospective effects of sea level rise. The rates for pre-FIRM (subsidized) structures are also modified every year based in part on a determination of what is known as the "Historical Average Loss Year". The goal of the NFIP is for subsidized policyholders to pay premiums that are sufficient, when combined with the premium paid by actuarially priced (post-FIRM) policyholders, to provide the NFIP sufficient revenue to pay losses associated with the historical average loss year.

[15] In economics, "market signal" refers to information passes indirectly or unintentionally between participants in a market. For example, higher flood insurance rates convey the information that a property is viewed as being riskier than previously thought.

[16] Under the NFIP grandfathering policy, whenever FEMA revises the flood risk maps used to calculate the premium for specific homes, a policy holder can choose between the new map and the old map, whichever results in the lower rate (NFIP, 2007).

[17] Although rates for individual policies may be grandfathered, rates for the entire A or V Zone (or any flood zone) can still increase each year up to a maximum of 10 percent; therefore a grandfathered policy may still see annual rate increases. For example, a post-FIRM structure might be originally constructed in an A Zone at 30 cm (1 ft) above base flood elevation. If shore erosion, sea-level rise, or a revised mapping procedure leads to a new map that shows the same property to be in the V Zone and 60 cm (2 ft) below base flood elevation, the policy holder can continue to pay as if the home was 30 cm above base flood elevation in the A Zone. However, the entire class of A Zone rates could still increase as a result of annual class-wide rate adjustments based on the annual "Actuarial Rate Review". Those class-wide increases could be caused by long-term erosion, greater flooding from sea-level rise, increased storm severity, higher reconstruction or administrative costs, or any other factors that increase the cost of paying claims by policyholders.

[18] This calculation assumes a storm-wave height adjustment of 90 cm and no sea-level rise (see NFIP, 2008).

[19] Note that: (1) the NFIP insures against damages caused by flood-related-erosion; (2) the probability of flood-related erosion is considered in defining the landward limit of V Zones; and (3) flood insurance rates in the V

Zone are generally much higher than A Zone rates. Part of the reason for this is consideration of the potential for flood-related erosion.

[20] The year 2005 (Hurricanes Katrina, Rita, and Wilma) is excluded from such calculations.

[21] Pre-FIRM owners of homes a few feet *below* the base flood elevation could achieve modest saving by elevating homes a few feet *above* the base flood elevation; but those savings are small compared to the savings available to the owner of a post-FIRM home at the same elevation relative to base flood elevation.

[22] Appendix 1, however, does examine such policies.

[23] 06-096 Code of Maine Rules §355(3)(B)(1) (2007).

[24] 310 Code Mass Regulations § 10.30 (2005).

[25] Rhode Island Coastal Resource Management Program §210.3(B)(4) and §300.7(D) (2007).

[26] Maryland House Bill 273-2008.

[27] MD Code Environment § 16-201(c)

[28] Maryland House Bill 273-2008.

[29] 23 U.S.C. §135(f)(1) (2008).

[30] Cognitive dissonance is a feeling of conflict or anxiety caused by holding two contradictory ideas simultaneously, especially when there is a discrepancy between one's beliefs or actions and information that contradicts those beliefs or actions. When confronted with information (*e.g.*, about risk) that contradicts one's pre-existing beliefs or self-image (*e.g.*, that they are acting reasonably), people often respond by discounting, denying, or ignoring the information (*e.g.*, Festinger [1957], Harmon-Jones and Mills, [1999]).

[31] 33 USC §2213.

[32] 44 Code of Federal Regiulations §60.3(e)(4)

[33] 44 CFR §206.226(j)

[34] See 61 Federal Register 65,873, 65,915 (December 13, 1996) (reissuing Nationwide Wetland Permit 13, Bank Stabilization activities necessary for erosion prevention). *See also* Reissuance of Nationwide Permits, 72 Fed. Reg. 11,1108-09, 11183 (March 12, 2007) (reissuing Nationwide Wetland Permit 13 and explaining that construction of erosion control structures along coastal shores is authorized).

[35] See The Clean Water Act of 1977, § 404, 33 U.S.C. § 1344; The Rivers and Harbors Act of 1899, § 10, 33 U.S.C. §§ 403, 409 (1994).

[36] Department of Interior Secretarial Order 3226

[37] Reissuance of Nationwide Permits, 72 Federal Register 11,1108-09, 11183 ((March 12, 2007) (reissuing Nationwide Wetland Permit 13 and explaining that construction of erosion control structures along coastal shores is authorized). See also Nationwide Permits 3 (Maintenance), 31 (Maintenance of Existing Flood Control Facilities) and 45 (Repair of Uplands Damaged by Discrete Events). 72 Federal Register 11092-11198 (March 12, 2007).

[38] Reissuance of Nationwide Permits, 72 Federal Register 11, 11183, 11185 ((March 12, 2007) (explaining that permit 13 requires fill to be minimized and that permit 27 does not allow conversion of open to water to another habitat such as beach or tidal wetlands)

[39] 16 U.S.Code. §3501 *et seq.*

[40] NFIP only covers the first $250,000 in flood losses (44 CFR 61.6) For homes with a construction cost greater than $250,000, federal insurance reduces a property owner's risk, but to a lesser extent.

[41] 44 CFR §60.3(c)(2)

[42] 44 CFR §61.5(a)

[43] Communities are eligible for emergency beach nourishment after a storm, provided that the beach had been previously nourished (GAO, 2007).

[44] The other mid-Atlantic states each have less than 6 square kilometers within the CBRA system. A small area within the system in Delaware is intensely developed (see Box 9.2).

[45] "Most current onsite wastewater system codes require minimum separation distances of at least 18 inches from the seasonally high water table or saturated zone irrespective of soil characteristics. Generally, 2- to 4- foot separation distances have proven to be adequate in removing most fecal coliforms in septic tank effluent" U.S. EPA (2002).

PART IV OVERVIEW. NATIONAL IMPLICATIONS AND A SCIENCE STRATEGY FOR THE WAY FORWARD

Authors: S. Jeffress Williams, USGS; E. Robert Thieler, USGS; Benjamin T. Gutierrez, USGS; Dean B. Gesch; USGS; Donald R. Cahoon, USGS; K. Eric Anderson, USGS

Climate change and effects such as sea-level rise have global implications and will increasingly affect the entire nation. While this Product focuses primarily on the mid- Atlantic region of the United States, many of the issues discussed in earlier chapters are relevant at the national scale.

Chapter 13 draws on findings from the mid-Atlantic focus area that have relevance to other parts of the United States, provides an overview of coastal environments and landforms in the United States, and describes the issues faced in understanding how these environments may be impacted and respond to sea-level rise. The diversity of U.S. coastal settings includes bedrock coasts in Maine, glacial bluffs in New York, barrier islands in the mid-Atlantic and Gulf of Mexico, coral reefs in Florida, the Caribbean, and Hawaii, one of the world's major delta systems in Louisiana, a wide variety of pocket beaches and cliffed coasts along the Pacific coast, Pacific atolls, and a number of arctic coastline types in Alaska. In addition, the large bays and estuaries around the country also exhibit a diverse range of shoreline types, large wetland systems, and extensive coastal habitats.

Understanding how the different coastal environments of the United States will respond to future climate and sea-level change is a major challenge. In addition, as highlighted in earlier Parts of this Product, human actions and policy decisions also substantially influence the evolution of the coast. The knowledge gaps and data limitations identified in this Product focusing on the mid-Atlantic have broad relevance to the rest of the U.S. Chapter 14 identifies opportunities for increasing the scientific understanding of future sea-level rise impacts. This includes basic and applied research in the natural and the social sciences. A significant emphasis is placed on developing linkages between scientists, policy makers, and stakeholders at all levels, so that information can be shared and utilized efficiently and effectively as sea-level rise mitigation and adaptation plans evolve.

In: Coastal Sensitivity to Sea Level Rise-Focusing on the Mid-... ISBN: 978-1-60741-440-7
Editor: Melvin C. Urajner

Chapter 13

IMPLICATIONS OF SEA-LEVEL RISE TO THE NATION

S. Jeffress Williams[1], Benjamin T. Gutierrez[1], James G. Titus[2], Eric Anderson[3], Stephen K. Gill[4], Donald R. Cahoon[1], E. Robert Thieler[1] and Dean B. Gesch[1]

[1]USGS
[2]USEPA
[3]USGS (retired)
[4]NOAA

Key Findings

- Nationwide, more than one-third of the United States population currently lives in the coastal zone and movement to the coast and development continues, along with the current and growing vulnerability to coastal hazards such as storms and sea-level rise. Fourteen of the 20 largest U.S. urban centers are located along the coast. With the very likely accelerated rise in sea level and increased storm intensity, the conflicts between people and development at the coast and the natural processes will increase, causing economic and societal impacts.
- For much of the U.S., shores comprised of barrier islands, dunes, spits, and sandy bluffs, erosion processes will dominate at highly variable rates in response to sea-level rise and storms over the next century and beyond. Some coastal landforms in the U.S. may undergo large changes in shape and location if the rate of sea-level rise increases as predicted. Increased inundation and more frequent flooding will affect estuaries and low-lying coastal areas. The response to these driving forces will vary depending on the type of coastal landform and local conditions, but will be more extreme, more variable and less predictable than the changes observed over the last century.
- For higher sea-level rise scenarios, some barrier island coasts and wetlands may cross thresholds and undergo significant and irreversible changes. These changes

include rapid landward migration and segmentation of some barrier islands and disintegration and drowning of wetlands.

- Nationally, tidal wetlands already experiencing submergence by sea-level rise and associated land loss, in concert with other factors, will continue to deteriorate in response to changing climate.
- Coastal change is driven by complex and interrelated processes. Over the next century and beyond, with an expected acceleration in sea-level rise, the potential for coastal change is likely to be greater than has been observed in historic past. These changes to coastal regions will have especially large impacts on urban centers and developed areas. Some portions of the U.S. coast, however, will be subject primarily to inundation from sea-level rise over the next century. A substantial challenge remains to quantify the various effects of sea-level rise and to identify the dominant coastal change processes for each region of the U.S. coast.
- Many coastal areas in the United States will likely experience an increased frequency and magnitude of storm-surge flooding and coastal erosion due to storms over the next century, in response to sea-level rise. The impacts from these storm events are likely to extend farther inland from the coast than those that would be affected by sea-level rise alone.
- Understanding, predicting, and responding to the environmental and societal effects of sea-level rise would benefit from a national program of integrated research that includes the natural and social sciences. Research on adaptation, mitigation, and avoidance-of-risk measures would enable improved understanding of the many and varied potential societal impacts of sea-level rise that would benefit the U.S as well as coastal nations around the world.

13.1. Introduction

As defined in the SAP 4.1 Prospectus and discussed in earlier chapters, this Product focuses on assessing potential impacts to the mid-Atlantic region; however, some discussion of impacts to other regions and the nation as a whole is warranted. The mid- Atlantic region is highly vulnerable to sea-level rise, but regions like the central Gulf Coast (Louisiana, Texas) are just as vulnerable or more so. The challenge in carrying out a national assessment is that nationwide data bases and scientific publications of national scale and scope are limited. Modest efforts at monitoring and observations for national-scale assessments of coastal change and hazards are underway by various organizations, but more effort is needed. The discussion in ection 13.3 is largely the expert opinions of the lead authors, informed by results of the two expert science panel reports (Reed *et al.*, 2008, Gutierrez, *et al.*, 2007) and available scientific literature. Because of the relative lack of adequate background literature and high reliance on expert opinion, the likehood statements as used in other chapters are not included in this discussion of potential impacts to the nation.

A large and expanding proportion of the U.S. population and related urban development is located along the Atlantic, Gulf of Mexico, and Pacific coasts and increasingly conflicts with the natural processes associated with coastal change from storms and sea- level rise (see a review in Williams *et al.,* 1991). Development in low-lying regions (*i.e.*, New Orleans) and islands (*e.g.*, in the Chesapeake Bay, Caribbean, Pacific Ocean) are particularly at risk (see Gibbons and Nicholls, 2006) In the future, as the effects of climate change intensify, these interactions will become more frequent and more challenging to society. Currently, more than one-third of the population lives in the coastal zone and movement to the coast and development continues, along with the growing vulnerability to coastal hazards. Fourteen of the 20 largest U.S. urban centers are located along the coast (Crossett *et al.*, 2004; Crowell *et al.*, 2007). With the likely accelerated rise in sea level and increased storm intensity, the conflicts between people and development at the coast and the natural processes will increase, affecting all parts of society (Leatherman, 2001; FitzGerald *et al.,* 2008).

Global sea-level rise associated with climate change is likely to be in the range of 19 cm to as much as 1 m over the next century and possibly as much as 4 to 6 m over the next several centuries (IPCC, 2007; Rahmstorf, 2007; Rahmstorf, e*t al.*, 2007; Overpeck *et al.*, 2006). The expected rise will increase erosion and the frequency of flooding and coastal areas will be at increasing risk. For some regions, adaptation using engineering means may be effective; for other coastal areas, however, adaption by relocation landward to higher elevated ground may be appropriate for longer-term sustainability (NRC, 1987).

Coastal landforms reflect the complex interaction between the natural physical processes that act on the coast, the geologic characteristics of the coast, and human activities. Spatial and temporal variations in these physical processes and the geology along the coast are responsible for the wide variety of landforms around the United States (Williams, 2003). With future sea-level rise, portions of the U.S. ocean coast are likely to undergo long-term net erosion, at rates higher than those that have been observed over the past century (see Chapter 3). The exact manner and rates at which these changes are likely to occur depend on the character of coastal landforms (*e.g.*, barrier islands, cliffs) and the physical processes (*e.g.*, waves and winds) that shape these landforms (see Chapters 3 and 4). Low-relief coastal regions, areas undergoing land subsidence and land subject to frequent storm landfalls, such as the northern Gulf of Mexico, Florida, Hawaii, Puerto Rico, the San Francisco-Sacramento Delta region, and the Mid-Atlantic region, are particularly vulnerable.

13.2. Types of Coasts

Coasts are dynamic junctions of the oceans, atmosphere, and land and differ greatly in physical character and vulnerability to erosion, storms, and sea-level rise (NRC, 1990). The principal coastal types are described in Chapters 3 and 4, and summarized below. With future sea-level rise, all of these landforms will become more dynamic (Nicholls *et al.*, 2007), but predicting and quantifying changes that are likely to occur with high confidence is currently scientifically challenging.

13.2.1. Cliff and Bluff Shorelines

Substantial portions of the U.S. coast are comprised of coastal cliffs and bluffs that vary greatly in height, morphology, and sedimentary composition. These occur predominantly along the New England and Pacific coasts, Hawaii, and Alaska. Coastal cliff is a general term that refers to steep slopes along the shoreline that commonly form in response to long-term rise in sea-level. The term "bluff" also can refer to escarpments eroded into unlithified material, such as glacial till, along the shore (Hampton and Griggs, 2004). The terms "cliff" and "bluff" are often used interchangeably. Coastal cliffs erode in response to a variety of both marine and terrestrial processes. Cliff retreat can be fairly constant, but can also be episodic. In contrast to sandy coasts, which may erode landward or accrete seaward, cliffs retreat only in a landward direction. Because rocky cliff coasts are composed of resistant materials, erosion can can occur more slowly than for those comprised of unconsolidated sediments and response times to sea-level rise much longer than for sandy coasts (NRC, 1987), but land slumping due to wave action or land surface water runoff can result in rapid retreat. Hampton and Griggs (2004) provide a review of the origin, U.S. distribution, evolution, and regional issues associated with coastal cliffs. Predicting the response of coastal cliffs to future sea-level rise is a topic of active research (Trenhaile, 2001; Walkden and Hall, 2005; Dickson *et al.*, 2007; Walkden and Dickson, 2008)

13.2.2. Sandy Shores, Pocket Beaches, Barrier Beaches, Spits, and Dunes

Sandy beaches are often categorized into a few basic types which commonly include mainland, pocket, and barrier beaches (Wells, 1995; Davis and FitzGerald, 2004). The sediments that comprise beaches are derived mainly from the erosion of the adjacent mainland and continental shelf, and sometimes from sediments supplied from coastal rivers. Mainland beaches occur where the land intersects the shore. Some mainland beaches occur in low-relief settings and are surrounded by coastal dunes, while others occur along steep portions of the coast and are backed by bluffs. Examples of mainland beaches include the shores of eastern Long Island, northern New Jersey (Oertel and Kraft, 1994), and parts of Delaware, (Kraft, 1971). Pocket beaches form in small bays, often occurring between rocky headlands and are common along parts of the southern New England coast, portions of California and Oregon (Hapke *et al.,* 2006), and in parts of the Hawaiian Islands. Barrier beaches and spits are the most abundant coastal landforms along the Atlantic and Gulf of Mexico coasts. In general, it is expected that accelerations in sea-level rise will enhance beach erosion globally, but on a local scale this response will depend on the sediment budget (Nicholls *et al.,* 2007).

13.2.3. Coastal Marshes, Mangroves, and Mud Flat Shorelines

Coastal wetlands include swamps and tidal flats, salt and brackish marshes, mangroves, and bayous. They form in low-relief, low-energy sheltered coastal environments, often in conjunction with river deltas, landward of barrier islands, and along the flanks of estuaries (*e.g.*, Delaware Bay, Chesapeake Bay, Everglades, Lake Pontchatrain, Galveston Bay, San Francisco Bay, and Puget Sound). Most coastal wetlands are in Louisiana, North and South Carolina, south Florida, and Alaska (Dahl, 1990; NRC, 1995a). Wetlands are extremely vulnerable to sea-level rise and can maintain their elevation and viability only if sediment accumulation (both mineral and organic matter) keeps pace with sea-level rise (Cahoon *et al.*, 2006; Nyman *et al.,* 2006; Morris *et al.,* 2002; Rybczyk and Cahoon, 2002). Future wetland area will also be determined, in part, by the amount of space (*e.g.*, mud flat or tidal flat area) available for landward migration and the rates of lateral erosion of the seaward edge of the marsh (see Chapter 4; Poulter, 2005). Wetlands will be especially vulnerable to the higher projected rates of future sea- level rise (*e.g.*, greater than 70 cm by the year 2100), but some will survive a 1 meter rise (Morris *et al.,* 2002). Even under lower accelerated sea-level rise rates, wetlands may be sustained only where conditions are optimal for vertical wetland development (*e.g.*, abundant sediment supply and low regional subsidence rate) (Rybczyk and Cahoon, 2002).

Mud flat shorelines represent a relatively small portion of U.S. coasts, but are important in providing the foundation for wetlands and marshes (Mitsch and Gosselink, 1986). They are frequently associated with wetlands, and occur predominately in low-energy, low-relief regions with high inputs of fine-grained river-born sediments and organic materials and large tidal ranges. These shoreline types are common in western Louisiana (*i.e.*, Chenier Plain) and along northeastern parts of the Gulf Coast of Florida. Muddy coasts may be drowned with sea-level rise unless sediment inputs are sufficiently large, such as the Atchafalaya River delta region of southwestern Louisiana, and the flats are able to be colonized by plants.

13.2.4. Tropical Coral Reef Coasts

Tropical coral reefs, made up of living organisms very sensitive to ocean temperature and chemistry, are found in the U.S. along the south coast of Florida; around the Hawaiian Islands, Puerto Rico, the Virgin Islands, and many of the U.S. territories in the Pacific (Riegl and Dodge, 2008). In tropical environments, living coral organisms build reefs that are important ecological resources (Smith and Buddemeier, 1992; Boesch *et al.*, 2000). Most corals are able to tolerate rates of sea-level rise of 10 to 20 mm per year or more (Smith and Buddemeier, 1992; Bird, 1995; Wells, 1995; Hallock, 2005). Nonetheless, the ability of coral reef systems to survive future sea-level rise will depend heavily on other climate change impacts such as increase in ocean temperature and or acidity, sediment runoff from the land, as well as episodic storm erosion (Hallock, 2005; Nicholls *et al.*, 2007). In addition, human caused stresses such as over-fishing or pollution can contribute to the vulnerability of these systems to climate change (Buddemeier *et al.*, 2004; Mimura *et al.*, 2007).

13.3. Potential for Future Shoreline Change

Over the next century and beyond, with an expected acceleration in sea-level rise, the potential for coastal change will increase and coastal change is likely to be more widespread and variable than has been observed in historic past (NRC, 1987; Brown and McLachlan, 2002; Nicholls *et al.*, 2007). However, it is difficult at present to quantitatively attribute shoreline changes directly to sea-level rise (Rosenzweig *et al.*, 2007). The potential changes include increased coastal erosion, more frequent tidal and storm-surge flooding of low-relief areas, and wetland deterioration and losses. Many of these changes will occur in all coastal states. These changes to the coastal zone can be expected to have especially large impacts to developed areas (Nicholls *et al.*, 2007). Some portions of the U.S. coast will be subject principally to inundation from sea-level rise over the next century, including upper reaches of bays and estuaries (*e.g.*, Chesapeake and Delaware Bays, Tampa Bay, Lake Pontchartrain, San Francisco Bay), and hardened urban shorelines. Erosion, sediment transport, and sediment deposition in coastal environments are active processes and will drive coastal change in concert with the combined effects of future sea-level rise and storms (Stive, 2004).

Coastal landforms may become even more dynamic and that erosion will dominate changes in shoreline position over the next century and beyond (Nicholls *et al.*, 2007).

Wetlands with sufficient sediment supply and available land for inland migration may be able to maintain elevation, keeping pace with sea-level rise, but sediment starved wetlands and those constrained by engineering structures (*e.g.*, seawalls, revetments) or steep uplands are likely to deteriorate and convert to open water through vertical accretion deficits and lateral erosion (see Chapter 4). On barrier island shores, erosion is likely to occur on both the ocean front and the landward side of the island due to a combination of storm activity, changes in sediment budget, more frequent tidal flooding, and rising water levels (Nicholls *et al.,* 2007).

Sea-level rise is a particular concern for islands (Mimura *et al.*, 2007). Of particular concern are islands comprised of coral atolls (*e.g.*, Midway Atoll), which are typically low-lying and dependent on the health of coral reefs that fringe the atolls. Populated islands with higher elevations (*e.g.*, the Northern Mariana Islands) are also frequently at risk as the infrastructure is frequently located in low-lying coastal regions along the periphery of the islands.

Many coastal areas in the United States will likely experience an increased frequency and magnitude of storm-surge flooding, greater wave heights, and more erosion due to storms as part of the response to sea-level rise (NRC, 1987; Woodworth and Blackman, 2004; Nicholls *et al.*, 2007; Gutowski *et al.*, 2008). Impacts from these storm events may extend farther inland than those that would be affected by sea-level rise alone. Many regions may also experience large changes to coastal systems, such as increased rates of erosion, barrier island and dune landward migration, and potential barrier island collapse (Nicholls *et al.*, 2007; see also Chapters 1, 3, and 14 for discussion of geomorphic thresholds). The potential of crossing thresholds, potentially leading to barrier and wetland collapse, may increase with higher rates of sea-level rise.

The use of so called "soft" coastal engineering mitigation measures, such as beach nourishment, usually using sand dredged from offshore Holocene-age sand bodies, may reduce the risk of storm flooding and coastal erosion temporarily (NRC, 1987, 1995b). However, an important issue is whether or not these practices are able to be maintained into the future to provide sustainable and economical shoreline protection in the face of high cost, need for periodic re-nourishment, and limited sand resources of suitable quality for nourishment for many regions of the country (NRC, 1995b; Magoon et al., 2004). Results from offshore geologic mapping studies indicate that most continental shelf regions of the U.S. have relatively limited Holocene-age sediment that can be deemed available and suitable for uses such as beach nourishment (Schwab *et al.,* 2000; Gayes *et al.*, 2003; Pilkey *et al.*, 1981; Kraft, 1971). In some cases, potential sand volumes are reduced because of economic and environmental factors such as water depth, benthic environmrntal concerns, and concerns that sand removal may alter sediment exchange with the adjacent coast (Bliss *et al.*, 2009). The result is limited volumes of high-quality offshore sand resources readily available for beach nourishment. The issue of relying long term on using offshore sand for beach nourishment to mitigate erosion is important and needs to be addressed.

More widespread implementation of regional sediment or best sediment management practices to conserve valuable coastal sediments from offshore disposal of clean sandy dredged spoils will enhance the long term sustainability of sandy coastal landforms (NRC, 2007). The use of so called "hard" engineering structures (*e.g.*, seawalls, breakwaters) to protect property from erosion and flooding may be justified for urban coasts, but their use on sandy shores can further exacerbate erosion over time due to disruption of sediment transport processes. Alternatives, such as relocation landward, strategic removal of development or limiting redevelopment following storm disasters from highly vulnerable parts of the coast, may provide longer term sustainability of both coastal landforms and development, especially if the higher rates of sea-level rise are realized (NRC, 1987). An example of abandonment of an island in Chesapeake Bay due to sea-level rise is detailed in Gibbons and Nicholls (2006). If coastal development is relocated, those areas could be converted to marine protected areas, public open-space lands that would serve to buffer sea-level rise effects landward and also provide recreation benefits and wildlife habitat values (see Salm and Clark, 2000).

13.4. Conclusions

Global climate is changing, largely due to carbon emissions from human activities (IPCC, 2001; 2007). Sea-level rise is one of the impacts of climate change that will affect all coastal regions of the United States over the next century and beyond (NRC, 1987; Nicholls *et al.*, 2007). The scientific tools and techniques for assessing the effects of future sea-level rise on coastal systems are improving, but much remains to be done in order to develop useful forecasts of potential effects. Chapter 14 of this Product identifies research opportunities that, if implemented, would lead to better understanding and prediction of sea-level rise effects that are likely to further impact the United States in the near future. Planning for accelerating sea-level rise should include thorough evaluation of a number of alternatives, such as cost-effective and sustainable shore protection and strategic relocation of development within urban centers. Important decisions like these should ideally be based on the best available science and careful consideration of longterm benefits for a sustainable future, and the total economic, social, and environmental costs of various methods of shore protection, relocation, and adaptation.

References

Bird, E. C. F. (1995). Present and future sea-level: the effects of predicted global changes. In: *Climate Change: Impact and Coastal Habitation*, [Eisma, D. (ed.)]. CRC Press, Boca Raton, FL, pp. 29-56.

Bliss, J. D., Williams, S. J. & Bolm, K. S. (2009). Modeling and assessment of marine sand resources, New York Bight, USA. In: *Contributions to industrial minerals research.* [Bliss, J.D., P.R. Moyle, and K.R. Long, (eds.)]. U.S. Geological Survey bulletin 2209-M. U.S. Geological Survey, [Reston, VA], 22 pp.

Boesch, D. F., Field, J. C. & Scavia (eds.), D. (2000). *The Potential Consequences of Climate Variability and Change on Coastal Areas and Marine Resources.* NOAA Coastal Ocean Program decision analysis series #21. National Oceanic and Atmospheric Administration, Silver Spring, MD, 163 pp. <http://www.cop.noaa.gov/pubs/das/das2 1 .pdf>

Brown, A. C. & McLachlan, A. (2002). Sandy shore ecosystems and the threats facing them: some predictions for the year 2025. *Environmental Conservation, 29(1),* 62-77.

Buddemeier, R. W., Kleypas, J. A. & Aronson, R. B. (2004). *Coral Reefs and Gobal Climate Change.* Pew Center on Global Climate Change, Arlington, VA, 44 pp.

Cahoon, D. R., Hensel, P. F., Spencer, T., Reed, D. J., McKee, K. L. & Saintilan, N. (2006). Coastal wetland vulnerability to relative sea level rise: wetland elevation trends and process controls. In: *Wetlands and Natural Resource Management* [Verhoeven, J.T.A., B. Beltman, R. Bobbink, and D. Whigham (eds.)]. Ecological studies volume 190, Springer, Berlin and New York, pp. 27 1-292.

Crossett, K., Culliton, T. J., Wiley, P. & Goodspeed, T. R. (2004). *Population Trends along the Coastal United States, 1980–2008.* National Oceanic and Atmospheric Administration, Silver Spring, MD, 47 pp.

Crowell, M., Coulton, K. & McAfee, S. (2007). How many people live in coastal areas?. *Journal of Coastal Research, 23(5),* iii-vi, editorial.

Dahl, T. E. (1990). *Wetlands losses in the United States 1780's to 1980's*, U.S. Department of the Interior, Fish and Wildlife Service, Washington, DC, 21 pp.

Davis, R. A. & FitzGerald, D. M. (2004). *Beaches and Coasts.* Blackwell Publishing, Malden, MA, 419 pp.

Dickson, M. E., Walkden, M. J. A. & Hall, J.W. (2007). Systemic impacts of climate change on an eroding coastal region over the twenty-first century. *Climatic Change, 84(2),* 141-166.

FitzGerald, D. M., Fenster, M. S., Argow, B.A. & Buynevich, I.V. (2008). Coastal impacts due to sea-level rise. *Annual Reviews of Earth and Planetary Sciences, 36*, 601-647.

Gayes, P. T., Schwab, W. C., Driscoll, N. W., Morton, R. A., Baldwin, W. E., Denny, J. F. Wright, E. E., Harris, M. S. Katuna, M.P. Putney, T. R. & Johnstone, E. (2003). Sediment dispersal pathways and conceptual sediment budget for a sediment starved embayment; Long Bay, South Carolina. In: *Coastal Sediments '03, 5th Annual Symposium on Coastal Engineering and Science of Coastal Sediment Processes*, Clearwater Beach, FL, May 18-23, 2003. East Meets West Productions, Corpus Christi, TX, 14 pp.

Gibbons, S. J. A. & Nicholls, R. J. (2006). Island abandonment and sea-level rise: an historical analog from the Chesapeake Bay, USA. *Global Environmental Change, 16(1),* 40-47.

Gutierrez, B. T., Williams, S. J. & Thieler, E. R. (2007). *Potential for Shoreline Changes Due to Sea-level Rise along the U.S. Mid-Atlantic Region.* U.S. Geological Survey open file report 2007-1278. U.S. Geological Survey, Reston, VA, 26 pp. <http://pubs.usgs.gov/of/2007/1 278/>

Gutowski, W. J., Hegerl, G. C., Holland, G. J., Knutson, T. R., Mearns, L. O., Stouffer, R. J. , Webster, P. J., Wehner, M. F. & Zwiers, F. W. (2008). Causes of observed changes in extremes and projections of future changes. In: *Weather and Climate Extremes in a Changing Climate: Regions of Focus: North America, Hawaii, Caribbean, and U.S. Pacific Islands* [Karl, T.R., G.A. Meehl, C.D. Miller, S.J. Hassol, A.M. Waple, and W.L. Murray (eds.)]. Synthesis and Assessment Product 3.3. U.S. Climate Change Science Program, Washington, DC, pp. 81-116.

Hallock, P. (2005). Global change and modern coral reefs: new opportunities to understand shallow-water carbonate depositional processes. *Sedimentary Geology, 175(1-4),* 19-33.

Hampton, M.A. and G.B. Griggs (eds.), 2004: *Formation, Evolution, and Stability of Coastal Cliffs — Status and Trends.* U.S. Geological Survey professional paper 1693. U.S. Geological Survey, Reston, VA, 123 pp.

IPCC (Intergovernmental Panel on Climate Change) (2001). *Climate Change 2001: The Scientific Basis.* Contribution of Working Group I to the Third Assessment Report of the

Intergovernmental Panel on Climate Change [Houghton, J.T., Y. Ding, D.J. Griggs, M. Noguer, P.J. van der Linden, X. Dai, K. Maskell, and C.A. Johnson (eds.)]. Cambridge University Press, Cambridge, UK, and New York, 881 pp.

IPCC (Intergovernmental Panel on Climate Change) (2007). *Climate Change 2007: The Physical Science Basis.* Contribution of Working Group I to the Fourth Assessment Report of the Intergovernmental Panel on Climate Change [Solomon, S., D. Qin, M. Manning, Z. Chen, M. Marquis, K.B. Averyt, M. Tignor, and H.L. Miller (eds.)]. Cambridge University Press, Cambridge, UK and New York, 996 pp.

Leatherman, S. P. (2001). Social and economic costs of sea level rise. In: *Sea Level Rise: History and Consequences* [Douglas, B.C., M.S. Kearney, and S.P. Leatherman (eds.)]. Academic Press, San Diego, pp. 181-223.

Magoon, O. T., Williams, S. J., Lent, L. K. Douglass, S. L. Edge, B. L., Richmond, J. A., Treadwell, D. D., Ewing, L. C. & Pratt, A. P. (2004). Economic impacts of anthropogenic activities on coastlines of the U.S., In: *Coastal Engineering 2004,* Proceedings of the 29th International Conference, Lisbon, Portugal. World Scientific Publishing, New Jersey and London, pp. 3022-3035.

Mimura, N. L., Nurse, L., McLean, R. F., Agard, J., Briguglio, L., Lefale, P., Payet, R. & Sem, G. (2007). Small islands. In: *Climate Change 2007: Impacts, Adaptation and Vulnerability.* Contribution of Working Group II to the Fourth Assessment Report of the Intergovernmental Panle on Climate Change [Parry, M.L., O.F. Canziani, J.P., Palutikof, P.J. van der Linden, and C.E. Hanson (eds.)]. Cambridge University Press, Cambridge, UK, and New York, pp. 688-7 16.

Mitsch, W. J. & Gosselink, J. G. (1986). *Wetlands.* Van Nostrand, New York, 537 pp.

Morris, J. T., Sundareshwar, P. V., Nietch, C. T., Kjerfve, B. & Cahoon, D. R. (2002). Responses of coastal wetlands to rising sea level. *Ecology, 83(10),* 2869-2877.

NRC (National Research Council) (1987). *Responding to Changes in Sea Level.* National Academy Press, Washington, DC, 148 pp.

NRC (National Research Council) (1990). *Managing Coastal Erosion.* National Academy Press, Washington, DC, 182 pp.

NRC (National Research Council) (1995a). *Wetlands: Characteristics and Boundaries.* National Academy Press, Washington, DC, 306 pp.

NRC (National Research Council) (1995b). *Beach Nourishment and Protection.* National Academy Press, Washington, DC, 334 pp.

NRC (National Research Council) (2007). *Mitigating Shore Erosion Along Sheltered Coasts.* National Academies Press, Washington, DC, 174 pp.

Nicholls, R. J., Wong, P. P., Burkett, V. R., Codignotto, J. O., Hay, J. E., McLean, R. F., Ragoonaden, S. & Woodroffe, C. D. (2007). Coastal systems and low-lying areas. In: *Climate Change 2007: Impacts, Adaptation and Vulnerability.* Contribution of Working Group II to the Fourth Assessment Report of the Intergovernmental Panel on Climate Change [Parry, M.L., O.F. Canziani, J.P. Palutikof, P.J. van der Linden, and C.E. Hanson (eds.)]. Cambridge University Press, Cambridge, UK, and New York, pp. 315-356.

Nyman, J. A., Walters, R. J., Delaune, R. D. & Patrick Jr., W. H. (2006). Marsh vertical accretion via vegetative growth. *Estuarine Coastal and Shelf Science, 69(3-4),* 370-380.

Oertel, G. F. & Kraft, J. C. (1994). New Jersey and Delmarva barrier islands. In: *Geology of Holocene Barrier Island Systems* [Davis, R.A. (ed.)]. Springer-Verlag, New York, pp. 207-232.

Overpeck, J. T., Otto-Bliesner, B. L., Miller, G. H., Muhs, D. R., Alley, R. B. & Keihl, J.T. (2006). Paleo-climatic evidence for the future ice-sheet instability and rapid sea level rise. *Science, 311(5768),* 1747-1750.

Pilkey, O. H., Blackwekder, B. W., Knebel, H. J.& Ayers, M. W. (1981). The Georgia Embayment continental shelf: Stratigraphy of a submergence. *Geological Society of America Bulletin, 92(1),* 52-63.

Rahmstorf, S. (2007). A semi-empirical approach to projecting future sea-level rise. *Science, 315(5810),* 368-370.

Rahmstorf, S., Cazenave, A., Church, J.A., Hansen, J. E., Keeling, R. F., Parker, D. E. & Somerville, R. C. J. (2007). Recent climate observations compared to projections. *Science, 316(5825),* 709.

Reed, D. J., Bishara, D., Cahoon, D., Donnelly, J. Kearney, M., Kolker, A. Leonard, L., Orson, R. A. & Stevenson, J.C. (2008). Site-specific scenarios for wetlands accretion as sea level rises in the mid-Atlantic region. Section 2.1 in: *Background Documents Supporting Climate Change Science Program Synthesis and Assessment Product 4.1: Coastal Elevations and Sensitivity to Sea Level Rise* [Titus, J.G. and E.M. Strange (eds.)]. EPA 430R07004. U.S. Environmental Protection Agency, Washington, DC, pp. 134-174. <http://epa.gov/climatechange/effects/coastal/background.html>

Riegl, B. & Dodge (eds.), E. (2008). *Coral Reefs of the USA.* Coral reefs of the worldvolume 1. Springer-Verlag, Dordrecht and London, 806 pp.

Rosenzweig, C., Casassa, G., Karoly, D. J., Imeson, A., Liu, C., Menzel, A., Rawlins, S., Root, T. L., Seguin, B. Tryjanowski, P. & Hanson, C. E. (2007). Assessment of observed changes and responses in natural and managed systems. In: *Climate Change 2007: Impacts, Adaptation and Vulnerability. Contribution of Working Group II to the Fourth Assessment Report of the Intergovernmental Panel on Climate Change.* [Parry, M.L., O.F. Canziani, J.P. Palutikof, and P.J. van der Linden, (eds.)]. Cambridge University Press, Cambridge, UK, and New York, pp. 79-131.

Rybczyk, J. M. & Cahoon, D. R. (2002). Estimating the potential for submergence for two subsiding wetlands in the Mississippi River delta. *Estuaries, 25(5),* 985-998.

Salm, R. V. & Clark, J.R. (2000). *Marine and Coastal Protected Areas.* International Union for Conservation of Nature and Natural Resources, Switzerland, 370 pp.

Schwab, W. C., Thieler, E. R., Allen, J. R., Foster, D. S., Swift, B. A. & Denny, J. F. (2000). Influence of inner-continental shelf geologic framework on the evolution and behavior of the barrier island system between Fire Island inlet and Shinnecock inlet, Long Island, New York. *Journal of Coastal Research, 16(2),* 408-422.

Smith, S. V. & Buddemeier, R.W. (1992). Global change and coral reef ecosystems. *Annual Reviews of Ecological Systematics, 23,* 89-118.

Stive, M. J. F. (2004). How important is global warming for coastal erosion? An editorial comment. *Climatic Change, 64(1-2),* 27-39.

Trenhaile, A. S. (2001). Modeling the effect of late Quaternary interglacial sea levels on wave-cut shore platforms. *Marine Geology, 172(3-4),* 205-223.

Walkden, M. and M. Dickson (2008). Equilibrium erosion of soft rock shores with a shallow or absent beach under increased sea level rise. *Marine Geology, 251(1-2),* 75-84.

Walkden, M. J. A. & Hall, J. W. (2005). A predictive Mesoscale model of the erosion and profile development of soft rock shores. *Coastal Engineering, 52(6),* 535-563.

Wells, J. T. (1995). Effects of sea level rise on coastal sedimentation and erosion. In: *Climate Change: Impact and Coastal Habitation*, [Eisma, D. (ed.)]. CRC Press, Boca Raton, FL, pp. 111-136.

Williams, S. J. (2003). Coastal and marine processes. Chapter 1.1.3.2. In: *Our Fragile World: Challenges and Opportunities for Sustainable Development*: Encyclopedia of Life Support Systems (EOLSS), [Cilek, V. (ed.)]. Developed under the auspices of the UNESCO, EOLSS Publishers, Oxford, UK, 13 pp. <http://www.eolss.net>

Williams, S. J. Dodd, K. & Gohn, K. K. (1991). *Coasts in Crisis*. U.S. Geological Survey circular 1075. U.S. Geological Survey, [Reston, VA], 30 pp.

Woodworth, P. L & Blackman, D. L. (2004). Evidence for systematic changes in extreme high waters since the mid-1970s. *Journal of Climate, 17(6),* 1190-1197.

In: Coastal Sensitivity to Sea Level Rise-Focusing on the Mid-… ISBN: 978-1-60741-440-7
Editor: Melvin C. Urajner

Chapter 14

A SCIENCE STRATEGY FOR IMPROVING THE UNDERSTANDING OF SEA-LEVEL RISE AND ITS IMPACTS ON U.S. COASTS

E. Robert. Thieler, K. Eric Anderson, Donald R. Cahoon, S. Jeffress Williams and Benjamin T. Gutierrez
USGS

Key Findings

- Understanding, predicting, and responding to the environmental and human effects of sea-level rise requires an integrated program of research that includes natural and social sciences.
- Monitoring of modern processes and environments could be improved by expanding the network of basic observations and observing systems, developing time series data on environmental and landscape changes, and assembling baseline data for the coastal zone.
- The historic and geologic record of coastal change should be used to improve the understanding of natural and human-influenced coastal systems, increase knowledge of sea-level rise and coastal change over the past few millennia, identify thresholds or tipping points in coastal systems, and more closely relate past changes in climate to coastal change.
- Increases in predictive capabilities can be achieved by improving quantitative assessment methods and integrating studies of the past and present into predictive models.
- Research on adaptation, mitigation, and avoidance measures will enable better understanding of the societal impacts of sea-level rise.
- Decision making in the coastal zone can be supported by providing easy access to data and resources, transferring knowledge of vulnerability and risk that affect decision making, and educating the public about consequences and alternatives.

14.1. Introduction

Chapter 14 identifies several major themes that present opportunities to improve the scientific understanding of future sea-level rise and its impacts on U.S. coastal regions. Advances in scientific understanding will enable the development of higher quality and more reliable information for planners and decision makers at all levels of government, as well as the public.

A number of recent studies have focused specifically on research needs in coastal areas. Two National Research Council (NRC) studies, *Science for Decision-making* (NRC, 1999) and *A Geospatial Framework for the Coastal Zone* (NRC, 2004) contain recommendations for science activities that can be applied to sea-level rise studies. Other relevant NRC reports include *Responding to Changes in Sea Level* (NRC, 1987), *Sea Level Change* (NRC, 1 990b) and *Abrupt Climate Change* (NRC, 2002). The Marine Board of the European Science Foundation's Impacts of Climate Change on the European Marine and Coastal Environment (Philippart *et al.*, 2007) identified numerous research needs, many of which have application to the United States. Recent studies on global climate change by the Pew Charitable Trusts also included the coastal zone (*e.g.*, Neumann *et al.*, 2000, Panetta, 2003; Kennedy *et al.*, 2002). Other studies by the NRC (1990a, 1990b, 1990c, 2001, 2006a, 2007) and the Heinz Center (2000, 2002a, 2002b, 2006) have addressed issues relevant to the impacts of sea-level rise on the coastal zone. These reports and related publications have helped guide the development of the potential research and decision-support activities described in the following sections.

14.2. A Science Strategy to Address Sea-Level Rise

An integrated scientific program of sea level studies that seeks to learn from the historic and geologic past, and monitors ongoing physical and environmental changes, will improve the level of knowledge and reduce the uncertainty about potential responses of coasts, estuaries, and wetlands to sea-level rise. Outcomes of both natural and social scientific research will support decision making and adaptive management in the coastal zone. The main elements of a potential science strategy and their interrelationships are shown in Figure14.1.

Building on and complementing ongoing efforts at federal agencies and universities, a research and observation program could incorporate new technologies to address the complex scientific and societal issues highlighted in this Product. These studies could include further development of a robust monitoring program for all coastal regions, leveraging the existing network of site observations, as well as the growing array of coastal observing systems. Research should also include studies of the historic and recent geologic past to understand how coastal systems evolved in response to past changes in sea level. The availability of higher resolution data collected over appropriate time spans, coupled with conceptual and numerical models of coastal evolution, will provide the basis for improved quantitative assessments and the development of predictive models useful for decision making. Providing ready access to interpretations from scientific research—as well as the underlying data—by means of publications, data portals, and decision-support systems will allow coastal managers to evaluate alternative strategies for mitigation, develop appropriate responses to sea-level rise, and practice adaptive management as new information becomes available.

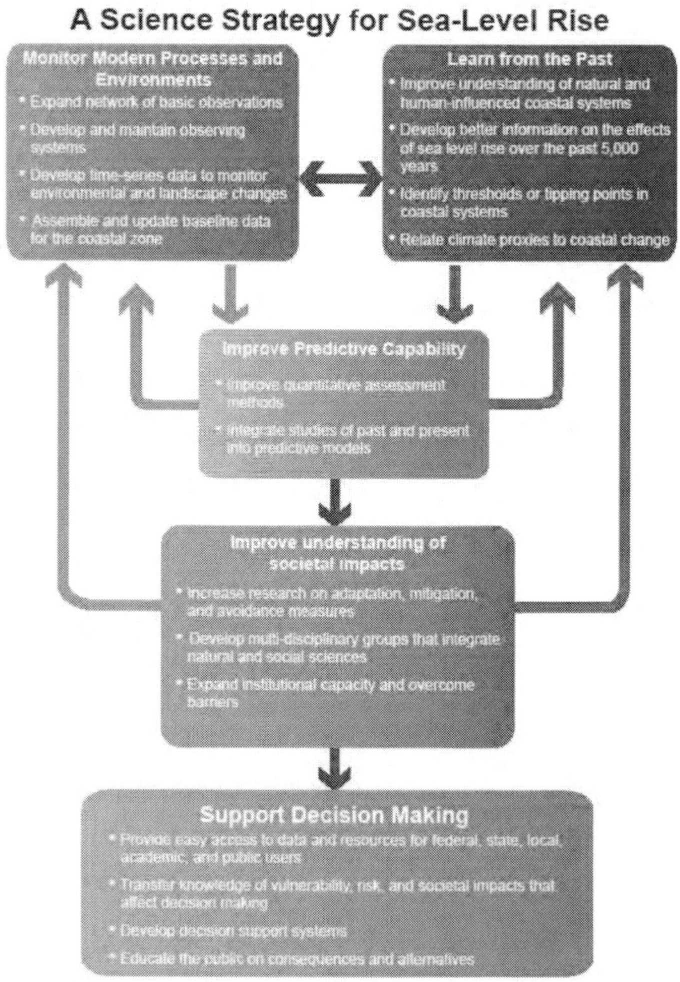

Figure 14.1. Schematic flow diagram summarizing a science strategy for improvement of scientific knowledge and decision-making capability needed to address the impacts of future sea-level rise.

14.2.1. Learn from the Historic and Recent Geologic Past

Studies of the recent geologic and historical record of sea-level rise and coastal and environmental change are needed to improve the state of knowledge of the key physical and biological processes involved in coastal change. As described throughout this Product, particularly in Chapters 1 through 5, significant knowledge gaps exist that inhibit useful prediction of future changes. The following research activities will help refine our knowledge of past changes and their causes.

Improve understanding of natural and human-influenced coastal systems

Significant opportunities exist to improve predictions of coastal response to sea-level rise. For example, scientists' understanding of the processes controlling rates of sediment flux in both natural and especially in human-modified coastal systems is still evolving. This is particularly true at the regional (littoral cell) scale, which is often the same scale at which management decisions are made. As described in Chapters 3 and 6, the human impact on coastal processes at management scales is not well understood. Shoreline engineering such as bulkheads, revetments, seawalls, groins, jetties, and beach nourishment can fundamentally alter the way a coastal system behaves by changing the transport, storage, and dispersal of sediment. The same is true of development and infrastructure on mobile landforms such as the barrier islands that comprise much of the mid-Atlantic coast.

Develop better information on the effects of sea-level rise over the past 5,000 years

The foundation of modern coastal barrier island and wetland systems has evolved over the past 5,000 years as the rate of sea-level rise slowed significantly (see Chapters 1, 3, and 4). More detailed investigation of coastal sedimentary deposits is needed to understand the rates and patterns of change during this part of the recent geologic past. Advances in methods to obtain samples of the geologic record, along with improvements in analytical laboratory techniques since the early 1990s, have significantly increased the resolution of the centennial-to-millennial scale record of sea-level rise and coastal environmental change (*e.g.*, Gehrels, 1994; Gehrels *et al.*, 1996; van de Plassche *et al.,* 1998; Donnelly *et al.*, 2001; Horton *et al.*, 2006) and provide a basis for future work. Archaeological records of past sea-level change also exist in many locales, and provide additional opportunities to understand coastal change and impacts on human activity.

Understand thresholds in coastal systems that, if crossed, could lead to rapid changes to coastal and wetland systems

Several aspects of climate change studies, such as atmosphere-ocean interactions, vegetation change, sea ice extent, and glacier and ice cap responses to temperature and precipitation, involve understanding the potential for abrupt climate change or "climate surprises" (NRC, 2002; Meehl *et al.*, 2007). Coastal systems may also respond abruptly to changes in sea-level rise or other physical and biological processes (see Chapter 3, Box 3.1). Coastal regions that may respond rapidly to even modest changes in future external forcing need to be identified, as well as the important variables driving the changes. For example, limited sediment supply, and/or permanent sand removal from the barrier system, in combination with an acceleration in the rate of sea-level rise, could result in the development of an unstable state for some barrier island systems (*i.e.*, a behavioral threshold or tipping point, as described in Chapters 1 and 3). Coastal responses could result in landward migration or roll-over, or barrier segmentation. Understanding and communicating the potential for such dramatic changes in the form and rate of coastal change will be crucial for the development of adaptation, mitigation, and other strategies for addressing sea-level rise.

The future evolution of low-elevation, narrow barriers will likely depend in part on the ability of salt marshes in back-barrier lagoons and estuaries to keep pace with sea-level rise (FitzGerald *et al.*, 2004, 2008; Reed *et al.*, 2008). It has been suggested that a reduction of salt marsh in back-barrier regions could change the hydrodynamics of back- barrier systems, altering local sediment budgets and leading to a reduction in sandy materials available to sustain barrier systems (FitzGerald *et al.*, 2004, 2008).

Relate climate proxies to coastal change

Links between paleoclimate proxies (*e.g.*, atmospheric gases in ice cores, isotopic composition of marine microfossils, tree rings), sea-level rise, and coastal change should be explored. Previous periods of high sea level, such as those during the last several interglacial periods, provide tangible evidence of higher-than-present sea levels that are broadly illustrative of the potential for future shoreline changes. For example, high stands of sea level approximately 420,000 and 125,000 years ago left distinct shoreline and other coastal features on the U.S. Atlantic coastal plain (Colquhoun *et al.*, 1991; Baldwin *et al.*, 2006). While the sedimentary record of these high stands is fragmentary, opportunities exist to relate past shoreline positions with climate proxies to improve the state of knowledge of the relationships between the atmosphere, sea level, and coastal evolution. Future studies may also provide insight into how coastal systems respond to prolonged periods of high sea level and rapid sea-level fluctuations during a high stand. Examples of both exist in the geologic record and have potential application to understanding and forecasting future coastal evolution.

14.2.2. Monitor Modern Coastal Conditions

The status and trends of sea-level change, and changes in the coastal environment, are monitored through a network of observation sites, as well as through coastal and ocean observing systems. Monitoring of modern processes and environments could be improved by expanding the network of basic observations, as well as the continued development of coastal and ocean observing systems. There are numerous ongoing efforts that could be leveraged to contribute to understanding patterns of sea-level rise over space and time and the response of coastal environments.

Expand the network of basic observations

An improvement in the coverage and quality of the U.S. network of basic sea-level observations could better inform researchers about the rate of sea-level rise in various geographic areas. Tide gauges are a primary source of information for sea-level rise data at a wide range of time scales, from minutes to centuries. These data contribute to a multitude of studies on local to global sea-level trends. Tide gauge data from the United States include some of the longest such datasets in the world and have been especially valuable for monitoring long-term trends. A denser network of high-resolution gauges would more rigorously assess regional trends and effects. The addition of tide gauges along the open ocean coast of the United States would be valuable in some regions. These data can be used in concert with satellite altimetry observations.

Tide-gauge observations also provide records of terrestrial elevation change that contributes to relative sea-level change, and can be coupled with field- or model-based measurements or estimates of land elevation changes. Existing and new gauges should be co-located with continuously operating Global Positioning System (GPS) reference stations (CORS) or surveyed periodically using GPS and other Global Navigation Satellite System technology. This will enable the coupling of the geodetic (earth-based) reference frame and the oceanographic reference frame at the land-sea interface. Long time series from CORS can provide precise local vertical land movement information in the ellipsoidal frame (*e.g.*, Snay *et al.*, 2007; Woppelmann *et al.*, 2007). Through a combined effort of monitoring ellipsoid heights and the geoid, as well as through gravity field monitoring, changes to coastal elevations can be adequately tracked.

Develop and maintain coastal observing systems

Observing systems have become an important tool for examining environmental change. They can be place-based (*e.g.*, specific estuaries or ocean locations) or consist of regional aggregations of data and scientific resources (*e.g.*, the developing network of coastal observing systems) that cover an entire region. Oceanographic observations also need to be integrated with observations of the physical environment, as well as habitats and biological processes.

An example of place-based observing systems is the National Estuarine Research Reserve System (NERRS: <http://www.nerrs.noaa.gov>), a network of 27 reserves for long-term research, monitoring, education, and resource stewardship. Targeted experiments in such settings can potentially elucidate impacts of sea-level rise on the physical environment, such as shoreline change or impacts to groundwater systems, or on biological processes, such as species changes or ecosystem impacts. Important contributions can also be made by the Long Term Ecological Research sites (<http://www.lternet.edu>) such as the Virginia Coast Reserve in the mid-Atlantic area (part of the focus of this Product). The sites combine long-term data with current research to examine ecosystem change over time. Integration of these ecological monitoring networks with the geodetic and tide gauge networks mentioned previously would also be an important enhancement.

The Integrated Ocean Observing System (IOOS) (<http://www.ocean.us>) will bring together observing systems and data collection efforts to understand and predict changes in the marine environment. Many of these efforts can contribute to understanding changes in sea-level rise over space and time. These observing systems incorporate a wide range of data types and sources, and provide an integrated approach to ocean studies. Such an approach should enable sea-level rise-induced changes to be distinguished from the diverse processes that drive changes in the coastal and marine environment.

A new initiative began in 2005 with a worldwide effort to build a Global Earth Observation System of Systems (GEOSS) (<http://www.earthobservations.org>) over the next 10 years. GEOSS builds upon existing national, regional, and international systems to provide comprehensive, coordinated Earth observations from thousands of instruments worldwide, which have broad application to sea-level rise studies.

Develop time series data to monitor environmental and landscape changes

Observations of sea level using satellite altimetry (*e.g.*, TOPEX/Poseidon and Jason-1) have provided new and important insights into the patterns of sea-level change across space and time. Such observations have allowed scientists to examine sea-level trends and compare them to the instrumental record (Church *et al.*, 2001, 2004), as well as predictions made by previous climate change assessments (Rahmstorf, 2007). The satellite data provide spatial coverage not available with ground-based methods such as tide gauges, and provide an efficient means for making global observations. Plans for future research could include a robust satellite observation program to ensure comprehensive coverage.

Studies of environmental and landscape change can also be expanded across larger spatial scales and longer time scales. Examples include systematic mapping of shoreline changes and coastal barriers and dunes around the United States (*e.g.*, Morton and Miller, 2005), and other national mapping efforts to document land-use and land-cover changes (*e.g.*, the NOAA Coastal Change Analysis Program: <http://www.csc.noaa.gov/crs/lca/ccap.html>). It is also important to undertake a rigorous study of land movements beyond the point scale of tide gauges and GPS networks. For example, the application of an emerging technology—Interferometric Synthetic Aperture Radar (InSAR)—enables the development of spatially-detailed maps of land-surface displacement over broad areas (Brooks *et al.*, 2007).

Determining wetland sustainability to current and future sea-level rise requires a broader foundation of observations if they are to be applied with high confidence at regional and national scales. In addition, there is a significant knowledge gap concerning the viability or sustainability of human-impacted and restored wetlands in a time of accelerating sea- level rise. The maintenance of a network of sites that utilize surface elevation tables and soil marker horizons for measuring marsh accretion or loss will be essential in understanding the impacts on areas of critical wetland habitat. The addition of sites to the network would aid in delineating regional variations (Cahoon *et al.*, 2006). Similar longterm studies for coastal erosion, habitat change, and water quality are also essential.

Coastal process studies require data to be collected over a long period of time in order to evaluate changes in beach and barrier profiles and track morphological changes over a time interval where there has been a significant rise in sea level. These data will also reflect the effects of storms and the sediment budget that frequently make it difficult to extract the coastal response to sea-level change. For example, routine lidar mapping updates to track morphological changes and changes in barrier island area above mean high water (*e.g.*, Morton and Sallenger, 2003), as well as dune degradation and recovery, and shore-face profile and near-shore bathymetric evolution may provide insight into how to distinguish various time and space scales of coastal change and their relationship to sea-level rise.

Time series observations can also be distributed across the landscape and need not be tied to specific observing systems or data networks. They do, however, need a means to have their data assimilated into a larger context. For example, development of new remote sensing and *in-situ* technologies and techniques would help fill critical data gaps at the land-water interface.

Assemble and update baseline data for the coastal zone

Baseline data for the coastal zone, including elevation, bathymetry, shoreline position, and geologic composition of the coast, as well as biologic and ecologic parameters such as vegetation and species distribution, and ecosystem and habitat boundaries, should be collected at high spatial resolution. As described in Chapter 2, existing 30-m (100-ft) digital elevation models are generally inadequate for meaningful mapping and analyses in the coastal zone. The use of lidar data, with much better horizontal and vertical accuracy, is essential. While some of these mapping data are being collected now, there are substantial areas around the United States that would benefit from higher quality data. More accurate bathymetric data, especially in the near-shore, is needed for site-specific analyses and to develop a complete topographic-bathymetric model of the coastal zone to be able to predict with greater confidence wave and current actions, inundation, coastal erosion, sediment transport, and storm effects.

To improve confidence in model predictions of wetland vulnerability to sea-level rise, more information is needed on: (1) maximum accretion rates (*i.e.*, thresholds) regionally and among vegetative communities; (2) wetland dynamics across larger landscape scales; (3) the interaction of feedback controls on flooding with other accretion drivers (*e.g.*, nutrient supply and soil organic matter accumulation); (4) fine-grained, cohesive sediment supplies; and (5) changing land use in the watershed (*i.e.*, altered river flows and accommodation space for landward migration of wetlands). In addition, population data on different species in near shore areas are needed to accurately judge the effects of habitat loss or transformation. More extensive and detailed areas of habitat mapping will enable preservation efforts to be focused on the most important areas.

14.2.3. Predict Future Coastal Conditions

Studies of the past history of sea-level rise and coastal response, combined with extensive monitoring of present conditions, will enable more robust predictions of future sea-level rise impacts. Substantial opportunities exist to improve methods of coastal impact assessment and prediction of future changes.

Develop quantitative assessment methods that identify high-priority areas needing useful predictions

Assessment methods are needed to identify both geographic and topical areas most in need of useful predictions of sea-level rise impacts. For example, an assessment technique for objectively assessing potential effects of sea-level rise on open coasts, the Coastal Vulnerability Index (CVI), has been employed in the United States and elsewhere (*e.g.*, Gornitz *et al.*, 1997; Shaw *et al.*, 1998; Thieler and Hammar-Klose, 1999, 2000a, 2000b). Although the CVI is a fairly simplistic technique, it can provide useful insights and has found application as a coastal planning and management tool (Thieler *et al.*, 2002). Such assessments have also been integrated with socioeconomic vulnerability criteria to yield a more integrative measure of community vulnerability (Boruff *et al.*, 2005).

Projecting long-term wetland sustainability to future sea-level rise requires data on accretionary events over sufficiently long time scales that include the return periods of major storms, floods, and droughts, as well as information on the effects of wetland elevation feedback on inundation and sedimentation processes that affect wetland vertical accretion. Numerical models can be applied to predict wetland sustainability at the local scale, but there is not sufficient data to populate these models at the regional or national scale (see Chapter 4). Given this data constraint, current numerical modeling approaches will need to improve or adapt such that they can be applied at broader spatial scales with more confidence.

Integrate studies of past and present coastal behavior into predictive models

Existing shoreline-change prediction techniques are typically based on assumptions that are either difficult to validate or too simplistic to be reliable for many real-world applications (see Appendix 2). As a result, the usefulness of these modeling approaches has been debated in the coastal science community (see Chapter 3). Newer models that include better representations of real-world settings and processes (*e.g.*, Cowell *et al.*, 1992; Stolper *et al.*,

2005; Pietrafesa *et al.*, 2007) have shown promise in predicting coastal evolution. Informing these models with improved data on past coastal changes should result in better predictions of future changes.

The process of marine transgression across the continental shelf has left an incomplete record of sea-level and environmental change. An improved understanding of the rate and timing of coastal evolution will need to draw on this incomplete record, however, in order to improve models of coastal change. Using a range of techniques, such as high-resolution seafloor and geologic framework mapping coupled with geochronologic and paleoenvironmental studies, the record of coastal evolution during the Pleistocene (1.8 million to 11,500 years ago) and the Holocene (the last 11,500 years) can be explored to identify the position and timing of former shorelines and coastal environments.

14.2.4. Improve Understanding of Societal Impacts

Research in the social sciences will be critical to understanding the potential effects on society and social systems resulting from sea-level rise.

Increase research on adaptation, mitigation, and avoidance measures

This Product describes a wide variety of potential impacts of sea-level rise, including the effects on the physical environment, biological systems, and coastal development and infrastructure. While the ability to predict future changes is currently inadequate for many decisions, adaptation, mitigation, and avoidance strategies must evolve as scientific knowledge and predictive ability increase. For example, expanded research and assessments of the economic and environmental costs of present and future actions are needed to allow a more complete analysis of the tradeoffs involved in sea-level rise decision making. In addition, opportunities to engage stakeholders such as federal agencies, states, counties, towns, non-government organizations, and private landowners in the design and implementation of sea-level rise impact and response planning should be created.

Develop multi-disciplinary groups that integrate natural and social sciences

Interdisciplinary research that combines natural and social sciences will be crucial to understanding the interplay of the physical, environmental, and societal impacts of sea-level rise. Development of programs that facilitate such collaborations should be encouraged.

Expand institutional capacity and overcome barriers

Substantial opportunities exist to expand and improve upon the ability of institutions to respond to sea-level rise (see Chapter 10, 11, and 12). Research is needed to define the capacity needed for decision making, as well as the methods that can be best employed (*e.g.*, command and control, economic incentive) to achieve management goals. Overcoming the institutional barriers described in Chapter 12 is also necessary for effective response to the management challenges presented by sea-level rise.

14.2.5. Develop Coastal Decision Support Systems for Planning and Policy Making

For coastal zone managers in all levels of government, there is a pressing need for more scientific information, a reduction in the ranges of uncertainty for processes and impacts, and new methods for assessing options and alternatives for management strategies. Geospatial information on a wide range of themes such as topography, bathymetry, land cover, population, and infrastructure, that is maintained on regular cycle will be a key component of planning for mitigation and adaptation strategies. For example, specialized themes of data such as hydric (abundantly moist) soils may be critical to understanding the potential for wetland survival in specific areas. Developing and maintaining high-resolution maps that incorporate changes in hazard type and distribution, coastal development, and societal risk will be critical. Regularly conducting vulnerability assessments and reviews will be necessary in order to adapt to changing conditions.

Provide easy access to data and information resources for federal, state, local, academic, and public users

Understanding and acting on scientific information about sea-level rise and its impacts will depend upon common, consistent, shared databases for integrating knowledge and providing a basis for decision making. Thematic data and other value-added products should adhere to predetermined standards to make them universally accessible and transferable through internet portals. All data should be accompanied by appropriate metadata describing its method of production, extent, quality, spatial reference, limitations of use, and other characteristics (NRC, 2004).

An opportunity exists to develop a national effort to develop and apply data integration tools to combine terrestrial and marine data into a seamless geospatial framework. This would involve the collection of real-time tide data and the development of more sophisticated hydrodynamic models for the entire U.S. coastline, as well as the establishment of protocols and tools for merging bathymetric and topographic datasets (NRC, 2004). Modern and updated digital flood insurance rate maps (DFIRM) that incorporate future sea-level rise are needed in the coastal zone (see Chapter 9).

Transfer scientific knowledge to studies of vulnerability, risk, and societal impacts

In addition to basic scientific research and environmental monitoring, a significant need exists to integrate the results of these efforts into comprehensive vulnerability and risk assessments. Tools are needed for mapping, modeling, and communicating risk to help public agencies and communities understand and reduce their vulnerability to, and risk of, sea-level rise hazards. Social science research activities are also needed that examine societal consequences and economic impacts of sea-level rise, as well as identify institutional frameworks needed to adapt to changes in the coastal zone. For example, analyses of the economic costs of armoring shores at risk of erosion and the expected lifespan of such efforts will be required, as will studies on the durability of armored shorefronts under different sea-level rise scenarios. The physical and biological consequences of armoring shores will need to be quantified and the tradeoffs communicated. Effective planning for sea-level rise will also require integrated economic assessments on the impact to fisheries, tourism, and commerce.

Applied research in the development of coastal flooding models for the subsequent study of ecosystem response to sea-level rise is underway in coastal states such as North Carolina (Feyen *et al.*, 2006). There is also a need for focused study on the ecological impacts of sea-level rise and in how the transfer of this knowledge can be made to coastal managers for decision-making.

Develop decision support systems

County and state planners need tools to analyze vulnerabilities, explore the implications of alternative response measures, assess the costs and benefits of options, and provide decision-making support. These might take the form of guidelines, checklists, or software tools. In addition, there is a need to examine issues in a landscape or ecosystem context rather than only administrative boundaries.

In addition to new and maintained data, models, and research, detailed site studies are needed to assess potential impacts on a site-specific basis and provide information that allows informed decision making. Appropriate methodologies need to be developed and made available. These will have to look at a full range of possible impacts including aquifer loss by saltwater intrusion, wetland loss, coastal erosion, and infrastructure implications, as well as the impact of adaptation measures themselves. Alternative strategies of adaptive management will be required. Each locality may need a slightly different set of responses to provide a balanced policy of preserving ecosystems, protecting critical infrastructure, and adjusting to property loss or protection. Providing a science-based set of decision support tools will provide a sound basis for making these important decisions.

Educate the public on consequences and alternatives

Relative to other natural hazards such as earthquakes, volcanic eruptions, and severe weather (*e.g.*, hurricanes, tornadoes) that typically occur in minutes to days, sea-level rise has a long time horizon over which effects become clear. Thus, it is often difficult to communicate the consequences of this sometimes slow process that occurs over many years. The impacts of sea-level rise, however, are already being felt across the United States (see Chapter 13). Public education will be crucial for adapting to physical, environmental, economic, and social changes resulting from sea-level rise. Research activities that result in effective means to conduct public education and outreach concerning sea-level rise consequence and alternatives should be encouraged.

REFERENCES

* Indicates non-peer reviewed literature; available upon request

Baldwin, W. E., Morton, R. A., Putney, T. R., Katuna, M. P., Harris, M. S., Gayes, P.T. Driscoll, N. W., Denny, J. F. & Schwab, W. C. (2006). Migration of the Pee Dee River system inferred from ancestral paleochannels underlying the South Carolina Grand Strand and Long Bay inner shelf. *Geological Society of America Bulletin*, *118(5/6)*, 533-549.

Boruff, B. J., Emrich, C. & Cutter, S. L. (2005). Erosion hazard vulnerability of US coastal counties. *Journal of Coastal Research*, *21(5)*, 932-942.

Brooks, B. A., Merrifield, M. A., Foster, J., Werner, C. L., Gomez, F., Bevis, M. & Gill, S. (2007). Space geodetic determination of spatial variability in relative sea level change, Los Angeles Basin. *Geophysical Research Letters*, 34, L01611, doi: 10.1 029/2006GL028 171.

Cahoon, D. R., Hensel, P. F. Spencer, T., Reed, D. J., McKee, K. L. & Saintilan, N. (2006). Coastal wetland vulnerability to relative sea-level rise: wetland elevation trends and process controls. In: *Wetlands and Natural Resource Management* [Verhoeven, J.T.A., B. Beltman, R. Bobbink, and D. Whigham (eds.)]. Ecological studies volume 190. Springer, Berlin and New York, pp. 271-292.

Church, J. A., Gregory, J. M., Huybrechts, P., Kuhn, M., Lambeck, K., Nhuan, M. T., Qin, D. & Woodworth, P. L. (2001). Changes in sea level. In: *Climate Change 2001: The Scientific Basis.* Contribution of Working Group 1 to the Third Assessment Report of the Intergovernmental Panel on Climate Change [Houghton, J.T., Y. Ding, D.J. Griggs, M. Noguer, P. J. van der Linden, X. Dai, K. Maskell, and C.A. Johnson (eds.)]. Cambridge University Press, Cambridge UK, and New York, pp. 639-693.

Church, J. A., White, N. J., Coleman, R., Lambeck, K. & Mitrovica, J. X. (2004). Estimates of the regional distribution of sea-level rise over the 1950-2000 period. *Journal of Climate, 17(13),* 2609-2625.

Colquhoun, D. J., Johnson, G. H., Peebles, P. C., Huddlestun, P. F. & Scott, T. (1991). Quaternary geology of the Atlantic coastal plain. In: *Quaternary Nonglacial Geology: Conterminous U.S.* [Morrison, R.B. (ed.)]. The Geology of North America v. K-2. Geological Society of America, Boulder, CO, pp. 629-650.

Cowell, P. J., Roy, P.S. & Jones, R. A. (1992). Shoreface translation model: computer simulation of coastal-sand-body response to sea-level rise. *Mathematics and Computers in Simulation, 33(5-6),* 603-608.

Donnelly, J. P., Bryant, S. S., Butler, J., Dowling, J., Fan, L., Hausmann, N., Newby, P., Shuman, B., Stern, J., Westover, K. & Webb III, T. (2001). A 700-year sedimentary record of intense hurricane landfalls in southern New England. *Geological Society of America Bulletin, 113(6),* 7 14-727.

* Feyen, J., Hess, K., Spargo, E., Wong, A., White, S., Sellars, J. & Gill, S. (2006). Development of continuous bathymetric/topographic unstructured coastal flooding model to study sea-level rise in North Carolina. In: *Proceedings of the 9th International Conference on Estuarine and Coastal Modeling*, Charleston, SC, October 31 – November 2, 2005. [Spaulding, M. (ed.)]. American Society of Civil Engineers, Reston, VA, pp. 338-356.

FitzGerald, D. M., Buynevich, I.V. & Argow, B. A. (2004). Model of tidal inlet and barrier island dynamics in a regime of accelerated sea-level rise. *Journal of Coastal Research, Special issue 39,* 789-795.

FitzGerald, D. M., Fenster, M. S., Argow, B.A. & Buynevich, I.V. (2008). Coastal impacts due to sea-level rise. *Annual Reviews of Earth and Planetary Sciences, 36,* 601-647.

Gehrels, W. R. (1994). Determining relative sea level change from salt-marsh foraminifera and plant zones on the coast of Maine, U.S.A. *Journal of Coastal Research, 10(4),* 990-1009.

Gehrels, W. R., Belknap, D.F. & Kelley, J. T. (1996). Integrated high-precision analyses of Holocence relative sea level changes: lessons from the coast of Maine. *Geological Society of America Bulletin, 108(9),* 1073-1088.

* Gornitz, V. T., Beaty, W. & Daniels, R.C. (1997). *A Coastal Hazards Database for the U.S. West Coast.* ORNL/CDIAC-81, NDP-043C. Oak Ridge National Laboratory, Oak Ridge, TN, 162 pp. <http://cdiac.ornl.gov/ndps/ndp043c.html>

* Heinz Center (2000). *Evaluation of Erosion Hazards.* The H. John Heinz III Center for Science, Economics, and the Environment, Washington, DC, 203 pp.

* Heinz Center (2002a). *Human Links to Coastal Disasters.* The H. John Heinz III Center for Science, Economics, and the Environment, Washington, DC, 139 pp.

* Heinz Center (2002b). *The State of the Nation's Ecosystems: Measuring the Lands, Waters, and Living Resources of the United States.* Cambridge University Press, New York, 270 pp.

* Heinz Center (2006). *Filling the Gap--Priority Data Needs and Key Management Challenges for National Reporting on Ecosystem Condition.* The H. John Heinz III Center for Science, Economics, and the Environment, Washington, DC, 110 pp.

Horton, B. P., Corbett, R., Culver, S. J., Edwards, R. J. & Hillier, C. (2006). Modern salt marsh diatom distributions of the Outer Banks, North Carolina, and the development of a transfer function for high resolution reconstructions of sea level. *Estuarine, Coastal, and Shelf Science, 69(3-4),* 381-394.

* Kennedy, V. S., Twilley, R. R., Kleypas, J.A., Cowan, J.H. & Hare, S. R. (2002). *Coastal and Marine Ecosystems & Global Climate Change: Potential Effects on U.S. Resources.* Pew Center on Global Climate Change, Arlington, VA, 52 pp.

Meehl, G. A., Stocker, T. F., Collins, W.D., Friedlingstein, P., Gaye, A.T., Gregory, J. M., Kitoh, A., Knutti, R., Murphy, J. M., Noda, A., Raper, S. C. B., Watterson, I. G., Weaver, A. J. & Zhao, Z. C. (2007). Global climate projections. In: *Climate Change 2007: The Physical Science Basis.* Contribution of Working Group I to the Fourth Assessment Report of the Intergovernmental Panel on Climate Change [Solomon, S., D. Qin, M. Manning, Z. Chen, M. Marquis, K.B. Averyt, M. Tignor, and H.L. Miller (eds.)]. Cambridge University Press, Cambridge, UK, and New York, pp. 747-845.

Morton, R. A. & Miller, T. L. (2005). *National Assessment of Shoreline Change: Part 2, Historical Shoreline Changes and Associated Coastal Land Loss Along the U.S. Southeast Atlantic Coast.* U.S. Geological Survey open-file report 2005-1401, U.S. Geological Survey, [St. Petersburg, FL], 35 pp. <http://pubs.usgs.gov/of/2005/1 401>

Morton, R.A. and A.H. Sallenger Jr., 2003: Morphological impacts of extreme storms on sandy beaches and barriers. *Journal of Coastal Research, 19(3),* 560-573.

* Neumann, J.E., G. Yohe, R. Nicholls, and M. Manion, 2000: *Sea-level Rise and Global Climate Change: A Review of Impacts to U.S. Coasts.* Pew Center on Global Climate Change, Arlington, VA, 38 pp. <http://www.pewclimate.org/globalwarming-in-depth/all_reports/sea_level_rise>

NRC (National Research Council) (1987). *Responding to Changes in Sea Level: Engineering Implications.* National Academy Press, Washington, DC, 148 pp.

NRC (National Research Council) (1990a). *Managing Coastal Erosion.* National Academy Press, Washington, DC, 182 pp.

NRC (National Research Council) (1990b). *Sea Level Change.* National Academy Press, Washington, DC, 256 pp.

NRC (National Research Council) (1990c). *Spatial Data Needs: The Future of the National Mapping Program.* National Academy Press, Washington, DC, 78 pp.

NRC (National Research Council) (1999). *Science for Decisionmaking: Coastal and Marine Geology at the U.S. Geological Survey.* National Academy Press, Washington, DC, 113 pp.

NRC (National Research Council) (2001). *Sea Level Rise and Coastal Disasters: Summary of a Forum.* National Academies Press, Washington, DC, 24 pp.

NRC (National Research Council) (2002). *Abrupt Climate Change: Inevitable Surprises.* National Academy Press, Washington, DC, 230 pp.

NRC (National Research Council) (2004). *A Geospatial Framework for the Coastal Zone: National Needs for Coastal Mapping and Charting.* National Academies Press, Washington, DC, 149 pp.

NRC (National Research Council) (2006). *Beyond Mapping: Meeting National Needs Through Enhanced Geographic Information Science.* National Academies Press, Washington, DC, 100 pp.

NRC (National Research Council) (2007). *Mitigating Shore Erosion on Sheltered Coasts.* National Academies Press, Washington, DC, 174 pp.

* Panetta, L. E. (2003). *America's Living Oceans: Charting a Course for Sea Change: A Report to the Nation: Recommendations for a New Ocean Policy.* Pew Oceans Commission, Arlington, VA, 145 pp. <http://www.pewtrusts.org/pdf/env_ pew_oceans>

Philippart, C. J. M. Anadón, R., Danovaro, R., Dippner, J. W., Drinkwater, K. F., Hawkins, S. J., O'Sullivan, G., Oguz, T. & Reid, P.C. (2007). *Impacts of Climate Change on the European Marine and Coastal Environment.* Marine Board position paper 9, European Science Foundation, Strasbourg, France, 84 pp. <http ://www. sesame 1 .pdf>

Pietrafesa, L. J., Kelleher, K., Karl, T., Davidson, M., Peng, M., Bao, S., Dickey, D., Xie, L., Liu, H. & Xia, M. (2007). A new architecture for coastal inundation and flood warning prediction. *Marine Technology Society Journal*, *40(4),* 71-77.

Rahmstorf, S. (2007). A semi-empirical approach to projecting future sea-level rise. *Science*, *315(5810),* 368-370.

Reed, D. J., Bishara, D., Cahoon, D., Donnelly, J., Kearney, M., Kolker, A., Leonard, L., Orson, R.A. & Stevenson, J.C. (2008). Site-specific scenarios for wetlands accretion as sea level rises in the mid-Atlantic region. Section 2.1 in: *Background Documents Supporting Climate Change Science Program Synthesis and Assessment Product 4.1: Coastal Elevations and Sensitivity to Sea Level Rise* [Titus, J.G. and E.M. Strange (eds.)]. EPA 430R07004. U.S. Environmental Protection Agency, Washington, DC, pp. 134-174. <http://epa.gov/climatechange/effects/coastal/background.html>

Shaw, J., Taylor, R.B., Solomon, S., Christian, H.A. & Forbes, D.L. (1998). Potential impacts of global sea-level rise on Canadian coasts. *The Canadian Geographer*, *42(4),* 365-379.

Snay, R., Cline, M., Dillinger, W. Foote, R., Hilla, S., Kass, W., Ray, J., Rohde, J., Sella, G. & Soler, T. (2007). Using global positioning system-derived crustal velocities to estimate rates of absolute sea level change form North American ride gauge records. *Journal of Geophysical Research*, *112*, B04409, doi: 10.1 029/2006JB004606.

Stolper, D., List, J. H. & Thieler, E. R. (2005). Simulating the evolution of coastal morphology and stratigraphy with a new morphological-behavior model (GEOMBEST). *Marine Geology*, *218(1-4),* 17-36.

Thieler, E. R. & Hammar-Klose, E.S. (1999). *National Assessment of Coastal Vulnerability to Sea-Level Rise: Preliminary Results for the U.S. Atlantic Coast.* U.S. Geological Survey open-file report 99-593. U.S. Geological Survey, Reston, VA, 1 sheet. <http://pubs.usgs.gov/of/of99-593/>

Thieler, E. R. & Hammar-Klose, E.S. (2000a). *National Assessment of Coastal Vulnerability to Sea-Level Rise: Preliminary Results for the U.S. Pacific Coast.* U.S. Geological

Survey open-file report 00-178. U.S. Geological Survey, Reston, VA, 1 sheet. <http://pubs.usgs.gov/of/2000/of00- 178/>

Thieler, E. R. & Hammar-Klose, E.S. (2000b). *National Assessment of Coastal Vulnerability to Sea-Level Rise: Preliminary Results for the U.S. Gulf of Mexico Coast.* U.S. Geological Survey open-file report 00-179. U.S. Geological Survey, Reston, VA, 1 sheet. <http://pubs.usgs.gov/of/2000/of00- 179/>

Thieler, E. R., Williams, S. J. & Beavers, R. (2002). *Vulnerability of U.S. National Parks to Sea-Level Rise and Coastal Change.* U.S. Geological Survey fact sheet FS 095- 02. [U.S. Geological Survey, Reston, VA], 2 pp. <http://pubs.usgs.gov/fs/fs095- 02/>

van de Plassche, O., van der Borg, K. & de Jong, A. F. M. (1998). Sea level-climate correlation during the past 1400 years. *Geology*, *26(4),* 319-322.

Woppelmann, G., Miguez, B.M., Bouin, M. N. & Altamimi, Z. (2007). Geocentric sea-level trend estimates from GPS analyses at relevant tide gauges world wide, *Global and Planetary Change*, *57(3-4),* 396-406.

In: Coastal Sensitivity to Sea Level Rise-Focusing on the Mid-… ISBN: 978-1-60741-440-7
Editor: Melvin C. Urajner

Appendix 1. State and Local Information on Vulnerable Species and Coastal Policies in the Mid-Atlantic

Overview

Appendix 1 discusses many of the species that depend on potentially vulnerable habitat in specific estuaries, providing local elaboration of the general issues examined in Chapter 5. It also describes key statutes, regulations, and other policies that currently define how state and local governments are responding to sea level rise, providing support for some of the observations made in Part III. This set of information was not developed as a quantitative nor analytical assessment and therefore is not intended as a complete or authoritative basis for decision-making; rather, it is a starting point for those seeking to discuss local impacts and to examine the types of decisions and potential policy responses related to sea-level rise.

The sections concerning species and habitat are largely derived from a U.S. EPA report developed in support of this Synthesis and Assessment Product (U.S. EPA, 2008), with additional input from stakeholders as well as expert and public reviewers. That report synthesized what peer-reviewed literature was available, and augmented that information with reports by organizations that manage the habitats under discussion, databases, and direct observations by experts in the field. The sections that concern state and local policies are based on statutes, regulations, and other official documents published by state and local governments.

Characterizations of likelihood in this Product are largely based on the judgment of the authors and on published peer-reviewed literature and existing policies, rather than a formal quantification of uncertainty. Data on how coastal ecosystems and specific species may respond to climate change is limited to a small number of site-specific studies, often carried out for purposes unrelated to efforts to evaluate the potential impact of sea level rise. Although being able to characterize current understanding—and the uncertainty associated with that information—is important, quantitative and qualitative assessments of likelihood are not unavailable for the site-specific issues discussed in this Appendix. Unlike the main body of the Product, any likelihood statements in this Appendix regarding specific habitat or species reflect likelihood as expressed in particular reports being cited. Statements about the implications of coastal policies in this Appendix are based on the authors qualitative assessment of available published literature and of the governmental policies. Published information, data, and tools are evolving to further examine sea-level rise at this scale.

The synthesis was compiled by the following authors for the specific areas of focus and edited by K. Eric Anderson, USGS; Stephen K. Gill, NOAA; Daniel Hudgens, Industrial Economics, Inc.; and James G. Titus, U.S. EPA:

A Long Island
Lead Authors: Daniel Hudgens, Industrial Economics Inc.; Ann Shellenbarger Jones, Industrial Economics Inc.; James G. Titus, EPA.
Contributing Authors: Elizabeth M. Strange, Stratus Consulting Inc.; J. Tanski, New York Sea Grant; G. Sinha, University of Ohio;

B. New York Metropolitan Area
Lead Author: Elizabeth M. Strange, Stratus Consulting Inc;
Contributing Authors: Daniel Hudgens, Industrial Economics Inc.; Ann Shellenbarger Jones, Industrial Economics Inc.

C. New Jersey Shore
Lead Author: James G. Titus, U.S. EPA
Contributing Author: Elizabeth M. Strange, Stratus Consulting Inc.

D. Delaware Estuary
Lead Author: James G. Titus, U.S. EPA
Contributing Authors: Chris Linn, Delaware Valley Regional Planning Commission; Danielle Kreeger, Partnership for the Delaware Estuary, Inc.; Michael Craghan, Middle Atlantic Center for Geography & Environmental Studies; Michael Weinstein, New Jersey Marine Sciences Consortium (NJMSC) and Director, New Jersey Sea Grant College Program

E. The Atlantic Coast of Virginia, Maryland, and Delaware
Lead Author: James G. Titus, U.S. EPA
Contributing Author: Elizabeth M. Strange, Stratus Consulting Inc.

F. Chesapeake Bay
Lead Author: James G. Titus, U.S. EPA
Contributing Authors: Ann Shellenbarger Jones, Industrial Economics Inc.; Peter Conrad, City of Baltimore; Elizabeth M. Strange, Stratus Consulting; Zoe Johnson, Maryland Department of Natural Resources; Michael Weinstein, New Jersey Sea Grant

G. North Carolina
Lead Authors: James G. Titus, U.S. EPA; Rebecca L. Feldman, NOAA; Ben Poulter, Potsdam Institute for Climate Impact Research
Contributing Authors: Jeff DeBlieu, The Nature Conservancy; Ann Shellenbarger Jones, Industrial Economics Inc.

A1.A. Long Island

The North Shore of Long Island is generally characterized by high bluffs of glacial origin, making this area less susceptible to problems associated with increased sea level. The

South Shore, by contrast, is generally low lying and fronted by barrier islands, except for the easternmost portion. As a result, there are already major planning efforts underway in the region to preserve the dry lands under threat of inundation. A brief discussion of these efforts, especially on the South Shore, is provided in Section A1.A.2 of this Appendix. Maps and estimates of the area of land close to sea level are provided in Titus and Richman (2001). Further information on portions of the South Shore can be found in Gornitz *et al.* (2002).

A1.A.1. Environmental Implications

North Shore and Peconic Bay

Of the 8,426 hectares (ha) (20,820 acres [ac]) of tidal wetlands in Long Island Sound, about 15 percent are in New York, primarily along the shores of Westchester and Bronx counties (Holst *et al.*, 2003). Notable areas of marsh are in and around Stony Brook Harbor and West Meadow, bordering the Nissequogue River and along the Peconic Estuary (NYS DOS, 2004). In general, tidal wetlands along the North Shore are limited; the glacial terminal moraine[1] resulted in steep uplands and bluffs and more kettle-hole[2] wetlands along the eastern portion (LISHRI, 2003). In the eastern portion, there has already been a significant loss of the historical area of vegetated tidal wetlands (Holst *et al.*, 2003; Hartig and Gornitz, 2004), which some scientists partially attribute to sea-level rise (Mushacke, 2003).

The loss of vegetated low marsh reduces habitat for several rare bird species (*e.g.*, seaside sparrow) that nest only or primarily in low marsh (see Section 5.2). Low marsh also provides safe foraging areas for small resident and transient fishes (*e.g.*, weakfish, winter flounder). Diamondback terrapin live in the creeks of the low marsh, where they feed on plants, mollusks, and crustaceans (LISF, 2008). Some wetlands along Long Island Sound may be allowed to respond naturally to sea-level rise, including some in the Peconic Estuary. Where migration is possible, preservation of local biodiversity as well as some regionally rare species is possible (Strange *et al.*, 2008).

Beaches are far more common than tidal wetlands in the Long Island Sound study area. Several notable barrier beaches exist. For example, the sandy barrier-beach system fronting Hempstead Harbor supports a typical community progression from the foreshore to the bay side, or backshore (LISHRI, 2003). The abundant invertebrate fauna provide forage for sanderling, semipalmated plovers, and other migrating shorebirds (LISHRI, 2003). The maritime beach community between the mean high tide and the primary dune provides nesting sites for several rare bird species, including piping plover, American oystercatcher, black skimmer, least tern, common tern, roseate tern, the Northeastern beach tiger beetle, and horseshoe crab (LISHRI, 2003) (see Box A1.1). Diamondback terrapin use dunes and the upper limit of the backshore beach for nesting (LISHRI, 2003).

Since nearly all of the Long Island Sound shoreline is densely populated and highly developed, the land may be armored in response to sea-level rise, raising the potential for beach loss. The Long Island Sound Habitat Restoration Initiative cautions: "Attempts to alter the natural cycle of deposition and erosion of sand by construction of bulkheads, seawalls, groins, and jetties interrupt the formation of new beaches" (LISHRI, 2003).

Shallow water habitats are a major ecological feature in and around the Peconic Estuary. Eelgrass beds provide food, shelter, and nursery habitats to diverse species, including worms, shrimp, scallops and other bivalves, crabs, and fish (PEP, 2001). Horseshoe crabs forage in the eelgrass beds of Cedar Point/Hedges Bank, where they are prey for loggerhead turtles (federally listed as threatened), crabs, whelks, and sharks (NYS DOS, 2004). Atlantic silverside spawn here; silverside eggs provide an important food source for seabirds, waterfowl, and blue crab, while adults are prey for bluefish, summer flounder, rainbow smelt, white perch, Atlantic bonito, and striped bass (NYS DOS, 2004). The Cedar Point/Hedges Bank Shallows eelgrass beds are known for supporting a bay scallop fishery of statewide importance (NYS DOS, 2004).

Other noteworthy habitats that could be affected by sea-level rise include the sea-level fen vegetation community that grows along Flanders Bay (NYS DOS, 2004), and the Long Island's north shore tidal flats, where longshore drift carries material that erodes from bluffs and later deposits it to form flats and barrier spits or shoals (LISHRI, 2003). One of the largest areas of tidal mudflats on the North Shore is near Conscience Bay, Little Bay, and Setauket Harbor west of Port Jefferson (NYS DOS, 2004). Large beds of hard clams, soft clams, American oysters, and ribbed mussels are found in this area (NYS DOS, 2004).

South Shore

Extensive back-barrier salt marshes exist to the west of Great South Bay in southern Nassau County (USFWS, 1997). These marshes are particularly notable given widespread marsh loss on the mainland shoreline of southern Nassau County (NYS DOS and USFWS, 1998; USFWS, 1997). To the east of Jones Inlet, the extensive back-barrier and fringing salt marshes are keeping pace with current rates of sea-level rise, but experts predict that the marshes' ability to keep pace is likely to be marginal if the rate of sea-level rise increases moderately, and that the marshes are likely to be lost under higher sea-level rise scenarios (Strange *et al.*, 2008, interpreting the findings of Reed *et al.*, 2008). Opportunities for marsh migration along Long Island's South Shore would be limited if the mainland shores continue to be bulkheaded. Outside of New York City, the state requires a minimum 22.9-meter (m) (75-foot [ft]) buffer around tidal wetlands to allow marsh migration, but outside of this buffer, additional development and shoreline protection are permitted[3] (NYS DEC, 2006). Numerous wildlife species could be affected by salt marsh loss. For example, the Dune Road Marsh west of Shinnecock Inlet provides nesting sites for several species that are already showing significant declines, including clapper rail, sharp-tailed sparrow, seaside sparrow, willet, and marsh wren (USFWS, 1997). The salt marshes of Gilgo State Park provide nesting sites for northern harrier, a species listed by the state as threatened (NYS DOS, 2004).

Of the extensive tidal flats along Long Island's southern shoreline, most are found west of Great South Bay and east of Fire Island Inlet, along the bay side of the barrier islands, (USFWS, 1997) in the Hempstead Bay–South Oyster Bay complex, (USFWS, 1997) and around Moriches and Shinnecock Inlets (NYS DOS and USFWS, 1998). These flats provide habitat for several edible shellfish species, including soft clam, hard clam, bay scallop, and blue mussel. The tidal flats around Moriches and Shinnecock Inlets are particularly important foraging areas for migrating shorebirds. The South Shore Estuary Reserve Council asserts that "because shorebirds concentrate in just a few areas during migration, loss or degradation of key sites could devastate these populations" (NYS DOS and USFWS, 1998).

The back-barrier beaches of the South Shore also provide nesting sites for the endangered roseate tern and horseshoe crabs (USFWS, 1997). Shorebirds, such as the red knot, feed preferentially on horseshoe crab eggs during their spring migrations.

Increased flooding and erosion of marsh and dredge spoil islands will reduce habitat for many bird species that forage and nest there, including breeding colonial waterbirds, migratory shorebirds, and wintering waterfowl. For example, erosion on Warner Island is reducing nesting habitat for the federally endangered roseate tern and increasing flooding risk during nesting (NYS DOS and USFWS, 1998). The Hempstead Bay–South Oyster Bay complex includes a network of salt marsh and dredge spoil islands that are important for nesting by herons, egrets, and ibises. Likewise, Lanes Island and Warner Island in Shinnecock Bay support colonies of the state-listed common tern and the roseate tern (USFWS, 1997).

Box A1.1. Effects on the Piping Plover

Piping Plover *Charadrius melodus*

Habitat: The piping plover, federally listed as threatened, is a small migratory shorebird that primarily inhabits open sandy barrier island beaches on Atlantic coasts (USFWS, 1996). Major contributing factors to the plover's status as threatened are beach recreation by pedestrians and vehicles that disturb or destroy plover nests and habitat, predation by mammals and other birds, and shoreline development that inhibit the natural renewal of barrier beach and overwash habitats (USFWS, 1996). In some locations, dune maintenance for protection of access roads associated with development appears to be correlated with absence of piping plover nests from former nesting sites (USFWS, 1996).

Locations: The Atlantic population of piping plovers winters on beaches from the Yucatan Peninsula to North Carolina. In the summer, they migrate north and breed on beaches from North Carolina to Newfoundland (CLO, 2004). In the mid-Atlantic region, breeding pairs of plovers can be observed on coastal beaches and barrier islands, although suitable habitat is limited in some areas. In New York, piping plovers breed more frequently on Long Island's sandy beaches, from Queens to the Hamptons, in the eastern bays and in the harbors of northern Suffolk County. New York's Breezy Point barrier beach, at the mouth of Jamaica Bay, consistently supports one of the largest piping plover nesting sites in the entire New York Bight coastal region (USFWS, 1997). New York has seen an increase in piping plover breeding pairs in the last decade from less than 200 in 1989 to near 375 in recent years (2003 to 2005), representing nearly a quarter of the Atlantic coast's total breeding population (USFWS, 2004a). Despite this improvement, piping plovers remain state listed as endangered in New York (NYS DEC, 2007).

Impact of Sea-Level Rise: Where beaches are prevented from migrating inland by shoreline armoring, sea-level rise will negatively impact Atlantic coast piping plover populations. To the degree that developed shorelines result in erosion of ocean beaches, and to the degree that stabilization is undertaken as a response to sea-level rise, piping plover habitat will be lost. In contrast, where beaches are able to migrate landward, plovers may find newly available habitat. For example, on Assateague Island, piping plover populations increased after a storm event that created an overwash area on the north of the island (Kumer, 2004). This suggests that if barrier beaches are allowed to migrate in response to sea-level rise, piping plovers might adapt to occupy new inlets and beaches created by overwash events.

Beach nourishment, the anticipated protection response for much of New York's barrier beaches such as Breezy Point, can benefit piping plovers and other shorebirds by increasing available nesting habitat in the short term, offsetting losses at eroded beaches, but may also be detrimental, depending on timing and implementation (USFWS, 1996). For instance, a study in Massachusetts found that plovers foraged on sandflats created by beach nourishment (Cohen *et al*,, 2005). However, once a beach is built and people spread out to enjoy it, many areas become restricted during nesting season. Overall, throughout the Mid- Atlantic, coastal development and shoreline stabilization projects constitute the most serious threats to the continuing viability of storm-maintained beach habitats and their dependent species, including the piping plover (USFWS, 1996).

Photograph credit: USFWS, New Jersey Field Office /Gene Nieminen, 2006.

A1.A.2. Development, Shore Protection, and Coastal Policies

New York State does not have written policies or regulations pertaining specifically to sea-level rise in relation to coastal zone management, although sea-level rise is becoming recognized as a factor in coastal erosion and flooding by New York State Department of State (DOS) in the development of regional management plans.

Policies regarding management and development in shoreline areas are primarily based on three laws. Under the Tidal Wetlands Act program, the Department of Environmental Conservation (DEC) classifies various wetland zones and adjacent areas where human activities may have the potential to impair wetland values or adversely affect their function; permits are required for most activities that take place in these areas. New construction greater than 9.3 square meters (sq m) (100 square feet [sq ft]), excluding docks, piers, and bulkheads) as well as roads and other infrastructure must be set back 22.9 m (75 ft) from any tidal wetland, except within New York City where the setback is 9.1 m (30 ft)[4].

The Waterfront Revitalization and Coastal Resources Act (WRCRA) allows the DOS to address sea-level rise indirectly through policies regarding flooding and erosion hazards (NOAA, 1982). Seven out of 44 written policies related to management, protection, and use of the coastal zone address flooding and erosion control. These polices endeavor to move development away from areas threatened by coastal erosion and flooding hazards, to ensure that development activities do not exacerbate erosion or flooding problems and to preserve natural protective features such as dunes. They also provide guidance for public funding of coastal hazard mitigation projects and encourage the use of nonstructural erosion and flood control measures where possible (NYS DOS, 2002).

Under the Coastal Erosion Hazard Areas Act program, the DEC identified areas subject to erosion and established two types of erosion hazard areas (structural hazard and natural protective feature areas) where development and construction activities are regulated[5]. Permits are required for most activities in designated natural protective feature areas. New development (*e.g.*, building, permanent shed, deck, pool, garage) is prohibited in nearshore areas, beaches, bluffs, and primary dunes. These regulations, however, do not extend far inland and therefore do not encompass the broader area vulnerable to sea-level rise.

New York State regulates shore protection structures along estuaries and the ocean coast differently. The state's Coastal Erosion Hazard Law defines coastal erosion hazard areas as those lands with an average erosion rate of at least 30 cm (1 ft) per year.[6] Within those erosion hazard areas, the local governments administer the programs to grant or deny permits, generally following state guidelines.[7] Those guidelines requires that individual property owners first evaluate non-structural approaches; but if they are unlikely to be effective, hard structures are allowed (New York State, 2002a).

Shoreline structures, which by definition include beach nourishment in New York State, are permitted only when it can be shown that the structure can prevent erosion for at least thirty years and will not cause an increase in erosion or flooding at the local site or nearby locations (New York State, 2002a). Setbacks, relocation, and elevated walkways are also encouraged before hardening.

Currently, all of the erosion hazard areas are along the open coast. Therefore, the state does not directly regulate shore protection structures along estuarine shores. However, under the federal Coastal Zone Management Act, New York's coastal management program reviews federal agency permit applications, to ensure consistency with policies of the State's coastal management program (NOAA, 2008; USACE, 2007). The state has objected to nationwide permit 13 issued by the Corps of Engineer's wetlands regulatory program (see Section 12.2.2 in Chapter 12), which provides a general authorization for erosion control structures (NYS DOS, 2006). The effect of that objection is that nationwide permit 13 does not automatically provide a property owner with a permit for shore protection unless the state concurs with such an application (NYS DOS, 2006). The state has also objected to the application of nationwide permits 3 (which includes maintenance of existing shore protection structures) and 31 (maintenance of existing flood control activities) within special management areas (NYS DOS, 2006).

Similar to the New York metropolitan area, the policies for Long Island reflect the fact that the region is intensely developed in the west and developing fast in the east. Much of the South Shore, particularly within Nassau County, is already developed and has already been protected, primarily by bulkheads. The Long Island Sound Management Program estimates that approximately 50 percent of the Sound's shoreline is armored (NYS DOS, 1999).

Some of the South Shore's densely developed communities facing flooding problems, such as Freeport and Hempstead, have already implemented programs that call for elevating buildings and infrastructure in place and installing bulkheads for flood protection. The Town of Hempstead has adopted the provisions of the state's Coastal Erosion Hazards Area Act, described in Section A1 .B of this Appendix, because erosion and flooding along Nassau County's ocean coast have been a major concern. The Town of Hempstead has also been actively working with the U.S. Army Corps of Engineers (USACE) to develop a long-term storm damage reduction plan for the heavily developed Long Beach barrier island (USACE, 2003).

Beach nourishment and the construction of flood and erosion protection structures are also common on the island. For example, in the early 1990s USACE constructed a substantial revetment around the Montauk Lighthouse at the eastern tip of Long Island and after a new feasibility study has proposed construction of a larger revetment (Bleyer, 2007). USACE is also reformulating a plan for the development of long-term storm damage prevention projects along the 134 kilo (km) 83 mile [mi]) portion of the South Shore of Suffolk County. As part of this effort, USACE is assessing at-risk properties within the 184 square kilometer (sq km)

(71 square miles [sq mi]) floodplain, present and future sea-level rise, restoration and preservation of important coastal landforms and processes, and important public uses of the area (USACE, 2008b).

To obtain state funding for nourishment, communities must provide public access every 800 m (0.5 mi) (New York State, 2002b). In 1994, as terms of a legal settlement between federal, state, and local agencies cooperating on the rebuilding of the beach through nourishment, the community of West Hampton provided six walkways from the shorefront road to allow public access to the beach (Dean, 1999). In communities that have not had such state-funded projects, however, particularly along portions of the bay shore communities in East Hampton, South Hampton, Brookhaven, and Islip, public access to tidal waters can be less common (NYS DOS, 1999).

The Comprehensive Coastal Management Plan (CCMP) of the Peconic Bay National Estuary Program Management Plan calls for "no net increase of hardened shoreline in the Peconic Estuary". The intent of this recommendation is to discourage individuals from armoring their coastline; yet this document is only a management plan and does not have any legal authority. However, towns such as East Hampton are trying to incorporate the plan into their own programs. In 2006, the town of East Hampton adopted and is now enforcing a defined zoning district overlay map that prevents shore armoring along much of the town's coastline (Town of East Hampton, 2006). Despite such regulations, authorities in East Hampton and elsewhere recognize that there are some areas where structures will have to be allowed to protect existing development.

The New York Department of State (DOS) is also examining options for managing erosion and flood risks through land use measures, such as further land exchanges. For example, there is currently an attempt to revise the proposed Fire Island to Montauk Point Storm Damage Reduction Project to consider a combination of nourishment and land use measures. One option would be to use beach nourishment to protect structures for the next few decades, during which time development could gradually be transferred out of the most hazardous locations. Non-conforming development could eventually be brought into conformance as it is reconstructed, moved, damaged by storms or flooding, or other land use management plans are brought into effect.

A1.B. New York Metropolitan Area

The New York metropolitan area has a mixture of elevated and low-lying coastlines. Low-lying land within 3 m of mean sea level (Gornitz *et al.*, 2002) include the borough of Queens' northern and southeastern shore, respectively (where New York's two major airports, LaGuardia and John F. Kennedy International Airport, are located); much of the recreational lands along Jamaica Bay's Gateway National Recreation Area (*e.g.*, Floyd Bennett Field, Jamaica Bay Wildlife Refuge, Fort Tilden, Riis Park); and the Staten Island communities of South Beach and Oakwood Beach. In New Jersey, the heavily developed coast of Hudson County (including Hoboken, Jersey City, and Bayonne) is also within 3 m, as is much of the area known as the Meadowlands (area around Giants Stadium). Other areas with sections of low-lying lands are found in Elizabeth and Newark, New Jersey (near Newark Airport). The area also includes the ecologically-significant Raritan Bay-Sandy Hook habitat complex at the apex of the New York region (also known as the New York Bight), where the east-west oriented coastline of New England and Long Island intersects the north-south oriented coastline of the mid-Atlantic at Sandy Hook.

Given its large population, the effects of hurricanes and other major storms combined with higher sea levels could be particularly severe in the New York metropolitan area. With much of the area's transportation infrastructure at low elevation (most at 3 m or less), even slight increases in the height of flooding could cause extensive damage and bring the thriving city to a relative standstill until the flood waters recede (Gornitz *et al.*, 2002).

Comprehensive assessments of the vulnerability of the New York City metropolitan area are found in Jacob *et al.* (2007) and Gornitz *et al.* (2002). Jacob *et al.* summarize vulnerability, coastal management and adaptation issues. Gornitz *et al.* details the methodology and results of a study that summarizes vulnerability to impacts of climate change, including higher storm surges, shoreline movement, wetland loss, beach nourishment and some socioeconomic implications. These assessments use sea-level rise estimates from global climate models available in 2002. Generalized maps depicting lands close to sea level are found in Titus and Richman (2001) and Titus and Wang (2008).

If sea-level rise impairs coastal habitat, many estuarine species would be at risk. This Section provides additional details on the possible environmental implications of sea- level rise for the greater New York metropolitan area, including New York City, the lower Hudson River, the East River, Jamaica Bay, the New Jersey Meadowlands, Raritan Bay and Sandy Hook Bay. The following subsections discuss tidal wetlands, beaches, tidal flats, marsh and bay islands, and shallow waters. (Sections A1.A.2 and A1.D.2 discuss the statewide coastal policies of New York and New Jersey.)

Tidal Wetlands. Examples of this habitat include:

- *Staten Island*: The Northwest Staten Island/Harbor Herons Special Natural Waterfront Area is an important nesting and foraging area for herons, ibises, egrets, gulls, and waterfowl (USFWS, 1997). Several marshes on Staten Island, such as Arlington Marsh and Saw Mill Creek Marsh, provide foraging areas for the birds of the island heronries. Hoffman Island and Swinburne Island, east of Staten Island, provide important nesting habitat for herons and cormorants, respectively (Bernick, 2006).

- *Manhattan*: In the marsh and mudflat at the mouth of the Harlem River at Inwood Hill Park (USFWS, 1997) great blue herons are found along the flat in winter, and snowy and great egrets are common from spring through fall (NYC DPR, 2001).

- *Lower Hudson River*: The Piermont Marsh, a 412 hectare (ha) (1,017 acre [ac]) brackish wetland on the western shore of the lower Hudson River has been designated for conservation management by New York State and the National Oceanic and Atmospheric Administration (NOAA) (USFWS, 1997). The marsh supports breeding birds, including relatively rare species such as Virginia rail, swamp sparrow, black duck, least bittern, and sora rail. Anadromous and freshwater fish use the marsh's tidal creeks as a spawning and nursery area. Diamondback terrapin reportedly nest in upland areas along the marsh (USFWS, 1997).

- *Jamaica Bay*: Located in Brooklyn and Queens, this bay is the largest area of protected wetlands in a major metropolitan area along the U.S. Atlantic Coast. The bay includes the Jamaica Bay Wildlife Refuge, which has been protected since 1972 as part of the Jamaica Bay Unit of the Gateway National Recreation Area. Despite extensive disturbance from dredging, filling, and development, Jamaica Bay remains one of the most important migratory shorebird stopover sites in the New York Bight (USFWS, 1997). The bay provides overwintering habitat for many duck species, and mudflats support foraging migrant species (Hartig *et al.*, 2002). The refuge and Breezy Point, at the tip of the Rockaway Peninsula, support populations of 214 species that are state or federally listed or of special emphasis, including 48 species of fish and 120 species of birds (USFWS, 1997). Salt marshes such as Four Sparrow Marsh provide nesting habitat for declining sparrow species and serve 326 species of migrating birds (NYC DPR, undated). Wetlands in some parts of the bay currently show substantial losses (Hartig *et al.*, 2002).

- *Meadowlands*: The Meadowlands contain the largest single tract of estuarine tidal wetland remaining in the New York/New Jersey Harbor Estuary and provide critical habitat for a diversity of species, including a number of special status species. Kearney Marsh is a feeding area for the state-listed endangered least tern, black skimmer, and pied-billed grebe. Diamondback terrapin, the only turtle known to occur in brackish water, is found in the Sawmill Wildlife Management Area (USFWS, 1997).

- *Raritan Bay-Sandy Hook*: The shorelines of southern Raritan Bay include large tracts of fringing salt marsh at Conaskonk Point and from Flat Creek to Thorn's Creek. These marshes are critical for large numbers of nesting and migrating bird species. The salt marsh at Conaskonk Point provides breeding areas for bird species such as green heron, American oystercatcher, seaside sparrow, and saltmarsh sharp- tailed sparrow, as well as feeding areas for herons, egrets, common tern, least tern, and black skimmer. In late May and early June, sanderlings, ruddy turnstones, semipalmated sandpipers, and red knots feed on horseshoe crab eggs near the mouth of Chingarora Creek. Low marsh along the backside of Sandy Hook spit provides forage and protection for the young of marine fishes, including winter flounder, Atlantic menhaden, bluefish, and striped bass, and critical habitat for characteristic bird species of the low marsh such as clapper rail, willet, and marsh wren (USFWS, 1997).

Estuarine Beaches. Relatively few areas of estuarine beach remain in the New York City metropolitan area, and most have been modified or degraded (USFWS 1997; Strange 2008a). In Jamaica Bay, remaining estuarine beaches occur off Belt Parkway (*e.g.*, on Plumb Beach) and on the bay islands (USFWS, 1997). Sandy beaches are still relatively common along the shores of Staten Island from Tottenville to Ft. Wadsworth. The southern shoreline of Raritan

Bay includes a number of beaches along Sandy Hook Peninsula and from the Highlands to South Amboy, some of which have been nourished. There are also beaches on small islands within the Shrewsbury-Navesink River system (USFWS, 1997).

Although limited in area, the remaining beaches support an extensive food web. Mud snails and wrack-based species (*e.g.*, insects, isopods, and amphipods) provide food for shorebirds including the piping plover, federally listed as threatened (USFWS, 1997). The beaches around Sandy Hook Bay have becoming important nestling places in winter for several species of seals (USFWS, 1997). The New Jersey Audubon Society reports that its members have observed gulls and terns at the Raritan Bay beach at Morgan on the southern shore, including some rare species such as black-headed gull, little gull, Franklin's gull, glaucous gulls, black tern, sandwich tern, and Hudsonian godwit. Horseshoe crabs lay their eggs on area beaches, supplying critical forage for shorebirds (Botton *et al.*, 2006). The upper beach is used by nesting diamondback terrapins; human- made sandy trails in Jamaica Bay are also an important nest site for terrapins in the region, although the sites are prone to depredations by raccoons (Feinberg and Burke, 2003).

Tidal flats. Like beaches, tidal flats are limited in the New York City metropolitan region, but the flats that remain provide important habitat, particularly for foraging birds. Tidal flats are also habitat for hard and soft shell clams, which are important for recreational and commercial fishermen where not impaired by poor water quality. Large concentrations of shorebirds, herons, and waterfowl use the shallows and tidal flats of Piermont Marsh along the lower Hudson River as staging areas for both spring and fall migrations (USFWS, 1997). Tidal flats in Jamaica Bay are frequented by shorebirds and waterfowl, and an intensive survey of shorebirds in the mid-1980s estimated more than 230,000 birds of 31 species in a single year, mostly during the fall migration (Burger, 1984). Some 1,460 ha (3,600 ac) of intertidal flats extend offshore an average of 0.4 km (0.25 mi) from the south shore of the Raritan and Sandy Hook Bays, from the confluence of the Shrewsbury and Navesink Rivers, west to the mouth of the Raritan River. These flats are important foraging and staging areas for migrating shorebirds, averaging over 20,000 birds, mostly semipalmated plover, sanderling, and ruddy turnstone. The flats at the mouth of Whale Creek near Pirate's Cove attract gulls, terns, and shorebirds year round. Midwinter waterfowl surveys indicate that an average of 60,000 birds migrate through the Raritan Bay-Sandy Hook area in winter (USFWS, 1997). Inundation with rising seas will eventually make flats unavailable to short-legged shorebirds, unless they can shift feeding to marsh ponds and pannes (Erwin *et al.*, 2004). At the same time, disappearing saltmarsh islands in the area are transforming into intertidal mudflats. This may increase habitat for shorebirds at low tide, but it leaves less habitat for refuge at high tide (Strange 2008a).

Shallow water habitat. This habitat is extensive in the Hudson River, from Stony Point south to Piermont Marsh, just below the Tappan Zee Bridge (USFWS, 1997). This area features the greatest mixing of ocean and freshwater, and concentrates nutrients and plankton, resulting in a high level of both primary and secondary productivity. Thus, this part of the Hudson provides key habitat for numerous fish and bird species. It is a major nursery area for striped bass, white perch, tomcod, and Atlantic sturgeon, and a wintering area for the federally endangered shortnose sturgeon. Waterfowl also feed and rest here during spring and fall migrations. Some submerged aquatic vegetation (SAV) is also found here, dominated by water celery, sago pondweed, and horned pondweed (USFWS, 1997).

Marsh and bay islands. Throughout the region, these islands are vulnerable to sea-level rise (Strange 2008a). Between 1974 and 1994, the smaller islands of Jamaica Bay lost nearly 80 percent of their vegetative cover (Strange 2008a, citing Hartig *et al.*, 2002). Island marsh deterioration in Jamaica Bay has led to a 50 percent decline in area between 1900 and 1994 (Gornitz *et al.*, 2002). Marsh loss has accelerated, reaching an average annual rate of 18 ha (45 ac) per year between 1994 and 1999 (Hartig *et al.*, 2002). The islands provide specialized habitat for an array of species:

- Regionally important populations of egrets, herons, and ibises are or have been located on North and South Brother islands in the East River and on Shooter's Island, Prall's Island, and Isle of Meadows in Arthur Kill and Kill van Kull (USFWS, 1997).
- North and South Brother Islands have the largest black crowned night heron colony in New York State, along with large numbers of snowy egret, great egret, cattle egret, and glossy ibis (USFWS, 1997).
- Since 1984, an average of 1,000 state threatened common tern have nested annually in colonies on seven islands of the Jamaica Bay Wildlife Refuge (USWFS, 1997).
- The heronry on Carnarsie Pol also supports nesting by great black-backed gull, herring gull, and American oystercatcher (USFWS, 1997).
- The only colonies of laughing gull in New York State, and the northernmost breeding extent of this species, occur on the islands of East High Meadow, Silver Hole Marsh, Jo Co Marsh, and West Hempstead Bay (USFWS, 1997).
- Diamondback terrapin nest in large numbers along the sandy shoreline areas of the islands of Jamaica Bay, primarily Ruler's Bar Hassock (USFWS, 1997).

A1.C. New Jersey Shore

The New Jersey shore has three types of ocean coasts (see Chapter 3 of this Product). At the south end, Cape May and Atlantic Counties have short and fairly wide "tide-dominated" barrier islands. Behind the islands, 253 sq km (97 sq mi) of marshes dominate the relatively small open water bays. To the north, Ocean County has "wave dominated" coastal barrier islands and spits. Long Beach Island is 29 km (18 mi) long and only two to three blocks wide in most places; Island Beach to the north is also long and narrow. Behind Long Beach Island and Island Beach lie Barnegat and Little Egg Harbor Bays. These shallow estuaries range from 2 to 7 km (about 1 to 4 mi) wide, and have 167 sq km (64 sq mi) of open water (USFWS, 1997) with extensive eelgrass, but only 125 sq km (48 sq mi) of tidal marsh (Jones and Wang, 2008). Monmouth County's ocean coast is entirely headlands, with the exception of Sandy Hook at the northern tip of the Jersey Shore. Non-tidal wetlands are immediately inland of the tidal wetlands along most of the mainland shore[8].

A1.C.1. Environmental Implications

There have been many efforts to conserve and restore species and habitats in the barrier island and back-barrier lagoon systems in New Jersey. Some of the larger parks and wildlife areas in the region include Island Beach State Park, Great Bay Boulevard State Wildlife Management Area, and the E.B. Forsythe National Wildlife Refuge (Forsythe Refuge) in Ocean and Atlantic counties. Parts of the Cape May Peninsula are protected by the Cape May National Wildlife Refuge (USFWS, undated[a]), the Cape May Point State Park (NJDEP, undated) and The Nature Conservancy's (TNC's) Cape May Migratory Bird Refuge (TNC, undated).

Tidal and Nearshore Nontidal Marshes. There are 18,440 ha (71 sq mi), 29,344 ha (113 sq mi), and 26,987 ha (104 sq mi) of tidal salt marsh in Ocean, Atlantic, and Cape May counties, respectively (Jones and Wang, 2008). The marshes in the study area are keeping pace with current local rates of sea-level rise of 4 millimeters (mm) per year, but are likely to become marginal with a 2 mm per year acceleration and be lost with a 7 mm per year acceleration, except where they are near local sources of sediments (*e.g.*, rivers such as the Mullica and Great Harbor rivers in Atlantic County) (Strange 2008b, interpreting the findings of Reed *et al.*, 2008).

There is potential for wetland migration in Forsythe Refuge, and other lands that preserve the coastal environment such as parks and wildlife management areas. Conservation lands are also found along parts of the Mullica and Great Egg Harbor Rivers in Atlantic County. However, many estuarine shorelines in developed areas are hardened, limiting the potential for wetland migration (Strange, 2008b).

As marshes along protected shorelines experience increased tidal flooding, there may be an initial benefit to some species. If tidal creeks become wider and deeper fish may have increased access to forage on the marsh surface (Weinstein, 1979). Sampling of larval fishes in high salt marsh on Cattus Island, Beach Haven West, and Cedar Run in Ocean County showed that high marsh is important for mummichog, rainwater killifish, spotfin killifish, and sheepshead minnow (Talbot and Able, 1984). The flooded marsh surface and tidal and nontidal ponds and ditches appear to be especially important for the larvae of these species (Talbot and Able, 1984). However, as sea level rises, and marshes along hardened shorelines convert to open water, marsh fishes will lose access to these marsh features and the protection from predators, nursery habitat, and foraging areas provided by the marsh (Strange 2008b).

Loss of marsh area would also have negative implications for the dozens of bird species that forage and nest in the region's marshes. Initially, deeper tidal creeks and marsh pools will become inaccessible to short-legged shorebirds such as plovers (Erwin *et al.*, 2004). Long-legged waterbirds such as the yellow-crowned night heron, which forages almost exclusively on marsh crabs (fiddler crab and others), will lose important food resources (Riegner, 1982). Eventually, complete conversion of marsh to open water will affect the hundreds of thousands of shorebirds that stop in these areas to feed during their migrations. The New Jersey Coastal Management Program estimates that some 1.5 million migratory shorebirds stopover on New Jersey's shores during their annual migrations (Cooper *et al.*, 2005). Waterfowl also forage and overwinter in area marshes. Mid-winter aerial waterfowl counts in Barnegat Bay alone average 50,000 birds (USFWS, 1997). The tidal marshes of the Cape May Peninsula provide stopover areas for hundreds of thousands of shorebirds, songbirds, raptors, and waterfowl during their seasonal migrations (USFWS, 1997). The peninsula is also an important staging area and overwintering area for seabird populations. Surveys conducted by the U.S. Fish and Wildlife Service from July through December 1995 in Cape May County recorded more than 900,000 seabirds migrating along the coast (USFWS, 1997).

As feeding habitats are lost, local bird populations may no longer be sustainable (Strange, 2008b). For example, avian biologists suggest that if marsh pannes and pools continue to be lost in Atlantic County as a result of sea-level rise, the tens of thousands of shorebirds that feed in these areas may shift to feeding in impoundments in the nearby Forsythe Refuge. Such a shift would increase shorebird densities in the refuge ten-fold and reduce population sustainability due to lower per capita food resources and disease from crowding (Erwin *et al.*, 2006).

Local populations of marsh nesting bird species will also be at risk where marshes drown. This will have a particularly negative impact on rare species such as seaside and sharp- tailed sparrows, which may have difficulty finding other suitable nesting sites. According to a synthesis of published studies in Greenlaw and Rising (1994) and Post and Greenlaw (1994), densities in the region ranged from 0.3 to 20 singing males per hectare and 0.3 to 4.1 females per hectare for the seaside and sharp-tailed sparrows, respectively (Greenlaw and Rising, 1994). Loss and alteration of suitable marsh habitats are the primary conservation concerns for these and other marsh-nesting passerine birds (BBNEP, 2001).

Shore protection activities (nourishment and vegetation control) are underway to protect the vulnerable freshwater ecosystems of the Cape May Meadows (The Meadows), which are located behind the eroding dunes near Cape May Point (USACE, 2008a). Freshwater coastal ponds in The Meadows are found within about one hundred meters (a few hundred feet) of the shoreline and therefore could easily be inundated as seas rise. The ponds provide critical foraging and resting habitat for a variety of bird species, primarily migrating shorebirds (NJDEP, undated). Among the rare birds seen in The Meadows by local birders are buff-breasted sandpipers, arctic tern, roseate tern, whiskered tern, Wilson's phalarope, black rail, king rail, Hudsonian godwit, and black-necked stilt (Kerlinger, undated). The Nature Conservancy, the United States Army Corps of Engineers (USACE), and the New Jersey Department of Environmental Protection (NJDEP) have undertaken an extensive restoration project in the Cape May Migratory Bird Refuge, including beach replenishment to protect a mile-long stretch of sandy beach that provides nesting habitat for the piping plover (federally listed as threatened), creation of plover foraging ponds, and creation of island nesting sites for terns and herons (TNC, 2007).

Estuarine Beaches. Estuarine beaches are largely disappearing in developed areas where shoreline armoring is the preferred method of shore protection. The erosion or inundation of bay islands would also reduce the amount of beach habitat. Many species of invertebrates are found within or on the sandy substrate or beach wrack (seaweed and other decaying marine plant material left on the shore by the tides) along the tide line of estuarine beaches (Bertness, 1999). These species provide a rich and abundant food source for bird species. Small beach invertebrates include isopods and amphipods, blood worms, and beach hoppers, and beach macroinvertebrates include soft shell clams, hard clams, horseshoe crabs, fiddler crabs, and sand shrimp (Shellenbarger Jones, 2008a).

Northern diamondback terrapin nest on estuarine beaches in the Barnegat Bay area (BBNEP, 2001). Local scientists consider coastal development, which destroys terrapin nesting beaches and access to nesting habitat, to be one of the primary threats to diamondback terrapins, along with predation, road kills, and crab trap bycatch (Strange, 2008b, citing Wetland Institute [undated]).

Loss of estuarine beach could also have negative impacts on various beach invertebrates, including rare tiger beetles (Strange, 2008b). Two sub-species likely exist in coastal New Jersey: *Cicindela dorsalis dorsalis*, the northeastern beach tiger beetle, which is a federally listed threatened species and a state species of special concern and regional priority, and *Cicindela dorsalis media*, the southeastern beach tiger beetle, which is state- listed as rare (NJDEP, 2001). In the mid-1990s, the tiger beetle was observed on the undeveloped ocean beaches of Holgate and Island Beach. Current surveys do not indicate whether this species is also found on the area's estuarine beaches, but it feeds and nests in a variety of habitats (USFWS, 1997). The current abundance and distribution of the northeastern beach tiger beetle in the coastal bays is a target of research (State of New Jersey, 2005). At present, there are plans to reintroduce the species in the study region at locations where natural ocean beaches remain (State of New Jersey, 2005).

Tidal Flats. The tidal flats of New Jersey's back-barrier bays are critical foraging areas for hundreds of species of shorebirds, passerines, raptors, and waterfowl (BBNEP, 2001). Important shorebird areas in the study region include the flats of Great Bay Boulevard Wildlife Management Area, North Brigantine Natural Area, and the Brigantine Unit of the Forsythe Refuge (USFWS, 1997). The USFWS estimates that the extensive tidal flats of the Great Bay alone total 1,358 ha (3,355 ac). Inundation of tidal flats with rising seas would eliminate critical foraging opportunities for the area's abundant avifauna. As tidal flat area declines, increased crowding in remaining areas could lead to exclusion and mortality of many foraging birds (Galbraith *et al.*, 2002; Erwin *et al.*, 2004). Some areas may become potential sea grass restoration sites, but whether or not "enhancing" these sites as eelgrass areas is feasible will depend on their location, acreage, and sediment type (Strange, 2008b).

Shallow Nearshore Waters and Submerged Aquatic Vegetation (SAV). The Barnegat Estuary is distinguished from the lagoons to the south by more open water and SAV and less emergent marsh. Within the Barnegat Estuary, dense beds of eelgrass are found at depths under 1 m, particularly on sandy shoals along the backside of Long Beach Island and Island Beach, and around Barnegat Inlet, Manahawkin Bay, and Little Egg Inlet. Eelgrass is relatively uncommon from the middle of Little Egg Harbor south to Cape May, particularly locations where water depths are more than 1 m, such as portions of Great South Bay (USFWS, 1997).

Seagrass surveys from the 1 960s through the 1 990s indicate that there has been an overall decline in seagrass beds in Barnegat Estuary, from 6,823 ha (16,847 ac) in 1968 to an average of 5,677 ha (14,029 ac) during the period 1996 to 1998 (BBNEP, 2001). Numerous studies indicate that eelgrass has high ecological value as a source of both primary (Thayer *et al.*, 1984) and secondary production (Jackson *et al.*, 2001) in estuarine food webs. In Barnegat Estuary eelgrass beds provide habitat for invertebrates, birds, and fish that use the submerged vegetation for spawning, nursery, and feeding (BBNEP, 2001). Shallow water habitat quality may also be affected by adjacent shoreline protections. A Barnegat Bay study found that where shorelines are bulkheaded, SAV, woody debris, and other features of natural shallow water habitat are rare or absent, with a resulting reduction in fish abundance (Byrne, 1995).

Marsh and Bay Islands. Large bird populations are found on marsh and dredge spoil islands of the New Jersey back-barrier bays. These islands include nesting sites protected from predators for a number species of conservation concern, including gull-billed tern, common tern, Forster's tern, least tern, black skimmer, American oystercatcher, and piping plover (USFWS, 1997). Diamondback terrapins are also known to feed on marsh islands in the bays (USFWS, 1997).

Some of the small islands in Barnegat Bay and Little Egg Harbor extend up to about 1 m above spring high water (Jones and Wang, 2008), but portions of other islands are very low, and some low islands are currently disappearing. Mordecai (MLT, undated) and other islands (Strange, 2008b) used by nesting common terns, Forster's terns, black skimmers, and American oystercatchers are vulnerable to sea-level rise and erosion (MLT, undated). With the assistance of local governments, the Mordecai Land Trust is actively seeking grants to halt the gradual erosion of Mordecai Island, an 18-ha (45-ac) island just west of Beach Haven on Long Beach Island (MLT, undated). Members of the land trust have documented a 37 percent loss of island area since 1930. The island's native salt marsh and surrounding waters and SAV beds provide habitat for a variety of aquatic and avian species. NOAA National Marine Fisheries Service considers the island and its waters Essential Fish Habitat for spawning and all life stages of winter flounder as well as juvenile and adult stages of Atlantic sea herring, bluefish, summer flounder, scup, and black sea bass (MLT, undated). The island is also a strategically-located nesting island for many of New Jersey's threatened and endangered species, including black skimmers, least terns, American bitterns, and both yellow-crowned and black-crowned night herons (MLT, 2003).

Sea-level fens. New Jersey has identified 12 sea-level fens, encompassing 126 acres. This rare ecological community is restricted in distribution to Ocean County, New Jersey, between Forked River and Tuckerton, in an area of artesian groundwater discharge from the Kirkwood-Cohansey aquifer. Additional recent field surveys have shown possible occurrences in the vicinity of Tuckahoe in Cape May and Atlantic counties (Walz *et al.*, 2004). These communities provide significant wetland functions in the landscape as well as supporting 18 rare plant species, of which one is state-listed as endangered. (Walz *et al.*, 2004).

A1.C.2. Development, Shore Protection, and Coastal Policies

At least five state policies affect the response to sea-level rise along New Jersey's Atlantic Coast: the Coastal Facility Review Act, the Wetlands Act, the State Plan, an unusually strong public trust doctrine, and the state's strong support for beach nourishment— and opposition to both erosion-control structures and shoreline retreat— along ocean shores. This section discusses the latter policy; the first four are discussed in Section A1.D.2 of this Appendix.

In 1997, then-Governor Whitman promised coastal communities that "there will be no forced retreat," and that the government would not force people to leave the shoreline. That policy does not necessarily mean that there will always be government help for shore protection. Nevertheless, although subsequent administrations have not expressed this view so succinctly, they have not withdrawn the policy either. In fact, the primary debate in New Jersey tends to be about the level of public access required before a community is eligible to receive beach nourishment, not the need for shore protection itself (see Chapter 8 of this Product).

With extensive development and tourism along its shore, New Jersey has a well-established policy in favor of shore protection along the ocean[9]. The state generally prohibits new hard structures along the ocean front; but that was not always the case. A large portion of the Monmouth County shoreline was once protected with seawalls, with a partial or total loss of beach (Pilkey *et al.*, 1981). Today, beach nourishment is the preferred method for reversing beach erosion and providing ocean front land with protection from coastal storms (Mauriello, 1991). The entire Monmouth County shoreline now has a beach in front of the old seawalls. Beach nourishment has been undertaken or planned for at least one community in every coastal county from Middlesex along Raritan Bay, to Salem along the Delaware River. Island Beach State Park, a barrier spit along the central portion of Barnegat Bay just north of Long Beach Island, is heavily used by New Jersey residents and includes the official beach house of the Governor. Although it is a state park, it is currently included in the authorized USACE Project for beach nourishment from Manasquan to Barnegat Inlet. In the case of Cape May Meadows[10], environmental considerations have prompted shore protection efforts (USACE, 2008a). The area's critical freshwater ecosystem is immediately behind dunes that have eroded severely as a result of the jetties protecting the entrance to the Cape May Canal.

Some coastal scientists have suggested the possibility of disintegrating barrier islands along the New Jersey shore (see Chapter 3). Although the bay sides of these islands are bulkheaded, communities are unlikely to seriously consider the option of being encircled by a dike as sea level rises (see Box A1.2). Nevertheless, Avalon uses a combination of floodwalls and checkvalves to prevent tidal flooding; and Atlantic City's stormwater management system includes underground tanks with checkvalves. These systems have been implemented to address current flooding problems; but they would also be a logical first step in a strategy to protect low-lying areas with structural solutions as sea level rises[11]. Other authors have suggested that a gradual elevation of barrier islands is more likely (see Box A1.2).

Wetlands along the back-barrier bays of New Jersey's Atlantic coast are likely to have some room to migrate inland, because they are adjacent to large areas of non-tidal wetlands. One effort at the state level to preserve such coastal resources is the state's Stormwater Management Plan, which establishes a special water resource protection area that limits development within 91.4 m (300 ft) along most of its coastal shore (NJDEP DWM, 2004). Although the primary objective of the regulation is to improve coastal water quality and reduce potential flood damage, it serves to preserve areas suitable for the landward migration of wetlands.

BOX A1.2. SHORE PROTECTION ON LONG BEACH ISLAND

The effects of sea-level rise can be observed on both the ocean and bay sides of this 29-km (18-mi) long barrier island. Along the ocean side, shore erosion has threatened homes in Harvey Cedars and portions of Long Beach township. During the 1990s, a steady procession of dump trucks brought sand onto the beach from inland sources. In 2007, the USACE began to restore the beach at Surf City and areas immediately north. The beach had to be closed for a few weeks, however, after officials discovered that munitions (which had been dumped offshore after World War II) had been inadvertently pumped onto the beach.

High tides regularly flood the main boulevard in the commercial district of Beach Haven, as well as the southern two blocks of Central Avenue in Ship Bottom. Referring to the flooded parking lot during spring tides, the billboard of a pizza parlor in Beach Haven Crest boasts "Occasional Waterfront Dining."

U.S. EPA's 1989 Report to Congress used Long Beach Island as a model for analyzing alternative responses to rising sea level, considering four options: a dike around the island, beach nourishment and elevating land and structures, an engineered retreat which would include the creation of new bayside lands as the ocean eroded, and making no effort to maintain the island's land area (U.S. EPA, 1989; Titus *et al.*, 1991). Giving up the island was the most expensive option (Weggel *et al.*, 1989; Titus, 1990). The study concluded that a dike would be the least expensive in the short run, but unacceptable to most residents due to the lost view of the bay and risk of being on a barrier island below sea level (Titus, 1991). In the long run, fostering a landward migration would be the least expensive, but it would unsettle the expectations of bay front property owners and hence require a lead time of a few generations between being enacted and new bayside land actually being created. Thus, the combination of beach nourishment and elevating land and structures appeared to be the most realistic, and U.S. EPA used that assumption in its nationwide cost estimate (U.S. EPA, 1989; Titus *et al.*, 1991).

Long Beach Township, Ship Bottom, Harvey Cedars, and Beach Haven went through a similar thinking process in considering their preferred response to sea-level rise. In resolutions enacted by their respective councils, they concluded that a gradual elevation of their communities would be preferable to either dikes or the retreat option. In the last ten years, several structural moving companies have had ongoing operations, continually elevating homes (see Figure 12.5).

Box Figure A1.2. Street flooding in Long Beach Island, New Jersey at one of the higher tides.

A1.D. Delaware Estuary

A1.D.1. Environmenal Implications

On both sides of Delaware Bay, most shores are either tidal wetlands or sandy beaches with tidal wetlands immediately behind them. In effect, the sandy beach ridges are similar to the barrier islands along the Atlantic, only on a smaller scale. Several substantial communities with wide sandy beaches on one side and marsh on the other side are along Delaware Bay—especially on the Delaware side of the bay. Although these communities are potentially vulnerable to inundation, shoreline erosion has been a more immediate threat to these communities. Detailed discussions of the dynamics of Delaware shorelines are found in Kraft and John (1976).

Delaware Bay is home to hundreds of species of ecological, commercial, and recreational value (Dove and Nyman, 1995). Unlike other estuaries in the Mid-Atlantic, the tidal range is greater than the ocean tidal range, generally about 2 m. In much of Delaware Bay, tidal marshes appear to be at the low end of their potential elevation range, increasing their vulnerability to sea-level rise (Kearney *et al.*, 2002). Recent research indicates that 50 to 60 percent of Delaware Bay's tidal marsh has been degraded, primarily because the surface of the marshes is not rising as fast as the sea (Kearney *et al.*, 2002). One possible reason is that channel deepening projects and consumptive withdrawals of fresh water have changed the sediment supply to the marshes (Sommerfield and Walsh, 2005). Many marsh restoration projects are underway in the Delaware Bay (*cf.* Teal and Peterson, 2005): dikes have been removed to restore tidal flow and natural marsh habitat and biota; however, in some restoration areas invasion by common reed (*Phragmites australis)* has been a problem (Abel and Hagan, 2000; Weinstein *et al.*, 2000).

The loss of tidal marsh as sea level rises would harm species that depend on these habitats for food and shelter, including invertebrates, finfish, and a variety of bird species (Kreeger and Titus, 2008). Great blue herons, black duck, blue and green-winged teal, Northern harrier, osprey, rails, red winged blackbirds, widgeon, and shovelers all use the salt marshes in Delaware Bay. Blue crab, killifish, mummichog, perch, weakfish, flounder, bay anchovy, silverside, herring, and rockfish rely on tidal marshes for feeding on the mussels, fiddler crabs, and other invertebrates and for protection from predators (Dove and Nyman, 1995).

Delaware Bay is a major stopover area for six species of migratory shorebirds, including most of the Western Hemisphere's population of red knot (USFWS, 2003). On their annual migrations from South America to the Arctic, nearly a million shorebirds move through Delaware Bay, where they feed heavily on invertebrates in tidal mudflats, and particularly on horseshoe crab eggs on the bay's sandy beaches and foreshores (Walls *et al.*, 2002). Horseshoe crabs have been historically abundant on the Delaware Bay shores. A sea-level rise modeling study estimated that a 6-centimeter (cm) (2-ft) rise in relative sea level over the next century could reduce shorebird foraging areas in Delaware Bay by 57 percent or more by 2100 (Galbraith *et al.*, 2002).

Invertebrates associated with cordgrass stands in the low intertidal zone include grass shrimp, ribbed mussel, coffee-bean snail, and fiddler crabs (Kreamer, 1995). Blue crab, sea turtles, and shorebirds are among the many species that prey on ribbed mussels; fiddler crabs are an important food source for bay anchovy and various species of shorebirds (Kreamer, 1995). Wading birds such as the glossy ibis feed on marsh invertebrates (Dove and Nyman, 1995). Waterfowl, particularly dabbling ducks, use low marsh areas as a wintering ground.

Sandy beaches and foreshores account for the majority of the Delaware and New Jersey shores of Delaware Bay. As sea level rises, beaches can be lost if either shores are armored or if the land behind the existing beach has too little sand to sustain a beach as the shore retreats (Nordstrom, 2005). As shown in Table A1.1, so far only 4 percent (Delaware) and 6 percent (New Jersey) of the natural shores have been replaced with shoreline armoring. Another 15 percent (Delaware) and 4 percent (New Jersey) of the shore is developed. Although conservation areas encompass 58 percent of Delaware Bay's shores, they include only 32 percent of beaches that are optimal or suitable habitat for horseshoe crabs (Kreeger and Titus, 2008).

Beach nourishment has been relatively common along the developed beach communities on the Delaware side of the bay. Although beach nourishment can diminish the quality of habitat for horseshoe crabs, nourished beaches are more beneficial than an armored shore; most beach nourishment along the New Jersey shore of Delaware Bay has been justified by environmental benefits (Kreeger and Titus, 2008; USACE 1 998b,,c).

Many Delaware Bay beaches have a relatively thin layer of sand. Although these small beaches currently have enough sand to protect the marshes immediately inland from wave action, some beaches may not be able to survive accelerated sea-level rise even in areas without shoreline armoring, unless artificial measures are taken to preserve them (Kreeger and Titus (2008). For example, Delaware has already nourished beaches with the primary purpose of restoring horseshoe crab habitat (Smith *et al.*, 2002) (see Box A1.3.).

Table A1.1. The Shores of Delaware Bay: Habitat Type and Conservation Status of Shores Suitable for Horseshoe Crabs

Shoreline length	Delaware		New Jersey		New Jersey and Delaware
By Habitat Type (percent of bay shoreline	km	%	km	%	%
Beach	68	74	62	42	54
Armored Shore	3.7	4	8.3	6	5
Organic	20	22	78	53	41
Total Shoreline	91	100	148	100	100
By Indicator of Future Shore Protection (kin)					
Shore Protection Structures	2.7	2.9	5.1	3.4	3
Development	13	15	5.7	3.8	8
By Suitability for Horseshoe Crab (percent of Bay shoreline					
Optimal Habitat	31.3	34	26.0	18	24
Suitable Habitat	10.5	12	5.1	3.5	6.6
Less Suitable Habitat	29.0	32	49.0	33	33
Unsuitable Habitat	20.0	22	67.0	46	37
Within Conservations Lands by Suitability for Horseshoe Crab (percent of equally suitable lands)					
Optimal Habitat	12.9	41	9.6	37	39
Optimal and Suitable Habitat	13.6	33	9.8	32	32
Optimal, Suitable, and Less Suitable Habitat	32.2	46	43.3	54	50
All Shores	44.7	49	92.7	63	58

Source: Kreeger and Titus (2008), compiling data developed by Lathrop et al. (2006).

BOX A1.3. HORSESHOE CRABS AND ESTUARINE BEACHES

The Atlantic horseshoe crab (*Limulus polyphemus)*, an ancient species that has survived virtually unchanged for more than 350 million years, enters estuaries each spring to spawn along sandy beaches. The species has experienced recent population declines, apparently due to overharvesting as well as habitat loss and degradation (Berkson and Shuster, 1999).

Population Status and Sea-Level Rise

In Delaware Bay, as elsewhere along its range, horseshoe crabs depend on narrow sandy beaches and the alluvial and sand bar deposits at the mouths of tidal creeks for essential spawning habitat. A product of wave energy, tides, shoreline configuration, and over longer periods, sea-level rise, the narrow sandy beaches utilized by horseshoe crabs are diminishing at sometimes alarming rates due to beach erosion as a product of land subsidence and sea-level increases (Nordstrom, 1989; Titus *et al.*, 1991). At Maurice Cove in Delaware Bay, for example, portions of the shoreline have eroded at a rate of 4.3 m per year between 1842 and 1992 (Weinstein and Weishar, 2002); an estimate by Chase (1979) suggests that the shoreline retreated 150 m landward in a 32-year period, exposing ancient peat deposits that are believed to be suboptimal spawning habitat (Botton *et al.*, 1988). If human infrastructure along the coast leaves estuarine beaches little or no room to transgress inland as sea level rises, concomitant loss of horseshoe crab spawning habitat is likely (Galbraith *et al.*, 2002). Kraft *et al.* (1992) estimated this loss, along with wetland "drowning", as greater than 90 percent in Delaware Bay (about 33,000 ha).

Horseshoe Crab Spawning and Shorebird Migrations

Each spring, horseshoe crab spawning coincides with the arrival of hundreds of thousands of shorebirds migrating from South America to their sub-Arctic nesting areas. While in Delaware Bay, shorebirds feed extensively on horseshoe crab eggs to increase their depleted body mass before continuing their migration (Castro and Myers, 1993; Clark, 1996). Individual birds may increase their body weight by nearly one-third before leaving the area. There is a known delicate relationship between the horseshoe crab and red knots (Baker *et al.*, 2004). How other shorebirds might be affected by horseshoe crab population decline is uncertain (Smith *et al.*, 2002).

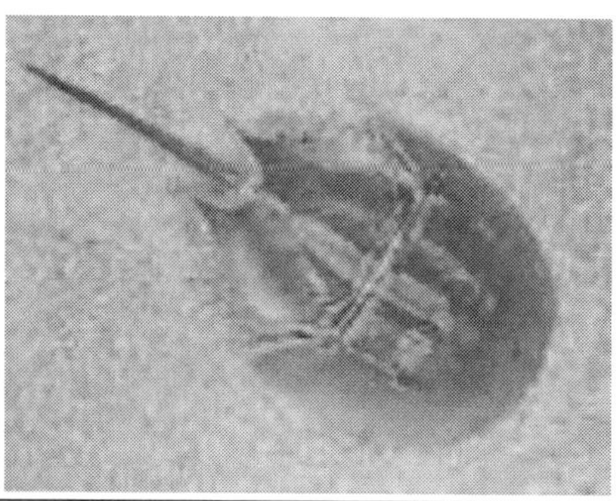

Numerous other animals, including diamondback terrapins, and Kemp's ridley sea turtles, rely on the sandy beaches of Delaware Bay to lay eggs or forage on invertebrates such as amphipods and clams. When tides are high, numerous fish also forage along the submerged sandy beaches, such as killifish, mummichog, rockfish, perch, herring, silverside, and bay anchovy (Dove and Nyman, 1995).

A1.D.2. Development, Shore Protection, and Coastal Policies

A1.D.2.1. New Jersey

Policies that may be relevant for adapting to sea-level rise in New Jersey include policies related to the Coastal Facility Review Act (CAFRA), the (coastal) Wetlands Act of 1970, the State Plan, an unusually strong public trust doctrine, and strong preference for beach nourishment along the Atlantic Ocean over hard structures or shoreline retreat. This Section discusses the first four of these policies (nourishment of ocean beaches is discussed in Section A1.C of this Appendix).

CAFRA applies to all shores along Delaware Bay and the portion of the Delaware River south of Killcohook National Wildlife Area, as well as most tidal shores along the tributaries to Delaware Bay. The act sometimes limits development in the coastal zone, primarily to reduce runoff of pollution into the state's waters (State of New Jersey, 2001). Regulations promulgated under the Wetlands Act of 1970 prohibit development in tidal wetlands unless the development is water-dependent and there is no prudent alternative (NJAC 7:7E-2.27 [c]). Regulations prohibit development of freshwater wetlands under most circumstances (NJAC 7:7E-2.27 [c]). The regulations also prohibit development within 91.4 m (300 ft of tidal wetland, unless the development has no significant adverse impact on the wetlands (NJAC 7:7-3.28 [c]). These regulations, like Maryland's Critical Areas Act (see Section A1 .E.2), may indirectly reduce the need for shore protection by ensuring that homes are set back farther from the shore than would otherwise be the case (NOAA 2007, see Section 6.2 in Chapter 6). For the same reason, existing restrictions of development in nontidal wetlands (see Section 10.3) may also enable tidal wetlands to migrate inland.

The New Jersey state plan provides a statewide vision of where growth should be encouraged, tolerated, and discouraged—but local government has the final say. In most areas, lands are divided into five planning areas. The state encourages development in (1) metropolitan and (2) suburban planning areas, and in those (3) fringe planning areas that are either already developed or part of a well-designed new development. The state discourages development in most portions of (4) rural planning areas and (5) land with valuable ecosystems, geologic features, or wildlife habitat, including coastal wetlands and barrier spits/islands (State of New Jersey, 2001). However, even these areas include developed enclaves, known as "centers" where development is recognized as a reality (State of New Jersey, 2001). The preservation of rural and natural landscapes in portions of planning areas (4) and (5) is likely to afford opportunities for wetlands to migrate inland as sea level rises. Nevertheless, New Jersey has a long history of building dikes along Delaware Bay and the Delaware River to convert tidal wetlands to agricultural lands (see Box 5.1) and dikes still protect some undeveloped lands.

BOX A1.4. THE GIBBSTOWN LEVEE, NEW JERSEY

The Gibbstown Levee along the Delaware River in New Jersey once served a function similar to the dikes in Cumberland County, preventing tidal inundation and lowering the water table to a level below mean sea level. When the dike was built 300 years ago (USACE, undated[a]), the tides were 1 m lower and the combination dike and tide gate kept the water levels low enough to permit cultivation. But rising sea level and land subsidence have left this land barely above low tide, and many lands drain too slowly to completely drain during low tide. Hence, farmland has converted to non-tidal wetland.

By keeping the creek a meter or so lower than it would be if it rose and fell with the tides, the levee improves drainage during rainstorms for Greenwich Township. Nevertheless, it is less effective today than when the sea was 50 to 100 cm lower. During extreme rainfall, the area can flood fairly easily because the tide gates have to be closed most of the day. Heavy rain during a storm surge is even more problematic because for practical purposes there is no low tide to afford the opportunity to get normal drainage by opening the tide gate. Evacuations were necessary during hurricane Floyd when part of this dike collapsed as a storm tide brought water levels of more than ten feet above mean low water (NCDC, 1999).

Officials in Greenwich Township are concerned that the dikes in Gloucester County are in danger of failing. "The Gibbstown Levee was repaired in many places in 1962 by the U.S. Army Corps of Engineers under Public Law 84-99" (USACE, 2004). Part of the problem appears to be that most of these dikes are the responsibility of meadow companies originally chartered in colonial times. These companies were authorized to create productive agricultural lands from tidal marshes. Although harvests of salt hay once yielded more than enough revenue to maintain the dikes, this type of farming became less profitable during the first half of the twentieth century. Moreover, as sea level has continued to rise, the land protected by the dikes has mostly reverted to marsh (Weinstein *et al.*, 2000; Abel *et al.*, 2000). Revenues from these lands, if any, are insufficient to cover the cost of maintaining the dikes (DiMuzio, 2006). As a result, the dikes are deteriorating, leading officials to fear a possible catastrophic dike failure during storm, or an increase in flood insurance rates (DELO, 2006). The officials hope to obtain federal funding (DELO, 2006).

Even if these dikes and their associated tide gates are fortified, the dry land will gradually be submerged unless pumping facilities are installed (see Section 6.2 in Chapter 6), because much of the area is barely above low tide even today (Titus and Wang 2008). Although freshwater marshes in general seem likely to be able to keep pace with rising sea level (Reed *et al.*, 2008), wetlands behind dikes do not always fare as well as those exposed to normal tidal currents (Reed *et al.*, 2008). Over longer periods of time, increases in salinity of the Delaware River resulting from rising sea level and reduced river flows during droughts could enable saltwater to invade these fresh marshes (Hull and Titus, 1986), which would convert them to open water ponds.

If pumping facilities are not sufficient for a daily pumping of all the very low lands protected by the dikes, the primary impact of the dikes could be to prevent flooding from storm surges and ordinary tides. For the isolated settlements along Marsh Dike Road and elsewhere, elevating homes and land surfaces may be possible; although property values are less than along the barrier islands, sources for fill material are closer. One could envision that Gibbstown, Bridgetown, and other more populated communities could be encircled with a ring dike with a pumping system that drains only the densely developed area; or they too may elevate land as the sea rises.

In Cumberland County, salt marsh has been reclaimed for agricultural purposes for more than 200 years (Sebold, 1992 and references therein). Over the last few decades, many of the dikes that were constructed have been dismantled. Some have failed during storms. Others have been purchased by conservation programs seeking to restore wetlands, most notably Public Service Enterprise Group (PSEG) in its efforts to offset possible environmental effects of a nuclear power plant. Although the trend is for dike removal, the fact that diked farms have been part of the landscape for centuries leads one to the logical inference that dikes may be used to hold back a rising sea once again. Cumberland County has relatively little coastal development, yet the trend in coastal communities that have not become part of a conservation program has been for a gradual retreat from the shore. Several small settlements along Delaware Bay are gradually being abandoned.

The state plan contemplates a substantial degree of agricultural and environmental preservation along the Delaware River and its tidal tributaries in Salem and lower Gloucester County. An agricultural easement program in Gloucester County reinforces that expectation. Farther up the river, in the industrial and commercial areas, most of the shoreline is already bulkheaded, to provide the vertical shore that facilitates docking—but the effect is also to stop coastal erosion. The eventual fate of existing dikes, which protect lightly developed areas, is unclear (Box A1.4).

The public trust doctrine in New Jersey has two unique aspects. First, the public has an easement along the dry beach between mean high water and the vegetation line. Although other states have gradually acquired these easements in most recreational communities, few states have general access along the dry beach. As a result, people are entitled to walk along river and bay beaches. The laws of Delaware and Pennsylvania, by contrast, grant less public access along the shore. In most states, the public owns the land below mean high water. In these two states, the public owns the land below mean low water. The public has an easement along the wet beach between mean low and mean high water, but only for navigation, fishing, and hunting—not for recreation (see Chapter 8 of this Product for additional details)

Second, the New Jersey Supreme Court has held that the public is entitled to perpendicular access to the beach[12]. The holding does not mean that someone can indiscriminately walk across any landowner's property to get to the water, but it does require governments to take prudent measures to ensure that public access to the water accompanies new subdivisions[13].

As trustee, the New Jersey Department of Environmental Protection has promulgated rules preserving the public trust rights to parallel and perpendicular access. The regulations divide new construction (including shore protection structures) into three classes: single family homes (or duplexes); development with two or three homes; and all other residential and nonresidential development. Along most of the tidal Delaware River, any development other than a single family home requires a public walkway at least 3 m (10 ft) wide along the shore. By contrast, along Delaware Bay, areas where one might walk along the beach rather than require a walkway, the regulations have a more general requirement for public access. (See Table A1.2).

A1.D.2.2. Delaware

Kent County does not permit subdivisions—and generally discourages most development—in the 100-year coastal floodplain, as does New Castle County south of the Chesapeake and Delaware Canal[14]. Because the 100-year floodplain for storm surge extends about 2 m above spring high water, which is often more than one kilometer inland, the floodplain regulations often require a greater setback than the erosion-hazard (see *e.g.*, A1.G.2) and environmental (*e.g.*, A1.E.2 and A1.F.2) setbacks elsewhere in the mid-Atlantic. Thus, a greater amount of land may be available for potential wetland migration (see Section 6.2 in Chapter 6). Nevertheless, if sea level continues to rise, it is logical to assume that this buffer would not last forever.

Preservation easements and land purchases have also contributed to a major conservation buffer (DDA, 2004), which would leave room for wetlands to migrate inland as sea level rises (see Chapter 6.2). The state is purchasing agricultural preservation easements in the coastal zone, and a significant portion of the shore is in Prime Hook or Bombay Hook National Wildlife Refuge. The majority of the shore south of the canal is part of some form of preservation or conservation land.

A1.D.2.3. Pennsylvania

Pennsylvania[15] is the only state in the nation along tidal water without an ocean coast[16]. As a result, the state's sensitivity to sea-level rise is different than other states. Floods in the tidal Delaware River are as likely to be caused by extreme rainfall over the watershed as storm surges. The Delaware River is usually fresh along almost all of the Pennsylvania shore. Because Philadelphia relies on freshwater intakes in the tidal river, the most important impact may be the impact of salinity increases from rising sea level on the city's water supply (Hull and Titus, 1986).

The state of Pennsylvania has no policies that directly address the issue of sea-level rise[17]. Nevertheless, the state has several coastal policies that might form the initial basis for a response to sea-level rise, including state policies on tidal wetlands and floodplains, public access, and redeveloping the shore in response to the decline of water-dependent industries.

Tidal Wetlands and Floodplains

Pennsylvania's Dam Safety and Waterway Management Rules and Regulations[18] require permits for construction in the 100-year floodplain or wetlands. The regulations do not explicitly indicate whether landowners have a right to protect property from erosion or rising water level. A permit for a bulkhead or revetment seaward of the high-water mark can be awarded only if the project will not have a "significant adverse impact" on the "aerial extent of a wetland" or on a "wetland's values and functions". A bulkhead seaward of the high-water mark, however, eliminates the tidal wetlands on the landward side. If such long-term impacts were viewed as "significant," permits for bulkheads could not be awarded except where the shore was already armored. But the state has not viewed the elimination of mudflats or beaches as "significant" for purposes of these regulations; hence it is possible to obtain a permit for a bulkhead.

Table A1.2. New Jersey Regulatory Requirements for (Parallel) Access along, and (Perpendicular) to the Shore for New Development or Shore Protection Structures Along Delaware Estuary

	Single Family[4]	Two or Three Residential Structures[5]	All other Development[6]
Designated Urban Rivers[1]	No requirement	*Along the shore*: 20-ft preservation buffer, including 10-ft wide walkway *To the shore*: 10-ft wide walkway every half mile.	*Along the Shore*: 30-ft presse-rvation buffer, including 16-ft wide walkway *To the Shore*: 20-ft wide presser-vation buffer, including 10-ft wide walkway, every half mile
Beaches along Major Bodies of Water[2]	Access along and to the bea-ch is required.	Access along and to the beach is required.	Access along and to the beach is required.
All other coas-tal areas (exce-pt Hudson River)	No requirement	Alternative access on site or nearby.	Access along the beach and shore is required.

[1] Within this region, Cohansey River within Bridgeton, Maurice River within Millville, and Delaware River from the CAFRA boundary upstream to the Trenton Makes Bridge (Trenton). Also applies to Arthur Kill, Kill Van Kull west of Bayonne Bride, Newark Bay, Elizabeth Riber, Hackwnsask River, Rahway River, and Raritan River.

[2] Delaware Bay within this region. Also Atlantic Ocean, Sandy Hook Bay, and Raritan Bay.

[3] See Section B of this Appendix for Hudson River requirements.

[4] NJAC 7:7E-8. 11 (f)(6-7)

[5] NJAC 7:7E-8. 11 (f)(4-5)

[6] NJAC 7:7E-8.11 (d-e)

The rules do not restrict construction of bulkheads or revetments landward of the high water mark. However, they do prohibit permits for any "encroachment located in, along, across or projecting into a wetland, unless the applicant affirmatively demonstrates that...the ... encroachment will not have an adverse impact on the wetland..."[19].

Therefore, shoreline armoring can eliminate coastal wetlands (or at least prevent their inland expansion[20]) as sea level rises by preventing their landward migration. Like the shore protection regulations, Pennsylvania's Chapter 105 floodplains regulations consider only existing floodplains, not the floodplains that would result as the sea rises.

Public Access

Public access for recreation is an objective of the Pennsylvania Coastal Zone Management program. This policy, coupled with ongoing redevelopment trends in Pennsylvania, may tend to ensure that future development includes access along the shore. If the public access is created by setting development back from the shore, it may tend to also make a gradual retreat possible. If keeping public access is a policy goal of the governmental authority awarding the permit for shore protection, then public access need not be eliminated, even if shores are armored (see Titus, 1998 and Table A1.2).

Development and Redevelopment

Industrial, commercial, residential, recreational, wooded, vacant, transportation, and environmental land uses all occupy portions of Pennsylvania's 100-km coast. Generally speaking, however, the Pennsylvania coastal zone is consistently and heavily developed. Only about 18 percent of the coastal area is classified as undeveloped (DVRPC, 2003). Much of the shoreline has been filled or modified with bulkheads, docks, wharfs, piers, bulkheads, revetments and other hard structures over the past two centuries (DVRPC, 2000).

The Pennsylvania coast is moving from an industrial to a post-industrial landscape. The coastal zone is still dominated by manufacturing and industrial land uses, but a steady decline in the industrial economy over the past 60 years has led to the abandonment of many industrial and manufacturing facilities. Some of these facilities sit empty and idle; others have been adapted for uses that are not water dependent.

A majority of Pennsylvania's Delaware River shore is classified as developed, but sizable expanses (especially near the water) are blighted and stressed (DVRPC, 2003). Because of the decaying industrial base, many residential areas along the Delaware River have depressed property values, declining population, high vacancy rates, physical deterioration, and high levels of poverty and crime (DVRPC, 2003). Many—perhaps most—of the refineries, chemical processing plants, and other manufacturing facilities that operate profitably today may close in the next 50 to 100 years. (DVRPC, 2003)

New paradigms of waterfront development have emerged that offer fresh visions for southeastern Pennsylvania's waterfront. In late 2001, Philadelphia released the Comprehensive Redevelopment Plan for the North Delaware Riverfront—a 25-year redevelopment vision for a distressed ten-mile stretch of waterfront led by the design firm Field Operations. Delaware County, meanwhile, developed its Coastal Zone Compendium of Waterfront Provisions (1998) to guide revitalization efforts along its coast. Likewise, Bucks County just finished a national search for a design firm to create a comprehensive plan outlining the revitalization of its waterfront. Meanwhile, the Schuylkill River Development Corporation produced the Tidal Schuylkill River Master Plan.

All of these plans and visions share common elements. They view the region's waterfronts as valuable public amenities that can be capitalized on, and they view the estuary as something for the region to embrace, not to turn its back on. They emphasize public access along the water's edge, the creation of greenways and trails, open spaces, and the restoration of natural shorelines and wetlands where appropriate (DRCC, 2006).

A1.E. The Atlantic Coast of Virginia, Maryland, and Delaware (including coastal bays)

Between Delaware and Chesapeake Bays is the land commonly known as the Delmarva Peninsula. The Atlantic coast of the Delmarva consists mostly of barrier islands separated by tidal inlets of various sizes (Theiler and Hammar-Klose, 1999; Titus *et al.*, 1985). Behind these barrier islands, shallow estuaries and tidal wetlands are found. The large area of tidal wetlands behind Virginia's barrier islands to the south are mostly mudflats; marshes and shallow open water are more common in Maryland and adjacent portions of Virginia and Delaware. The barrier islands themselves are a small portion of the low land in this region

(Titus and Richman, 2001). The northern portion of the Delaware shore consists of headlands, rather than barrier islands (see Chapter 3 of this Product).

A1.E.1. Environmental Implications

Tidal Marshes and Marsh Islands. The region's tidal marshes and marsh-fringed bay islands provide roosting, nesting, and foraging areas for a variety of bird species, both common and rare, including shorebirds (piping plover, American oystercatcher, spotted sandpiper), waterbirds (gull-billed, royal, sandwich, and least terns and black ducks) and wading birds such as herons and egrets (Conley, 2004). Particularly at low tide, the marshes provide forage for shorebirds such as sandpipers, plovers, dunlins, and sanderlings (Burger *et al.*, 1997). Ducks and geese, including Atlantic brants, buffleheads, mergansers, and goldeneyes, overwinter in the bays' marshes (DNREC, undated). The marshes also provide nesting habitat for many species of concern to federal and state agencies, including American black duck, Nelson's sparrow, salt marsh sharp- tailed sparrow, seaside sparrow, coastal plain swamp sparrow, black rail, Forster's tern, gull-billed tern, black skimmers, and American oystercatchers (Erwin *et al.*, 2006).

The marshes of the bay islands in particular are key resources for birds, due to their relative isolation and protection from predators and to the proximity to both upland and intertidal habitat. For example, hundreds of horned grebes prepare for migration at the north end of Rehoboth Bay near Thompson's Island (Ednie, undated). Several bird species of concern in this region nest on shell piles (shellrake) on marsh islands, including gull-billed terns, common terns, black skimmers, royal tern, and American oystercatchers (Erwin, 1996; Rounds *et al.*, 2004). Dredge spoil islands in particular are a favorite nesting spot for the spotted sandpiper, which has a state conservation status of vulnerable to critically imperiled in Maryland, Delaware, and Virginia (Natureserve, 2008). However, marsh islands are also subject to tidal flooding, which reduces the reproductive success of island-nesting birds (Eyler *et al.*, 1999).

Sea-level rise is considered a major threat to bird species in the Virginia Barrier Island/Lagoon Important Bird Area (IBA) (Watts, 2006). Biologists at the Patuxent Wildlife Research Center suggest that submergence of lagoonal marshes in Virginia would have a major negative effect on marsh-nesting birds such as black rails, seaside sparrows, saltmarsh sharp-tailed sparrows, clapper rails, and Forster's terns (Erwin *et al.*, 2004). The U.S. Fish and Wildlife Service considers black rail and both sparrow species "birds of conservation concern" because populations are already declining in much of their range (USFWS, 2002). The number of bird species in Virginia marshes was found to be directly related to marsh size; the minimum marsh size found to support significant marsh bird communities was 4.1 to 6.7 ha (10 to 15 ac) (Watts, 1993).

The region's tidal marshes also support a diversity of resident and transient estuarine and marine fish and shellfish species that move in and out of marshes with the tides to take advantage of the abundance of decomposing plants in the marsh, the availability of invertebrate prey, and refuge from predators (Boesch and Turner, 1984; Kneib, 1997). Marine transients include recreationally and commercially important species that depend on the marshes for spawning and nursery habitat, including black drum, striped bass, bluefish, Atlantic croaker, sea trout, and summer flounder. Important forage fish that spawn in local marsh areas include spot, menhaden, silver perch, and bay anchovy. Shellfish species found in the marshes include clams, oysters, shrimps, ribbed mussels, and blue crabs (Casey and Doctor, 2004).

Salt Marsh Adaptation to Sea-level Rise. Salt marshes occupy thousands of acres in eastern Accomack and Northampton counties (Fleming *et al.*, 2006). Marsh accretion experts believe that most of these marshes are keeping pace with current rates of sea-level rise, but are unlikely to continue to do so if the rate of sea-level rise increases by another 2 mm per year (Strange 2008c, interpreting the findings of Reed *et al.*, 2008). However, some very localized field measurements indicate that accretion rates may be insufficient to keep pace even with current rates of sea-level rise (Strange, 2008d). For instance, accretion rates as low as 0.9 mm per year (Phillips Creek Marsh) and as high as 2.1 mm per year (Chimney Pole Marsh) have been reported (Kastler and Wiberg, 1996), and the average relative sea-level rise along the Eastern Shore is estimated as 2.8 to 4.2 mm per year (May, 2002).

In some areas, marshes may be able to migrate onto adjoining dry lands. For instance, lands in Worcester County that are held for the preservation of the coastal environment might allow for wetland migration. Portions of eastern Accomack County that are opposite the barrier islands and lagoonal marshes owned by The Nature Conservancy are lightly developed today, and in some cases already converting to marsh. In unprotected areas, marshes may be able to migrate inland in low-lying areas. From 1938 to 1990 mainland salt marshes on the Eastern Shore increased in area by 8.2 percent, largely as a result of encroachment of salt marsh into upland areas (Kastler and Wiberg, 1996).

The marsh islands of the coastal bays are undergoing rapid erosion; for example, Big Piney Island in Rehoboth Bay experienced erosion rates of 10 m (30 ft) per year between 1968 and 1981, and is now gone (Swisher, 1982; Strange *et al.*, 2008). Seal Island in Little Assawoman Bay is eroding rapidly after being nearly totally devegetated by greater snow geese (Strange, 2008c). Island shrinking is also apparent along the Accomack County, Virginia shore; from 1949 to 1990, Chimney Pole marsh showed a 10 percent loss to open water (Kastler and Wiberg, 1996). The United States Army Corps of Engineers (USACE) has created many small dredge spoil islands in the region, many of which are also disappearing as a result of erosion (Federal Register, 2006).

Sea-Level Fens. The rare sea-level fen vegetation community is found in a few locations along the coastal bays, including the Angola Neck Natural Area along Rehoboth Bay in Delaware and the Mutton Hunk Fen Natural Area Preserve fronting Gargathy Bay in eastern Accomack County (VA DCR, undated [a][b]). The Division of Natural Heritage within the Virginia Department of Conservation and Recreation believes that chronic sea- level rise with intrusions of tidal flooding and salinity poses "a serious threat to the longterm viability" of sea-level fens (VA DCR, 2001).

Shallow Waters and Submerged Aquatic Vegetation (SA V). Eelgrass beds are essential habitat for summer flounder, bay scallop, and blue crab, all of which support substantial recreational and commercial fisheries in the coastal bays (MCBP, 1999). Various waterbirds feed on eelgrass beds, including brant, canvasback duck, and American black duck (Perry and Deller, 1996). Shallow water areas of the coastal bays that can maintain higher salinities also feature beds of hard and surf clams (DNREC, 2001).

Tidal Flats. Abundant tidal flats in this region provide a rich invertebrate food source for a number of bird species, including whimbrels, dowitchers, dunlins, black-bellied plovers, and semi-palmated sandpipers (Watts and Truitt, 2002). Loss of these flats could have significant impacts. For example, 80 percent of the Northern Hemisphere's whimbrel population feeds on area flats, in large part on fiddler crabs (TNC, 2006). The whimbrel is considered a species "of conservation concern" by the United States Fish and Wildlife Service, Division of Migratory Bird Management (USFWS, 2002).

Beaches. Loss of beach habitat due to sea-level rise and erosion below protective structures could have a number of negative consequences for species that use these beaches:

- Horseshoe crabs rarely spawn unless sand is at least deep enough to nearly cover their bodies, about 10 cm (4 inches [in]) (Weber, 2001). Shoreline protection structures designed to slow beach loss can also block horseshoe crab access to beaches and can entrap or strand spawning crabs when wave energy is high (Doctor and Wazniak, 2005).
- The rare northeastern tiger beetle depends on beach habitat (USFWS, 2004b).
- *Photuris bethaniensis* is a globally rare firefly located only in interdunal swales on Delaware barrier beaches (DNREC, 2001).
- Erosion and inundation may reduce or eliminate beach wrack communities of the upper beach, especially in developed areas where shores are protected (Strange, 2008c). Beach wrack contains insects and crustaceans that provide food for many species, including migrating shorebirds (Dugan *et al.*, 2003).
- Many rare beach-nesting birds, such as piping plover, least tern, common tern, black skimmer, and American oystercatcher, nest on the beaches of the coastal bays (DNREC, 2001)

Coastal Habitat for Migrating Neotropical Songbirds. Southern Northampton County is one of the most important bird areas along the Atlantic Coast of North America for migrating neotropical songbirds such as indigo buntings and ruby-throated hummingbirds (Watts, 2006). Not only are these birds valued for their beauty but they also serve important functions in dispersing seeds and controlling insect pests. It is estimated that a pair of warblers can consume thousands of insects as they raise a brood (Mabey *et al.*, undated). Migrating birds concentrate within the tree canopy and thick understory vegetation found within the lower 10 km (6 mi) of the peninsula within 200 m (650 ft) of the shoreline. Loss of this understory vegetation as a result of rising seas would eliminate this critical stopover area for neotropical migrants, many of which have shown consistent population declines since the early 1 970s (Mabey *et al.*, undated)

A1.E.2. Development, Shore Protection, and Coastal Policies

A1.E.2.1. Atlantic Coast

Less than one-fifth of the Delmarva's ocean coast is developed, and the remaining lands are owned by private conservation organizations or government agencies. Almost all of the Virginia Eastern Shore's 124-km (77-mi) ocean coast is owned by the U.S. Fish and Wildlife

Service, NASA, the State, or The Nature Conservancy[21]. Of Maryland's 51 km (32 mi) of ocean coast, 36 km (22 mi) are along Assateague Island National Seashore. The densely populated Ocean City occupies approximately 15 km (9 mi). More than three-quarters of the barrier islands and spits in Delaware are part of Delaware Seashore State Park, while the mainland coast is about evenly divided between Cape Henlopen State Park and resort towns such as Rehoboth, Dewey Beach, and Bethany Beach. With approximately 15 km of developed ocean coast each, Maryland and Delaware have pursued beach nourishment to protect valuable coastal property and preserve the beaches that make the property so valuable (Hedrick *et al.*, 2000).

With development accounting for only 15 to 20 percent of the ocean coast, the natural shoreline processes are likely to dominate along most of these shores. Within developed areas, counteracting shoreline erosion in developed areas with beach nourishment may continue as the primary activity in the near term. A successful alternative to beach nourishment, as demonstrated by a USACE (2001a) and National Park Service project to mitigate jetty impacts along Assateague Island, is to restore sediment transport rates by mechanically bypassing sand from the inlet and tidal deltas into the shallow nearshore areas that have been starved of their natural sand supply. Beginning in 1990, the USACE and the Assateague Island National Seashore partnered to develop a comprehensive restoration plan for the northern end of Assateague Island. The "North End Restoration Project" included two phases. The first phase, completed in 2002, provided a one-time placement of sand to replace a portion of sand lost over the past 60 years due to the formation of the inlet and subsequent jetty stabilization efforts. The second phase is focused on re-establishing a natural sediment supply by mechanically bypassing sand from the inlet and tidal deltas into the shallow nearshore areas[22].

A1.E 2.2. Coastal Bay Shores

The mainland along the back-barrier bays has been developed to a greater extent than the respective ocean coast in all three states (MRLCC, 2002; MDP, 2001; DOSP, 1997). Along the coastal bays, market forces have led to extensive development at the northern end of the Delmarva due to the relatively close proximity to Washington, Baltimore, and Philadelphia. Although connected to the densely populated Hampton Roads area by the Chesapeake Bay Bridge-Tunnel, southern portions of the Delmarva are not as developed as the shoreline to the north. Worcester County, Maryland, reflects a balance between development and environmental protection resulting from both recognition of existing market forces and a conscious decision to preserve Chincoteague Bay. Development is extensive along most shores opposite Ocean City and along the bay shores near Ocean City Inlet. In the southern portion of the county, conservation easements or the Critical Areas Act preclude development along most of the shore. Although the Critical Areas Act encourages shore protection, and conservation easements in Maryland preserve the right to armor the shore (MET, 2006), these low-lying lands are more vulnerable to inundation than erosion (*e.g.*, Titus *et al.*, 1991) and are therefore possible candidates for wetland migration.

Of the three states, Maryland has the most stringent policies governing development along coastal bays Under the Chesapeake and Atlantic Coastal Bays Critical Areas Protection Program, new development must be set back at least 100 ft from tidal wetlands or open water[23]. In most undeveloped areas, the statute also limits future development density to one home per 20 ac within 305 m (1000 ft) of the shore[24] and requires a 61-m (200-ft) setback.[25]

In Virginia, new development must be set back at least 30 m (100 ft). (see Section A1.F.2 in this Appendix for additional discussion of the Maryland and Virginia policies.) The Delaware Department of Natural Resources has proposed a 30-m (100-ft) setback along the coastal bays (DNREC, 2007); Sussex County currently requires a 1 5-m (50-ft) setback[26].

While shore protection is currently more of a priority along the Atlantic Coast, preventing the inundation of low-lying lands may eventually be necessary as well. Elevating these low areas appears to be more practical than erecting a dike around a narrow barrier island (Titus, 1990). Most land surfaces on the bayside of Ocean City were elevated during the initial construction of residences (McGean, 2003). In an appendix for U.S. EPA's 1989 Report to Congress, Leatherman (1989) concluded that the only portion of Fenwick Island where bayside property would have to be elevated with a 50 cm rise in sea level would be the portion in Delaware (*i.e.*, outside of Ocean City). He also concluded that Wallops Island, South Bethany, Bethany, and Rehoboth Beach are high enough to avoid tidal inundation for the first 50 to 100 cm of sea-level rise. The Town of Ocean City has begun to consider how to respond to address some of the logistical problems of elevating a densely developed barrier island (see Box A1.5).

The Maryland Coastal Bays Program considers erosion (due to sea-level rise) and shoreline hardening major factors contributing to a decline in natural shoreline habitat available for estuarine species in the northern bays (MCBP, 1999). Much of the shoreline of Maryland's northern coastal bays is protected using bulkheads or stone riprap, resulting in unstable sediments and loss of wetlands and shallow water habitat (MCBP, 1999). Armoring these shorelines will prevent inland migration of marshes, and any remaining fringing marshes will ultimately be lost (Strange 2008c). The Coastal Bays Program estimated that more than 600 ha (1,500 ac) of salt marshes have already been lost in the coastal bays as a result of shoreline development and stabilization techniques (MCBP, 1999). If shores in the southern part of Maryland's coastal bays remain unprotected, marshes in low-lying areas would be allowed to potentially (see Chapter 4) expand inland as sea level rises (Strange 2008c).

Box A1.5. Elevating Ocean City as Sea Level Rises

Logistically, the easiest time to elevate low land is when it is still vacant, or during a coordinated rebuilding. Low parts of Ocean City's bay side were elevated during the initial construction. As sea level rises, the town of Ocean City has started thinking about how it might ultimately elevate.

Ocean City's relatively high bay sides make it much less vulnerable to inundation by spring tides than other barrier islands. Still, some streets are below the 10-year flood plain, and as sea level rises, flooding will become increasingly frequent.

However, the town cannot elevate the lowest streets without considering the implications for adjacent properties. A town ordinance requires property owners to maintain a 2 percent grade so that yards drain into the street. The town construes this rule as imposing a reciprocal responsibility on the town itself to not elevate roadways above the level where yards can drain, even if the road is low enough to flood during minor tidal surges. Thus, the lowest lot in a given area dictates how high the street can be.

As sea level rises, failure by a single property owner to elevate could prevent the town from elevating its streets, unless it changes this rule. Yet public health reasons require drainage, to prevent standing water in which mosquitoes breed. Therefore, the town has an interest in ensuring that all property owners gradually elevate their yards so that the streets can be elevated as the sea rises without causing public health problems.

The Town of Ocean City (2003) has developed draft rules that would require that, during any significant construction, yards be elevated enough to drain during a 10-year storm surge for the life of the project, considering projections of future sea-level rise. The draft rules also state that Ocean City's policy is for all lands to gradually be elevated as the sea rises.

A1.F. Chesapeake Bay

The Chesapeake Bay region accounts for more than one-third of the lowland in the Mid-Atlantic (see Titus and Richman 2001). Accordingly, the first subsection (A1.F.1) on vulnerable habitat, development, and shore protection) divides the region into seven subregions. Starting with Hampton Roads, the subsections proceed clockwise around the Bay to Virginia's Middle Peninsula and Northern Neck, then up the Potomac River to Washington, D.C., then up Maryland's Western Shore, around to the Upper Eastern Shore, and finally down to the Lower Eastern Shore. The discussions for Virginia are largely organized by planning district; the Maryland discussions are organized by major section of shore. The second subsection compares the coastal policies of Maryland and Virginia that are most relevant to how these states respond to rising sea level[27].

A1.F.1. Inundation, Development and Shore Protection, and Vulnerable Habitat

A1.F.1.1. Hampton Roads

Most of the vulnerable dry land in the Hampton Roads region is located within Virginia Beach and Chesapeake. These low areas are not, however, in the urban portions of those jurisdictions. Most of Virginia Beach's very low land is either along the back-barrier bays near the North Carolina border, or along the North Landing River. Most of Chesapeake's low land is around the Northwest River near the North Carolina border, or the along the Intracoastal Waterway. The localities located farther up the James and York Rivers have less low land. An important exception is historic Jamestown Island, which has been gradually submerged by the rising tides since the colony was established 400 years ago (see Box 10.1 in chapter 10).

Development and Shore Protection

Norfolk is home to the central business district of the Hampton Roads region. Newport News has similar development to Norfolk along its southern shores, with bluffs giving rise to less dense residential areas further north along the coast. The city of Hampton is also highly developed, but overall has a much smaller percentage of commercial and industrial development than Norfolk or Newport News.

Outside of the urban core, localities are more rural in nature. These localities find themselves facing mounting development pressures and their comprehensive plans outline how they plan to respond to these pressures (*e.g.*, Suffolk, 1998; York County, 1999; James City County, 2003; Isle of Wight County, 2001). Overall, however, the makeup of these outlying localities is a mix of urban and rural development, with historic towns and residential development dotting the landscape.

Virginia Beach has sandy shores along both the Atlantic Ocean and the mouth of Chesapeake Bay. Dunes dominate the bay shore, but much of the developed ocean shore is protected by a seawall, and periodic beach nourishment has occurred since the mid-1950s (Hardaway *et al.*, 2005). Along Chesapeake Bay, by contrast, the Virginia Beach shore has substantial dunes, with homes set well back from the shore in some areas. Although the ground is relatively high, beach nourishment has been required on the bay beaches at Ocean Park (Hardaway *et al.*, 2005). Norfolk has maintained its beaches along Chesapeake Bay mostly with breakwaters and groins. Shores along other bodies of water are being armored. Of Norfolk's 269 km (167 mi) of shoreline, 113 km (70 mi) have been hardened (Berman *et al.*, 2000).

Overall trends in the last century show the dunes east of the Lynn Haven inlet advancing into the Bay (Shellenbarger Jones and Bosch, 2008c). West from the inlet, erosion, beach nourishment, and fill operations as well as condominium development and shoreline armoring have affected the accretion and erosion patterns (Hardaway *et al.*, 2005). Along the shores of Norfolk, the rate of erosion is generally low, and beach accretion occurs along much of the shore (Berman *et al.*, 2000). Most of the shore along Chesapeake Bay is protected by groins and breakwaters, and hence relatively stable (Hardaway *et al.*, 2005). On the other side of the James River, the bay shoreline is dominated by marshes, many of which are eroding (Shellenbarger Jones and Bosch, 2008c).

Since 1979, Virginia Beach has had a "Green Line", south of which the city tries to maintain the rural agricultural way of life. Because development has continued, Virginia Beach has also established a "Rural Area Line," which coincides with the Green Line in the eastern part of the city and runs 5 km (3 mi) south of it in the western portion. Below the Rural Area Line, the city strongly discourages development and encourages rural legacy and conservation easements (VBCP, 2003). In effect, the city's plan to preserve rural areas will also serve to preserve the coastal environment as sea level rises throughout the coming century and beyond (see Sections 6.1.3, 6.2, 10.3). To the west, by contrast, the City of Chesapeake is encouraging development in the rural areas, particularly along major corridors. Comprehensive plans in the more rural counties such as Isle of Wight and James City tend to focus less on preserving open space and more on encouraging growth in designated areas (Isle of Wight, 2001; James City County, 2003). Therefore, these more remote areas may present the best opportunity for long-range planning to minimize coastal hazards and preserve the ability of ecosystems to migrate inland.

Vulnerable Habitat

Much of the tidal wetlands in the area are within Poquoson's Plum Tree Island National Wildlife Refuge. Unlike most mid-Atlantic wetlands, these wetlands are unlikely to keep pace with the current rate of sea-level rise (Shellenbarger Jones and Bosch, 2008c, interpreting the findings of Reed *et al.*, 2008). The relative isolation of the area has made it a haven for over 100 different species of birds. The refuge has substantial forested dune hummocks (CPCP,

1999), and a variety of mammals use the higher ground of the refuge. Endangered sea turtles, primarily the loggerhead, use the near shore waters. Oyster, clams, and blue crabs inhabit the shallow waters and mudflats, and striped bass, mullet, spot, and white perch have been found in the near shore waters and marsh (USFWS, undated[b]).

The wetlands in York County appear able to keep pace with the current rate of sea-level rise. Assuming that they are typical of most wetlands on the western side of Chesapeake Bay, they are likely to become marginal with a modest acceleration and be lost if sea- level rise accelerates to 1 cm per year (Shellenbarger Jones and Bosch, 2008c, interpreting the findings of Reed *et al.*, 2008). Bald eagles currently nest in the Goodwin Islands National Estuarine Research Reserve (Watts and Markham, 2003). This reserve includes intertidal flats, 100 ha (300 ac) of eelgrass and widgeon grass (VIMS, undated), and salt marshes dominated by salt marsh cordgrass and salt meadow hay.

A1.F.1.2. York River to Potomac River

Two planning districts lie between the York and Potomac rivers. The Middle Peninsula Planning District includes the land between the York and Rappahannock rivers. The Northern Neck is between the Rappahannock and Potomac rivers.

Development and Shore Protection

A large portion of the necks along Mobjack Bay has a conservation zoning that allows only low-density residential development "in a manner which protects natural resources in a sensitive environment[28.] The intent is to preserve contiguous open spaces and protect the surrounding wetlands[29]. The county also seeks to maintain coastal ecosystems important for crabbing and fishing. As a result, existing land use would not prevent wetlands and beaches along Mobjack Bay from migrating inland as sea level rises.

Gloucester County also has suburban country side zoning, which allows for low density residential development, including clustered sub-developments[30] along part of the Guinea Neck and along the York River between Carter Creek and the Catlett islands. These developments often leave some open space that might convert to wetlands as sea level rises even if the development itself is protected. The county plan anticipates development along most of the York River. Nevertheless, a number of areas are off limits to development. For example, the Catlett islands are part of the Chesapeake Bay National Estuarine Research Reserve in Virginia, managed as a conservation area[31].

Along the Northern Neck, shoreline armoring is already very common, especially along Chesapeake Bay and the Rappahannock Rivers shores of Lancaster County. Above Lancaster County, however, development is relatively sparse along the Rappahannock River and shoreline armoring is not common. Development and shoreline armoring are proceeding along the Potomac River.

Vulnerable Habitat

Like the marshes of Poquoson to the south, the marshes of the Guinea Neck and adjacent islands are not keeping pace with the current rates of sea-level rise (Shellenbarger Jones and Bosch, 2008a, interpreting the findings of Reed *et al.*, 2008). For more than three decades, scientists have documented their migration onto farms and forests (Moore, 1976). Thus, the continued survival of these marshes depends on land-use and shore protection decisions.

Upstream from the Guinea Neck, sea-level rise is evident in the York River's tributaries, not because wetlands are converting to open water but because the composition of wetlands is changing. Along the Pamunkey and Mattaponi Rivers, dead trees reveal that tidal hardwood swamps are converting to brackish or freshwater marsh as the water level rises (Rheinhardt, 2007). Tidal hardwood swamps provide nesting sites for piscivorous (fish eating) species such as ospreys, bald eagles, and double-crested cormorants (Robbins and Blom, 1996).

In Mathews County, Bethel Beach (a natural area preserve separating Winter Harbor from Chesapeake Bay) is currently migrating inland over an extensive salt marsh area (Shellenbarger Jones and Bosch, 2008a). The beach is currently undergoing high erosion (Berman *et al.*, 2000), and is home to a population of the Northeastern beach tiger beetle (federally listed as threatened) and a nesting site for rare least terns, which scour shallow nests in the sand (VA DCR, 1999). In the overwash zone extending toward the marsh, a rare plant is present, the sea-beach knotweed (*Polygonum glaucum*) (VA DCR, 1999). The marsh is also one of few Chesapeake Bay nesting sites for northern harriers (*Circus cyaneus*), a hawk that is more commonly found in regions further north (VA DCR, 1999). As long as the shore is able to migrate, these habitats will remain intact; but eventually, overwash and inundation of the marsh could reduce habitat populations (Shellenbarger Jones and Bosch, 2008a).

A1.F.1.3. The Potomac River

Virginia Side. Many coastal homes are along bluffs, some of which are eroding (Bernd-Cohen and Gordon, 1999). Lewisetta is one of the larger vulnerable communities along the Potomac. Water in some ditches rise and fall with the tides, and some areas drain through tide gates. With a fairly modest rise in sea level, one could predict that wetlands may begin to take over portions of people's yards, the tide gates could close more often, and flooding could become more frequent. Somewhat higher in elevation than Lewisetta, Old Town Alexandria and Belle Haven (Fairfax County) both flood occasionally from high levels in the Potomac River.

Maryland Side. Much of the low-lying land is concentrated around St. George Island and Piney Point in St. Mary's County, and along the Wicomico River and along Neal Sound opposite Cobb Island in Charles County. Relatively steep bluffs, however, are also common.

Development and Shore Protection

West of Chesapeake Bay, the southwestern shoreline of the Potomac River is the border between Maryland and Virginia[32]. As a result, islands in the Potomac River, no matter how close they are to the Virginia side of the river, are part of Maryland or the District of Columbia. Moreover, most efforts to control erosion along the Virginia shore take place partly in Maryland (or the District of Columbia) and thus could potentially be subject to Maryland (or Washington, D.C.) policies[33].

Development is proceeding along approximately two-thirds of the Potomac River shore. Nevertheless, most shores in Charles County, Maryland are in the resource conservation area defined by the state's Critical Areas Act (and hence limited to one home per 20 ac) (MD DNR, 2007). A significant portion of Prince George's County's shoreline along the Potomac and its tributaries are owned by the National Park Service and other conservation entities that seek to preserve the coastal environment (MD DNR, 2000).

In Virginia, parks also account for a significant portion of the shore (ESRI, 1999). In King George County, several developers have set development back from low-lying marsh areas, which avoids problems associated with flooding and poor drainage. Water and sewer regulations that only apply for lot sizes less than 10 acres may provide an incentive for larger lot sizes. In Stafford County, the CSX railroad line follows the river for several miles, and is set back to allow shores to erode, but not so far back as to allow for development between the railroad and the shore (ADC, 2008a).

Vulnerable Habitat

The Lower Potomac River includes a diverse mix of land uses and habitat types. *Freshwater tidal marshes* in the Lower Potomac are found in the upper reaches of tidal tributaries. In general, freshwater tidal marshes in the Lower Potomac are keeping pace with sea-level rise through sediment and peat accumulation, and are likely to continue to do so, even under higher sea-level rise scenarios (Strange and Shellenbarger Jones, 2008a, interpreting the findings of Reed *et al.*, 2008).

Brackish tidal marshes are a major feature of the downstream portions of the region's rivers. In general, these marshes are keeping pace with sea-level rise today, but are likely to be marginal if sea level rise accelerates by 2 mm per year, and be lost if sea level accelerates 7 mm per year (Strange and Shellenbarger Jones 2008a, interpreting the findings of Reed *et al.*, 2008). Loss of brackish tidal marshes would eliminate nesting, foraging, roosting, and stopover areas for migrating birds (Strange and Shellenbarger Jones, 2008a). Significant concentrations of migrating waterfowl forage and overwinter in these marshes in fall and winter. Rails, coots, and migrant shorebirds are transient species that feed on fish and invertebrates in and around the marshes and tidal creeks. (Strange and Shellenbarger Jones, 2008a). The rich food resources of the tidal marshes also support rare bird species such as bald eagle and northern harrier (White, 1989).

Unnourished *beaches and tidal flats* of the Lower Potomac are likely to erode as sea levels rise. Impacts on beaches are highly dependent on the nature of shoreline protection measures selected for a specific area. For example, the developed areas of Wicomico Beach and Cobb Island are at the mouth of the Wicomico River in Maryland. Assuming that the shores of Cobb Island continue to be protected, sea-level rise is likely to eliminate most of the island's remaining beaches and tidal flats (Strange and Shellenbarger Jones, 2008a).

Finally, where the *cliffs and bluffs* along the Lower Potomac are not protected (*e.g.*, Westmoreland State Park, Caledon Natural Area), natural erosional processes will generally continue, helping to maintain the beaches below (Strange and Shellenbarger Jones, 2008a).

Above Indian Head, the Potomac River is fresh. Tidal wetlands are likely to generally keep pace with rising sea level in these areas (see Chapter 4 of this Product). Nevertheless, the Dyke Marsh Preserve faces an uncertain future. Its freshwater tidal marsh and adjacent mud flats are one of the last major remnants of the freshwater tidal marshes of the Upper Potomac River (Johnston, 2000). A recent survey found 62 species of fish, nine species of amphibians, seven species of turtles, two species of lizards, three species of snakes, 34 species of mammals, and 76 species of birds in Dyke Marsh (Engelhardt *et al.*, 2005). Many of the fish species present (*e.g.*, striped bass, American shad, yellow perch, blueback herring) are important for commercial and recreational fisheries in the area (Mangold *et al.*, 2004).

Parklands on the Mason Neck Peninsula are managed for conservation, but shoreline protection on adjacent lands may result in marsh loss and reduced abundance of key bird species (Strange and Shellenbarger Jones, 2008b). The Mason Neck National Wildlife Refuge hosts seven nesting bald eagle pairs and up to 100 bald eagles during winter, has one of the largest great blue heron colonies in Virginia, provides nesting areas for hawks and waterfowl, and is a stopover for migratory birds.

A1.F.1.4. District of Columbia

Within the downtown area, most of the lowest land is the area filled during the 1 870s, such as Hains Point and the location of the former Tiber and James Creeks, as well as the Washington City Canal that joined them (See Box 5.2 in Chapter 5). The largest low area is the former Naval Air Station, now part of Bolling Air Force Base, just south of the mouth of the Anacostia River, which was part of the mouth of the Anacostia River during colonial times. A dike protects this area, where most of the low land between Interstate- 295 and the Anacostia River was open water when the city of Washington was originally planned.

Development and Shore Protection

The central city is not likely to be given up to rising sea level; city officials are currently discussing the flood control infrastructure necessary to avoid portions of the downtown area from being classified as part of the 100-year floodplain. Nevertheless, natural areas in the city account for a substantial portion of the city's shore, such as Roosevelt Island and the shores of the Potomac River within C&O Canal National Historic Park.

As part of the city's efforts to restore the Anacostia River, District officials have proposed a series of environmental protection buffers along the Anacostia River with widths between 15 and 90 m (50 and 300 ft). Bulkheads are being removed except where they are needed for navigation, in favor of natural shores in the upper part of the river and bioengineered "living shorelines" in the lower portion (DCOP, 2003).

Vulnerable Habitat

The Washington, D.C. area features sensitive wetland habitats potentially vulnerable to sea-level rise. Several major areas are managed for conservation or are the target of restoration efforts, making ultimate impacts uncertain. The wetlands around the Anacostia River are an example. Local organizations have been working to reverse historical modifications and restore some of the wetlands around several heavily altered lakes. Restoration of the 13-ha (32-ac) Kenilworth Marsh was completed in 1993; restoration of the Kingman Lake marshes began in 2000 (USGS, undated). Monitoring of the restored habitats demonstrates that these marshes can be very productive. A recent survey identified 177 bird species in the marshes, including shorebirds, gulls, terns, passerines, and raptors as well as marsh nesting species such as marsh wren and swamp sparrow (Paul *et al.*, 2004).

Roosevelt Island is another area where sea-level rise effects are uncertain. Fish in the Roosevelt Island marsh provide food for herons, egrets, and other marsh birds (NPS, undated). The ability of the tidal marshes of the island to keep pace with sea-level rise will depend on the supply of sediment, and increased inundation of the swamp forest could result in crown dieback and tree mortality (Fleming *et al.*, 2006).

A1.F.1.5. Western Shore: Potomac River to Susquehanna River

The Western Shore counties have relatively little low land, unlike the low counties across the Bay. The Deal/Shady Side peninsula (Anne Arundel County) and Aberdeen Proving Grounds (Harford County) are the only areas with substantial amounts of low-lying land. The block closest to the water, however, is similarly low in many of the older communities, including parts of Baltimore County, Fells Point in Baltimore (see Box A1.6), downtown Annapolis, North Beach, and Chesapeake Beach, all of which flooded during Hurricane Isabel.

Between the Potomac and the Patuxent Rivers, the bay shore is usually a sandy beach in front of a bank less than 3 m (10 ft) high. Cliffs and bluffs up to 35 m (115 ft) above the water dominate the shores of Calvert County (Shellenbarger Jones and Bosch, 2008b). The shores north of Calvert County tend to be beaches; but these beaches become narrower as one proceeds north, where the wave climate is milder.

Box A1.6. Planning for Sea-Level Rise in Baltimore

Only 3.2 percent of the City of Baltimore's 210 sq km (81 sq mi) of land is currently within the coastal floodplain. This land, however, includes popular tourist destinations such as Inner Harbor and the Fells Point Historic District, as well as industrial areas, some of which are being redeveloped into mixed use developments with residential, commercial, and retail land uses. The map below depicts the areas that the city expects to be flooded by category 1, 2, 3 and 4 hurricanes, which roughly correspond to water levels of 1.78 m (6 ft), 3.0 m (10 ft), 4.2 m (14 ft), and 5.5 m (18 ft) above North Amercan Vertical Datum (NAVD88). Approximately 250 homes are vulnerable to a category 1, while 700 homes could be flooded by a category 2 hurricane (Baltimore, 2006). As Hurricane Isabel passed in September 2003, water levels in Baltimore Harbor generally reached approximately 2.4 m (8 ft) above NAVD, flooding streets and basements, but resulting in only 16 flood insurance claims (Baltimore, 2006).

The city's All Hazards Plan explicitly includes rising sea level as one of the factors to be considered in land use and infrastructure planning[34]. The All Hazards Plan has as an objective to "develop up-to-date research about hazards" and a strategy under that objective to "study the threat, possible mitigation and policy changes for sea-level rise." As a first step toward accurate mapping of possible sea-level rise scenarios, the city is exploring options for acquiring lidar. Policies developed for floodplain management foreshadow the broad methods the city is likely to use in its response.

Property values are high, and there is a long-standing practice of armoring shores to facilitate port-related activities and more recently, protect waterfront structures from shore erosion. In most areas, there is not enough room between the harbor and waterfront buildings to fit a dike. Even where there is room, the loss of waterfront views would be unacceptable in tourist and residential areas (see Section 6.5 in Chapter 6; Titus, 1990). In addition, storm sewers, which drain by gravity into the harbor, would have to be fit with pumping systems.

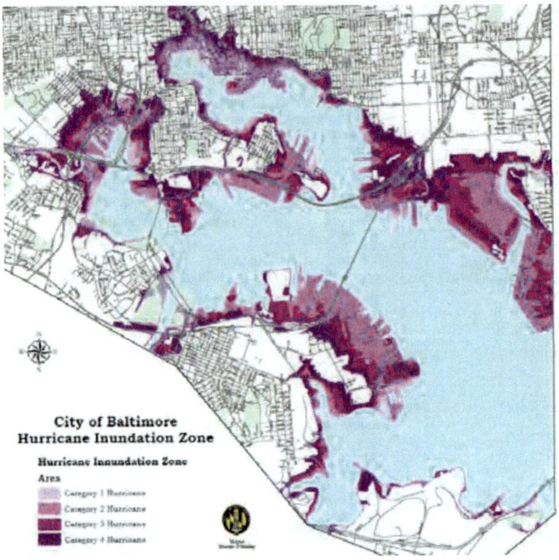

Box Figure A1.6. Inundation Zone for Baltimore Harbor under category 1,2,3, and 4 hurricanes.

Fells Point Historic District This historic community has 24 ha (60 ac) within the 100-year flood plain. Fells Point is a Federal Historic District and pending approval as a Local Historic District. The row houses here were built predominantly in the early to mid-nineteenth century and cannot be easily elevated. Elevating brick and stone structures is always more difficult than elevating a wood frame structure. But because row houses are, by definition, attached to each other, elevating them one at a time is not feasible.

Many of these homes have basements, which already flood. FEMA regulations do not permit basements in new construction in the floodplain and treats (44 CFR §60.3[c] [2]) existing basements as requiring mitigation. Possible mitigation for basements includes relocation of utilities, reinforcement of walls, and filling.

In theory, homes could be remodeled to add stairways and doors to convert what is now the second floor to a first floor and convert the first floors to basements. But doing so would reduce the livable space. Moreover, federal and local preservation laws, as well as community sensibilities, preclude adding third stories to these homes. Elevating streets is also problematic because below-grade utilities need to be elevated. In the last decade only one street (one block of Caroline Street) has been elevated specifically to reduce flooding.

FEMA Flood Hazard Mapping and Sea-level Rise

Baltimore City is a participating jurisdiction in the National Flood Insurance Program through its regulation of development in the floodplain and through overall floodplain management. The city is currently funded through the Cooperative Technical Partnership (CTP) to update its flood maps. Federal flood mapping policies require that Flood Insurance Rate Maps be based on existing conditions (see Section 10.7.5.3 in Chapter 10). Therefore, the floodplain maps do not consider future sea level rise. As a result, the city will be permitting new structures with effective functional lifespan of 50 to 100 years but elevated only to current flood elevations. One strategy to surmount this limitation is to add "freeboard," or additional elevation to the effective BFE. Baltimore already requires one additional foot of freeboard.

The City of Baltimore is concerned, however, that 0.3 to 0.6 additional meters of freeboard is inequitable and inefficient. If flood levels will be, for example, 1 meter higher than the flood maps currently assume, then lands just outside the current flood boundary are also potentially vulnerable. If the city were to add 1 meter of freeboard to property in the floodplain, without addressing adjacent properties outside the floodplain, then adjacent property owners would have divergent requirements that city officials would find difficult to justify. (see Figure 10.6.)

Infrastructure

Baltimore has two regional sewerage plants. One of them, the Patapsco Wastewater Treatment Plant, sits on ground that is less than two meters above mean sea level and floods occasionally (see Box Figure A1 .6). The facility itself is elevated and currently drains by gravity into the Patapsco River (USGS 7.5-minute map series). With a significant rise in sea level, however, pumping will be needed and possibly additional protections against storms (Smith, 1998: Titus *et al.*, 1987, . Numerous streets, with associated conduits and utility piping, are within the existing coatal floodplain and would potentially be affected by sea-level rise (see Box Figure A1.6).

Development and Shore Protection

The Western Shore was largely developed before the Maryland's Critical Areas Act was passed. Stone revetments are common along the mostly developed shores of Anne Arundel and Baltimore Counties. Yet Calvert County has one of the only shore protection policies in the nation that prohibits shore protection along an estuary, even when the prohibition means that homes will be lost. Calvert County's erosion policy is designed to preserve unique cliff areas that border Chesapeake Bay.

The county allows shoreline armoring in certain developed areas to protect property interests, but also bans armoring in other areas to protect endangered species and the unique landscape[35]. Cliffs in Calvert County are separated into categories according to the priority for preservation of the land. Although a county policy prohibiting shore protection would appear to run counter to the state law granting riparian owner the right to shore protection, to date no legal challenges to the cliff policy have been made. The state has accepted the county's policy, which is embodied in the county's critical areas plan submitted to the state under the Critical Areas Act. Recognizing the potential environmental implications, living shoreline protection is becoming increasingly commonplace along the Western Shore.

Vulnerable Habitat

A range of sea-level rise impacts are possible along the western shore of Chesapeake Bay, including potential loss of key habitats. First, marshes are expected to be marginal with mid-range increases in sea-level rise, and to be lost with high-range increases in sea- level rise (Shellenbarger Jones and Bosch, 2008b, interpreting the findings of Reed *et al.*, 2008). The ability to migrate is likely to determine coastal marsh survival as well as the survival of the crustaceans, mollusks, turtles, and birds that depend on the marshes. In upper reaches of tributaries, however, marsh accretion is likely to be sufficient to counter sea-level rise (Shellenbarger Jones and Bosch, 2008b, interpreting the findings of Reed *et al.*, 2008). Several key locations warrant attention:

- In the Jug Bay Sanctuary, along the upper Patuxent River, marsh inundation is causing vegetation changes, compounding stress on local bird species (Shellenbarger Jones and Bosch, 2008b).
- Cove Point Marsh in Calvert County is a 60-ha (150-ac) freshwater, barrier-beach marsh. Numerous state-defined rare plant species are present, including American frog's-bit, silver plumegrass, various ferns, and unique wetland communities (Steury, 2002), as well as several rare or threatened beetle species. With current rates of sea-level rise, the marsh is continuing to migrate, but will soon hit the northern edge of local residential development.
- The potential loss of the wide mudflats at Hart-Miller Island would eliminate major foraging and nesting areas for several high conservation priority species (Shellenbarger Jones and Bosch, 2008b).
- Given the extent of development and shoreline armoring in Anne Arundel County and Baltimore City/County, both intertidal areas and wetlands are likely to be lost with even a modest acceleration in sea-level rise (Shellenbarger Jones and Bosch, 2008b).
-

Beach loss, particularly in St. Mary's, Calvert, and Anne Arundel Counties along Chesapeake Bay, may occur in areas without nourishment. In general, beach loss will lead to habitat loss for resident insects (including the Northeastern beach tiger beetle, federally listed as threatened) and other invertebrates, as well as forage loss for larger predators such as shorebirds (Lippson and Lippson, 2006)[36].

The Calvert County cliffs represent unique habitat that could be degraded by sea-level rise; however, the cliffs are not likely to be lost entirely. The Puritan tiger beetle and Northeastern beach tiger beetle, both federally listed, are present in the area (Shellenbarger Jones and Bosch, 2008b). While natural erosion processes are allowed to continue in the protected cliff areas in the southern portion of the county, shoreline protections in the more northern developed areas are increasing erosion rates in adjacent areas (Wilcock *et al.*, 1998).

A1.F.1.6. Upper Eastern Shore

The Eastern Shore above Rock Hall is dominated by bluffs and steep slopes rising to above 6 m (20 ft). Tolchester Beach, Betterton Beach, and Crystal Beach are typical in that regard. From Rock Hall south to around the middle of Kent Island, all of the land within a few kilometers of the Chesapeake Bay or its major tributaries consists of low-lying land.

Between the Choptank River and Ocohannock Creek along the Eastern Shore of Chesapeake Bay lies one of the largest concentrations of land close to sea level. Water levels in roadside ditches rise and fall with the tides in the areas west of Golden Hill in Dorchester County and several necks in Somerset County. Many farms abut tidal wetlands, which are gradually encroaching onto those farms. Some landowners have responded by inserting makeshift tide gates over culverts, decreasing their own flooding but increasing it elsewhere. Throughout Hoopers Island, as well as the mainland nearby, there are: numerous abandoned driveways that once led to a home but are now ridges flooded at high tide and surrounded by low marsh or open water; recently abandoned homes that are still standing but surrounded by marsh; and dead trees still standing in areas where marsh has invaded a forest.

Development and Shore Protection

Along the Chesapeake Bay, recent coastal development has not placed a high value on the beach. The new bayfront subdivisions often provide no public access to the beach, and as shores erode, people erect shore-protection structures that eventually eliminate the beach (see Chapter 6 of this Product; Titus, 1998). Some traditional access points have been closed (Titus, 1998). Maintaining a beach remains important to some of the older bay resort communities where residents have long had a public beach—but even communities with "beach" in the name are seeing their beaches replaced with shore protection structures.

Maryland's Critical Areas Act, however, is likely to restrict the extent of additional development along the Eastern Shore of Chesapeake Bay to a greater extent than along the Western Shore. The resource conservation areas where development is discouraged include half of the Chesapeake Bay shoreline between the Susquehanna and Choptank Rivers. Among the major tributaries, most of the Sassafras, Chester, and Choptank Rivers are similarly preserved; the Act did not prevent development along most of the Wye, Elk, and North East Rivers. Existing development is most concentrated in the northern areas near Interstate-95, Kent Island, and the various necks near Easton and St. Michaels.

Vulnerable Habitat

Above Kent Island. The environmental implications of sea-level rise effects in the upper Chesapeake Bay are likely to be relatively limited. The Susquehanna River provides a large (though variable) influx of sediment to the upper Chesapeake Bay, as well as almost half of Chesapeake Bay's freshwater input (CBP, 2002). This sediment generally is retained above the Chesapeake Bay Bridge and provides material for accretion in the tidal wetlands of the region (CBP, 2002). The other upper Chesapeake Bay tributaries characteristically have large sediment loads as well, and currently receive sufficient sediment to maintain wetlands and their ecological function. As such, the upper Chesapeake Bay will continue to provide spawning and nursery habitat for crabs and fish, as well as nesting and foraging habitat for migratory and residential birds, including bald eagles and large numbers of waterfowl. Likewise, while some of the beaches may require nourishment for retention, the general lack of shoreline protections will minimize interferences with longshore sediment transport. Hence, beaches are likely to remain intact throughout much of the region (Shellenbarger Jones, 2008b).

Two areas in the upper bay—Eastern Neck and Elk Neck—appear most vulnerable to sea-level rise effects. First, Eastern Neck Wildlife Refuge lies at the southern tip of Maryland's Kent County. Ongoing shoreline protection efforts seek to reduce erosion of habitats supporting many migratory waterfowl and residential birds, as well as turtles, invertebrates, and the Delmarva fox squirrel, federally listed as endangered. In many marsh locations, stands of invasive common reed are the only areas retaining sufficient sediment (Shellenbarger Jones, 2008b). Local managers have observed common reed migrating upland into forested areas as inundation at marsh edges increases, although widespread marsh migration of other species has not been observed (Shellenbarger Jones, 2008b). The three-square bulrush marshes on Eastern Neck have been largely inundated, as have the black needle rush marshes on Smith Island and other locations, likely causes of reductions in black duck counts (Shellenbarger Jones, 2008b).

Other sea-level rise impacts are possible in Cecil County, in and around the Northeast and Elk Rivers. The headwaters of the rivers are tidal freshwater wetlands and tidal flats, spawning and nursery areas for striped bass and a nursery area for alewife, blueback herring, hickory shad, and white perch, as well as a wintering and breeding area for waterfowl (USFWS, 1980). Accretion is likely to be sufficient in some areas due to the large sediment inputs in the Upper Bay. Where accretion rates are not sufficient, wetland migration would be difficult due to the upland elevation adjacent to the shorelines. These conditions increase the chances of large tidal fresh marsh losses (Shellenbarger Jones, 2008b). Other sensitive Cecil County habitats exist such as the cliffs at Elk Neck State Park and the Sassafras River Natural Resource Management Area, which will be left to erode naturally losses (Shellenbarger Jones, 2008b). Finally, marsh loss is possible in and around the Aberdeen Proving Ground in Harford County. The Proving Ground is primarily within 5 m of sea level and contains 8000 ha (20,000 ac) of tidal wetlands.

Kent Island to Choptank River. The central Eastern Shore region of Chesapeake Bay contains diverse habitats, and sea-level rise holds equally diverse implications, varying greatly between subregions. Large expanses of marsh and tidal flats are likely to be lost, affecting shellfish, fish, and waterfowl populations (Shellenbarger Jones, 2008c). Several subregions merit consideration:

- Marshes along the Chester River are likely to be marginal with moderate sea-level rise rate increases (Shellenbarger Jones, 2008c, interpreting the findings of Reed *et al.*, 2008; see Chapter 4 of this Product).
- Loss of the large tidal flats exist at the mouth of the Chester River (Tiner and Burke, 1995) may result in a decline in the resident invertebrates and fish that use the shallow waters as well as the birds that feed on the flats (Shellenbarger Jones, 2008c; Robbins and Blom, 1996).
- The Eastern Bay side of nearby Kent Island has several tidal creeks, extensive tidal flats, and wetlands. Existing marshes and tidal flats are likely to be lost (see Chapter 4) (although some marsh may convert to tidal flat). Increasing water depths are likely to reduce the remaining SAV; a landward migration onto existing flats and marshes will depend on sediment type and choice of shoreline structure (Shellenbarger Jones, 2008c).
- Portions of the Wye River shore are being developed. If these shores are protected and the marshes and tidal flats in these areas are lost, the juvenile fish nurseries will be affected and species that feed in the marshes and SAV will lose an important food source (MD DNR, 2004).

Certain key marsh areas are likely to be retained. The upper reaches of tributaries, including the Chester and Choptank Rivers, are likely to retain current marshes and the associated ecological services. Likewise, Poplar Island will provide a large, isolated marsh and tidal flat area (USACE, undated[b]). In addition, the marshes of the Wye Island Natural Resource Management Area support a large waterfowl population (MD DNR, 2004). Maryland DNR will manage Wye Island to protect its biological diversity and structural integrity, such that detrimental effects from sea-level rise acceleration are minimized (MD DNR, 2004).

Beach loss is also possible in some areas. The Chesapeake Bay shore of Kent Island historically had narrow sandy beaches with some pebbles along low bluffs, as well as some wider beaches and dune areas (*e.g.*, Terrapin Park). As development continues, however, privately owned shores are gradually being replaced with stone revetments. The beaches will be unable to migrate inland, leading to habitat loss for the various resident invertebrates, including tiger beetles, sand fleas, and numerous crab species (Shellenbarger Jones, 2008c). Shorebirds that rely on beaches for forage and nesting will face more limited resources (Lippson and Lippson, 2006). Likewise, on the bay side of Tilghman Island, the high erosion rates will tend to encourage shoreline protection measures, particularly following construction of waterfront homes (MD DNR, undated). Beach loss, combined with anticipated marsh loss in the area, will eliminate the worms, snails, amphipods, sand fleas, and other invertebrates that live in the beach and intertidal areas and reduce forage for their predators (Shellenbarger Jones, 2008c).

A1.F.1.7. Lower Eastern Shore

Approximately halfway between Crisfield on the Eastern Shore and the mouth of the Potomac River on the Western Shore, are the last two inhabited islands in Chesapeake Bay unconnected by bridges to the mainland: Smith (Maryland) and Tangier (Virginia). Both islands are entirely below the 5-ft elevation contour on a USGS topographic map. Along the Eastern Shore of Northampton County, by contrast, elevations are higher, often with bluffs of a few meters.

Development and Shore Protection

Along Chesapeake Bay, islands are threatened by a combination of erosion and inundation. Wetlands are taking over portions of Hoopers and Deal Islands, but shore erosion is the more serious threat. During the middle of the nineteenth century, watermen who made their living by fishing Chesapeake Bay made their homes on various islands in this region. Today, Bloodsworth and Lower Hoopers Islands are uninhabitable marsh, and the erosion of Barren and Poplar Islands led people to move their homes to the mainland (Leatherman, 1992). Smith Island is now several islands, and has a declining population. Hoopers and Deal Islands are becoming gentrified, as small houses owned by watermen are replaced with larger houses owned by wealthier retirees and professionals.

Virtually all of the beaches along Chesapeake Bay are eroding. Shore erosion of beaches and clay shores along the Chester, Nanticoke, and Chester Rivers is slower than along the Bay but enough to induce shoreline armoring along most developed portions. The lower Eastern Shore has a history of abandoning lowlands to shore erosion and rising sea level to a greater extent than other parts of the state (Leatherman, 1992).

Today, Smith and Tangier are the only inhabited islands without a bridge connection to the mainland. Government officials at all levels are pursuing efforts to prevent the loss of these lands, partly because of their unique cultural status and—in the case of Tangier—a town government that works hard to ensure that the state continues to reinvest in schools and infrastructure. The USACE has several planned projects for halting shore erosion, but to date, no efforts are underway to elevate the land (USACE, 2001b; Johnson, 2000). The replacement of traditional lifestyles with gentrified second homes may increase the resources available to preserve these islands.

The mainland of Somerset County vulnerable to sea-level rise is mostly along three necks. Until recently, a key indicator of the cost-effectiveness of shore protection was the availability of a sewer line[37]. As sea level rises, homes without sewer may be condemned as septic systems fail. The incorporated town of Crisfield, in the southernmost neck, has long had sewer service, which has been recently expanded to nearby areas. The town itself is largely encircled by an aging dike. Deal Island, no longer the thriving fishing port of centuries gone by, still has moderate density housing on most of the dry land.

Wicomico County's low-lying areas are along both the Wicomico and Nanticoke Rivers. Unlike Somerset, Wicomico has a large urban/suburban population, with the Eastern Shore's largest city, Salisbury. Planners accept the general principals of the state's Critical Areas Act, which discourages development along the shore.

Much of coastal Dorchester County is already part of Blackwater Wildlife Refuge. The very low land south of Cambridge that is not already part of the refuge is farmland. Because most of the low-lying lands west of Cambridge are within Resource Conservation Areas (CBCAC, 2001), significant development would be unlikely under the state's Critical Areas Act (see Section A1 .F.2). On the higher ground along the Choptank River, by contrast, many waterfront parcels are being developed. In July 2008, the State of Maryland Board of Public Works approved the purchase of 295 ha (729 ac) of land along the Little Blackwater River, near the town to Cambridge in Dorchester County. Funded by the state's Program Open Space, the purchase will allow for the preservation and restoration of more than two-thirds of a 434-ha (1,072-ac) parcel that was previously slated for development[38].

Vulnerable Habitat

On the lower Eastern Shore of Chesapeake Bay in Maryland, habitats vulnerable to sea-level rise are diverse and include beaches, various types of tidal marsh, non-tidal marshes, and upland pine forests.

Narrow sandy beaches exist along discrete segments of shoreline throughout the region, particularly in Somerset County. Given the gradual slope of the shoreline, one might infer that these habitats could accommodate moderate sea-level rise by migrating upslope, assuming no armoring or other barriers exist. Many of the beaches provide critical nesting habitat for the diamondback terrapin (*Malaclemys terrapin*), and proximity of these nesting beaches to nearby marshes provides habitat for new hatchlings (see Box A1.7)

BOX A1.7. THE DIAMONDBACK TERRAPIN, *MALACLEMYS TERRAPIN*

The diamondback terrapin, *Malaclemys terrapin*, comprising seven subspecies, is the only turtle that is fully adapted to life in the brackish salt marshes of estuarine embayments, lagoons, and impoundments (Ernst and Barbour, 1972). Its range extends from Massachusetts to Texas in the narrowest of coastal strips along the Atlantic and Gulf coasts of the United States (Palmer and Cordes, 1988). Extreme fishing pressure on the species resulted in population crashes over much of their range so that by 1920 the catch in Chesapeake Bay had fallen to less than 900 pounds. The Great Depression put a halt to the fishery, and during the mid-twentieth century, populations began to recover (CBP, 2006). Although a modest fishery has been reestablished in some areas, stringent harvest regulations are in place in several states. In some instances, states have listed the species as endangered (Rhode Island), threatened (Massachusetts), or as a "species of concern" (Georgia, Delaware, New Jersey, Louisiana, North Carolina, and Virginia). In Maryland, the status of the northern diamondback subpopulation is under review (MD DNR, 2006).

Effects of Sea-level Rise

The prospect of sea-level rise, along with land subsidence at many coastal locations, increasing human habitation of the shore zone and shoreline stabilization, places the habitat of terrapins at increasing risk. Loss of prime nesting beaches remains a major threat to the diamondback terrapin population in Chesapeake Bay (MD DTTF, 2001). Because human infrastructure (*i.e.*, roadways, buildings, and impervious surfaces) leaves tidal salt marshes with little or no room to transgress inland, one can infer that the ecosystem that terrapins depend on may be lost with concomitant extirpation of the species.

Of the 87,000 ha (340 sq mi) of tidal marsh in the Chesapeake Bay, a majority is located in the three-county lower Eastern Shore region (Darmondy and Foss, 1979). The marshes are critical nursery grounds for commercially important fisheries (*e.g.*, crabs and rockfish); critical feeding grounds for migratory waterfowl; and home to furbearers (*e.g.*, muskrat and nutria).

Areas of Virginia's Eastern Shore are uniquely vulnerable to sea-level rise since large portions of Northampton and Accomack counties lie near sea level. Because most of the land in the two counties is undeveloped or agricultural, the area also has a high potential for wetland creation relative to other Virginia shorelines.

Most notably, the bay side of northern Accomack County is primarily tidal salt marsh, with low-lying lands extending several kilometers inland. Unprotected marshes are already migrating inland in response to sea-level rise, creating new wetlands in agricultural areas at a rate of 16 ha (40 ac) per year (Strange, 2008e). Given the anticipated lack of shoreline protection and insufficient sediment input, the seaward boundaries of these tidal wetlands are likely to continue retreating (Strange, 2008e, interpreting the findings of Reed *et al.*, 2008). The upland elevations are higher in southern than northern Accomack County, however, making wetland migration more difficult.

The salt marshes of Accomack County support a variety of species, including rare bird species such as the seaside sparrow, sharp-tailed sparrow, and peregrine falcon (VA DCR, undated [a][b]). Growth and survival of these species may be reduced where shores are hardened, unless alternative suitable habitat is available nearby. Furthermore, longterm tidal flooding will decrease the ability of nekton (*i.e.*, free-swimming finfish and decapod crustaceans such as shrimps and crabs) to access coastal marshes.

A1.F.2. Baywide Policy Context

Chesapeake Bay's watershed has tidal shores in Virginia, Maryland, the District of Columbia, and Delaware. Because the shores of Delaware and the District of Columbia. account for a small portion of the total, this subsection focuses on Virginia and Maryland. (The federal Coastal Zone Management Act's definition of "coastal state" excludes the District of Columbia[39].)

Coastal management officials of Maryland have cooperated with the U.S. EPA since the 1 980s in efforts to learn the ramifications of accelerated sea-level rise for their activities (AP, 1985). Increased erosion from sea-level rise was one of the factors cited for the state's decision in 1985 to shift its erosion control strategy at Ocean City from groins to beach nourishment (AP, 1985). The state also developed a planning document for rising sea level (Johnson, 2000), and sea-level rise was a key factor motivating Maryland to become the second mid-Atlantic state to obtain lidar elevation data for the entire coastal floodplain.

Neither Maryland nor Virginia has adopted a comprehensive policy to explicitly address the consequences of rising sea level. Nevertheless, the policies designed to protect wetlands, beaches, and private shorefront properties are collectively an implicit policy. Both states prevent new buildings within 30.5 m (100 ft) of most tidal shores; Maryland also limits the density of new development in most areas to one home per 8 ha (20 ac) within 305 m (1,000 ft) of the shore. Virginia allows most forms of shore protection. Maryland encourages shore protection[40], but discourages new bulkheads in favor of revetments or nonstructural measures (MD DNR, 2006). Both states have programs to inform property owners of nonstructural options and have created programs and educational outreach efforts to train marine contractors on "living shoreline" design and installation techniques. Both states work with the federal government to obtain federal funds for beach nourishment along their respective ocean resorts (Ocean City and Virginia Beach); Virginia also assists local governments in efforts to nourish public beaches along Chesapeake Bay and its tributaries. Summaries of these land use, wetlands, and beach nourishment policies follow.

During 2007, both states established climate change commissions to inform policy makers about options for responding to sea level rise and other consequences of changing climate[41]. The Maryland Commission on Climate Change is charged with developing a climate action plan to address both the causes and consequences of climate change[42]. Its interim report (MCCC, 2008) recommends that the state (1) protect and restore natural shoreline features (*e.g.*, wetlands) and (2) reduce growth and development in areas vulnerable to sea-level rise and its ensuing coastal hazards. The Virginia commission has an Adaptation Subgroup.

A1.F.2.1. Land Use

The primary state policies related to land use are Maryland's Chesapeake and Atlantic Coastal Bays Critical Area Protection Act, Virginia's Chesapeake Bay Preservation Act, and Virginia's Coastal Primary Sand Dunes & Beaches Act.

Maryland Chesapeake Bay and Atlantic Coastal Bays Critical Area Protection Act. The Maryland General Assembly enacted the Chesapeake Bay Critical Area Protection Act in 1984 to reverse the deterioration of the Bay[43]. (The statute now applies to Atlantic coastal bays as well; see Section A1.E.2) The law seeks to control development in the coastal zone and preserve a healthy bay ecosystem. The jurisdictional boundary of the Critical Area includes all waters of Chesapeake and Atlantic Coastal Bays, adjacent wetlands[44], dry land within 305 m (1,000 ft) of open water[45], and in some cases dry land within 305 m inland of wetlands that are hydraulically connected to the bays[46].

The act created a Critical Areas Commission to set criteria and approve local plans[47]. The commission has divided land in the critical area into three classes: intensely developed areas (IDAs), limited development areas (LDAs), and resource conservation areas (RCAs)[48]. Within the RCAs, new development is limited to an average density of one home per 8 ha (20 ac)[49] and set back at least 61 m (200 ft)[50], and the regulations encourage communities to "consider cluster development, transfer of development rights, maximum lot size provisions, and/or additional means to maintain the land area necessary to support the protective uses"[51]. The program limits future intense development activities to lands within the IDAs, and permits some additional low-intensity development in the LDAs. However, the statute allows up to 5 percent of the RCAs in a county to be converted to an IDA[52], although a 61-m (200-ft) buffer applies in those locations.

The three categories were originally delineated based on the land uses of 1985. Areas that were dominated by either agriculture, forest, or other open space, as well as residential areas with densities less than one home in 2 ha (5 acres), were defined as RCAs[53]. Thus, the greatest preservation occurs in the areas that had little development when the act was passed, typically lands that are far from population centers and major transportation corridors—particularly along tributaries (as opposed to the Bay itself). The boundary of the critical area was based on wetland maps created in 1972. MCCC (2008) pointed out that rising sea level and shoreline erosion had made that boundary obsolete in some locations. As a result, the Legislature directed the Critical Areas Commission to update the maps based on 2007 to 2008 imagery, and thereafter at least once every 12 years[54].

The Critical Areas Program also established a 30.5-m (100-ft) natural buffer adjacent to tidal waters, which applies to all three land categories[55]. No new development activities are allowed within the buffer[56], except water-dependent facilities. By limiting development in the buffer, the program prevents additional infrastructure from being located in the areas most vulnerable to sea-level rise. In some cases, the 30.5-m buffer provides a first line of defense against coastal erosion and flooding induced by sea-level rise. But the regulations also encourage property owners to halt shore erosion[57]. Nonstructural measures are preferred, followed by structural measures[58], with an eroding shore the least preferable (Titus, 1998).

Virginia Chesapeake Bay Preservation Act. The Chesapeake Bay Preservation Act[59] seeks to limit runoff into the bay by creating a class of land known as Chesapeake Bay Preservation Areas. The act also created the Chesapeake Bay Local Assistance Board to implement[60] and enforce[61] its provisions. Although the act defers most site-specific development decisions to local governments[62], it lays out the broad framework for the preservation areas[63] and provides the Board with rulemaking authority to set overall criteria[64]. The Board has issued regulations[65] defining the programs that local governments must develop to comply with the act[66].

All localities must create maps that define the locations of the preservation areas, which are subdivided into resource management areas[67] and resource protection areas (RPAs)[68]. RPAs include areas flooded by the tides, as well as a 30.5 m (100-ft) buffer inland of the tidal shores and wetlands[69]. Within the buffer, development is generally limited to water dependent uses, redevelopment, and some water management facilities. Roads may be allowed if there is no practical alternative. Similarly, for lots subdivided before 2002, new buildings may encroach into the 30.5 m buffer if necessary to preserve the owner's right to build; but any building must still be at least 15.2 m (50 ft) from the shore[70]. Property owners, however, may still construct shoreline defense structures within the RPA. The type of shoreline defense installed is not regulated (beyond certain engineering considerations). Consequently, hard structures can be installed anywhere along Virginia's shoreline.

Virginia Coastal Primary Sand Dunes & Beaches Act. Virginia's Dunes and Beaches Act preserves and protects coastal primary sand dunes while accommodating shoreline development. The act identifies eight counties and cities that can adopt a coastal primary sand dune zoning ordinance, somewhat analogous to a Tidal Wetlands ordinance: Accomack, Northampton, Virginia Beach, Norfolk, Hampton, Mathews, Lancaster, and Northumberland (Hardaway *et al.*, 2001); all but Hampton and Accomack have done so. The act defines beaches as (1) the shoreline zone of unconsolidated sandy material; (2) the land extending from mean low water landward to a marked change in material composition or in physiographic form (*e.g.*, a dune, marsh, or bluff); and (3) if a marked change does not occur, then a line of woody vegetation or the nearest seawall, revetment, bulkhead or other similar structure.

A1.F.2.2. Wetlands and Erosion Control Permits

Virginia. The Tidal Wetlands Act seeks to "...preserve and prevent the despoliation and destruction of wetlands while accommodating necessary economic development in a manner consistent with wetlands preservation" (VA Code 28.2-1302). It provides for a Wetlands Zoning ordinance that any county, city, or town in Virginia may adopt to regulate the use and development of local wetlands. Under the ordinance, localities create a wetlands board consisting of five to seven citizen volunteers. The jurisdiction of these local boards extends from mean low water (the Marine Resources Commission has jurisdiction over bottom lands seaward of mean low water) to mean high water where no emergent vegetation exists, and slightly above spring high water[71] where marsh is present. The board grants or denies permits for shoreline alterations within their jurisdiction (Trono, 2003). The Virginia Marine Resources Commission has jurisdiction over the permitting of projects within state-owned subaqueous lands and reviews projects in localities that have no local wetlands board by virtue of not having adopted a wetland zoning ordinance[72].

Maryland. The Wetlands and Riparian Rights Act[73] gives the owner of land bounding on navigable water the right to protect their property from the effects of shore erosion. For example, property owners who erect an erosion control structure in Maryland can obtain a permit to fill vegetated wetlands[74] and fill beaches and tidal waters up to 3 m (10 ft) seaward of mean high water[75]. In addition, Maryland's statute allows anyone whose property has eroded to fill wetlands and other tidal waters to reclaim any land that the owner has lost since the early 1970s[76]. (USACE has delegated most wetland permit approval to the state[77].) Although the state has long discouraged bulkheads, much of the shore has been armored with stone revetments (Titus, 1998).

Shore protection structures tend to be initially constructed landward of mean high water, but neither Virginia nor Maryland[78] require their removal once the shore erodes to the point where the structures are flooded by the tides. Nor has either state prevented construction of replacement bulkheads within state waters, although Maryland encourages revetments.

For the last several years, Maryland has encouraged the "living shorelines" approach to halting erosion (*e.g.*, marsh planting and beach nourishment) over hard structures and revetments over bulkheads[79]. Few new bulkheads are built for erosion control, and existing bulkheads are often replaced with revetments. Nevertheless, obtaining permits for structural options has often been easier (NRC, 2007; Johnson and Luscher, 2004). For example, in the aftermath of Hurricane Isabel, many property owners sought expedited permits to replace shore protection structures that had been destroyed by storms. Maryland wanted to make obtaining a permit to replace a destroyed bulkhead with a living shoreline as easy as obtaining a permit to rebuild the bulkhead; but the state was unable to obtain federal approval. The permits issued by USACE authorized replacement of the damaged structures with new structures of the same kind, but they did not authorize owners to replace lost revetments and bulkheads with living shoreslines, or even to replace lost bulkheads with revetments (Johnson and Luscher, 2004).

Table A1.3. Selected State Funded Beach Nourishment Projects along Estuarine Shores in Maryland and Virginia

Location	City or County	$Cost (Millions)
Maryland (2001 to 2008)		
North Beach	Calvert	n.a.
Sandy Point	Anne Arundel	n.a.
PT Lookout State Park	St Mary's	n.a.
Choptank River Fishing Pier	Talbot	n.a.
Jefferson Island.	St. Mary's	n.a.
Tanners Creek	St. Mary's	n.a.
Bay Ridge	Anne Arundel	n.a.
Hart and MIllers Island.	Baltimore County Co.	n.a.
Rock Hall Town Park	Kent	n.a.
Claiborne Landing	Talbot	n.a.
Terrapin Beach,	Queen Anne's	n.a.
Jefferson Is. Club - St Catherine Island	St. Mary's	n.a.
Elms Power Plant Site	St. Mary's	n.a.
Virginia (1995 to 2005)		
Bay Shore	Norfolk	5.0
Parks along James River	Newport News	1.0
Buckroe Beach	Hampton	1.3
Cape Charles	Northampton	0.3
Colonial Beach	Westmoreland	0.3
Aquia Landing	Stafford	0.2

Source: Maryland Department of Natural Resources; Virginia Board on Conservation and Development of Public Beaches

Recognizing the environmental consequences of continued shoreline armoring, the Living Shoreline Protection Act of 2008[80]. Under the act, the Department of Environment will designate certain areas as appropriate for structural shoreline measures (*e.g.*, bulkheads and revetments). Outside of those areas, only nonstructural measures (*e.g.*, marsh creation, beach nourishment) will be allowed unless the property owner can demonstrate that nonstructural measures are infeasible[81].

A1.F.2.3. Beach Nourishment and Other Shore Protection Activities

Virginia. Until 2003, the Board on Conservation and Development of Public Beaches promoted maintenance, access, and development along the public beaches of Virginia. The largest beach nourishment projects have been along the 21 km (13 mi) of public beach along the Atlantic Ocean in Virginia Beach. During the last 50 years, the state has provided three percent of the funding for beach nourishment at Virginia Beach, with the local and federal shares being 67 percent and 30 percent, respectively (VA PBB, 2000).

Virginia has made substantial efforts to promote beach nourishment (and public use of beaches) along Chesapeake Bay and its tributaries. Norfolk's four guarded beaches serve 160,000 visitors each summer (VA PBB, 2000). When shore erosion threatened property, the tourist economy, and local recreation, the Beach Board helped the city construct a series of breakwaters with beachfill and a terminal groin at a cost of $5 million (VA PBB, 2000). State and local partnerships have also promoted beach restoration projects in several other locations along Chesapeake Bay and the Potomac and York rivers. (see Table A1.3).

Maryland. Maryland's primary effort to protect shores along the bay is through the Department of Natural Resource's Shore Erosion Control Program. Until 2008, the program provided interest-free loans and technical assistance to Maryland property owners to resolve erosion problems through the use of both structural and nonstructural shore erosion control projects; the program is now limited to "living shoreline" (see Box 5.3 in Chapter 5) approaches. The program provides contractor and homeowner training to support the installation of "living shorelines". The Department of Natural Resources has been involved in several beach nourishment projects along Chesapeake Bay (see Table A1.3), many of which include breakwaters or groins to retain sand within the area nourished.

The Maryland Port Administration and the USACE have also used dredge spoils to restore Poplar and Smith Islands (USACE, 2001b). Preliminary examinations are under way to see if dredged materials can be used to restore other Chesapeake Bay islands such as James and Barren Islands (Federal Register, 2006), or to protect valuable environmental resources such as the eroding lands of the U.S. Fish and Wildlife Service (USFWS) Blackwater National Wildlife Refuge (USFWS, 2008).

A1.G. North Caroina

A1.G.1. Introduction

North Carolina's coastline is outlined by a barrier island system, with approximately 500 km (300 mi) of shoreline along the Atlantic Ocean. North Carolina's winding estuarine shorelines extend a total of approximately 10,000 linear km (6,000 mi) (Feldman, 2008).

There are three well-known capes along the coastline: Cape Hatteras, Cape Lookout, and Cape Fear, in order from north to south. The "Outer Banks" of North Carolina include the barrier islands and barrier spits from Cape Lookout north to the Virginia state line. Much of this land is owned by the federal government, including Cape Lookout National Seashore, Cape Hatteras National Seashore, Pea Island National Wildlife Refuge, and Currituck National Wildlife Refuge. The Outer Banks also include several towns, including Kitty Hawk, Nags Head, Rodanthe, and Ocracoke (see Section A1.G.4.2). North and east of Cape Lookout, four rivers empty into the Albemarle and Pamlico Sounds. Albemarle Sound, Pamlico Sound, and their tidal tributaries, sometimes collectively called the Albemarle-Pamlico Estuarine System, comprise the second largest estuarine system in the United States.

Previous assessments of North Carolina's estuarine regions have divided the state's coastal regions into two principal provinces (geological zones), each with different characteristics (*e.g.*, Riggs and Ames, 2003). The zone northeast of a line drawn between Cape Lookout and Raleigh (located about 260 km west of the cape) is called the Northern Coastal Province, which includes the Outer Banks and most of the land bordering the Albemarle and Pamlico Sounds. It has gentle slopes, three major and three minor inlets, and long barrier islands with a moderately low sediment supply, compared to barrier islands worldwide (Riggs and Ames, 2003). The rest of the state's coastal zone—the Southern Coastal Province—has steeper slopes, an even lower sediment supply, short barrier islands, and many inlets.

The Albemarle-Pamlico Peninsula is the land between Albemarle and Pamlico Sounds, to the west of Roanoke Island. The potential vulnerability of this 5,500 sq km (2,100 sq mi) peninsula (Henman and Poulter, 2008) is described in Box A1.8. The majority of Dare and Hyde counties are less than 1 m (approximately 3 ft) above sea level, as is a large portion of Tyrell County (Poulter and Halpin, 2007). Along the estuarine shorelines of North Carolina, wetlands are widespread, particularly in Hyde, Tyrell, and Dare counties. North Carolina's Division of Coastal Management mapped a total of more than 11,000 sq km (4,400 sq mi) of wetlands in the 20 coastal counties in North Carolina (Sutter, 1999). Wetlands types present include marshes, swamps, forested wetlands, pocosins (where evergreen shrubs and wetland trees occupy peat deposits), and many other types (Sutter, 1999).

Where the land is flat, areas a few meters above sea level drain slowly—so slowly that most of the lowest land is nontidal wetland (Richardson, 2003). Because rising sea level decreases the average slope between nearby coastal areas and the sea, it slows the speed at which these areas drain. Some of the dry land within a few meters above the tides could convert to wetland from even a small rise in sea level; and nontidal wetlands at these elevations would be saturated more of the time (McFadden *et al.*, 2007; Moorhead and Brinson, 1995). Wetland loss could occur if dikes and drainage systems are built to prevent dry land from becoming wet (McFadden *et al.*, 2007).

The very low tide range in some of the sounds is another possible source of vulnerability. Albemarle Sound, Currituck Sound, and much of Pamlico Sound have a very small tide range because inlets to the ocean are few and far between (NOAA, 2008b). Some of the inlets are narrow and shallow as well. Although Oregon and Ocracoke inlets are more than 10 m (over 30 ft) deep, the inlets are characterized by extensive shoals on both the ebb and flood sides, and the channels do not maintain depth for long distances before they break into shallower finger channels. Like narrow channels, this configuration limits the flow of water between the ocean and sounds (NOAA, 2008c). Thus, although the astronomic tide range at the ocean entrances is approximately 90 cm (3 ft), it decreases to 30 cm (1 ft) just inside the inlets, and a few centimeters in the centers of the estuaries. It is possible that rising sea level combined with storm-induced erosion will cause more, wider, and/or deeper inlets in the future (Riggs and Ames, 2003; see Chapter 3 of this Product). If greater tide ranges resulted, more lands would be tidally inundated.

The configuration of the few inlets within the Northern Coastal Province reduces tidal flushing and keeps salinity levels relatively low in most of the estuaries in this area (Riggs and Ames, 2003). Salinity is relatively high at the inlets, but declines as one proceeds upstream or away from the inlets. Also, there can be a strong seasonal variation with lower salinities during the periods of maximum river discharge and higher salinities during periods of drought (Buzzelli *et al.*, 2003). The salinity in Albemarle-Pamlico Sound generally ranges from 0 to 20 parts per thousand (ppt), with the upper reaches of the Neuse and Pamlico Rivers, Albemarle Sound and Currituck Sound having salinities usually below 5 ppt (Caldwell, 2001; Tenore, 1972). (The typical salinity of the ocean is 35 ppt [Caldwell, 2001]). Some tidal marshes (which are irregularly flooded by the winds rather than regularly flooded by astronomical tides) are thus unable to tolerate salt water (Bridgham and Richardson, 1993; Poulter, 2005). In some areas, the flow of shallow groundwater to the sea is also fresh, so the soils are unaccustomed to salt water, and hence potentially vulnerable to increased salinity.

Box A1.8. Vulnerability of the Albemarle-Pamlico Peninsula and Emerging Stakeholder Response

Vulnerability to sea-level rise on the diverse Albemarle-Pamlico Peninsula is very high: about two-thirds of the peninsula is less than 1.5 m (5 ft) above sea level (Heath, 1975), and approximately 30 percent is less than 0.9 m (3 ft) above sea level (Poulter, 2005). Shoreline retreat rates in parts of the peninsula are already high, up to 7.6 m (25 ft) per year (Riggs and Ames, 2003). The ecosystems of the Albemarle-Pamlico Peninsula have long been recognized for their biological and ecological value. The peninsula is home to four national wildlife refuges, the first of which was established in 1932. In all, about one-third of the peninsula has been set aside for conservation purposes.

The Albemarle-Pamlico Peninsula is among North Carolina's poorest areas. Four of its five counties are classified as economically distressed by the state, with high unemployment rates, along with low average household incomes (NC Department of Commerce, 2008). However, now that undeveloped waterfront property on the Outer Banks is very expensive and very scarce, developers have discovered the small fishing villages on the peninsula and begun acquiring property in several areas—including Columbia (Tyrrell County), Engelhard (Hyde County) and Bath (Beaufort County). The peninsula is being marketed as the "Inner Banks" (Washington County, 2008). Communities across the peninsula are planning infrastructure, including wastewater treatment facilities and desalination plants for drinking water, to enable new development. Columbia and Plymouth (Washington County) have become demonstration sites in the North Carolina Rural Economic Development Center's STEP (Small Towns Economic Prosperity) Program, which is designed to support revitalization and provide information vital to developing public policies that support long-term investment in small towns (NC REDC, 2006).

There are already signs that sea-level rise is causing ecosystems on the Albemarle-Pamlico Peninsula to change. For example, at the Buckridge Coastal Reserve, a 7,547-ha (18,650-ac) area owned by the North Carolina Division of Coastal Management, dieback is occurring in several areas of Atlantic white cedar. Other parts of the cedar community are beginning to show signs of stress. Initial investigations suggest the dieback is associated with altered hydrologic conditions, due to canals and ditches serving as conduits that bring salt and brackish water into the peat soils where cedar usually grows. Storms have pushed estuarine water into areas that are naturally fresh, affecting water chemistry, peatland soils, and vegetation intolerant of saline conditions (Poulter and Pederson, 2006). There is growing awareness on the part of residents and local officials about potential vulnerabilities across the landscape (Poulter, *et al.*, 2009). Some farmers acknowledge that salt intrusion and sea- level rise are affecting their fields (Moorhead and Brinson, 1995). Researchers at North Carolina State University are using Hyde County farms to experiment with the development of new varieties of salt-tolerant soybeans (Lee *et al.*, 2004). Hyde County is building a dike around Swan Quarter, the county seat (Hyde Country, 2008).

A variety of evidence has suggested to some stakeholders that the risks to the Albemarle-Pamlico Peninsula merit special management responses. In fact, because so much of the landscape across the peninsula has been transformed by humans, some have expressed concern that the ecosystem may be less resilient and less likely to be able to adapt when exposed to mounting stresses (Pearsall *et al.*, 2005). Thus far, no comprehensive long-term response to the effects of sea-level rise on the peninsula has been proposed. In 2007, The Nature Conservancy, U.S. Fish and Wildlife Service, National Audubon Society, Environmental Defense, Ducks Unlimited, the North Carolina Coastal Federation and others began working to build an Albemarle-Pamlico Conservation and Communities Collaborative (AP3C) to develop a long-term strategic vision for the peninsula. Although this initiative is only in its infancy, sea-level rise will be one of the first and most important issues the partnership will address (Nature Conservancy, 2008).

The Nature Conservancy and other stakeholders have already identified several adaptive responses to sea-level rise on the Peninsula. Many of these approaches require community participation in conservation efforts, land protection, and adaptive management (Pearsall and Poulter, 2005). Specific management strategies that The Nature Conservancy and others have recommended include: plugging drainage ditches and installing tide gates in agricultural fields so that sea water does not flow inland through them, establishing cypress trees where land has been cleared in areas that are expected to become wetlands in the future, reestablishing brackish marshes in hospitable areas that are likely to become wetlands in the future, creating conservation corridors that run from the shoreline inland to facilitate habitat migration, reducing habitat fragmentation, banning or restricting hardened structures along the estuarine shoreline, and establishing oyster reefs and submerged aquatic vegetation beds offshore to help buffer shorelines (Pearsall and DeBlieu, 2005; Pearsall and Poulter, 2005).

More than other areas in the Mid-Atlantic, the Albemarle-Pamlico Sound region appears to be potentially vulnerable to the possibility that several impacts of sea-level rise might compound to produce an impact larger than the sum of the individual effects (Poulter and

Halpin, 2007; Poulter *et al.*, 2008). If a major inlet opened, increasing the tide range and salinity levels, it is possible that some freshwater wetlands that are otherwise able to keep pace with rising sea level would be poisoned by excessive salinity and convert to open water. Similarly, if a pulse of salt water penetrated into the groundwater, sulfate reduction of the organic-rich soil and peat that underlies parts of the region could cause the land surfaces to subside (Hackney and Yelverton, 1990; Henman and Poulter, 2008; Mitsch and Gosselink, 2000; Portnoy and Giblin, 1997). Moreover, a substantial acceleration in the rate of sea-level rise storms of the type described below could cause barrier islands to be breached (see Chapter 3). Pamlico Sound (and potentially Albemarle Sound) could be transformed from a protected estuary into a semi-open embayment with saltier waters, regular astronomical tides, and larger waves (Riggs and Ames, 2003).

A1.G.2. Shore Processes

A1.G.2.1. Ocean Coasts

North Carolina receives the highest wave energy along the entire east coast of the United States and the northwest Atlantic margin (Riggs and Ames 2003). The coast of North Carolina has shifted significantly over time due to storms, waves, tides, currents, rising sea level, and other natural and human activities. These factors have caused variable sediment transport, erosion, and accretion, along with the opening and closing of inlets (see, *e.g.*, Everts *et al.*, 1983).

The North Carolina Division of Coastal Management (NCDCM) has calculated long-term erosion rates along the coastline adjacent to the ocean by comparing the location of shorelines in 1998 with the oldest available maps of shoreline location, mostly from the 1940s. The average erosion rate was 0.8 m (2.6 ft) per year. Approximately 18 percent of the ocean coastline retreated by more than 1.5 m per year (5 ft per year), 20 percent eroded at an annual rate of 0.6 to 1.5 m (2 to 5 ft) per year, and 30 percent of the coastaline eroded by 0.6 m (2 ft) per year or less. However, 32 percent of the coastline accreted (NC DCM, 2003). The NCDCM recalculates long-term erosion rates about every five years to better track the dynamic shoreline trends and establish the setback line that determines where structures may be permitted on the oceanfront (NC DCM, 2005).

An analysis of shoreline change between approximately 1850 and 1980 in the area between the northern border of North Carolina and the point 8 km west of Cape Hatteras has been published. Data were averaged over 2 km reaches (stretches of coastline). Across the areas where data were available during this time period, approximately 68 percent of the ocean shoreline retreated towards the mainland, while approximately 28 percent advanced (or accreted) away from the mainland, and 4 percent did not change position (Everts *et al.*, 1983). On average, the parts of the coastline between Ocracoke Inlet and Cape Hatteras eroded an average of 4.5 m (14.8 ft) per year over 1852 to 1917, 8.3 m (27.2 ft) per year over 1917 to 1949, and 2.0 m (6.6 ft) per year over 1949 to 1980. The average erosion rate over the study period along the parts of the coastline facing east (between Cape Hatteras and Cape Henry, in Virginia) was 0.8 m (2.6 ft) per year. However, the study indicates that the coastline from Cape Hatteras to Oregon Inlet accreted slightly (an average of 0.4 m [1.3 ft] per year) over 1852 to 1917, eroded an average of 2.9 m (9.5 ft) per year over 1917 to 1949, and eroded an average of 1.3 m (4.3 ft) per year over 1949 to 1980. North of Oregon Inlet, the coastline was stable, on average, over 1852 to 1917; however, there was an average of 1.2 m (3.9 ft) per year of erosion over 1917 to 1949 and an average of 0.3 m (1.0 ft) per year of erosion in 1949 to1980 (Everts *et al.*, 1983).

The report cautions against predicting future shoreline change based on the limited data available from surveys conducted since 1850. The authors observe that shoreline change can be influenced by local features, such as inlets, capes, and shoals (Everts *et al.*, 1983). For example, shorelines north of the ridges of three offshore shoals intersecting North Carolina's ocean coast have retreated, whereas shorelines south of the ridges have generally advanced (Everts *et al.*, 1983). Everts *et al.* also point out that while geological evidence indicates that the barrier islands have migrated landward over thousands of years, the islands are presently narrowing from both sides, in part because overwash processes cannot carry sand to the estuarine side due to island width and development (Everts *et al.*, 1983).

More recently, researchers have used models to predict the amount of shoreline change that might result from future sea-level rise, above and beyond the shoreline change caused by other factors. For example, one analysis of statewide erosion rates over the past 100 years led researchers to estimate that a 1 m sea-level rise would cause the shore to retreat an average of 88 m (289 ft), in addition to the erosion caused by other factors (excluding inlets) (Leatherman *et al.*, 2000a). Another study estimated that a rise in sea level of 0.52 m between 1996 and 2050 would cause the shoreline at Nags Head to retreat between 33 and 43 m, or between 108 and 144 ft (Daniels, 1996).

Some researchers are concerned that the barrier islands themselves may be in jeopardy if sea-level rise accelerates. According to Riggs and Ames (2003), about 40 km (25 mi) of the Outer Banks are so sediment-starved that they are already in the process of "collapsing". Within a few decades, they estimate, portions of Cape Hatteras National Seashore could be destroyed by: (1) sea-level rise (at current rates or higher); (2) storms of the magnitude experienced in the 1990s; or (3) one or more Category 4 or 5 hurricanes hitting the Outer Banks (Riggs and Ames, 2003). Most of the Outer Banks between Nags Head and Ocracoke is vulnerable to barrier island segmentation and disintegration over the next century if the rate of sea-level rise accelerates by 2 mm per year—and portions may be vulnerable even at the current trend (see Chapter 3.)

A1.G.3. Vulnerable Habitats and Species

Some wetland systems are already at the limit of their ability to vertically keep pace with rising sea level, such as the remnants of the tidal marshes that connected Roanoke Island to the mainland of Dare County until the nineteenth century. The pocosin wetlands can vertically accrete by about 1 to 2 mm per year with or without rising sea level—when they are in their natural state (Craft and Richardson, 1998; Moorhead and Brinson, 1995). The human-altered drainage patterns, however, appear to be limiting their vertical accretion—and saltwater intrusion could cause subsidence and conversion to open water (Pearsall and Poulter, 2005).

A1.G.3.1. Estuarine Shoreline Retreat

The Pamlico and Albemarle Sounds, North Carolina's smaller sounds, and the lower reaches of the Chowan, Roanoke, Tar, and Neuse Rivers are affected by rising sea level

(Brinson *et al.*, 1985). Rising sea level is not the primary cause of shoreline retreat along estuarine shores in North Carolina. Storm waves cause shorelines to recede whether or not the sea is rising. A study of 21 sites estimated that shoreline retreat—caused by "the intimately coupled processes of wave action and rising sea level"—is already eliminating wetlands at a rate of about 3 sq km (800 ac) per year, mostly in zones of brackish marsh habitat, such as on the Albemarle-Pamlico Peninsula (Riggs and Ames, 2003).

Riggs and Ames (2003) compiled data collected across North Carolina shorelines, both those that are adjacent to wetlands and those that are not. These data show that the vast majority of estuarine shores in the region are eroding, except for the sound sides of barrier islands (which one might expect to advance toward the mainland). Shores have retreated almost 2 m (7 ft) per year, over periods as long as 30 years. Annual averages for most shoreline types are less than 1 m per year, (Table A1.4) but annual maxima exceed the average many-fold and can reach 8 m (26 ft) per year where the shoreline is characterized by sediment bluffs or high banks. One or a few individual storm events contribute disproportionately to average annual shoreline recession rates (Riggs and Ames, 2003).

An analysis of estuarine shoreline change is also included in Everts *et al.* (1983). The authors calculated average erosion rates for the periods around 1850 to 1915 and 1915 to 1980. Between Nags Head and Oregon Inlet, the estuarine points analyzed between 1850 and 1915 showed both advance rates greater than 4 m (13 ft) per year and retreat rates of close to 3 m (10 ft) per year. However, between 1915 and 1980, the estuarine points analyzed in this region showed a range of approximately 1 m per year of retreat to less than 1 m per year of advance. Study authors did not analyze the area adjacent to Oregon Inlet or along most of Pea Island. Just north of Rodanthe, the earlier dataset shows dramatic shoreline advance averaging 4 m per year, but the later dataset shows a relatively stable shoreline. Just south of Rodanthe, there was slow advance during the earlier period and slow retreat (of approximately 1 m per year or less) in the later period. Between Avon and Salvo, both datasets show shoreline retreat at rates not exceeding 2 m per year, with a slightly higher average rate of retreat in the later period than the earlier period (taken from Figure 34, Everts *et al.*, 1983).

The study indicates that the average retreat rate across all the estuarine points analyzed from 1852 to 1980 was 0.1 m (4 in) per year. However, this average masks an important trend seen both north and south of Oregon Inlet. The rate of shoreline change gradually changed from shoreline advance (movement towards the sounds) to shore retreat. The rate of advance was almost 2.0 m per year from 1852 to 1917. Shores were generally stable from 1917 to 1949, but they retreated over the period from 1949 to 1980. Erosion was greater along estuarine shores facing west (an average of 1.2 m per year over 1852 to 1980) than those facing north or south (averaging 0.1 m per year over 1852 to 1980). The authors observed that these data indicate that the North Carolina barrier islands in the study region did not appear to be migrating landward during the study period, but instead they narrowed from both sides. The present rate of island narrowing averages 0.9 m (3.0 ft) per year. Available data indicate that sand washed over the barrier islands to the estuarine side of islands (overwash) did not significantly affect shoreline change along the estuary, particularly after the artificial dunes were constructed, a process that might itself have caused erosion from the sound side because it removed sand from the estuarine system (Everts *et al.*, 1983). Away from the inlets connecting the Albemarle- Pamlico Estuarine System to the ocean, the authors conclude that the retreat of the estuarine shoreline "can be accounted for mostly by sea level rise" (Everts *et al.*, 1983).

A1.G.3.2. Potential for Wetlands to Keep Pace with Rising Sea Level

Sections 4.3, 4.4, and 4.6 in Chapter 4 discuss wetland vertical and horizontal development. In North Carolina, vertical accretion rates have, for the most part, matched the rate of sea-level rise (see Section 4.6.2 in Chapter 4; Cahoon, 2003; Erlich, 1980; Riggs *et al.*, 2000). Vertical accretion rates as high as 2.4 to 3.6 mm per year have been measured, but the maximum rate at which wetlands can accrete is not well understood (Craft e*t al.*, 1993). Further, relative sea-level rise in North Carolina in recent years has ranged from approximately 1.8 to 4.3 mm per year at different points along the North Carolina coast (Zervas, 2004). As discussed in Section 4.6.2.2, wetland drowning could result in some areas if rates of global sea-level rise increase by 2 mm per year and is likely if rates increase by 7 mm per year. Day *et al.* (2005) suggest that brackish marshes in the Mississippi Delta region cannot survive 10 mm per year of relative sea-level rise. Under this scenario, fringe wetlands of North Carolina's lower coastal plain would drown. However, swamp forest wetlands along the piedmont-draining rivers are likely to sustain themselves where there is an abundant supply of mineral sediments (*e.g.*, river floodplains, but not river mouths) (Kuhn and Mendelssohn, 1999). As sea level rises further and waters with higher salt content reach the Albemarle-Pamlico peninsula, the ability of peat-based wetlands to keep up is doubtful, where the peat, root map, and vegetation would first be killed by brackish water (Poulter, 2005; Portnoy and Giblin, 1997; Pearsall and Poulter, 2005).

Finally, as described in Chapter 3, in a scenario where there are high rates of sea-level rise, more inlets would likely be created and segmentation or disintegration of some of the barrier islands is possible. This would cause a state change from a non-tidal to tidal regime as additional inlets open, causing the Albemarle and Pamlico Sounds to have a significant tide range and increased salinity, which would greatly disrupt current ecosystems. In this scenario, wave activity in the sounds could change erosion patterns and could impact wetlands (Riggs and Ames, 2003).

Table A1.4. Estuarine Shoreline Erosion Rates by Shoreline Type and the Percent of Total Shoreline for Each Type. From Riggs and Ames (2003)

Shoreline type	Percent of shoreline	Maximum rate per year (m)	Average rate per year
Sediment Bank	38		
Low bank	30	2.7	1.0
Bluff/high bank	8	8.0	0.8
Back-barrier strandplain beach	<1	0.6	-0.2[1]
Organic Shoreline	62		
Mainland marsh	55	5.6	0.9
Back-barrier marsh	<1	5.8	0.4
Swamp forest	7	1.8	0.7
Human Modified	Unknown	2.0	0.2
Weighted Average[2]			2.7

[1] The negative erosion rate listed refers to this shoreline type, on average, accreting.

[2] This weighted average excludes strandplain beaches and human-modified shorelines.

A1.G.3.3. Environmental Implications of Habitat Loss and Shore Protection

Ecological/habitat processes and patterns. Some wetland functions are proportional to size. Other functions depend on the wetland's edges, that is, the borders between open water and wetland. Many irregularly flooded marshes in coastal North Carolina are quite large. In the absence of tidal creeks and astronomical tidal currents, pathways for fish and invertebrate movement are severely restricted, except when wind tides are unusually high or during storm events. By contrast, the twice-daily inundation of tidal marshes by astronomical tides increases connections across the aquatic-wetland edge, as does the presence of tidal creeks, which allow fish and aquatic invertebrates to exploit intertidal areas (Kneib and Wagner, 1994). Mobility across ecosystem boundaries is less prevalent in irregularly flooded marshes, where some fish species become marsh "residents" because of the long distances required to navigate from marshes to subtidal habitats (Marraro *et al.*, 1991). Where irregularly flooded marshes are inundated for weeks at a time, little is known about how resident species adapt. These include, among other species, several types of fish (*e.g.*, killifish and mummichogs), brown water snakes, crustaceans (various species of crabs), birds (yellowthroat, marsh wren, harrier, swamp sparrow, and five species of rails), and several species of mammals (nutria, cotton rat, and raccoon). North Carolina's coastal marshes are also home to a reintroduced population of red wolves, and sea-level rise could affect this population (see Box A1.9).

Effects of human activities. Levees associated with waterfowl impoundments have isolated large marsh areas in the southern Pamlico Sound from any connection with estuarine waters. Impoundments were built to create a freshwater environment conducive to migratory duck populations and thus eliminated most other habitat functions mentioned above for brackish marshes. Further, isolation from sea level influences has likely disconnected the impoundments from pre-existing hydrologic gradients that would promote vertical accretion of marsh soil. If the impoundments were opened to an estuarine connection after decades of isolation, they would likely become shallow, open-water areas incapable of reverting to wetlands (Day *et al.*, 1990).

Drainage ditches, installed to drain land so that it would be suitable for agriculture and timber harvesting, are prevalent in North Carolina. By the 1970s, on the Albemarle-Pamlico Peninsula, there were an estimated 32 km (20 mi) of streams and artificial drainage channels per square mile of land, while the ratio in other parts of North Carolina ranged from 1.4:1 to 2.8:1 (Heath, 1975). In Dare County, there are currently an estimated 4 km of drainage ditch features per sq km (Poulter *et al.*, 2008). In many cases, ditches, some of which were dug more than a century ago to drain farmland (Lilly, 1981), now serve to transport brackish water landward, a problem that could become increasingly prevalent as sea level rises. Saltwater intrusion into agricultural soils and peat collapse are major consequences of this process.

A number of tide gates have been installed on the Albemarle-Pamlico Peninsula to reduce brackish water intrusion, but these will serve their purpose only temporarily, given continued sea-level rise. One analysis indicates that plugging ditches in selected places to reduce saltwater flow inland would be effective for local stakeholders. Another option is to install new water control structures, such as tide gates, in selected locations (Poulter *et al.*, 2008). Plugging ditches would also help restore natural drainage patterns to the marshes.

A1.G.4. Development, Shore Protection, and Coastal Policies

A1.G.4.1. Statewide Policy Context

Several North Carolina laws and regulations have an impact on response to sea-level rise within the state. First, setback rules encourage retreat by requiring buildings being constructed or reconstructed to be set back a certain distance from where the shoreline is located when construction permits are issued. Second, North Carolina does not allow "hard" shoreline armoring[82] such as seawalls and revetments on oceanfront shorelines[83], preventing property owners from employing one possible method of holding back the sea to protect property[84]. Along estuarine shores, however, shoreline armoring is allowed landward of any wetlands. The North Carolina Coastal Resources Commission (CRC) is preparing new state regulations for the location and type of estuarine shoreline stabilization structures to help encourage alternatives to bulkheads (NC CRC, 2008b; Feldman, 2008). The goals are similar to the "living shorelines" legislation recently enacted in Maryland (see Section A1 .F.2.2). Adding sand to beaches (*i.e.*, beach nourishment) is the preferred method in North Carolina to protect buildings and roads along the ocean coastline.

The state's Coastal Area Management Act (CAMA) has fostered land use planning in the 20 coastal counties to which it applies. Regulations authorized by CAMA require local land use plans to "[d]evelop policies that minimize threats to life, property, and natural resources resulting from development located in or adjacent to hazard areas, such as those subject to erosion, high winds, storm surge, flowing, or sea level rise". However, the state's technical manual for coastal land use planning (NC DCM, 2002) does not mention sea-level rise. Accordingly, local land use plans either do not mention sea-level rise at all, mention it only in passing, or explicitly defer decisions about vulnerable areas until more information is available in the future (Feldman, 2008; Poulter *et al.*, 2009). Nevertheless, the regulatory requirement to consider sea-level rise may eventually encourage local jurisdictions to consider how the communities most vulnerable to sea-level rise should prepare and respond (Feldman, 2008). Land-use plans are updated regularly and are an important tool for increasing public awareness about coastal hazards.

CAMA and the state's Dredge and Fill Law authorize the CRC to regulate certain aspects of development within North Carolina's 20 coastal counties. For example, the CRC issues permits for development and classifies certain regions as Areas of Environmental Concern (AECs, *e.g.*, ocean hazard zones and coastal wetlands) where special rules governing development apply. Land use plans are binding in AECs. In response to the threat of damage to coastal structures from the waves, since 1980 North Carolina has required new development to be set back from the oceanfront. The setbacks are measured from the first line of stable natural vegetation[85]. Single-family homes of any size—as well as multi-family homes and non-residential structures with less than 5,000 sq ft of floor area—must be set back by 60 ft or 30 times the long-term rate of erosion as calculated by the state, whichever is greater. Larger multi-family homes and nonresidential structures must be set back by 120 ft or the erosion-based setback distance, whichever is greater. The setback distance for these larger structures is set as either 60 times the annual erosion rate or 105 ft plus 30 times the erosion rate, whichever is less[86]. North Carolina is considering changes to its oceanfront setback rules, including progressively larger setback factors for buildings with 10,000 sq ft of floor area or more (NC CRC, 2008a). Along estuarine shorelines, North Carolina has a 30-ft setback[87] and restricts development between 30 and 75 ft from the shore[88]. As the shore moves inland, these setback lines move inland as well.

As of 2000, the U.S. Army Corps of Engineers (USACE) participated in beach nourishment projects along more than 51 km (32 mi) of North Carolina's shoreline (including some nourishment projects that occurred as a result of nearby dredging projects), and nourishment along an additional 137 km (85 mi) of coastline had been proposed (USACE, 2000)[89]. If necessary, property owners can place large geotextile sandbags in front of buildings to attempt to protect them from the waves. Standards apply to the placement of sandbags, which is supposed to be temporary (to protect structures during and after a major storm or other short-term event that causes erosion, or to allow time for relocation)[90]. Buildings are supposed to be moved or removed within two years of becoming "imminently threatened" by shoreline changes[91].

North Carolina officials are in the process of reassessing certain state policies in light of the forces of shoreline change and climate change. Policy considerations have been affected by numerous studies that researchers have published on the potential effects of sea-level rise on North Carolina (Poulter *et al.*, 2009). The state legislature appointed a Legislative Commission on Global Climate Change to study and report on potential climate change effects and potential mitigation strategies, including providing recommendations that address impacts on the coastal zone[92]. The Commission's recommendations have not yet been finalized, but an initial draft version offered such suggestions as creating a mechanism to purchase land or conservation easements in low-lying areas at great risk from sea-level rise; providing incentives for controlling erosion along estuarine shorelines using ecologically beneficial methods; creating a commission to study adaptation to climate change and make recommendations about controversial issues; and inventorying, mapping, and monitoring the physical and biological characteristics of the entire shoreline (Feldman, 2008; Riggs *et al.*, 2007).

The CRC is also considering the potential effects of sea-level rise and whether to recommend any changes to its rules affecting development in coastal areas (Feldman, 2008). In addition, NCDCM is developing a Beach and Inlet Management Plan to define beach and inlet management zones and propose preliminary management strategies given natural forces, economic factors, limitations to the supply of beach-quality sand, and other constraints (Moffatt and Nichol, 2007).

A1.G.4.2. Current Land Use

Ocean Coast (from north to south). North Carolina's ocean coast, like the coasts of most states, includes moderate and densely developed communities, as well as undeveloped roadless barrier islands. Unlike other mid-Atlantic states, North Carolina's coast also includes a major lighthouse (at Cape Hatteras) that has been relocated landward, a roadless coastal barrier that is nevertheless being developed (described below), and densely populated areas where storms, erosion, and sea-level rise have caused homes to become abandoned or relocated.

The northern 23 km (14 mi) of the state's coastline is a designated undeveloped coastal barrier under the Coastal Barrier Resources Act (CBRA) and hence ineligible for most federal programs (USFWS, undated[c]) This stretch of barrier island includes two sections of Currituck National Wildlife Refuge, each about 2 km (1 mi) long, which are both off-limits to development. Nevertheless, the privately owned areas are gradually being developed, even though they are accessible only by boat or four-wheel drive vehicles traveling along the beach. The CBRA zones are ineligible for federal beach nourishment and flood insurance (USFWS, undated[c]).

Along the Dare County coast from Kitty Hawk south to Nags Head, federal legislation has authorized shore protection, and USACE (2006b) has concluded that the proposed project would be cost-effective. In some areas, homes have been lost to shoreline erosion (Pilkey *et al.*, 1998) (see Figure 12.6). Continued shore erosion has threatened some of the through-streets parallel to the shore, which had been landward of the lost homes. Given the importance of those roads to entire communities (see Section 12.2 in Chapter 12) small sand replenishment projects have been undertaken to protect the roads (Town of Kitty Hawk, 2005). The planned beach nourishment project does not extend along the coast to the north of Kitty Hawk. Those beaches are generally not open to the public and are currently ineligible for publicly funded beach nourishment.

From Nags Head to the southwestern end of Hatteras Island, most of the coast is part of Cape Hatteras National Seashore. A coastal highway runs the entire length, from which one can catch a ferry to Ocracoke Island, carrying through traffic to both Ocracoke and Carteret County. Therefore, the National Park Service must balance its general commitment to allowing natural shoreline processes to function (see Chapter 12.1; NRC 1988) with the needs to manage an important transportation artery. In most cases, the approach is a managed retreat, in which shores generally migrate but assets are relocated rather than simply abandoned to the sea. Congress appropriated \$9.8 million to move the Cape Hatteras Lighthouse 1,600 ft (468 m) inland in 1999 (NPS, 2000) (see Figure 11.1a). The coastal highway has been relocated inland in places. Because it is essential infrastructure, its protection would probably require maintaining the barrier island itself, for example, by filling inlets after severe storms. A possible exception is where the highway runs through Pea Island National Wildlife Refuge on the northern end of Hatteras Island, just south of the bridge over Oregon Inlet. The federal and state governments are considering the possibility that when a new bridge is built over Oregon Inlet, it would bypass the National Wildlife Refuge and extend over Pamlico Sound just west of Hatteras Island as far as Rodanthe (USDOI, 2007).

The undeveloped Portsmouth Island and Core Banks constitute Cape Lookout National Sea Shore and lack road access. Cape Lookout is located on Core Banks. Shackleford Banks, immediately adjacent to the southwest, is also roadless and uninhabited. Southwest of Cape Lookout, the coast consists mostly of developed barrier islands, conservation lands, and designated "undeveloped coastal barriers" that are nevertheless being developed. Bogue Banks includes five large communities with high dunes and dense forests (Pilkey *et al.*, 1998). Bogue Banks also receives fill to widen its beaches regularly.

To the west of Bogue Banks are the barrier islands of Onslow County and then Pender County. Some islands are only accessible by boat, and most of these are undeveloped. North Topsail Beach, on Topsail Island, has been devastated by multiple hurricanes, in part due to its low elevation and the island's narrow width. Erosion has forced multiple roads on the island to be moved. While some parts of North Topsail Beach are part of a unit under the CBRA system, making them ineligible for federal subsidies, development has occurred within them nonetheless (Pilkey *et al.*, 1998).

Further to the southwest are the barrier islands of New Hanover County, including Figure Eight Island, which is entirely privately-owned with no public access to the beach, and hence ineligible for public funding for beach nourishment (see Chapter 8). Wrightsville Beach, like many other communities southwest of Cape Lookout, has an inlet on each side. It is the site of a dispute to protect a hotel from being washed away due to inlet migration (Pilkey *et al.*, 1998). The USACE has made a long-term commitment to regular beach renourishment to maintain the place of the shoreline in Wrightsville Beach and Carolina Beach (USACE, 2006a). An exception to North Carolina's rules forbidding hardened structures has been granted in Kure Beach, west of Carolina Beach, where stone revetments have been placed on the oceanfront to protect Fort Fisher (which dates back to the Civil War). These structures also protect a highway that provides access to the area (Pilkey *et al.*, 1998). Most of the beach communities in New Hanover County are extensively developed.

Some of the barrier islands in Brunswick County, close to the South Carolina state line, are heavily forested with high elevations, making them more resilient to coastal hazards (Pilkey *et al.*, 1998). Holden Beach and Ocean Isle Beach, however, contain many dredge-and-fill finger canals. Historically, at least two inlets ran through Holden Beach; and storms could create new inlets where there are currently canals (Pilkey *et al.*, 1998).

Estuarine Shores. Significant urbanization was slow to come to this region for many reasons. Most of the area is farther from population centers than the Delaware and Chesapeake Estuaries. The Outer Banks were developed more slowly than the barrier islands of New Jersey, Delaware, and Maryland. Most importantly, the land is mostly low and wet.

Unlike the Delaware Estuary, North Carolina does not have a long history of diking tidal wetlands to reclaim land from the sea for agricultural purposes[93]. However, the state is starting to gain experience with dikes to protect agricultural lands from flooding. In Tyrrell County, the Gum Neck has been protected with a dike for four decades. A dike is under construction for the town and farms around Swan Quarter (Allegood, 2007), the county seat of Hyde County (which includes Ocracoke Island). Hurricanes Fran and Floyd led to federally-sponsored purchases of thousands of properties across North Carolina's eastern counties, facilitating the demolition or relocation of associated structures. Pamlico County has encouraged people to gradually abandon Goose Creek Island in the eastern portion of the county, by working with FEMA to relocate people rather than rebuild damaged homes and businesses (Barnes, 2001). By contrast, in other areas (*e.g.*, parts of Carteret County), people took the opposite approach and elevated homes.

Geography, coastal features, and community characteristics vary greatly along North Carolina's coast. Thus, one can assume that a variety of different planning and adaptation strategies related to shoreline change and sea-level rise would be needed, particularly over the long term. Scientists, managers, and community members in North Carolina have undertaken a variety of efforts to better understand and begin to address potential sea-level rise vulnerabilities and impacts. These research and collaborative efforts may increase awareness, receptivity, and readiness to make informed coastal management decisions in the future (Poulter *et al.*, 2009).

REFERENCES

* Indicates non-peer reviewed literature; available upon request

Abel, K. W. & Hagan, S. M. (2000). Effects of common reed (*Phragmites australis*) invasion on marsh surface macrofauna: Response of fishes and decapod crustaceans. *Estuaries, 23(5)*, 633-646.

Abel, K. W., Nemerson, D. M. Light, P. R. & Bush, R. O. (2000). Initial response of fishes to marsh restoration at a former salt hay farm bordering Delaware Bay. In: *Concepts and Controversies in Tidal Marsh Ecology*, [Weinstein, M.P. and D.A. Kreeger (eds.)]. Kluwer Academic Publishers, Dordrecht, the Netherlands,

* Allegood, J., (2007). Dike to protect Swan Quarter: Battered coastal village hopes to hold back flooding from the next hurricane. *Raleigh News and Observer*, May 29, 2007.

ADC (Alexandria Drafting Company), (2008). *Stafford County, VA Street Atlas*. Langennscheldt Publishing Group, Duncan, SC, 60 pp.

* AP (Associated Press), (1985). Doubled erosion seen for Ocean City. *Washington Post*, November 14th. (Maryland Section).

Ayers, R.A., (2005). *Human Impacts to Sensitive Natural Resources on the Atlantic Barrier Islands on the Eastern Shore of Virginia*. Virginia Department of Environmental Quality, Coastal Zone Management Program, Richmond, 14 pp. <http://www.deq.state.va.us/coastal/documents/task1 1 -07-04b.pdf>

Baker, A. J., Gonzalez, P. M., Piersma, T., Niles, L. J. & I. de Lima Serrano do Nacimento, (2004). Rapid population decline in Red Knots: fitness consequences of decreased refueling rates and late arrival in Delaware Bay. *Proceedings of the Royal Society of London, B, 271*, 875-882.

Barnes, J., (2001). *North Carolina's Hurricane History*. University of North Carolina Press, Chapel Hill, NC, 336 pp.

BBNEP (Barnegat Bay National Estuary Program, Scientific and Technical Advisory Committee), (2001). Chapter 7 of *The Barnegat Bay Estuary Program Characterization Report*. <http://www.bbep.org/Char _Rpt/Ch7/Chapter%207.htm>

Baltimore District, (1996). *Maryland State Programmatic General Permit (MDSPGP)*. U.S. Army Corps of Engineers, Baltimore.

Berkson, J. and Shuster C. N. Jr., (1999). The horseshoe crab: the battle for a true multiple-use resource. *Fisheries*, 24, 6-10.

Berman, M. R., Berquist, H., Dewing, S., Glover, J., Hershner, C. H., Rudnicky, T., Schatt, D. E. & Skunda, K. (2000). *Mathews County Shoreline Situation Report*. Special report in applied marine science and ocean engineering no. 364. Comprehensive Coastal Inventory Program, Virginia Institute of Marine Science, College of William and Mary, Gloucester Point, VA.

Bernd-Cohen, T. & Gordon, M. (1999). State coastal program effectiveness in protecting natural beaches, dunes, bluffs, and rock shores. *Coastal Management, 27*, 187-217.

Bernick, A. J., (2006). *New York City Audubon 's Harbor Herons Project: 2006 Interim Nesting Survey*. Report prepared for New York City Audubon, New York, 22 pp.

Bertness, M.D., (1999). *The Ecology of Atlantic Shorelines*. Sinauer Associates, Sunderland, MA, 417 pp.

* Bleyer, B., (2007). Erosion protection for Montauk lighthouse creates waves. *Lighthouse Digest.* <http://www.lighthousedigest.com/Digest/StoryPage.cfm?StoryKey=263 6>

Boesch, D. F. & Turner, R. E. (1984). Dependence of fishery species on salt marshes: the role of food and refuge. *Estuaries*, *7(4A)*, 460-468.

Botton, M. L., Loveland, R. E. & Jacobsen, T. R. (1988). Beach erosion and geochemical factors: influence on spawning success of horseshoe crabs (*Limulus polyphemus*) in Delaware Bay. *Marine Biology*, *99(3)*, 325-332.

Botton, M. L., Loveland, R. E., Tanacredi, J. T. & Itow, T. (2006). Horseshoe crabs (*Limulus polyphemus*) in an urban estuary (Jamaica Bay, New York) and the potential for ecological restoration. *Estuaries and Coasts*, *29(5)*, 820-830.

Bridgham, S. D. & Richardson, C. J. (1993). Hydrology and nutrient gradients in North Carolina peatlands. *Wetlands*, *13*, 207-218.

Brinson, M. M., Bradshaw, H. D. & Jones, M. N. (1985). Transitions in forested wetlands along gradients of salinity and hydroperiod. *Journal of the Elisha Mitchell Scientific Society*, *101*, 76-94.

Burger, J., (1984). Cited in New York State Department of State and the U.S. Fish and Wildlife Service, Southern New England–New York Bight Coastal Ecosystems Program. (1998). *Shorebirds*. South Shore Estuary Reserve, Technical Report Series. <http://www.nyswaterfronts.com/Final_Draft_HTML/Tech_Report_HTM/PDFs/C8A_Index_pdf.htm>

Burger, J., Niles, L. and Clark, K. E. (1997). Importance of beach, mudflat, and marsh habitats to migrant shorebirds in Delaware Bay. *Biological Conservation 79(2)*, 283-292.

Buzzelli, C., Ramus, J. R. and Paerl, H. W. (2003). Ferry-based monitoring of surface water quality in North Carolina estuaries. *Estuaries*, *26*, 975-984.

* Byrne, D. M., (1995). The effect of bulkheads on estuarine fauna: a comparison of littoral fish and macroinvertebrate assemblages at bulkheaded and non-bulkheaded shorelines in a Barnegat Bay Lagoon. In: *Second Annual Marine Estuarine Shallow Water Science and Management Conference, Atlantic City, NJ.* Environmental Protection Agency, Philadelphia, PA, pp. 53-56.

Cahoon, D. R., (2003). Storms as agents of wetland elevation change: their impact on surface and subsurface sediment processes. *Proceedings of the International Conference on Coastal Sediments 2003*, May 18-23, 2003, Clearwater Beach, Florida. World Scientific Publishing Corporation, Corpus Christi TX.

Caldwell, W. S., (2001). *Hydrologic and Salinity Characteristics of Currituck Sound and elected Tributaries in North Carolina and Virginia, 1998-99.* USGS water resources investigation report 01-4097. U.S. Geological Survey, Raleigh, NC, 36 pp.

Casey, J. & Doctor, S. (2004). Status of finfish populations in the Maryland Coastal Bays. In: *Maryland's Coastal Bays: Ecosystem Health Assessment 2004* [Wazniak, C.E. and M.R. Hall (eds.)]. DNR-12-1202-0009. Maryland Department of Natural Resources, Tidewater Ecosystem Assessment, Annapolis, chapter 8.4.

Castro, G. & Myers, J. P. (1993). Shorebird predation on eggs of horseshoe crabs during spring stopover on Delaware Bay. *Auk*, *110(4)*, 927-930.

CBCAC (Chesapeake Bay Critical Area Commission, Maryland Department of Natural Resources), (2001). Chesapeake Bay Critical Area Line. <http://www.marylandgis.net/metadataexplorer/full_metadata.j sp?docId=%7B480 9747B-1 DF0-463 5-92D1 - CE99AA1 A84C 1 %7D&loggedIn=false>

* CBP (Chesapeake Bay Program), (2002). *The Impact of Susquehanna Sediments on the Chesapeake Bay*. Scientific and Technical Advisory Committee Workshop Report. May 2002

* CBP (Chesapeake Bay Program), (2006). *Diamondback Terrapin*. [web site] <http://www. chesapeakebay.net/diamondback_terrapin.htm>

* Chase, C. M., (1979). *The Holocene Geologic History of the Maurice River Cove and its Marshes, Eastern Delaware Bay, New Jersey*. MS Thesis, University of Delaware, Newark, 129 pp.

Clark, K., (1996). Horseshoe crabs and the shorebird connection. In: *Proceedings of the Horseshoe Crab Forum: Status of the Resource* [Farrell, J. and C. Martin (eds.)].University of Delaware Sea Grant College Program, Lewes, pp. 23-25.

* CLO (Cornell Laboratory for Ornithology), 2004: *All About Birds: Piping Plover*. [web site] <http://www.birds >

Cohen, J. B., Wunker, E. H. & Fraser, J. D. (2005). Substrate and vegetation selection by piping plovers. *Wilson Journal of Ornithology, 120(2)*, 404-407

Conley, M., (2004). *Maryland Coastal Bays Aquatic Sensitive Areas Initiative Technical Report*. Prepared by the Maryland Department of Natural Resources, Coastal Zone Management Division.

Cooper, M. J. P., Beevers, M. D. & Oppenheimer, M. (2005). *Future Sea-level Rise and the New Jersey Coast*. Science, Technology, and Environmental Policy Program, Woodrow Wilson School of Public and International Affairs, Princeton University, Princeton, NJ, 37 pp. <http://www.princeton.edu/~step/people/Oppenheimer%20Future%20of%20Sea %20Level%20Rise.pdf>

CPCP (City of Poquoson Comprehensive Plan), 1999: *Environmental Element*. Poquoson, VA. <http://www.ci.poquoson.va.us/>

Craft, C. B. and Richardson, C. J. (1998). Recent and long-term organic soil accretion and nutrient accumulation in the everglades. *Soil Science Society of America Journal, 62*, 834-843.

Craft, C. B., Seneca, E. D. & Broome, S. W. (1993). Vertical accretion in microtidal regularly and irregularly flooded estuarine marshes. *Estuarine, Coastal, and Shelf Science, 37*, 371-386.

Daniels, R. C., (1996). An innovative method of model integration to forecast spatial patterns of shoreline change: A case study of Nags Head, North Carolina. *The Professional Geographer, 48(2)*, 195-209.

Darmondy, R. G. & Foss, J. E. (1979). Soil-landscape relationships of tidal marshes of Maryland. *Soil Science Society of America Journal, 43(3)*, 534-541.

Day, R. H., Holz, R. K. & Day J. W. Jr., (1990). An inventory of wetland impoundments in the coastal zone of Louisiana, USA: historical trends. *Environmental Management, 14(2)*, 229-240.

Day, J. W., Jr., Barras, J., Clairain, Johnston, J., Justic, D. & Kemp, G. P. J.-Y. Ko, Lane, R., Mitsch, Steyer, W. J., Templet, G. & Yañez-Arancibia, A. (2005). Implications of global climatic change and energy cost and availability for the restoration of the Mississippi delta. *Ecological Engineering, 24(4)*, 253-265.

DCOP (District of Columbia Office of Planning), (2003). Anacostia Riverparks Target Area Plan and Riverwalk Design Guidelines. District of Columbia Office of Planning, Washington, DC, 6 pp.

Dean, C., (1999). *Against the Tide: The Battle for America's Beaches*. Columbia University Press, New York, 279 pp.

DDA (Delaware Department of Agriculture), 2008: *Delaware Agricultural Preservation Program: Statewide District/Easement Maps*. Dover, DE. <http://dda.delaware.gov/aglands/ forms/2008/062008JuneMap.pdf>

DNREC (Delaware Department of Natural Resources and Environmental Control), undated: *Discover Delaware's Inland Bays*. Document No. 40-01-01/03/03/01. <www.dnrec.state.de.us/dnrec2000/Library/Misc/InlandBays>

DNREC (Delaware Department of Natural Resources and Environmental Control), (2001). *Inland Bays/Atlantic Ocean Basin Assessment*. Report. Document No. 40- 01/01/01/02. Published June 2001.

DNREC (Delaware Department of Natural Resources and Environmental Control), (2007). Inland Bays Pollution Control Strategy and Proposed Regulations. April 2007. DNREC, Division of Water Resources, Dover, DE.

Doctor, S. & Wazniak, C. E. (2005). Status of horseshoe crab, *Limulus polyp hem us,* populations in Maryland coastal bays. In: *Maryland's Coastal Bays: Ecosystem Health Assessment 2004* [Wazniak, C.E. and M.R. Hall, (eds.)]. DNR-12-1202- 0009. Maryland Department of Natural Resources, Tidewater Ecosystem Assessment, Annapolis, chapter 8.7.

DOSP (Delaware Office of State Planning), (1997). *Land Use/Land Cover*. Dover, DE.

Dove, L. E. & Nyman R. M. (eds.), (1995). *Living Resources of the Delaware Estuary*. Delaware Estuary Program Report Number 9 5-07. Partnership for the Delaware Estuary, Wilmington, DE.

Dugan, J. E., Hubbard, D. M., McCrary, M. D. & Pierson, M.O. (2003). The response of macrofauna communities and shorebirds to macrophyte wrack subsidies on exposed sandy beaches of southern California. *Estuarine, Coastal, and Shelf Science, 58S*, 25–40.

Ednie, A. P., Undated: *Birding Delaware's Prehistoric Past: Thompson's Island at Delaware Seashore State Park*. <http://www. dvoc.org/ DelValBirding /Places/ Thompsons Island DE.htm>

Engelhardt, K. A. M., Seagle, S. & Hopfensperger, K. N. (2005). *Should We Restore Dyke Marsh? A Management Dilemma Fcing George Washington Memorial Parkway*. Submitted to the George Washington Memorial Parkway, National Park Service, National Capital Region, McLean, VA.

* Erlich, R. N., (1980). *Early Holocene to Recent Development and Sedimentation of the Roanoke River Area, North Carolina*. M.S. thesis, Department of Geology, University of North Carolina, Chapel Hill, 83 pp.

Ernst, C. H. & Barbour, R. W. (1972). *Turtles of the United States*. University Press of Kentucky, Lexington, 347 pp.

Erwin, R. M., (1996). Dependence of waterbirds and shorebirds on shallow-water habitats in the mid-Atlantic coastal region: An ecological profile and management recommendations. *Estuaries, 19(2A)*, 213-219.

Erwin, R. M., Sanders, G. M. & Prosser, D. J. (2004). Changes in lagoonal marsh morphology at selected northeastern Atlantic Coast sites of significance to migratory waterbirds. *Wetlands, 24(4)*, 891-903.

Erwin, R. M., Sanders, G. M., Prosser, D. J. & Cahoon, D. R. (2006). High tides and rising seas: potential effects on estuarine waterbirds. In: *Terrestrial Vertebrates in Tidal*

Marshes: Evolution, Ecology, and Conservation [Greenberg, R. (ed.)]. Studies in avian biology no. 32. Cooper Ornithological Society, Camarillo, CA, pp. 2 14-228.

Everts, C. H., Battley J. P. Jr. & Gibson, P. N. (1983). *Shoreline Movements: Report 1, Cape Henry, Virginia, to Cape Hatteras, North Carolina, 1849-1980*. Technical report CERC-83-1. U.S. Army Corps of Engineers, Washington, DC, and National Oceanic and Atmospheric Administration, Rockville, MD, 111 pp.

Eyler, T. B., Erwin, R. M. Stotts, D. B. & Hatfield, J. S. (1999). Aspects of hatching success and chick survival in gull-billed terns in coastal Virginia. *Waterbirds*, 22, 54-59.

ESRI (Environmental Systems Research Institute), (1999). *Parks*. National Park Service Washington, DC.

Federal Register, (2006). Availability of a Draft Integrated Feasibility Report and Environmental Impact Statement for the Mid-Chesapeake Bay Island Ecosystem Restoration Project in Dorchester County, on Maryland's Eastern Shore. *Federal Register Notices*, *71(174)*, 53090-53091.

Feinberg, J. A. & Burke, R. L. (2003). Nesting Ecology and Predation of Diamondback Terrapins, *Malaclemys terrapin*, at Gateway National Recreation Area, New York. *Journal of Herpetology*, *37(3)*, 517-526.

Feldman, R. L., (2008). Recommendations for responding to sea level rise: lessons from North Carolina. In: *Proceedings of Solutions to Coastal Disasters 2008*, Oahu, HI. American Society of Civil Engineers, Reston VA, pp. 15-27.

Fleming, G. P., Coulling, P. P., K. D. Patterson & Taverna, K. (2006). *The Natural Communities of Virginia: Classification of Ecological Community Groups*. Second approximation, version 2.2. Virginia Department of Conservation and Recreation, Division of Natural Heritage, Richmond, VA. <http://www.dcr.virginia. gov/natural>

Galbraith, H., Jones, R., Park, P., Clough, J., Herrod-Julius, S., Harrington, B. & Page, G. (2002). Global climate change and sea-level rise: potential losses of intertidal habitat for shorebirds. *Waterbirds, 25(2),* 173-183.

Glick, P., Clough, J. & Nunley, B. (2008). *Sea Level Rise and Coastal Habitats in the Chesapeake Bay Region*. National Wildlife Federation, [Washington, DC], 121 pp.

Gornitz, V., Couch, S. & Hartig, E. K. (2002). Impacts of sea level rise in the New York City metropolitan area. *Global and Planetary Change*, *32(1)*, 61-88.

Greenlaw, J. S. & Rising, J. D. (1994). Sharp-tailed sparrow (*Ammodramus audacutus*). In: *The Birds of North America*, No. 127, [Poole, A. and F. Gill, (ed.)]. The Academy of Natural Sciences, Philadelphia and the American Ornithologists' Union, Washington, DC, as cited in Chapter 6 of *The Barnegat Bay Estuary Program Characterization Report*. Prepared by the Barnegat Bay National Estuary Program (Scientific and Technical Advisory Committee), January, 2001. <http://www.bbep.org/Char _Rpt/Ch6/Chapter%206.htm>

Hackney, C. T. & Yelverton, G. F. (1990). Effects of human activities and sea level rise on wetland ecosystems in the Cape Fear River Estuary, North Carolina, USA. In: *Wetland Ecology and Management: Case Studies* [Whigman, D.F., R.E. Good, and J. Kvet (eds).]. Kluwer Academic Publishers, Dordecht, the Netherlands, pp. 55-61.

Hardaway, C. S., Varnell, L. M., Milligan, D. A, Thomas, G. R. & Hobbs, C. H. (2001). *Chesapeake Bay Dune Systems: Evolution and Status*. Virginia Institute of Marine Science, 38 pp.

Hardaway, C. S., Jr., Milligan, D.A., Varnell, L. M., Wilcox, C., Thomas, G. R. & Comer, T. R. (2005). *Shoreline Evolution, Chesapeake Bay Shoreline, City of Virginia Beach, Virginia.*Virginia Institute of Marine Sciences, College of William and Mary (Gloucester Point) and Virginia Department of Environmental Quality (Richmond), 16 pp.<http://web each/Virginia_Beach_Shore_Evolution.pdf>

Hartig, E. K. & Gornitz, V. (2004). Salt marsh change, 1926-2003 at Marshlands Conservancy, New York. In: *7th Bienniel Long Island Sound Research Conference Proceedings*, November 4-5, 2004, Stony Brook, NY. Long Island Sound Foundation, Groton, CT, pp. 61-65. <http://lisfoundation.org/downloads/lisrc_proceedings2004.pdf>

Hartig, E. K., Gornitz, V., Kolker, A., Mushacke, F. & Fallon, D. (2002). Anthropogenic and climate-change impacts on salt marshes of Jamaica Bay, New York City. *Wetlands, 22(1)*, 71-89.

Heath, R., (1975). *Hydrology of the Albemarle-Pamlico Region, North Carolina: A Preliminary Report on the Impact of Agricultural Developments.* Water resources investigations 9-75. U.S. Geological Survey, Raleigh, NC, 98 pp.

Hedrick, C., Millhouser, W. & J. Lukens, (2000). *State, Territory, and Commonwealth Beach Nourishment Programs: A National Overview.* Office of Ocean & Coastal Resource Management Program policy series technical document 00-01. National Oceanic and Atmospheric Administration, National Ocean Service.

Henman, J. and B. Poulter, (2008). Inundation of freshwater peatlands by sea level rise: uncertainty and potential carbon cycle feedbacks. *Journal of Geophysical Research*, 113, G01011, doi:10.1029/2006JG000395.

Holst, L., Rozsa, R., Benoit, L., Jacobsen & Rilling, C. (2003). *Long Island Sound Habitat Restoration Initiative, Technical Support for Habitat Restoration, Section 1: Tidal Wetlands.* EPA Long Island Sound Office, Stamford, CT, 25 pp. <http://www.longislandsoundstudy.net/habitat>

Hull, C. H. J & J.G.Titus, I1986). *Greenhouse Effect, Sea Level Rise, and Salinity in the Delaware Estuary.* Delaware River Basin Commission, West Trenton, NJ. <www.risingsea.net/DE/DRBC.html>

Hyde County, (2008). Invitation for Bids (Construction and Vegetation of Phase XII, in the Swan Quarter Watershed Project). <http://www.hydecountync. gov/news/PRESS%20 RELEASES/Phase%20XII%20Final%20Bid%20Documents/Bid%20Documents%20&% 20Specs/II.%20%20Inv itation%20for%20Bids%20 1 2-06-08%20pdf.pdf>

Isle of Wight, County of., (2001) *Comprehensive Plan: Isle of Wight County, Virginia.* <http://www.co.isle-of-wight.va.us/index.php?option=com_content&task=view&id>

Jacob, K., V. Gornitz, and C. Rosenzweig, 2007: Vulnerability of the New York City metropolitan area to coastal hazards, including sea-level rise: Inferences for urban coastal risk management and adaptation policies. In: *Managing Coastal Vulnerability* [McFadden, L., R. Nicholls, and E. Penning-Rowsell (eds.)]. Elsevier, Amsterdam and Oxford, pp. 139-156.

Jackson, E.L., A.S. Rowden, M.J. Attrill, S. Bossey, & M. Jones, 2001: The importance of seagrass beds as habitat for fishery species. *Oceanography and Marine Biology Annual Review, 39*, 269-303.

James City County, (2003) *James City County Comprehensive Plan.* Land Use (pp. 101- 141) and Environment (pp. 42-67) Chapters. Williamsburg, Virginia.

Johnson, Z.P. (2000). *A Sea Level Rise Response Strategy for the State of Maryland.* Maryland Department of Natural Resources, Coastal Zone Management Division, Annapolis, 49 pp.

Johnson, Z. & A. Luscher (2004). Management, planning, and policy conference sessions. In: *Hurricane Isabel in Perspective* [Sellner, K.G. and N. Fisher (eds.)]. Chesapeake Research Consortium publication 05-160. Chesapeake Research Consortium, Edgewater, MD, pp. 221-232.

Johnston, D.W. (2000). The Dyke Marsh preserve ecosystem. *Virginia Journal of Science, 51*, 223-273.

Jones, R. & J. Wang, (2008). Interpolating elevations: proposed method for conducting overlay analysis of GIS data on coastal elevations, shore protection, and wetland accretion. Section 1.2 in: *Background Documents Supporting Climate Change Science Program Synthesis and Assessment Product 4.1: Coastal Elevations and Sensitivity to Sea-level Rise* [Titus, J.G. and E. Strange (eds.)]. EPA 430R07004, Environmental Protection Agency, Washington, DC, pp. 45-67. <http://epa.gov/climatechange/eEffects/coastal/background.html>

Kastler, J. A. & Wiberg, P. L. (1996). Sedimentation and boundary changes of Virginia salt marshes. *Estuarine, Coastal, and Shelf Science, 42(6)*, 683-700.

Kearney, M. S., Rogers, A. S., Townsend, J. R. G., Rizzo, E., Stutzer, D., Stevenson, J. C. & Sundborg, K. (2002). Landsat imagery shows decline of coastal marshes in Chesapeake and Delaware Bays. *EOS Transactions of the American Geophysical Union, 83(16)*, 173.

* Kerlinger, P., Undated: Cape May birding places: the Cape May Migratory Bird Refuge. *Cape May Times.* <http://www.capemaytimes.com/birds/ .htm>

Kneib, R. T., (1997). The role of tidal marshes in the ecology of estuarine nekton. *Oceanography and Marine Biology, an Annual Review*, *35*, 163-220.

Kneib, R. T. & Wagner, S. L. (1994). Nekton use of vegetated marsh habitats at different stages of tidal inundation. *Marine Ecology-Progress Series*, *106*, 227-23 8.

* Kozac, C., (2006). Alligator River National Wildlife Refuge threatened by global warming. *The Virginian-Pilot*, October 6, 2006.

Kraft, J. C. and John, C. J. (1976). *Introduction, the Geological Structure of the Shorelines of Delaware*. Delaware Sea Grant Technical Report, #DEL-SG-14-76.

Kraft, J. C., Hi-Il Y. & Khalequzzaman, H. I. (1992). Geologic and human factors in the decline of the tidal saltmarsh lithosome: the Delaware estuary and Atlantic coastal zone. *Sedimentology and Geology*, *80*, 23 3-246.

Kreamer, G. R. (1995). Saltmarsh invertebrate community. In: *Living Resources of the Delaware Estuary* [Dove, L.E. and R.M. Nyman (eds.)].The Delaware Estuary Program, pp. 81-90.

Kreeger, D. & Titus, J. G. (2008). Delaware Bay. Section 3.7 in: *Background Documents Supporting Climate Change Science Program Synthesis and Assessment Product 4.1: Coastal Elevations and Sensitivity to Sea Level Rise* [Titus, J.G. and E.M. Strange (eds.)]. EPA 430R07004. U.S. Environmental Protection Agency, Washington, DC, pp. 242-250.<http://epa.gov/climatechange/effects/coastal/background.html>

Kuhn, N. L. & Mendelssohn, I. A. (1999). Halophyte sustainability and sea level rise: Mechanisms of impact and possible solutions. In: *Halophyte Uses in Different Climates I: Ecological and Ecophysiological Studies* [Leith, H., (ed.)]. Backhuys Publishers, Leiden, Netherlands, pp. 113-126.

Kumer, J., (2004). Status of the endangered piping plover, *Charadrius melodus,* population in the Maryland Coastal Bays. In *Maryland's Coastal Bays: Ecosystem Health Assessment.* pp. 8-97.

Lathrop, R. G., Jr. & Love, A. (2007). *Vulnerability of New Jersey's Coastal habitats to Sea Level Rise.* Grant F. Walton Center for Remote Sensing and Spatial Analysis, Rutgers University, New Brunswick, NJ, and American Littoral Society, Highlands, NJ, 17 pp. <http://www.crssa.rutgers.edu>

Lathrop, R., Allen, M. & Love, A. (2006). *Mapping and Assessing Critical Horseshoe Crab Spawning Habitats in Delaware Bay.* Grant F. Walton Center for Remote Sensing and Spatial Analysis, Cook College, Rutgers University, New Brunswick, NJ, p.15 Table 8. <:http://deathstar.rutgers.edu/projects/delbay/>

Leatherman, S., (1989). National assessment of beach nourishment requirements associated with accelerated sea level rise. In*: The Potential Effects of Global Climate Change on the United States. Appendix: B, Sea Level Rise.* EPA 230-05- 89-052. U.S. Environmental Protection Agency, Washington, DC, pp. 2-1 to 2-30. <http://epa.gov/climatechange /effects/coastal/appB .html>

Leatherman, S. P. (1992). Coastal land loss in the Chesapeake Bay Region: An Historical Analogy approach to global climate analysis and response. In: *Regions and Global Warming: Impacts and Response Strategies* [Schmandt, J. (ed.)]. Oxford University Press.

Leatherman, S., Zhang, K. & Douglas, B. (2000). Sea level rise shown to drive coastal erosion: A reply. *EOS, Transactions of the American Geophysical Union*, *81(38)*, 437-441.

Lee, G. J., Carter, T. E. Jr., Villagarcia, M. R. Z. Li, X. Zhou, M.O. Gibbs & H.R. Boerma, (2004). A major QTL conditioning salt tolerance in S-100 soybean and descendent cultivars. *Theoretical and Applied Genetics*, *109(8)*, 1610-1619.

Lilly, J. P. (1981). A history of swamp land development in North Carolina. In: *Pocosin Wetlands* [Richardson, C.J., (ed.)]. Hutchinson Ross, Stroudsburg, PA, pp. 20-30.

Lippson, A. J. & Lippson, R. L. (2006). *Life in the Chesapeake Bay.* Johns Hopkins University Press, Baltimore, MD, 3rd edition, 324 pp.

* LISF (Long Island Sound Foundation), (2008). *Plants & Animals of Hammonasset.* Long Island Sound Foundation, Groton, CT. <http://www.lisfoundation.org/coastal_acces s/hamm_wildlife.html>

LISHRI (Long Island Sound Habitat Restoration Initiative), (2003). *Long Island Sound Habitat Restoration Initiative, Technical Support for Habitat Restoration, Section 5: Coastal Barriers, Beaches, and Dunes*. EPA Long Island Sound Office, Stamford, CT, 10 pp. <http://www.longislandsoundstudy.net/habitat>

Mabey, S., Watts, B. & McKay, L. Undated: *Migratory Birds of the Lower Delmarva: A Habitat Management Guide for Landowners.* The Center for Conservation Biology, College of William and Mary, Williamsburg, VA.

Mangold, M. F., Tipton, R. C., Eyler, S. M. & McCrobie, T. M. (2004). *Inventory of Fish Species within Dyke Marsh, Potomac River (2001-2004).* U.S. Fish and Wildlife Service in conjunction with Maryland Fishery Resources Office, Annapolis, MD.

Marraro, P. M, Thayer, G. W., LaCroix, M. L. & Colby, D. R. (1991). Distribution, abundance, and feeding of fish on a marsh on Cedar Island, North Carolina. In: *Ecology of a Nontidal Brackish Marsh in Coastal North Carolina* [Brinson, M.M., (ed.)]. NWRC open file report 9 1-03, U.S. Fish and Wildlife Service, Washington, DC, pp. 321-385.

Mauriello, M., (1991). Beach nourishment and dredging: New Jersey's policies. *Shore & Beach*, *59*, 3.

* May, M. K., (2002). *Pattern and Process of Headward Erosion in Salt Marsh Tidal Creeks*. Master's thesis, Department of Biology. Eastern Carolina University, Greenville, NC, 133 pp.

MCBP (Maryland Coastal Bays Program), (1999). *Today's Treasures for Tomorrow: Towards a Brighter Future. The Comprehensive Conservation and Management Plan for Maryland's Coastal Bays*. Maryland's Coastal Bays Program, Berlin, MD, 181 pp.

MCCC (Maryland Commission on Climate Change), (2008). *Interim Report to the Governor and the Maryland General Assembly: Climate Action Plan*. Maryland Department of Environment, Baltimore, MD.

McFadden, L., Spencer, T. & Nicholls, R. J. (2007). Broad-scale modelling of coastal wetlands: what is required? *Hydrobiologia*, *577*, 5-15.

* McGean, T., (2003). City Engineer, Town of Ocean City, Maryland. Presentation to Coastal Zone.

* MD DNR (Maryland Department of Natural Resources), Undated: Maryland Shoreline Changes Online, from the Maryland Department of Natural Resources. <http://shorelines.dnr.state.md.us/sc_online>

* MD DNR (Maryland Department of Natural Resources), (2000). *Maryland Atlas of Greenways, Water Trails, and Green Infrastructure*. Maryland Greenway Commission. <http://ww.dnr.state.md.us/greenways/counties/princegeorge.html>

* MD DNR (Maryland Department of Natural Resources), (2004). *Land and Water Conservation Service*. Wye Island NRMA Land Unit Plan.

* MD DNR (Maryland Department of Natural Resources), (2006). DNR receives approval for diamondback terrapin conservation, Press Release, 2 August, 2006.

* MD DNR (Maryland Department of Natural Resources), (2007). *Bay Smart: A Citizen's Guide to Maryland's Critical Area Program*. <http://www.dnr.state.md.us/ criticalarea/ download>.

MD DTTF (Maryland Diamondback Terrapin Task Force), (2001). Findings and Recommendations, Final Report to the Secretary of the MD DNR, September 2001, Executive Order 01.01.2001.05.

MET (Maryland Environmental Trust), (2006). *Model Conservation Easement*. Annapolis, Md. <http://www.dnr.state.md.us/MET/model.html>

Mitsch, W. J. & Gosselink, J. G. (2000). *Wetlands*. Van Nostrand Reinhold, New York, 3rd ed., 920 pp.

* MLT (Mordecai Land Trust), Undated: <http://www.mordecaimatters.org>

* MLT (Mordecai Land Trust), (2003). *Mordecai Island - Habitat Value*. <http://www.mordecaimatters.org/listing/Habitat _Value.pdf>

Moffatt and Nichol (2007) *Scope of Work for Development of a Beach and Inlet Management Plan for the State of North Carolina*. Raleigh, NC, 5 pp. <http://www.nccoas talmanagement. net/ Hazards/BIMP% 20Scope%20of%20Work %20Oct.%201 6,%202007.pdf>

Moore, K. (1976). *Gloucester County Tidal Marsh Inventory.* Special report No. 64 in applied science and ocean engineering. Virginia Institute of Marine Science, Gloucester Point, VA, 104 pp.

Moorhead, K. K. & Brinson, M. M. (1995). Response of wetlands to rising sea level in the lower coastal plain of North Carolina. *Ecological Applications*, *5(1)*, 261-271.

Mushacke, F. (2003). Wetland loss in the Peconic estuary. *Long Island Sound Tidal Wetland Loss Workshop*, June 24-25, 2003, Stony Brook, New York. Workshop Proceedings and Recommendations to the Long Island Sound Study. <http://www.longisland soundstudy.net /habitatrestoration/more.htm>

(MRLCC) Multi-Resolution Land Characteristics Consortium, 2002: *Land Cover 1992.* University of Virginia, Charlottesville.

Najjar, R. G., Walker, H. A., Anderson, P. J., Barro, E. J., Bord, R. J., Gibson, J. R., Kennedy, V. S., Knight, C. G., Megonigal, J. P., O'Connor, R. E. Polsky, C. D. Psuty, N. P. B., Richards, A., Sorenson, L. G., Steele, E. M. & Swanson, R.S. (2000). The potential impacts of climate change on the mid-Atlantic coastal region. *Climate Research, 14(3)*, 219-233.

Nature Conservancy, (2008). Adapting to Climate Change: Inundation on the Albemarle Sound. <http://www.nature.org/initiatives/climatechange/work/art261 97.html>

NatureServe, (2008). NatureServe Explorer: An online encyclopedia of life [web application]. Version 7.0. NatureServe, Arlington, VA. <http://www.natureserve.org/explorer>

NC CRC (North Carolina Coastal Resources Commission), (2008a). Draft rule language approved by CRC for public hearing on March 27, 2008, 15A NCAC 07H. 0306. <http://www.nccoastalmanagement.net/Hazards/7H0306 _March08.pdf>

NC CRC (North Carolina Coastal Resources Commission), (2008b) Memorandum CRC- 08-23 from Bonnie Bendell, DCM, to CRC re: Draft Amendments to the General Permit for Bulkheads and Riprap, May 1, 2008. <http://www.nccoastalmanagement.net /Hazards/CRC-08-23 .pdf>

NCDC (National Climatic Data Center), (1999). *Event Record Details: 16 Sept 1999, New Jersey.* <http://www4.ncdc.noaa.gov/cgi-win/wwcgi.dll?wwevent~ShowEvent~3 65151>

NC DCM (North Carolina Division of Coastal Management), (2002). *Technical Manual for Coastal Land Use Planning: A "How-To" Manual for Addressing the Coastal Resources Commission's 2002 Land Use Planning Guidelines.* North Carolina Division of Coastal Management, Raleigh, 63 pp. <http://www.nccoastalmanagement. net/Planning/ techmanual.pdf>

NC DCM (North Carolina Division of Coastal Management), (2003). Coastal Management News: *CAMAgram*: Spring 2003. <http://dcm2.enr.state.nc.us/ CAMAgram/Spring03/ rates.htm>

NC DCM (North Carolina Division of Coastal Management), (2005). *CAMA Handbook for Development in Coastal North Carolina.* NCDENR Division of Coastal Management, Morehead City, NC, section 2. <http://dcm2.enr. state.nc.us/Handbook/section2.htm>

NC (North Carolina) Department of Commerce, (2008). County Tier Designations. <http://www.nccommerce.com/en/BusinessServices/LocateYourBusiness/WhyN C/Incentives/CountyTierDesignations/>

NC REDC (North Carolina Rural Economic Development Center), (2006). Small Towns Initiative. NC Rural Economic Development Center, Raleigh, NC. <http://www. ncruralcenter.org/ smalltowns/initiative.htm>

New York State, (2002). *State Coastal Policies* (Policy 13, 14, 17 and 20). Coastal Management Program, Albany, NY. <http://nyswaterfronts. com/downloads/pdfs/ State_Coastal_ Policies.pdf>

NJDEP (New Jersey Department of Environmental Protection), Undated: Cape May Point State Park.<http://www. state.nj .us/dep/parksandforests/parks/capemay.html>

NJDEP (New Jersey Department of Environmental Protection), (1997). Coastal Report Task Force, New Jersey Department of Environmental Protection, 1997, New Jersey Coastal Report: A Framework Document for a Coastal Management Partnership.

NJDEP (New Jersey Department of Environmental Protection), (2001). Cape May County Rare Species and Natural Communities Presently Recorded in the New Jersey Natural Heritage Database. <http://www.nj .gov/dep/parksandforests/natural s/nj cape.txt>

NJDEP DWM (New Jersey Department of Environmental Protection Division of Watershed Management), (2004). *Stormwater Best Management Practices, Appendix D.* <http://www.nj stormwater>

NOAA (National Oceanic and Atmospheric Administration), (1982). *Ocean & Coastal Resource Management in New York.* <http://coastalmanagement. noaa.gov/mystate /ny.html>

NOAA (National Oceanic and Atmospheric Administration), (2007). *Construction Setbacks.* <http://coastalmanagement.noaa.gov/initiatives/shoreline_ppr_setbacks.html>

NOAA (National Oceanic and Atmospheric Administration), (2008a). Bench Mark Data Sheets: Duck Pier, NC.<http://tidesandcurrents.noaa.gov/station_retrieve.shtml?type = Bench + Mark+Data +Sheets>

NOAA (National Oceanic and Atmospheric Administration), (2008b). Make a Tide Prediction, State and Region Listing: North Carolina. <http://tidesandcurrents. noaa.gov/tides08/tpred2.html#NC>

NOAA (National Oceanic and Atmospheric Administration), (2008c). NOAA Nautical Charts for North Carolina: Charts 11536 to 12205.

Nordstrom, K. F. (1989). Erosion control strategies for bay and estuarine beaches. *Coastal Management*, *17*, 25-35.

Nordstrom, K. F., (2005). Beach nourishment and coastal habitats: research needs to improve compatibility. *Restoration Ecology, 13(1),* 2 15-222.

NPS (National Park Service), Undated: Description of Roosevelt Island.<http://www.nps.gov/ gwmp/pac/tri/backgrnd.html>

NPS (National Park Service), (2000). Synopsis: Cape Hatteras Lighthouse. [NPS Cape Hatteras National Seashore, Manteo, NC.] <http://www.nps.gov/archive/ caha/synopsis.htm>

NRC (National Research Council), (1988). *Saving Cape Hatteras Lighthouse from the Sea: Options and Policy Implications*. National Academy Press, Washington, DC, 136 pp.

NRC (National Research Council), (2007). *Mitigating Shore Erosion along Sheltered Coasts.* National Academies Press, Washington, DC, 174 pp.

NYC DPR (New York City Department of Parks and Recreation), Undated: *Four Sparrow Marsh Reserve.* <http://www.nycgovparks.org/sub_about/ parks_divisions/nrg/ forever_wild/site.ph p?FWID=2 1>

NYC DPR (New York City Department of Parks and Recreation), (2001). *Inwood Hill Park - Salt Marshes in New York City Parks*. Oct. 1. <http://www.nycgovparks. org/sub_your_ park/historical_signs. php?id=1 2864>

NYDOS (New York Department of State) (2006). Department of State review of US Army Corps of Engineers consistency determination for reissuance, modification, and issuance of new nationwide permits and conditions. Letter to USACE Buffalo and New York districts, December 8, (2006). New York Division of Coastal Resources, Albany. 5 pp.

NYS DEC (New York State Department of Environmental Conservation), (2006). *New York's Open Space Conservation Plan.* <http://www.dec.ny.us/opensp/index.html>

NYS DEC (New York State Department of Environmental Control), (2007). List of Endangered, Threatened and Special Concern Fish & Wildlife Species of New York State. Last revised August 8, 2007. <http://www.nynhp.org/>

NYS DOS (New York State Department of State), (1999). *Long Island Sound: Coastal Management Program.* 129pp. <http://www.nyswaterfronts. com/downloads/ pdfs/lis _ cmp/Combined_Chapters.p df>

NYS DOS (New York State Department of State), (2002). State Coastal Policies, Excerpted from the *State of New York Coastal Management Program and Final Environmental Impact Statement*, Section 6, August 1982.

NYS DOS (New York State Department of State), (2004). *Significant Coastal Fish and Wildlife Habitats.* <http://nyswaterfronts.com/ waterfront_natural> (Multiple location - specific pieces can be found on the site).

NYS DOS and USFWS (New York State Department of State and United States Fish and Wildlife Service), (1998). *Shorebirds.* South Shore Estuary Reserve, Technical Report Series, [NYS] Department of State, [Albany, NY], 28 pp. <http://www.nyswaterfronts. com/Final_Draft_HTML/Tech_Report_HTM/PDFs/ Chap3/Shorebirds.pdf>

Palmer, W. M. and Cordes, C. L. (1988). *Habitat Suitability Index Models: Diamondback Terrapin (Nesting) - Atlantic Coast.* Biological report 82(10.151). U.S. Fish and Wildlife Service, Washington, DC, 23 pp.

Paul, M., C. Krafft, and Hammerschlag, D. (2004). *Avian Comparisons Between Kingman and Kenilworth Marshes.* Final report 2001-2004. USGS Patuxent Wildlife Research Center, Laurel, MD.

* Pearsall, S. & DeBlieu, J. (2005). TNC's [The Nature Conservancy's] *Adaptation Efforts in Conservation Landscapes: Sentinel Ecosystem.* <http://www.climatescience.gov/ workshop2005/ presentations/EC 1. 9_Pearsall.pdf >

Pearsall, S. H., and B. Poulter, (2005). Adapting coastal lowlands to rising seas: a case study. In: *Principles of Conservation Biology* [Groom, M.J., G.K. Meffe, and C.R. Carroll, (eds.)]. Sinauer Press, Sunderland, MA, 3rd edition, pp. 3 66-370.

Pearsall, S., McCrodden, B. & Townsend, P. (2005). Adaptive management of flows in the Lower Roanoke River, North Carolina, USA. *Environmental Management, 35(4),* 353-367.

PEP (Peconic Estuary Program), (2001). *Peconic Estuary Comprehensive Conservation and Management Plan.* Sponsored by the United States Environmental Protection Agency under Sec. 320 of the Clean Water Act. Suffolk County Department of Health Services, Program Office, Riverhead, NY, 866 pp.

Perry, M. C. & Deller, A. S. (1996). Review of factors affecting the distribution and abundance of waterfowl in shallow-water habitats of Chesapeake Bay. *Estuaries, 19(2A),* 272-278.

Pilkey, D. F., Bullock, J. & Cowan, B. A. (1998). *The North Carolina Shore and Its Barrier Islands: Restless Ribbons of Sand.* Duke University Press, Durham, NC.

Pilkey, O. H., Howard, J. D., Brenninkmeyer, B., Frey, R., Hine, A., Kraft, J., Morton, R., D. Nummedal & Wanless, H. (1981). *Saving the American Beach: A Position Paper by Concerned Coastal Geologists.* Results of the Skidaway Institute of Oceanography Conference on America's Eroding Shoreline. Skidaway Institute of Oceanography, Savannah, GA.Portnoy, J.W. and A.E. Giblin, 1997: Biogeochemical effects of seawater restoration to diked salt marshes. *Ecological Applications*, *7(3)*, 1054-1063.

Post, W. & Greenlaw, J. S. (1994). Seaside sparrow *(Ammodramus maritimus).* In *The Birds of North America,* No. 127, Poole, A. and F. Gill, eds., The American Ornithologists' Union, Washington, DC.; The Academy of Natural Sciences, Philadelphia, as cited in Chapter 6 of *The Barnegat Bay Estuary Program Characterization Report.* Prepared by the Barnegat Bay National Estuary Program (Scientific and Technical Advisory Committee), 2001. <http://www.bbep.org/Char _Rpt/Ch6/Chapter%206.htm>

* Poulter, B., (2005). *Interactions Between Landscape Disturbance and Gradual Environmental Change: Plant Community Migration in Response to Fire and Sea Level Rise.* Ph.D. dissertation, Nicholas School of the Envrionment and Earth Sciences. Duke University, Durham, NC, 216 pp.

Poulter, B. & Halpin, P. N. (2007). Raster modeling of coastal flooding from sea-level rise. *International Journal of Geographic Information Sciences*, *22(2)*, 167-182.

Poulter, B. & Pederson, N. (2006). *Stand Dynamics and Climate Sensitivity of an Atlantic White Cedar (*Chamaecyparis thyiodes*) Forest: Implications for Restoration and Management.* Report to the North Carolina Division of Coastal Management. Eastern Kentucky University, Cumberland Laboratory of Forest Science Richmond, KY, 31 pp. <http://people.eku.edu/pedersonn/pub s/AWCreportPoulter.pdf>

Poulter, B., Goodall, J. & Halpin, P. N. (2008). Applications of network analysis for adaptive management of artificial drainage systems in landscapes vulnerable to sea level rise. *Journal of Hydrology*, *357(3-4),* 207-2 17.

Poulter, B., Feldman, R. L., Brinson, M. M., Horton, Orbach, M. K., Pearsall, E. Reyes, S. R. Riggs, and J. C. Whitehead, (2009). Sea level rise research and dialogue in North Carolina: creating windows for policy change. *Ocean and Coastal Management*, Article in Press, doi: 10.101 6/j .ocecoaman.2008.09.0 10.

Psuty, N. P. & Ofiara, D. D. (2002). *Coastal Hazard Management, Lessons and Future Direction from New Jersey.* Rutgers University Press, New Brunswick, NJ, 429 pp.

Reed, D. J., Bishara, D. Cahoon, D., Donnelly, J., Kearney, M., Kolker, A., Leonard, L. R.A. Orson, and J.C. Stevenson, (2008). Site-specific scenarios for wetlands accretion as sea level rises in the mid-Atlantic region. Section 2.1 in: *Background Documents Supporting Climate Change Science Program Synthesis and Assessment Product 4.1: Coastal Elevations and Sensitivity to Sea Level Rise* [Titus, J.G. and E.M. Strange (eds.)]. EPA 430R07004. U.S. Environmental Protection Agency, Washington, DC, pp. 134-174. <http://epa.gov/climatechange/effects/coastal/background.html>

Rheinhardt, R. (2007). Tidal freshwater swamps of a lower Chesapeake Bay Subestuary. In: *Ecology of Tidal Freshwater Forested Wetlands of the Southeastern United States* [Conner, W.H., T.W. Doyle, and K.W. Krauss (eds.)]. Springer Netherlands, Dordrecht, the Netherlands, 505 pp.

Richardson, C., (2003). Pocosins: hydrologically isolated or integrated wetlands on the landscape? *Wetlands*, *23(3)*, 563-576.

Riegner, M. F. (1982). The diet of yellow-crowned night-herons in the eastern and southern United States. *Colonial Waterbirds*, *5*, 173-176.

Riggs, S. R. & D. V. Ames, (2003). *Drowning the North Carolina Coast: Sea-level Rise and Estuarine Dynamics*. Publication number UNC-SG-03-04. North Carolina Sea Grant, Raleigh, NC, 152 pp.

Riggs, S. R., Rudolph, G. L. and Ames, D. V. (2000). *Erosional Scour and Geologic Evolution of Croatan Sound, Northeastern North Carolina*. Report no. FHWA/NC/2000-002. North Carolina Department of Transportation, Raleigh, NC, 115 pp.

Riggs, S. R., Stephenson, J. H. & Clark, W. (2007). Preliminary recommendations for mitigating the consequences of climate change within North Carolina. Prepared for the NC Legislative Commission on Global Climate Change, January 9, 2007. 3 pp.

Robbins, C. S. & Blom, E. A. T. (1996). *Atlas of the Breeding Birds of Maryland and the District of Columbia*. University of Pittsburgh Press, Pittsburgh, PA.

Rounds, R. A., Erwin, R. M. & Porter, J. H. (2004). Nest-site selection and hatching success of waterbirds in coastal Virginia: Some results of habitat manipulation. *Journal of Field Ornithology*, *75(4)*, 318.

Shellenbarger Jones, A., (2008a). Overview of mid-Atlantic coastal habitats and environmental implications of sea level rise. Section 3.1 in: *Background Documents Supporting Climate Change Science Program Synthesis and Assessment Product 4.1: Coastal Elevations and Sensitivity to Sea Level Rise* [Titus, J.G. and E.M. Strange (eds.)]. EPA 430R07004. U.S. Environmental Protection Agency, Washington, DC, pp. 188-210. <http://epa.gov/climatechange/effects/coastal/background.html>

Shellenbarger Jones, A., (2008b). Upper Chesapeake Bay shoreline. Section 3.17 in: *Background Documents Supporting Climate Change Science Program Synthesis and Assessment Product 4.1: Coastal Elevations and Sensitivity to Sea Level Rise* [Titus, J.G. and E.M. Strange (eds.)]. EPA 430R07004. U.S. Environmental Protection Agency, Washington, DC, pp. 290-293. <http://epa.gov/climatechange/ effects/coastal/ background.html>

Shellenbarger Jones, A., (2008c). The Chesapeake Bay shoreline of the central eastern shore. Section 3.18 in *Background Documents Supporting Climate Change Science Program Synthesis and Assessment Product 4.1: Coastal Elevations and Sensitivity to Sea Level Rise* [Titus, J.G. and E.M. Strange (eds.)]. EPA 430R07004. U.S. Environmental Protection Agency, Washington, DC, pp. 294- 297. <http://epa.gov/climatechange/ effects/coastal/background.html>

Shellenbarger Jones, A. & C. Bosch, (2008a). The Chesapeake Bay shoreline of middle peninsula. Section 3.12 in: *Background Documents Supporting Climate Change Science Program Synthesis and Assessment Product 4.1: Coastal Elevations and Sensitivity to Sea Level Rise* [Titus, J.G. and E.M. Strange (eds.)]. EPA 430R07004. U.S. Environmental Protection Agency, Washington, DC, pp. 269-272. <http://epa.gov/climatechange/effects/coastal/background.html>

Shellenbarger Jones, A. & Bosch, C. (2008b). Western shore Chesapeake Bay shoreline. Section 3.16 in: *Background Documents Supporting Climate Change Science Program Synthesis and Assessment Product 4.1: Coastal Elevations and Sensitivity to Sea Level Rise* [Titus, J.G. and E.M. Strange (eds.)]. EPA 430R07004. U.S. Environmental Protection Agency, Washington, DC, pp. 284- 289. <http://epa.gov/ climatechange/effects/ coastal/background.html>

Shellenbarger Jones, A. & Bosch, C. (2008c). The Chesapeake Bay shoreline near Hampton Roads. Section 3.11 in: *Background Documents Supporting Climate Change Science Program Synthesis and Assessment Product 4.1: Coastal Elevations and Sensitivity to Sea Level Rise* [Titus, J.G. and E.M. Strange (eds.)]. EPA 430R07004. U.S. Environmental Protection Agency, Washington, DC, pp. 265-268. <http://epa.gov/climatechange/effects/coastal/background.html>

Smith, D., Jackson, N., Love, S., Nordstrum, K., Weber, R. & Carter, D. (2002). Beach Nourishment on Delaware Shore Beaches to Restore Habitat for Horseshoe Crab Spawning and Shorebird Foraging. The Nature Conservancy: Wilmington, DE. 51 pp. Available on the USACE Threatened, Endangered, and Sensitive Species Proection and Management System web page:<http://el.erdc.usace.army.mil/ tessp/pdfs/New%20 Horseshoe% 20Crab%20Habi tat.pdf>

Sommerfield, C. K. & Walsh, D. R. (2005). Historical changes in the morphology of the subtidal Delaware estuary. In: *Proceedings of the First Delaware Estuary Science Conference*. 2005. [Kreeger, D.A. (ed.)]. Partnership for the Delaware Estuary, Report #05-01. 110 pp. <http://www.delawareestuary.org/pdf/ ScienceReportsbyPDEand DELEP/ PDEReport-05-01 -Proceedings 2005 SciConf.pdf>

State of New Jersey, (2001). New Jersey State Development and Redevelopment Plan. <http://www.nj .gov/dca/osg/plan/stateplan. shtml>

State of New Jersey, (2005). *New Jersey Comprehensive Wildlife Conservation Strategy for Wildlife of Greatest Conservation Need.* August 2005 Draft. 649 pp. Table C1. <http://www.njfishandwildlife.com/ensp/waphome.htm>

Steury, B., (2002). The vascular flora of Cove Point, Calvert County, Maryland. *The Maryland Naturalist 45(2),* 1-28.

Strange, E. M. (2008a). New York City, the Lower Hudson, and Jamaica Bay. Section 3.4 in: *Background Documents Supporting Climate Change Science Program Synthesis and Assessment Product 4.1: Coastal Elevations and Sensitivity to Sea Level Rise* [Titus, J.G. and E.M. Strange (eds.)]. EPA 430R07004. U.S. Environmental Protection Agency, Washington, DC, pp. 222-22 9. <http://epa.gov/climatechange/ effects/coastal /background.html>

Strange, E. M. (2008b). New Jersey's coastal bays. Section 3.6 in: *Background Documents Supporting Climate Change Science Program Synthesis and Assessment Product 4.1: Coastal Elevations and Sensitivity to Sea Level Rise* [Titus, J.G. and E.M. Strange (eds.)]. EPA 430R07004. U.S. Environmental Protection Agency, Washington, DC, pp. 235-241. <http://epa.gov/climatechange/effects/coastal/background.html>

Strange, E. M. (2008c). Maryland and Delaware coastal bays. Section 3.8 in: *Background Documents Supporting Climate Change Science Program Synthesis and Assessment Product 4.1: Coastal Elevations and Sensitivity to Sea Level Rise* [Titus, J.G. and E.M. Strange (eds.)]. EPA 430R07004. U.S. Environmental Protection Agency, Washington, DC, pp. 251-257. <http://epa.gov/climatechange/effects/coastal/background.html>

Strange, E. M. (2008d). The Atlantic side of the Virginia eastern shore. Section 3.9 in: *Background Documents Supporting Climate Change Science Program Synthesis and Assessment Product 4.1: Coastal Elevations and Sensitivity to Sea Level Rise* [Titus, J.G. and E.M. Strange (eds.)]. EPA 430R07004. U.S. Environmental Protection Agency, Washington, DC, pp. 258-261. <http://epa.gov/ climatechange/ effects/coastal/ background.html>

Strange, E. M. (2008e). The Virginia eastern shore of Chesapeake Bay. Section 3.19 in: *Background Documents Supporting Climate Change Science Program Synthesis and Assessment Product 4.1: Coastal Elevations and Sensitivity to Sea Level Rise* [Titus, J.G. and E.M. Strange (eds.)]. EPA 430R07004. U.S. Environmental Protection Agency, Washington, DC, pp. 298-300. <http://epa.gov/climatechange/ effects/coastal/ background. html>

Strange, E. M. & Shellenbarger A. Jones, (2008a). Lower Potomac. Section 3.14 in: *Background Documents Supporting Climate Change Science Program Synthesis and Assessment Product 4.1: Coastal Elevations and Sensitivity to Sea Level Rise* [Titus, J.G. and E.M. Strange (eds.)]. EPA 430R07004. U.S. Environmental Protection Agency, Washington, DC, pp. 275-279. <http://epa.gov/climatechange/ effects/coastal/ background.html>

Strange, E. M. & Shellenbarger Jones, A., (2008b). Upper Potomac. Section 3.15 in: *Background Documents Supporting Climate Change Science Program Synthesis and Assessment Product 4.1: Coastal Elevations and Sensitivity to Sea Level Rise* [Titus, J.G. and E.M. Strange (eds.)]. EPA 430R07004. U.S. Environmental Protection Agency, Washington, DC, pp. 280-283. <http://epa.gov/climatechange/ effects/ coastal/ background. html>

Strange, E. M., Shellenbarger Jones, A. Bosch, C. Jones, R., Kreeger, D. & J. G. Titus, (2008). Mid-Atlantic coastal habitats and environmental implications of sea level rise. Section 3 in: *Background Documents Supporting Climate Change Science Program Synthesis and Assessment Product 4.1: Coastal Elevations and Sensitivity to Sea Level Rise* [Titus, J. G. and Strange E. M. (eds.)]. EPA 430R07004. U.S. Environmental Protection Agency, Washington, DC, pp. 188- 342. <http://epa.gov/climatechange/effects/coastal/background.html>

Suffolk (City of Suffolk Department of Planning), (1998). *The Comprehensive Plan for 2018: City of Suffolk, Virginia.* Adopted March 25, 1998

Sutter, L. (1999). DCM wetland mapping in coastal North Carolina. North Carolina Division of Coastal Management. <http://dcm2.enr. state.nc.us/Wetlands/WTYPEMAPDOC.pdf>

* Swisher, M. L. (1982). *The Rates and Causes of Shore Erosion Around a Transgressive Coastal Lagoon, Rehoboth Bay, Delaware*. M.S. Thesis, College of Marine Studies. University of Delaware, Newark.

Talbot, C. W. & Able, K. W. (1984). Composition and distribution of larval fishes in New Jersey high marshes. *Estuaries, 7(4A)*, 434-443.

Teal, J. M. & Peterson, S. B. (2005). Introduction to the Delaware Bay salt marsh restoration. *Ecological Engineering*, *25(3),* 199-203.

Tenore, K. R. (1972). Macrobenthos of the Pamlico River Estuary, North Carolina, *Ecological Monographs*, *42(1),* 51-69.

Thayer, G. W., Kenworthy, W. J. & Fonseca, M.S. (1984). *The Ecology of Eelgrass Meadows of the Atlantic Coast: A Community Profile*. U.S. Fish and Wildlife Service, FWS/OBS-84/02.

Thieler, E. R. & Hammar-Klose, E.S. (1999). *National Assessment of Coastal Vulnerability to Sea-Level Rise: Preliminary Results for the U.S. Atlantic Coast*. U.S. Geological Survey open-file report 99-593. U.S. Geological Survey, Reston, VA, 1 sheet. <http://pubs.usgs.gov/of/1 999/of99-593/index.html>

Tiner, R. W. & Burke, D.G. (1995). *Wetlands of Maryland.* U.S. Fish and Wildlife Service, Region 5, Hadley, MA.

Titus, J. G. (1990). Greenhouse effect, sea-level rise, and barrier islands: case study of Long Beach Island, New Jersey. *Coastal Management, 18(1),* 65-90.

Titus, J. G. (1998). Rising seas, coastal erosion, and the takings clause: how to save wetlands and beaches without hurting property owners. *Maryland Law Review, 57(4),* 1279-1399.

Titus, J. G. & Richman, C. (2001). Maps of lands vulnerable to sea level rise: modeled elevations along the U.S. Atlantic and Gulf Coasts. *Climate Research, 18(3),* 205- 228.

Titus, J. G. & Wang, J. (2008). Maps of lands close to sea level along the middle Atlantic coast of the United States: an elevation data set to use while waiting for LIDAR. Section 1.1 in: *Background Documents Supporting Climate Change Science Program Synthesis and Assessment Product 4.1: Coastal Elevations and Sensitivity to Sea Level Rise* [Titus, J.G. and E.M. Strange (eds.)]. EPA 430R07004. U.S. Environmental Protection Agency, Washington, DC, pp. 2-44.

Titus, J. G., Leatherman, S. P., Everts, C. H., Kriebel, D. L. & Dean, R.G. (1985). *Potential Impacts of Sea Level Rise on the Beach at Ocean City, Maryland.* EPA 230-10- 85-013. Environmental Protection Agency, Washington, DC.

Titus, J. G., Park, R. A., Leatherman, S. P., Weggel, R. R, Greene, M. S., Mausel, P.W., Brown, S., Gaunt, G., Trehan, M. & Yohe, G. (1991). Greenhouse effect and sea level rise: the cost of holding back the sea. *Coastal Management, 19(2),* 171-204.

* TNC (The Nature Conservancy) Undated: *William D. and Jane C. Blair Jr. Cape May Migratory Bird Refuge.* <http://www.nature.org/wherewework/northamerica/states/newj ersey/work/art1 72 05 .html>

* TNC (The Nature Conservancy) (2006). Project profile for the Virginia Coast Reserve. Available online by searching on “field guides” at < http://www.nature.org/ wherewework>

* TNC (The Nature Conservancy) (2007). Press release: Migratory Bird Refuge: Re- Opened! <http://www.nature.org/wherewework/northamerica/states/newj ersey/work/art2 18 76.html>

Town of East Hampton (2006). *Coastal Erosion Overlay District Legislation, Resolution 2006-899.* <http://www.town.east-hampton.ny.us/coastal.pdf>

Town of Kitty Hawk, (2005). Council Circular 2: 5. <http://www.townofkittyhawk.org/ vertical/Sites/%7B99 1 BBDF3 -791 F-443 5-80BA-FEB 8F8D09BA4%7D/ uploads/%7B 1 A6EE852-D7D6-4A72-9DA6- B 1B576DDA266%7D.PDF>

Town of Ocean City, (2003). *Sea Level Rise Policy.* Coastal Legislation Committee, Ocean City Council, Town of Ocean City, Ocean City, MD.

* Trono, K. L. (2003). *An Analysis of the Current Shoreline Management Framework in Virginia: Focus on the Need for Improved Agency.* Virginia Shoreline Management Analysis Report from the Virginia Coastal Program’s publications web page.

USACE (U.S. Army Corps of Engineers), undated(a). *Timeline representing key dates ofGibbstown Levee and Repaupo Creek.* (Dike and floodgates constructed in late 1600s by Repaupo Meadow Company) < http://www.nap.usace.army.mil/Projects/ Repaupo/timeline.html>.

USACE (United States Army Corps of Engineers), Undated(b). Poplar Island Environmental Restoration Site. <http://www.nab.usace.army.mil/projects/Maryland>.

USACE (United States Army Corps of Engineers) (1998a). *Delaware Bay Coastline: Villas & Vicinity, NJ. final feasibility report and Environmental Impact Statement.* United States Army Corps of Engineers, Philadelphia District. Revised 1999.

USACE (United States Army Corps of Engineers) (1998b). *Delaware Bay Coastline: Reeds Beach and Pierces Poit, NJ. final integrated feasibility report and Environmental Impact Statement.* United States Army Corps of Engineers, Philadelphia District. Addendum 1999.

USACE (United States Army Corps of Engineers) (2000). Wilmington District. Attachment D: Dare County Beaches (Bodie Island Portion) Hurricane Protection and Beach Erosion Control Project, Dare County, North Carolina. Preliminary Compilation of Disposal/Nourishment Zones and Borrow Areas in Recovery from 5-Year Running Total. <www.saw.usace.army.mil/Dare%20County/Dare%20County%20Bodie%20Is.% 20%20FEIS%20Attachment%20D.pdf>

USACE (United States Army Corps of Engineers) (2001a). *Final Finding of No Significant Impact and Environmental Assessment, Ass ateague Island Short-term Restoration: Modifications to Proposed Project and Development of a Dredging Plan.* Worcester County, MD. Baltimore District, July 2001.

USACE (United States Army Corps of Engineers) (2001b). Smith Island, Maryland. Environmental Restoration and Protection. Final Integrated Feasibility Report and Environmental Assessment. Submitted by USACE Baltimore District in cooperation with Somerset County, Maryland, Maryland Department of Natural Resources, and Maryland Department of the Environment.

USACE (United States Army Corps of Engineers) (2003). *Regional Sediment Management (RSM) Demonstration Program Project Brief, New York District: Town of Hempstead, Long Island, New York.* <http://www.wes.army.mil/rsm/pubs/pdfs/rsm-db9.pdf>

USACE (United States Army Corps of Engineers) (2006a). *Annual Report, Fiscal Year 2006 of the Secretary of the Army on Civil Works Activities.* South Atlantic Division, Wilmington, N.C. District.

USACE (United States Army Corps of Engineers) (2006b). *Revised Record of Decision: Dare County Beaches (Bodie Island Portion), Hurricane Protection and Beach Erosion Control Project, Dare County, North Carolina.* U.S. Army Corps of Engineers. <http://www.saw.usace.army.mil/Dare%20County/Yelverton-RODDareCo.pdf>

USACE (United States Army Corps of Engineers), 2007: Activities Authorized by Nationwide Permit. <http://www.lrb.usace.army.mil/regulatory/nwp/NYNWP2007/NY NWP03 .doc>

USACE (United States Army Corps of Engineers), 2008a: Project Fact Sheet: *New Jersey Shore Protection, Lower Cape May Meadows – Cape May Point, NJ.* <http://www.nap.usace.army/mil/cenap-dp/projects/factsheets/NJLowe Cape May Meadows.pdf>

USACE (United States Army Corps of Engineers) (2008b). *Fire island Inlet to Montauk Point Reformulation Study, DRAFT.* <http://www.nan.usace.army.mil/fimp/>

USDOI (United States Department of the Interior) (2007). Letter from Willie Taylor, Director, Office of Environmental Policy and Compliance, DOI, to Gregory J. Thorpe, Ph.D., Manager, Project Development and Environmental Analysis Division, North Carolina Department of Transportation, re: Draft Environmental Impact Statement for

NC-12 Replacement of Herbert C. Bonner Bridge (No. 11) Over Oregon Inlet, Dare County, North Carolina. <http://www.fws.gov/peaisland/images 1 7-07.pdf>

U.S. EPA (Environmental Protection Agency) (1989). *The Potential Effects of Global Climate Change on the United States.* EPA 230-05-89-052. U.S. Environmental Protection Agency, Washington, DC.

U.S. EPA (Environmental Protection Agency) (2008). *Background Documents Supporting Climate Change Science Program Synthesis and Assessment Product 4.1: Coastal Elevations and Sensitivity to Sea Level Rise* [Titus, J.G. and E.M. Strange (eds.)]. EPA 430R07004. U.S. Environmental Protection Agency, Washington, DC, 354 pp. <http://epa.gov/climatechange/effects/coastal/background.html>

USFWS (United States Fish and Wildlife Service), Undated(a). Cape May National Wildlife Refuge. < http://www.fws.gov/northeast/capemay/>

USFWS (United States Fish and Wildlife Service), Undated(b). *Profile of the Plum Tree Island National Wildlife Refuge.* <http://www.fws.gov/refuges/profiles/index.cfm?id = 51512>

USFWS (United States Fish and Wildlife Service), Undated©: *John H. Chafee Coastal Barrier Resource System.* <http://www.fws.gov/habitatconservation/coastal_barrier>

USFWS (United States Fish and Wildlife Service) (1980). *Atlantic coast ecological inventory: Wilmington.* Washington, D.C.: No. 39074-A1-EI-250.

USFWS (United States Fish and Wildlife Service) (1993). *Species Account for the Red Wolf.* ><http://www.fws.gov/cookeville/docs/endspec/rwolfsa.html>

USFWS (United States Fish and Wildlife Service) (1996). *Piping Plover* (Charadrius melodus) *Atlantic Coast Population Revised Recovery Plan.* Prepared by the Atlantic Coast Piping Plover Recovery Team. U.S. Fish and Wildlife Service, Hadley, MA, 245 pp.

USFWS (United States Fish and Wildlife Service) (1997). *Significant Habitats and Habitat Complexes of the New York Bight Watershed.* U.S. Fish and Wildlife Service, Charlestown, RI. <http://training>

USFWS (United States Fish and Wildlife Service) (2002). Birds of Conservation Concern 2002. Division of Migratory Bird Management, Arlington, VA. Table 30. <http://www.fws.gov/migratorybirds/reports/reports.html>

USFWS (United States Fish and Wildlife Service) (2004a). *2002-2003 Status Update: U.S. Atlantic Coast Piping Plover Population.* U.S. Fish and Wildlife Service, Sudbury, MA, 8 pp. <http://www.fws.gov/northeast/pipingplover/status/index.html>

USFWS (United States Fish and Wildlife Service) (2004b). Eastern Shore of Virginia and Fisherman Island National Wildlife Refuges Comprehensive Conservation Plan. Chapter 3: Refuge and Resource Descriptions of Northeast Regional Office, Hadley, MA. <http://library.fws.gov/CCPs/eastshoreVA_index.htm>

USFWS (United States Fish and Wildlife Service) (2008). Blackwater National Wildlife Refuge. Wetland Restoration. <http://www.fws.gov/blackwater/restore.html>. Last updated January 22, 2008.

USGS (United States Geological Survey), Undated: *Anacostia Freshwater Tidal Reconstructed Wetlands.* <http://www.pwrc.usgs.gov/resshow/hammerschlag/anacostia.cfm>

VA DCR (Virginia Department of Conservation and Recreation), Undated(a). Parkers Marsh Natural Area Preserve Fact Sheet.<http://www.dcr.virginia.gov/natural s/parke rs. shtml>

VA DCR (Virginia Department of Conservation and Recreation), Undated(b) : Mutton Hunk Fen Natural Area Preserve. <http://www.dcr.virginia.gov/natural k. shtml>

VA DCR (Virginia Department of Conservation and Recreation) (1999). Bethel Beach Natural Area Preserve, fact sheet. <http://www.dcr.virginia.gov/natural>

VA DCR (Virginia Department of Conservation and Recreation) (2001). The Natural Communities of Virginia. Ecological Classification of Ecological Community Groups. First Approximation. Division of Natural Heritage Natural Heritage Technical Report 01-1, January 2001.

VA PBB (Virginia Public Beach Board) (2000). 20 Years of Coastal Management. Richmond, Virginia: Board on Conservation and Development of Public Beaches.

VBCP (Virginia Beach Comprehensive Plan) (2003). Introduction and General Strategy: Policy Document.

* VIMS (Virginia Institute for Marine Science), Undated: Chesapeake Bay National Estuarine Research Reserve in Virginia. Goodwin Islands. <http://www.vims.edu/cbnerr/reservesites/goodwin.htm>

Walls, E. A., Berkson, J. & Smith, S. A. (2002). The horseshoe crab, *Limulus polyp hem us*: 200 million years of existence, 100 years of study. *Reviews in Fisheries Science, 10(1),* 39-73.

Walz, K., Cronan, E., Domber, S., Serfes, M. Kelly, L. & Anderson, K. (2004). *The Potential Impacts of Open Marsh Management (OMWM) on a Globally Imperiled Sea-level Fen in Ocean County, New Jersey*. Prepared for the New Jersey Department of Environmental Protection, Coastal Management Office, 18 pp.

Washington County (2008). Return to the Heart of the Inner Banks. <http://www.originalwashington.com/>

Watts, B. D. (1993). *Effects of Marsh Size on Incidence Rates and Avian Community Organization within the Lower Chesapeake Bay*. Technical report CCBTR-93-03. Center for Conservation Biology, College of William and Mary, Williamsburg, VA, 53 pp.

Watts, B. D. (2006). *Synthesizing Information Resources for the Virginia Important Bird Area Program: Phase I, Delmarva Peninsula and Tidewater*. Technical Report CCBTR-06-05. Center for Conservation Biology, College of William and Mary, Williamsburg, VA.

Watts, B. D. & Markham, C. (2003). *The Influence of Salinity on Diet, Prey Delivery, and Nestling Growth in Bald Eagles in the Lower Chesapeake Bay: Progress Report*. Technical Report CCBTR-03-06. Center for Conservation Biology, College of William and Mary, Williamsburg, VA, 5pp.

Watts, B. D. & Truitt, B. R. (2002). Abundance of shorebirds along the Virginia barrier islands during spring migration. *The Raven, 71(2),* 33-3 9.

* Weber, R. G. (2001). Preconstruction horseshoe crab egg density monitoring and habitat availability at Kelly Island, Port Mahon, and Broadkill Beach Study areas. Prepared for the Philadelphia District Corps of Engineers, Philadelphia, PA. December 2001.

Weggel, J. R., Brown, S., Escajadillo, J., Breen, P. & Doheny, E. L. (1989). The cost of defending developed shorelines along sheltered water of the United States from a two meter rise in mean sea level. In: *The Potential Effects of Global Climate Change on the United States. Appendix: B, Sea Level Rise*. EPA 230-05-89-052. Environmental Protection Agency, Washington, DC, pp. 3-1 to 3-90. <http://epa.gov/climatechange/effects/coastal/appB .html>

Weinstein, M. P. (1979). Shallow marsh habitats as primary nurseries for fishes and shellfish, Cape Fear River, North Carolina, United States. *Fisheries Bulletin, 77(2),* 339-357.

Weinstein, M. P. & Weishar, L. L. (2002). Beneficial use of dredged material to enhance the restoration trajectories of formerly diked lands. *Ecological Engineering, 19(3),* 187-201.

Weinstein, M. P., Phillip, K. R. & Goodwin, P. (2000). Catastrophes, near-catastrophes, and the bounds of expectation: Success criteria for macroscale marsh restoration. In: *Concepts and Controversies in Tidal Marsh Ecology* [Weinstein, M.P. and D.A. Kreeger (eds.)]. Kluwer Academic Publishers, Dordrecht, the Netherlands, pp. 777-804.

* Wetlands Institute, Undated: Terrapin Conservation Program. <http://www.terrapinconservation.org>.

White, C. P. (1989). *Chesapeake Bay: Nature of the Estuary: A Field Guide.* Tidewater Publishers, Centreville, MD, 212 pp.

Wilcock, P. R., Miller, D.S., Shea, R.H. & Kerhin, R.T. (1998). Frequency of effective wave activity and the recession of coastal bluffs: Calvert Cliffs, Maryland. *Journal of Coastal Research, 14(1),* 256-268.

York (York County, Virginia) (1999). *Charting the Course to 2015: The York County Comprehensive Plan.* Yorktown, VA.

Zervas, C. (2004). North Carolina Bathymetry/Topography Sea Level Rise Project: Determination of Sea Level Trends. NOAA technical report NOS CO-OPS 041. NOAA National Ocean Service, Silver Spring, MD, 31 pp. <http://tidesandcurrents.noaa.gov/publications/techrpt4 1 .pdf>

End Notes

[1] A glacial terminal moraine is a glacial deposit landform that marks the limit of glacial advance.

[2] A kettle hole is a depression landform formed in glacial deposit sediments from a time when a large block of glacial ice remained and melted after a glacial retreat.

[3] The state has jurisdiction up to 300 feet beyond the tidal wetland boundary in most areas (but only 150 feet in New York City).

[4] Article 25, Environmental Conservation Law Implementing Regulations-6NYCRR PART 661.

[5] Environmental Conservation Law, Article 34

[6] § New York Environmental Conservation Law 34-0103 (3)(a)

[7] § New York Environmental Conservation Law 34-0105

[8] For comprehensive discussions of the New Jersey shore and the implications of sea level rise, see Cooper *et al.* (2005), Lathrop and Love (2007), Najjar *et al.* (2000) and Psuty and Ofiara (2002)

[9] For example, the primary coastal policy document during the Whitman administration suggested that even mentioning the term "retreat" would divide people and impede meaningful discussion of appropriate policies, in part because retreat can mean government restrictions on development or simply a decision by government not to fund shore protection (see NJDEP, 1997). Governor Whitman promised coastal mayors and residents that "there will be no forced retreat."

[10] The Meadows are within Cape May Point State Park and the Nature Conservancy's Cape May Migratory Bird Refuge.

[11] See Chapter 5 of this Product for explanation of structural mechanisms to combat flooding.

[12] Matthews v Bay Head Improvement Association, 471 A.2d 355. Supreme Court of NJ (1984).

[13] Federal law requires similar access before an area is eligible for beach nourishment.

[14] See Kent County Ordinances § 7.3 and New Castle Ordinance 40.10.313

[15] This section only addresses the Pennsylvania side of the river because Section C in this Appendix addressed the policy context for shore protection in New Jersey.

[16] This statement also applies to the District of Columbia.

[17] Philadelphia's flood regulations do consider sea level rise.

[18] These regulations were issued pursuant to the Dam Safety and Encroachment Act of 1978. Laws of Pennsylvania, The Dam Safety and Encroachments Act of November 26, 1978, P.L. 1375, No. 325.

[19] Pennsylvania Code, Chapter 105. Dam Safety and Waterway Management, Pennsylvania Department of Environmental Protection, 1997. Subchapter 105.1 8b.

[20] Chapter 3 of this Product concludes that most tidal wetlands in Pennsylvania are likely to keep pace with projected rates of sea level rise. However, that finding does not address erosion of wetlands at their seaward boundary. Even though wetlands can keep vertical pace with the rising water level, narrow fringing wetlands along rivers can be eliminated by shoreline armoring as their seaward boundaries erode and their landward migration is prevented. Moreover, even where the seaward boundary keeps pace, preventing an expansion of wetlands might be viewed as significant.

[21] A few residential structures are on Cedar Island, and Cobbs and Hog Islands have some small private inholdings (Ayers, 2005).

[22] See <http://www.nps.gov/asis/naturescience/resource-management>

[23] Maryland Natural Resources Code §8-1807(a). Code of Maryland Regulations §27.01.09.01 (C);

[24] Code of Maryland Regulations §27.01.02.05(C)(4).

[25] Maryland Natural Resources Code §8-1808.10

[26] Sussex County, DE. 2007. Buffer zones for wetlands and tidal and perennial non-tidal waters. § 115-193, Sussex County Code. Enacted July 19, 1988 by Ord. No. 521.

[27] As this report was being finalized, a comprehensive study of the impacts of sea level rise on the Chesapeake Bay region was completed by the National Wildlife Federation (Glick *et al.,* 2008).

[28] Gloucester County Code of Ordinances, accessed through Municode Online Codes: <http://livepublish.municode.com/22/lpext.dll?f = templates&fn = main-j.htm&vid = 10843>: "The intent of the SC-1 district is to allow low density residential development... .Cluster development is encouraged in order to protect environmental and scenic resources."

[29] Gloucester County Code of Ordinances, accessed through Municode Online Codes; <http://livepublish.municode.com/22/lpext.dll?f = templates&fn = main-j.htm&vid = 10843>

[30] Definition of suburban countryside in Gloucester County Code of Ordinances, accessed through Municode Online Codes: <http://livepublish.municode.com/22/lpext.dll?f = templates&fn = mainj.htm&vid = 10843>: "The intent of the SC-1 district is to allow low density residential development... .Cluster development is encouraged in order to protect environmental and scenic resources."

[31] See the Research Reserve's web page at <http://www.vims.edu/cbnerr/about/index.htm>.

[32] See Maryland v. Virginia, 540 US (2003)

[33] The Virginia Shore across from Washington, D.C. is mostly owned by the federal government, which would be exempt from District of Columbia policies.

[34] Baltimore (City of Baltimore Department of Planning) 2006 "All Hazards Plan for Baltimore City" Adopted by Baltimore City Planning Commission April 20, 2006 at 6, 10-11.

[35] See also Nationwide Permits 3 (Maintenance), 31 (Maintenance of Existing Flood Control Facilities) and 45 (Repair of Uplands Damaged by Discrete Events). 72 Federal Register 11092-11198 (March 12, 2007).

[36] For more detail on beach habitats and the species that occur in the mid-Atlantic region, see Shellenbarger Jones (2008a).

[37] The mounds systems have made it possible to inhabit low areas with high water tables.

[38] See <http://www.dnr.state.md.us/dnrnews/pressrelease2007/041 807.html>

[39] 16 USC §1453 (4)

[40] Code of Maryland Regulations§ 27.01.04.02.02-03

[41] Maryland Executive Order (01.01.2007.07); Virginia Executive Order 59 (2007).

[42] Maryland Executive Order (01.01.2007.07).

[43] Chesapeake Bay Critical Areas Protection Act, Maryland Code Natural Resources §8-1807.

[44] *i.e.* all state and private wetlands designated under Natural Resources Article, Title 9 (now Title 16 of the Environment Article).

[45] Maryland Code Natural Resources §8-1 807(c)(1)(i)(2).

[46] Lands that are less than 1,000 ft from open water *may* be excluded from jurisdiction if the lands are more than 1,000 ft from open water, and the wetlands between that land and the open water are highly functional and able to protect the water from adverse effects of developing the land. Maryland Code Natural Resources §8-1 807(c)(1)(i)(2) and §8-1 807(a)(2).

[47] Maryland Code Natural Resources §8-1808.

[48] Code of Maryland Regulations §27.01.02.02(A).

[49] Code of Maryland Regulations §27.0 1.02.05(C)(4).

[50] Maryland Code Natural Resources §8-1808.10 The required setback is only 100 feet for new construction on pre-existing lots.

[51] Code of Maryland Regulations §27.01.02.05(C)(4). activities to lands within the IDAs, and permits some additional low-intensity

[52] Code of Maryland Regulations §27.01.02.06.

[53] Code of Maryland Regulations §27.01.02.05.

[54] Maryland House Bill 1253 (2008) §3.

[55] Code of Maryland Regulations §27.0 1 .00.01 (C)(1).
[56] Code of Maryland Regulations §27.0 1 .00.01 (C)(2).
[57] Code of Maryland Regulations§ 27.01.04.02. 02
[58] Code of Maryland Regulations§ 27.01.04.02. 03.
[59] Code VA § 10.1-2100 et seq. As of August 8, 2003, the Act was posted on the Virginia Legislative Information System website as part of the Code of Virginia at: <http://legl .state.va.us/cgi-bin/legp504.exe?000+cod+TOC1 0010000021000000000000>.
[60] Code VA §10.1-2102.
[61] Code VA §10.1-2104.
[62] Code VA §10.1-2109.
[63] Code VA § 10.1-2107(B).
[64] Code VA §10.1-2107(A).
[65] Chesapeake Bay Preservation Area Designation and Management Regulations (9 VAC 10-20-10 et. seq.).
[66] 9 Virginia Administrative Code § 10-20-50.
[67] The act also provides for Resource Management Areas (RMAs) which are lands that, if improperly used or developed, have the potential to diminish the functional value of RPAs. Finally, areas in which development is concentrated or redevelopment efforts are taking place may be designated as Intensely Developed Areas (IDAs) and become subject to certain performance criteria for redevelopment. Private landowners are free to develop IDA and RMA lands, but must undergo a permitting process as well to prove that these actions will not harm the RPAs.
[68] 9 Virginia Administrative Code § 10-20-70.
[69] 9 Virginia Administrative Code § 10-20-80 (B).
[70] 9 Virginia Administrative Code § 10-20-130 (4).
[71] The act grants jurisdiction to an elevation equal to 1.5 times the mean tide range, above mean low water.
[72] Virginia Administrative Code §28.2.
[73] Maryland Environmental Code § 16-101 to § 16-503.
[74] See MD. CODE ANN., ENVIR. § 16-201 (1996); See Baltimore District (1996), app. at I-24, I-31. Along sheltered waters, the state encourages property owners to control erosion by planting vegetation. For this purpose, one can fill up to 35 feet seaward of mean high water. See MD. CODE ANN., ENVIR. § 16-202(c)(3)(iii) (Supp. 1997). Along Chesapeake Bay and other waters with significant waves, hard structures are generally employed.
[75] MD. CODE ANN., ENVIR. § 16-202(c)(2).
[76] MD. CODE ANN., ENVIR. § 16-201.
[77] See Baltimore District (1996) §§ 1-5
[78] The Maryland/Virginia border along the Potomac River is the low water mark. Courts have not ruled whether Maryland or Virginia environmental rules would govern a structure in Maryland waters attached to Virginia land.
[79] Maryland General Permit at 56, § A1(A)(1)(g).
[80] MD H.B. 273 (2008).
[81] MD Code Environment § 16-201(c)
[82] See Chapter 6 for an explanation of various shore protection options.
[83] 15A NCAC 07H.0101.
[84] Some hard structures exist along North Carolina's oceanfront shoreline (*e.g.*, adjacent to inlets). Many were built before 1985 when the statute was enacted to ban new hard structures, or were covered by exception in the rules. The Legislature regularly considers additional exceptions, such as terminal groins for beach nourishment projects and jetties for stabilizing inlets. *e.g.* North Carolina SB599 (2007-2008).
[85] Local governments can request that an alternative vegetation line be established under certain conditions. Additional rules also apply when there is a sand dune between the home and the shoreline, to protect the integrity of the dune.
[86] 15A NCAC 07H. 0305- 0306.
[87] 15A NCAC 07H.0306.
[88] 15A NCAC 07H.0209
[89] Although beach nourishment has been a common response to sea level rise in many areas along the coast, there has been a decline in the availability of suitable sand sources for nourishment, particularly along portions of the coast (Bruun, 2002). In addition, the availability of substantial federal funds allocated for beach nourishment has become increasingly questionable in certain areas, particularly in Dare County (Dare County, 2007; Coastal Science and Engineering, 2004).
[90] 15A NCAC 07H.0308
[91] 15A NCAC 7H.0306 (l)
[92] See the "North Carolina Global Warming Act," Session Law 2005-442.
[93] Nevertheless, it has had a few short-lived projects, most notably Lake Matamuskeet.

In: Coastal Sensitivity to Sea Level Rise-Focusing on the Mid-… ISBN: 978-1-60741-440-7
Editor: Melvin C. Urajner

APPENDIX 2. BASIC APPROACHES FOR SHORELINE CHANGE PROJECTIONS

Lead Author: Benjamin T. Gutierrez, USGS

Contributing Authors: S. Jeffress Williams, USGS; E. Robert Thieler, USGS

While the factors that influence changes in shoreline position in response to sea-level rise are well known, it has been difficult to incorporate this understanding into quantitative approaches that can be used to assess land loss over long time periods (*e.g.*, 50 to 100 years). The validity of some of the more common approaches discussed in this Appendix has been a source of debate in the scientific community (see Section 3.1). This Appendix reviews some basic approaches that have been applied to evaluate the potential for shoreline changes over these time scales.

The Bruun Model. One of the most widely known models developed for predicting shoreline change driven by sea-level rise on sandy coasts was formulated by Bruun (1962, 1988). This model is often referred to as the 'Bruun rule' and considers the two- dimensional shoreline response (vertical and horizontal) to a rise in sea level. A fundamental assumption of this model is that over time the cross-shore shape of the beach, or beach profile, assumes an equilibrium shape that translates upward and landward as sea level rises. Four additional assumptions of this model are that:

1. The upper beach is eroded due to landward translation of the profile.
2. The material eroded from the upper beach is transported offshore and deposited such that the volume eroded from the upper beach equals the volume deposited seaward of the shoreline.
3. The rise in the nearshore seabed as a result of deposition is equal to the rise in sea level, maintaining a constant water depth.
4. Gradients in longshore transport are negligible.

Mathematically, the model is depicted as:

$$R = \frac{L_*}{B + h_*} \cdot S \tag{A2.1}$$

where R is the horizontal retreat of the shore, $h*$ is the depth of closure or depth where sediment exchange between the shore face and inner shelf is assumed to be minimal, B is the height of the berm, $L*$ is the length of the beach profile to $h*$, and S is the vertical rise in sea level (Figure A2.1). This relationship can also be evaluated based on the slope of the shore face, Θ, as:

$$R = \frac{1}{\tan \Theta} \cdot S \tag{A2.2}$$

For most sites, it has been found that general values of Θ and R are approximately 0.01 to 0.02 and 50*S to 100*S, respectively (Wright, 1995; Komar, 1998; Zhang, 1998).

A few studies have been conducted to verify the Bruun Model (Schwartz, 1967; Hands, 1980; also reviewed in SCOR, 1991; Komar, 1998; and Dean and Dalrymple, 2002). In other cases, some researchers have advocated that there are several uncertainties with this approach, which limit its use in real-world applications (Thieler *et al.*, 2000; Cooper and Pilkey, 2004, also reviewed in Dubois, 2002). Field evaluations have also shown that the assumption of profile equilibrium can be difficult to meet (Riggs *et al.*, 1995; List *et al.*, 1997). Moreover, the Bruun relationship neglects the contribution of longshore transport, which is a primary mechanism of sediment transport in the beach environment (Thieler *et al.*, 2000) and there have been relatively few attempts to incorporate longshore transport rates into this approach (Everts, 1985).

A number of investigators have expanded upon the Bruun rule or developed other models that simulate sea-level rise driven shoreline changes. Dean and Maurmeyer (1983) adapted and modified the Bruun rule to apply to barrier islands (*e.g.*, the Generalized Bruun Rule). Cowell *et al.* (1992) developed the Shoreline Translation Model (STM), which incorporated several parameters that characterize the influence of the geological framework into sea-level rise driven shoreline change for barrier islands. Stolper *et al.* (2005) developed a rules-based geomorphic shoreline change model (GEOMBEST) that simulates barrier island evolution in response to sea-level rise. While these models can achieve results consistent with the current understanding of sea-level rise driven changes to barrier island systems, there is still need for more research and testing against both the geologic record and present-day observations to advance scientific understanding and inform management.

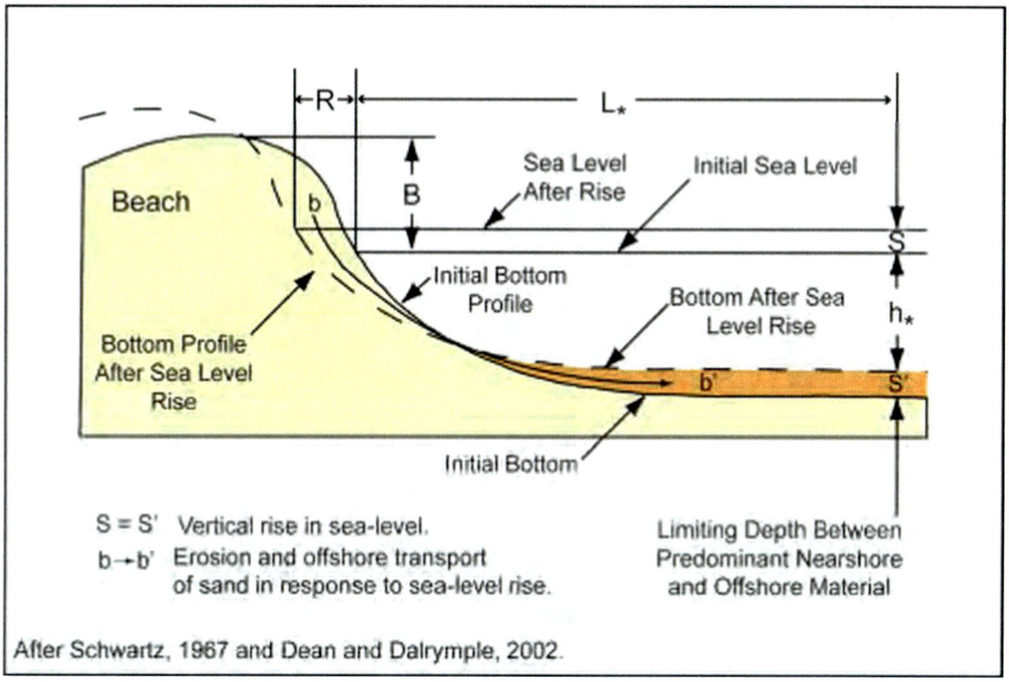

Figure A2.1. Illustration showing the Bruun Model and the basic dimensions of the shore that are used as model inputs.

Historical Trend Extrapolation. Another commonly used approach to evaluate potential shoreline change in the future relies on the calculation of shoreline change rates based on changes in shoreline position over time. In this approach, a series of shorelines from different time periods are assembled from maps for a particular area. In most cases, these shorelines are derived from either National Ocean Service T-sheets, aerial photographs, from Global Positioning System (GPS) surveys, or lidar surveys (Shalowitz, 1964; Leatherman, 1983; Dolan *et al.*, 1991; Anders and Byrnes, 1991; Stockdon *et al.*, 2002). The historical shorelines are then used to estimate rates of change over the time period covered by the different shorelines (Figure A2.2). Several statistical methods are used to calculate the shoreline change rates with the most commonly used being end-point rate calculations or linear regression (Dolan *et al.*, 1991; Crowell *et al.*, 1997). The shoreline change rates can then be used to extrapolate future changes in the shoreline by multiplying the observed rate of change by a specific amount of time, typically in terms of years (Leatherman, 1990; Crowell *et al.*, 1997). More specific assumptions can be incorporated that include other factors such as the rate of sea-level rise or geological characteristics of an area (Leatherman, 1990; Komar *et al.*, 1999).

Because past shoreline positions are readily available from maps that have been produced over time, the extrapolation of historical trends to predict future shoreline position has been applied widely for coastal management and planning (Crowell and Leatherman, 1999). In particular, this method is used to estimate building setbacks (Fenster, 2005). Despite this, relatively few studies have incorporated shoreline change rates into longterm shoreline change predictions to evaluate sea-level rise impacts, particularly for cases involving accelerated rates of sea-level rise (Kana *et al.*, 1984; Leatherman, 1984).

Historical trend analysis has evolved over the last few decades based on earlier efforts to investigate shoreline change (described in Crowell *et al.*, 2005). Since the early 1 980s, computer based Geographical Information System (GIS) software has been developed to digitally catalog shoreline data and facilitate the quantification of shoreline change rates (May *et al.*, 1982; Leatherman, 1983; Thieler *et al.*, 2005). At the same time, thorough review and critique of the procedures that are employed to make these estimates have been conducted (Dolan *et al.*, 1991; Crowell *et al.*, 1991, 1993, 1997; Douglas *et al.*, 1998; Douglas and Crowell, 2000; Honeycutt *et al.*, 2001; Fenster *et al.*, 2001; Ruggiero *et al.*, 2003; Moore *et al.*, 2006; Genz *et al.*, 2007).

Recently, a national scale assessment of shoreline changes that have occurred over the last century has been carried out by the U.S. Geological Survey (Gulf Coast: Morton *et al.*, 2004; southeastern U.S. coast: Morton and Miller, 2005; California coast: Hapke *et al.*, 2006). In addition, efforts are ongoing to complete similar analyses for the northeastern, mid-Atlantic, Pacific Northwest, and Alaskan coasts.

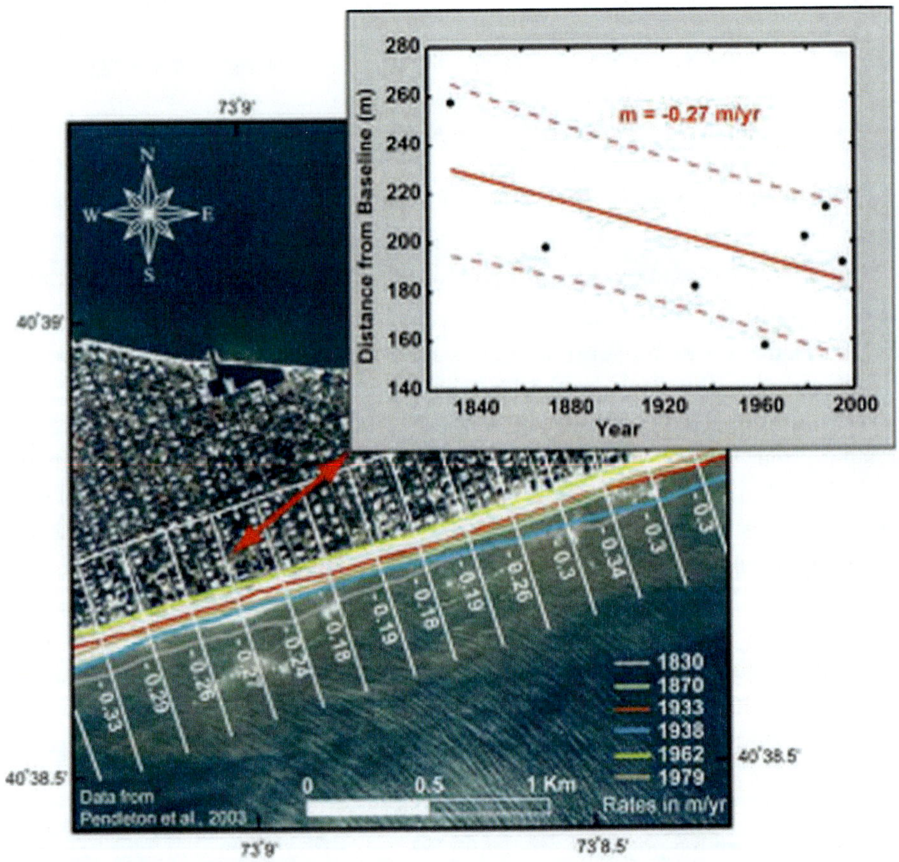

Figure A2.2. Aerial photograph of Fire Island, New York showing former shoreline positions and how these positions are used to calculate long-term shoreline change rates using linear regression. The inset box shows the shoreline positions at several points in time over the last 170 years. From the change in position with time, an average rate of retreat can be calculated. This is noted by the slope of the line, *m*. The red line in the inset box indicates the best fit line while the dashed lines specify the 95 percent confidence interval for this fit. Photo source: State of New York GIS.

The Sediment Budget. Another approach to shoreline change assessment involves evaluating the sediment mass balance, or sediment budget, for a given portion of the coast (Bowen and Inman, 1966; Komar, 1996; List, 2005; Rosati, 2005), as shown in Figure A2.3. Using this method, the gains and losses of sediment to a portion of the shore, often referred to as a control volume, are quantified and evaluated based on estimates of beach volume change. Changes in the volume of sand for a particular setting can be identified and evaluated with respect to adjacent portions of the shore and to changes in shoreline position over time. One challenge related to this method is obtaining precise measurements that minimize error since small vertical changes over these relatively low gradient shoreline areas can result in large volumes of material (NRC, 1987). To apply this approach, accurate measurements of coastal landforms, such as beach profiles, dunes, or cliff positions, are needed. Collection of such data, especially those on the underwater portions of the beach profile, is difficult. In addition, high- density measurements are needed to evaluate changes from one section of the beach to the next. While the results can be useful to understand where sediment volume changes occur, the lack of quality data and the expense of collecting the data limit the application of this method in many areas.

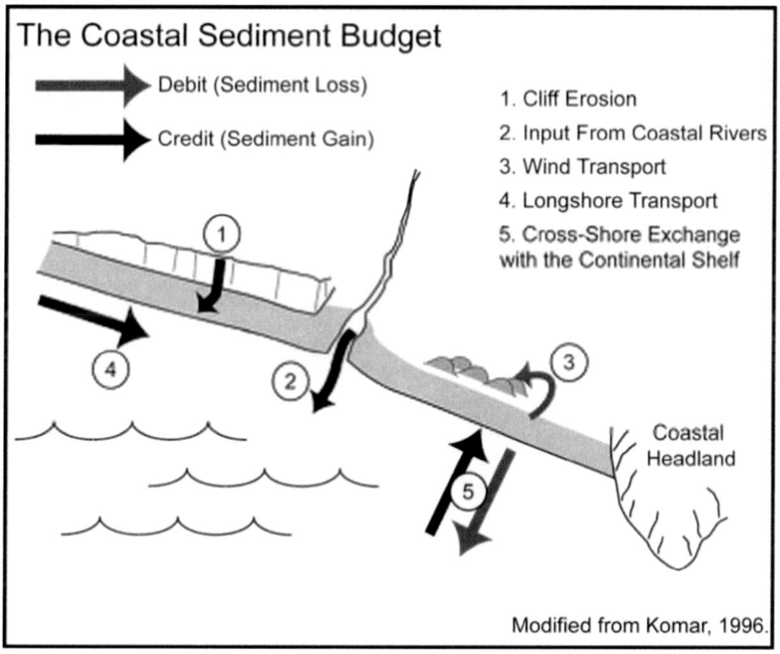

Figure A2.3. Schematic of the coastal sediment budget (modified from Komar, 1996). Using the sediment budget approach, the gains and losses of sediment from the beach and nearshore regions are evaluated to identify possible underlying causes for shoreline changes. In this schematic the main sediment gains are from: cliff erosion, coastal rivers, longshore transport, and cross-shore sediment transport from the continental shelf. The main sediment losses are due to: offshore transport from the beach to the shelf and wind transport from the beach to coastal dunes.

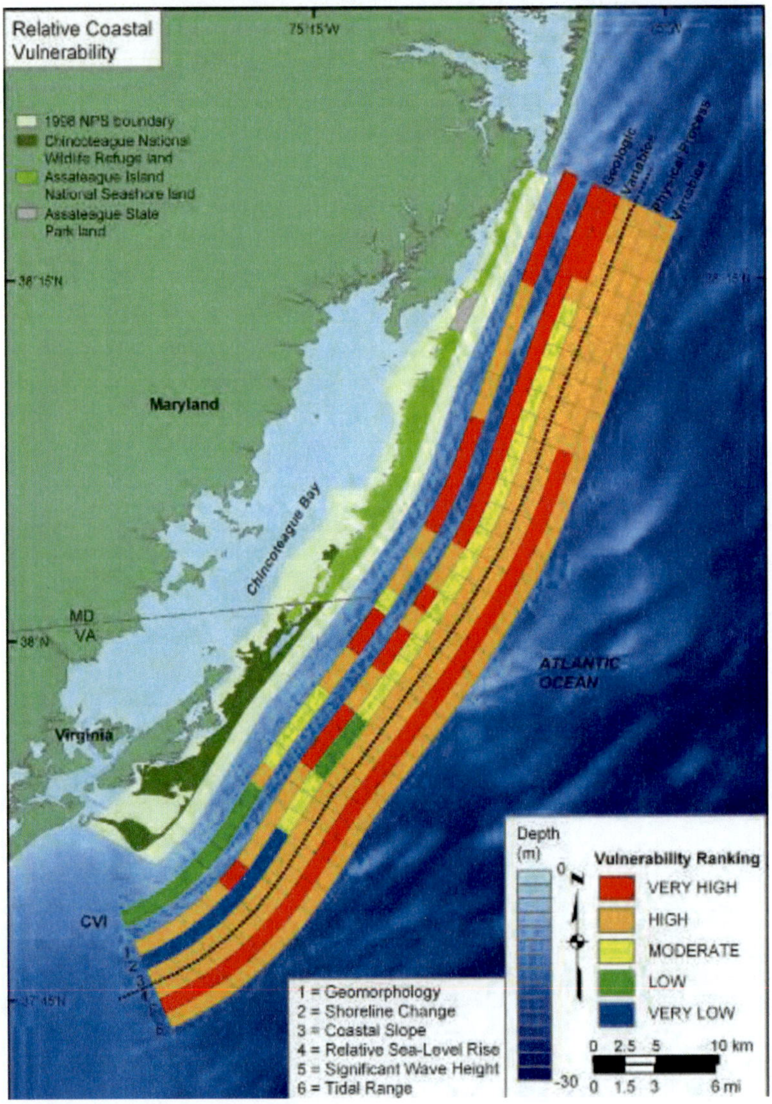

Figure A2.4. Coastal Vulnerability Index (CVI) calculated for Assateague Island National Seashore in Maryland. The inner most color-coded bar is the CVI estimate based on the other input factors (1 through 6). From Pendleton *et al.* (2004).

The Coastal Vulnerability Index. One approach that has been developed to evaluate the potential for coastal changes is through the development of a Coastal Vulnerability Index (CVI, Gornitz and Kanciruk, 1989; Gornitz, 1990; Gornitz *et al.*, 1994; Thieler and Hammar-Klose, 1999). Recently, the U.S. Geological Survey (USGS) used this approach to evaluate the potential vulnerability of the U.S. coastline on a national scale (Thieler and Hammar-Klose, 1999) and on a more detailed scale for the U.S. National Park Service (Thieler *et al.*, 2002). The USGS approach reduced the index to include six variables (geomorphology, shoreline change, coastal slope, relative sea-level change, significant wave height, and tidal range) which were considered to be the most important in determining a shoreline's susceptibility to sea-level rise (Thieler and Hammar-Klose, 1999). The CVI is calculated as:

$$CVI = \sqrt{\frac{a \times b \times c \times d \times e \times f}{6}} \quad \text{(A2.3)}$$

where *a* is the geomorphology, *b* is the rate of shoreline change, *c* is the coastal slope, *d* is the relative sea-level change, *e* is the mean significant wave height, and *f* is the mean tidal range.

The CVI provides a relatively simple numerical basis for ranking sections of coastline in terms of their potential for change that can be used by managers to identify regions where risks may be relatively high. The CVI results are displayed on maps to highlight regions where the factors that contribute to shoreline changes may have the greatest potential to contribute to changes to shoreline retreat (Figure A2.4).

REFERENCES

* Indicates non-peer reviewed literature; available upon request

Anders, F. J. & Byrnes, M. R. (1991). Accuracy of shoreline change rates as determined from maps and aerial photographs. *Shore and Beach*, 59(1), 17-26.

Bowen, A. J. & Inman, D. L. (1966). *Budget of Littoral Sands in the Vicinity of Point Arguello, California*. Coastal Engineering Research Center technical memoradum no. 19. [U.S. Army Corps of Engineers, Washington, DC], 56 pp.

Bruun, P., (1962). Sea-level rise as a cause of shore erosion. *Journal of Waterways and Harbors Division*, *88*, 117-130.

Bruun, P., (1988). The Bruun rule of erosion by sea-level rise: a discussion on large scale two- and three-dimensional usages. *Journal of Coastal Research*, *4(4)*, 627-648.

Cooper, J. A. G. & Pilkey, O. H. (2004). Sea-level rise and shoreline retreat: time to abandon the Bruun Rule. *Global and Planetary Change*, *43(3-4)*, 157-171.

Cowell, P. J., Roy, P. S. & Jones, R. A. (1992). Shoreface translation model: computer simulation of coastal-sand-body response to sea level rise. *Mathematics and Computers in Simulation*, *33(5-6)*, 603-608.

Crowell, M. & Leatherman S. P. (eds.), (1999). Coastal erosion mapping and management. *Journal of Coastal Research*, Special issue 28, 196 pp.

Crowell, M., Leatherman, S. P. & Buckley, M. K. (1991). Historical shoreline change; error analysis and mapping accuracy. *Journal of Coastal Research*, *7(3)*, 839-852.

Crowell, M., Leatherman, S. P. & Buckley, M. K. (1993). Shoreline change rate analysis: long-term versus short-term. *Shore and Beach*, *61(2)*, 13-20.

Crowell, M., Douglas, B. C. & Leatherman, S. P. (1997). On forecasting future U.S. shoreline positions: a test of algorithms. *Journal of Coastal Research*, *13(4)*, 1245-1255.

Crowell, M., Leatherman, S. P. & Douglas, B. (2005). Erosion: historical analysis a forecasting. In: *Encyclopedia of Coastal Science* [Schwartz, M.L. (ed.).] Springer, the Netherlands, pp. 846-850.

Dean, R. G. & Dalrymple, R. A. (2002). *Coastal Processes with Engineering Applications*. Cambridge University Press, New York, 475 pp.

Dean, R. G. & Maurmeyer, E. M. (1983). Models of beach profile response. In: *CRC Handbook of Coastal Processes* [Komar, P.D. (ed.)]. CRC Press, Boca Raton, FL, pp. 151-165.

Dolan, R., Fenster, M. S. & Holme, S. J. (1991). Temporal analysis of shoreline recession and accretion. *Journal of Coastal Research, 7(3)*, 723-744.

Douglas, B. C. & Crowell, M. (2000). Long-term shoreline position prediction and error propagation. *Journal of Coastal Research, 16(1)*, 145-152.

Douglas, B. C., Crowell, M. & Leatherman, S. P. (1998). Consideration for shoreline position prediction. *Journal of Coastal Research, 14(3)*, 1025-1033.

Dubois, R. N., (2002). How does a barrier shoreface respond to a sea-level rise? *Journal of Coastal Research, 18(2)*, iii-v.

Everts, C. H., (1985). Sea-level rise effects on shoreline position. *Journal of Waterway, Port, and Coastal Engineering, 111(6)*, 985-999.

Fenster, M. S., (2005). Setbacks. In: *Encyclopedia of Coastal Science* [Schwartz, M.L. (ed.)]. Springer, Dordrecht, the Netherlands, pp. 863-866.

Fenster, M.S., Dolan, R. & Morton, R. A. (2001). Coastal storms and shoreline change: signal or noise? *Journal of Coastal Research, 17(3)*, 714-720.

Genz, A. S., Fletcher, C. H., Dunn, R. A. Frazer, L. N. & Rooney, J. J. (2007). The predictive accuracy of shoreline change rate methods and alongshore beach variation on Maui, Hawaii. *Journal of Coastal Research, 23(1)*, 87-105.

Gornitz, V. M., (1990). Vulnerability of the east coast, USA to future sea-level rise. *Journal of Coastal Research*, Special issue 9, 201-237.

* Gornitz, V. M. & Kanciruk, P. (1989). Assessment of global coastal hazards from sea level rise. In: *Coastal Zone '89: Proceedings of the Sixth Symposium on Coastal and Ocean Management*, July 11-14, 1989, Charleston, SC. American Society of Civil Engineers, New York, pp. 1345-1359.

Gornitz, V. M., Daniels, R. C., White, T. W. & Birdwell, K. R. (1994). The development of a coastal risk assessment database: vulnerability to sea-level rise in the U.S. southeast. *Journal of Coastal Research*, Special issue *12*, 327-338.

Hands, E. B., (1980). *Prediction of Shore Retreat and Nearshore Profile Adjustments to Rising Water Levels on the Great Lakes.* Coastal Engineering Research Center technical paper TP 80-7. U.S. Army Corps of Engineers, Ft. Belvoir, VA, 199 pp.

Hapke, C. J., Reid, K. R., Richmond, B. M., Ruggiero, P. & List, J. (2006). *National Assessment of Shoreline Change: Part 3 Historical Shoreline Change and Associated Land Loss along Sandy Shorelines of the California Coast.* U.S. Geological Survey open-file report 2006-12 19. U.S. Geological Survey, Reston VA, 72 pp. <http://purl.access. gpo.gov/ GPO>

Honeycutt, M. G., Crowell, M. & Douglas, B.C. (2001). Shoreline position forecasting: impacts of storms, rate-calculation methodologies, and temporal scales. *Journal of Coastal Research, 17(3)*, 721-730.

Kana, T. W., Michel, J, Hayes, M. O. & Jensen, J. R. (1984). The physical impact of sea level rise in the area of Charleston, South Carolina. In: *Greenhouse Effect and Sea-level Rise: A Challenge for this Generation* [Barth, M. C. & J. G. Titus (eds.)]. Van Nostrand Reinhold, New York, pp. 105-150.

Komar, P. D., (1996). The budget of littoral sediments concepts and applications. *Shore and Beach, 64(3)*, 18-26.

Komar, P. D., (1998). *Beach Processes and Sedimentation.* Prentice Hall, Upper Saddle River, NJ, 2nd edition, 544 pp.

Komar, P. D., McDougal, W. G., Marra, J. J. & Ruggiero, P. (1999). The rational analysis of setback distances: applications to the Oregon coast. *Shore and Beach*, *67(1)*, 41-49.

Leatherman, S. P., (1983). Shoreline mapping: a comparison of techniques. *Shore and Beach*, *51(3)*, 28-33.

Leatherman, S. P., (1984). Coastal geomorphic responses to sea level rise: Galveston Bay, Texas. In: *Greenhouse Effect and Sea-level Rise: A Challenge for This Generation* [Barth, M. C. & J. G. Titus (eds.)]. Van Nostrand Reinhold, New York, pp. 151-178.

Leatherman, S. P., (1990). Modeling shore response to sea-level rise on sedimentary coasts. *Progress in Physical Geography*, *14(4)*, 447-464.

List, J. H., (2005). The sediment budget. In: *Encyclopedia of Coastal Science* [Schwartz, M.L. (ed.)]. Springer, Dordrecht, the Netherlands, pp. 846-850.

List, J. H., Sallenger, A. H., Hansen, M. E. & Jaffe, B. E. (1997). Accelerated relative sea-level rise and rapid coastal erosion: testing a causal relationship for the Louisiana barrier islands. *Marine Geology, 140(3-4)*, 347-365.

May, S., Kimball, W., Grandy, N. & Dolan, R. (1982). The coastal erosion information system (CEIS). *Shore and Beach*, 50(1), 19-25.

Moore, L., Ruggiero, P. & List, J. (2006). Comparing mean high water and high water line shorelines: Should proxy-datum offsets be incorporated in shoreline change analysis? *Journal of Coastal Research*, *22(4)*, 894-905.

Morton, R. A. & Miller, T. L. (2005). *National Assessment of Shoreline Change: Part 2, Historical Shoreline Changes and Associated Coastal Land Loss along the U.S. Southeast Atlantic Coast.* U.S. Geological Survey open-file report 2005-1401. U.S. Geological Survey, St. Petersburg, FL, 35 pp. <http://pubs.usgs.gov/of/2005/1 401/>

Morton, R. A., Miller, T. L. & Moore, L. J. (2004). *National Assessment of Shoreline Change: Part 1, Historical Shoreline Changes and Associated Coastal Land Loss along the U.S. Gulf of Mexico.* U.S. Geological Survey open file report 2004- 1043. U.S. Geological Survey, St. Petersburg, FL, 44 pp. <http://pubs.usgs.gov/of/2004/1 043/>

NRC (National Research Council), (1987). *Responding to Changes in Sea Level: Engineering Implications*. National Academy Press, Washington DC, 148 pp.

Pendleton, E. A., Williams, S. J. & Thieler, E. R. (2004). *Coastal Vulnerability Assessment of Fire Island National Seashore (ASIS) to Sea-level Rise.* U.S. Geological Survey open-file report 04-1020. [U.S. Geological Survey, Reston, VA], 20 pp. <http://pubs.usgs.gov/of/2004/1020/>

Riggs, S. R., Cleary, W. J. & Snyder, S. W. (1995). Influence of inherited geologic framework upon barrier beach morphology and shoreface dynamics. *Marine Geology*, *126(1-4)*, 213-234.

Rosati, J. D., (2005). Concepts in sediment budgets. *Journal of Coastal Research, 21(2)*, 307-322.

Ruggiero, P., Kaminsky, G. M. & Gelfenbaum, G. (2003). Linking proxy-based and datum-based shorelines on a high-energy coastline: Implications for shoreline change analyses. *Journal of Coastal Research*, Special issue *38*, 57-82.

Schwartz, M. L., (1967). The Bruun theory of sea-level rise as a cause of shore erosion. *Journal of Geology*, *75(1)*, 76-92.

SCOR (Scientific Committee on Ocean Research) Working Group 89, 1991: The response of beaches to sea level changes: a review of predictive models. *Journal of Coastal Research, 7(3)*, 895-921.

Shalowitz, A. L., (1964). *Shore and Sea Boundaries, With Special Reference to the Interpretation and Use of Coast and Geodetic Survey Data*. Publication 10-1. U.S. Department of Commerce, Coast and Geodetic Survey, Washington, DC, 749 pp.

Stockdon, H. F., Sallenger, A. H. List, J. H. & Holman, R. A. (2002). Estimation of shoreline position and change using airborne topographic Lidar data. *Journal of Coastal Research, 18(3)*, 502-513.

Stolper, D., List, J. H. & Thieler, E. R. (2005). Simulating the evolution of coastal morphology and stratigraphy with a new morphological-behavior model (GEOMBEST). *Marine Geology, 218(1-4)*, 17-36.

Thieler, E. R. & Hammar-Klose, E. (1999). *National Assessment of Coastal Vulnerability to Sea-level Rise--Preliminary Results for U.S. Atlantic Coast.* U.S. Geological Survey open-file report 99-593. U.S. Geological Survey, Reston, VA, 1 sheet. <http://purl.usgs.gov/of/1 999/of99-593/>

Thieler, E. R., Pilkey, O. H., Young, R. S., Bush, D. M. & Chai, F. (2000). The use of the mathematical models to predict beach behavior for U.S. coastal engineering: a critical review. *Journal of Coastal Research, 16(1)*, 48-70.

Thieler, E. R., Williams, S. J. & Beavers, R. (2002). *Vulnerability of U.S. National Parks to Sea-Level Rise and Coastal Change*. U.S. Geological Survey fact sheet FS 095- 02. [U.S. Geological Survey, Reston, VA], 2 pp. <http://pubs.usgs.gov/fs/fs095- 02/>

Thieler, E. R., Himmelstoss, E. A. Zichichi, J. L. & Miller, T. L. (2005). *Digital Shoreline Analysis System (DSAS) Version 3.0, An ArcGIS© Extension for Calculating Shoreline Change*. U.S. Geological Survey open-file report 2005-1304. U.S. Geological Survey, Reston, VA. <http://pubs.usgs.gov/of/2005/1 304/>

Wright, L. D., (1995). *Morphodynamics of Inner Continental Shelves*. CRC Press, Boca Raton, FL, 241 pp.

* Zhang, K., (1998). *Twentieth Century Storm Activity and Sea Level Rise along the U.S. East Coast and Their Impact on Shoreline Position*. Ph.D. dissertation, Department of Geography. University of Maryland, College Park, 266 leaves.

GLOSSARY

access, lateral
the right to walk or otherwise move along a shore, once someone has reached the shore

access, perpendicular
a legally permissible means of reaching the shore from dry land

access point
a place where anyone may legally gain access to the shore; usually a park, the end of a public street, or a public path; a place where perpendicular access is provided

accretion, lateral

the extension of land by natural forces acting over a long period of time, as on a beach by the washing-up of sand from the sea or on a floodplain by the accumulation of sediment deposited by a stream

accretion, vertical

the vertical accumulation of a sedimentary deposit; the increase in thickness of a sediment body as a result of sediment accumulation

active margin

a continental margin characterized by volcanic activity and earthquakes occurring where the edges of lithospheric plates are colliding; because these margins are largely confined to the rim of the Pacific, this type of margin is also referred to as a Pacific margin; compare with *passive margin*

armoring

the placement of fixed engineering structures, typically rock or concrete, on or along the shoreline to mitigate the effects of coastal erosion and protect structures; such structures include *seawalls*, *revetments*, *bulkheads*, and *rip-rap* (loose boulders)

astronomical tides

the alternating rise and fall of the ocean surface and connected waters, such as estuaries and gulfs, that result from the gravitational forces of the moon and sun

avulsion

a sudden cutting off or separation of land by a flood or by and abrupt change in the course of a stream, as by a stream breaking through a meander or by a sudden change in current whereby a stream deserts its old channel for a new one; OR rapid erosion of the shore by waves during a storm

barrier island, (or sometimes just barrier)

a long, narrow coastal sandy island that is above high tide and parallel to the shore, and that commonly has dunes, vegetated zones, and swampy terraces extending landward from the beach

barrier island roll-over

the landward migration or landward transgression of a *barrier island*, accomplished primarily over decadal or longer time scales through the process of storm overwash, periodic inlet formation, and wind-blown transport of sand

barrier migration

refers to the movement of an entire barrier island or barrier spit in response to sea-level rise, changes in sediment supply, storm surges or waves, or some combination of these factors

barrier raising

adding sediment to a barrier island or spit to increase its elevation; it is rarely done on a large scale (*e.g.*, the Galveston, Texas barrier island was raised after the hurricane of 1900) but individual lot owners sometimes import sediment to add elevation to their land, especially if the land is prone to flooding

barrier spit

a barrier island that is connected at one end to the mainland

bathymetry

the measurement of ocean depths and the mapping of the topography of the seafloor

beach

the unconsolidated material that covers a gently sloping zone extending landward from the low water line to the place where there is a definite change in material or physiographic form (such as a cliff), or to the line of permanent vegetation (usually the effective limit of the highest storm waves)

beach nourishment

the addition of sand, often dredged from offshore, to an eroding shoreline to enlarge or create a beach area, offering both temporary shore protection and recreational opportunities

berm

a commonly occurring, low, impermanent, nearly horizontal ledge or narrow terrace on the backshore of a beach, formed of material thrown up and deposited by storm waves

bluff

a high bank or bold headland with a broad, precipitous, sometimes rounded cliff face overlooking a plain or body of water

breakwater

an offshore structure (such as a wall or jetty) that, by breaking the force of the waves, protects a harbor, anchorage, beach or shore area

breach

(n.) a channel through a barrier spit or island typically formed by storm waves, tidal action, or river flow; breaches commonly occur during high storm surge cause by a hurricane or extra-tropical storm; (v.) to cut a deep opening in a landform

bulkhead

a structure or partition to retain or prevent sliding of the land; a secondary purpose is to protect uplands against damage from wave action

coastal plain

any lowland area bordering a sea or ocean, extending inland to the nearest elevated land, and sloping very gently seaward

coastal squeeze

the narrowing, potentially to the point of failure or elimination, of an environmental system (typically a beach or marsh) that is trapped between the transgressing sea on one side and an impassable barrier (*e.g.*, a sea wall or bulkhead) on the other

coastal zone

the area extending from the ocean inland across the region directly influenced by marine processes

coastline

the line that forms the boundary between the coast and the shore or the line that forms the boundary between the land and the water

continental shelf

the gently sloping underwater region at the edge of the continent that extends from the beach to where the steep continental slope begins, usually at depths greater than 300 feet

contour interval

the difference in elevations of adjacent contours on a topographic map

datum

a quantity, or a set of quantities, that serves as a basis for the calculation of other quantities; in terms of surveying and mapping, a datum is a point, line or surface used as a reference in measuring locations or elevations

delta

a low relief landform composed of sediments deposited at the mouth of a river that commonly forms a triangular or fan-shaped plain of considerable area crossed by many channels from the main river; forms as the result of accumulation of sediment supplied by the river in such quantity that it is not removed by tidal or wave-driven currents

DEM (digital elevation model)

the digital representation of the ground surface or terrain using a set of elevation data

deposition

the laying, placing, or throwing down of any material; typically refers to sediment

depth of closure

a theoretical depth below which sediment exchange between the nearshore (beach and shoreface) and the continental shelf is deemed to be negligible

dike

a wall generally of earthen materials designed to prevent the permanent submergence of lands below sea level, tidal flooding of lands between sea level and spring high water, or storm-surge flooding of the coastal floodplain

discount rate

downdrift

refers to the location of one section or feature along the coast in relation to another; often used to refer to the direction of net longshore sediment transport between two or more locations (*i.e.*, downstream)

dredge and fill

a process by which channels are dredged through wetlands or uplands to allow small boat navigation, and dredge spoil is placed on the adjacent land area to raise the land high enough to allow development; sometimes referred to as "lagoon development" or "canal estates"; used extensively before the 1970s

dredge spoil disposal (dredged material placement)

dredged material, or spoil, is material consisting of sediment or rock, excavated or dredged from an underwater location and removed to a placement site or disposal area; in the United States, designated areas must be coordinated with the Environmental Protection Agency and resource agencies such as the U.S. Fish and Wildlife Service and the National Marine Fisheries Service for environmental compliance, and with local interests for capacity and acceptability

dune

a low mound, ridge, bank or hill of loose, wind blown material (generally sand) either bare or covered with vegetation, capable of movement from place to place but typically retaining a characteristic shape

ebb current

the tidal current associated with the decrease in height of the tide, generally moving seaward or down a tidal river or estuary

ebb-tide delta

a large sand shoal commonly deposited at the mouths of tidal inlets formed by ebbing tidal currents and modified in shape by waves

erosion

the mechanical removal of sedimentary material by gravity, running water, moving ice, or wind; in the context of coastal settings erosion refers to the landward retreat of a shoreline indicator such as the water line, the berm crest, or the vegetation line; the loss occurs when sediments are entrained into the water column and transported from the source

erosion-based setback

a setback equal to an estimated annual erosion rate multiplied by a number of years set by statute or regulation (*e.g.*, 30 years)

estuary

a semi-enclosed coastal body of water which has a free connection with the open sea and within which sea water is measurably diluted with freshwater from land drainage; an inlet of the sea reaching into a river valley as far as the upper limit of tidal rise, usually being divisible into three sectors; (a) a marine or lower estuary, in free connection with the open sea; (b) a middle estuary subject to strong salt and freshwater mixing; and (c) an upper or fluvial estuary, characterized by fresh water but subject to daily tidal action; limits between these sectors are variable, and subject to constant changes in the river discharge

eustatic sea-level rise

refers to worldwide rise of sea level that affects all oceans; eustatic rise has various causes, but typically result from thermal expansion of ocean waters, and additions of water from glaciers, ice caps, and ice sheets

extra-tropical storm

refers to cyclonic weather systems, occurring in the middle or high latitudes (*e.g.*, poleward of the tropics) that are generated by colliding airmasses; these weather systems often spawn large storms occurring between late fall and early spring

fetch

the area of the open ocean over the surface of which the winds blow with constant speed and direction, generating waves

flood current

the tidal current associated with the increase in height of the tide, generally moving landward or up a tidal river or *estuary*

flood-tide delta

a large sand *shoal* commonly deposited on the landward side of a tidal inlet formed by flooding tidal currents

floodproofing

a technique that is intended to limit the amount of damage that will occur to a building or its contents during a flood (see also *dry floodproofing* and *wet floodproofing*)

geologic framework

refers to the underlying geological setting, structure, and *lithology* (rock/sediment type) in a given area

geomorphic or geomorphology

the external structure, form, and arrangement of rocks or sediments in relation to the development of the surface of the earth

glacial rebound

uplift of land following deglaciation due to the mass of the ice being removed from the land surface which causes an isostatic response of the *lithosphere*

global sea-level rise

the worldwide average rise in mean sea level (see *eustatic sea level*)

groin

an engineering structure oriented perpendicular to the coast, used to accumulate littoral sand by interrupting longshore transport processes; often constructed of concrete, timbers, steel, or rock

high marsh

the part of a marsh that lies between the *low marsh* and the marsh's upland border; this area can be expansive, extending hundreds of yards inland from the low marsh area; soils here are mostly saturated but only flooded during higher-than-average tides

hydrodynamic climate

the characteristics of nearshore or continental shelf currents in an area that typically result from waves, tides, and weather systems

inlet

a small, narrow opening, recess, indentation, or other entrance into a coastline or shore of a lake or river through which water penetrates landward, commonly refers to a waterway between two barrier islands that connects the sea and a *lagoon*

intertidal zone

see *littoral*

inundation

refers to the submergence of land by water

isostasy

equilibrium condition whereby portions of the Earth's crust are compensated (floating) by denser material below

jetty

an engineering structure built at the mouth of a river or tidal inlet stabilize a channel for navigation; designed to prevent shoaling of a channel by littoral materials and to direct and confine the stream or tidal flow

lagoon

a shallow coastal body of seawater that is separated form the open ocean by a barrier or coral reef; the term is commonly used to define the shore-parallel body of water behind a barrier island or barrier spit

levee

a wall, generally of earthen materials, designed to prevent riverine flooding after periods of exceptional rainfall

lidar (LIght Detection And Ranging)

a remote sensing instrument that uses laser light pulses to measure the elevation of the land surface with a high degree of accuracy and precision

lithology

the description of rocks on the basis of characteristics such as color, mineral composition, and grain size

lithosphere

the solid portion of the Earth, including the crust and part of the upper mantle

littoral

zone between high and low tide in coastal waters or the shoreline of a freshwater lake

littoral cell

a section of coast for which sediment transport processes can be isolated from the adjacent coast; within each littoral cell, a sediment budget can be defined that describes sinks, sources, and internal fluxes

littoral drift

the sedimentary material moved in the littoral zone under the influence of waves and currents

littoral transport

the movement of littoral drift in the littoral zone by waves and currents; includes movement parallel and perpendicular to the shore

littoral zone

a term describing the region on the shore occurring between high and low water marks

living shoreline

refers to a shore protection concept where some or all of the environmental characteristics of a natural shoreline are retained as the position of the shore changes

long lived

having a long lifetime or a long expected lifetime; long-lived infrastructure means infrastructure that is likely to be in service for a long time

longshore current

an ocean current in the littoral zone that moves parallel to the shoreline, produced by waves approaching at an angle to the shoreline

longshore transport

movement of sediment parallel to the shoreline in the surf zone by wave suspension and the longshore current

low marsh

the seaward edge of a salt marsh, usually a narrow band along a creek or ditch which is flooded at every high tide and exposed at low tide (also see *high marsh* for comparison)

marsh

a frequently or continually inundated wetland characterized by herbaceous vegetation adapted to saturated soil conditions (see also *salt marsh*)

mean high water

a tidal datum; the average height of high water levels observed over a 19-year period

mean higher high water

the average of the higher high water height of each tidal day observed over the national tidal datum epoch (see national tidal datum epoch

mean sea level

the 'still water level' (*i.e.*, the level of the sea with high frequency motions such as wind waves averaged out) averaged over a period of time such as a month or a year, such that periodic changes in sea level (*e.g.*, due to the tides) are also averaged out; the values of MSL are measured with respect to the level of marks on land (called 'benchmarks')

metadata

a file of information which captures the basic characteristics of a data or information resource; representing the who, what, when, where, why and how of the data resource; geospatial metadata are used to document geographic digital resources such as Geographic Information System (GIS) files, geospatial databases, and earth imagery

metes and bounds

the boundary lines and limits of a tract that is described and characterized by placing all data in the tract description as opposed to other references such as maps or plats

mixed energy coast

a coast in which the coastal landforms are shaped by a combination of wave and tidal currents

moral hazard

a circumstance in which insurance, lending practices, or subsidies designed to protect against a specified hazard induce people to take measures that increase the risk of that hazard

mudflat

a level area of fine silt and clay along a shore alternately covered and uncovered by the tide or covered by shallow water

national geodetic vertical datum of 1929 (NGVD29)

a fixed reference adopted as a standard geodetic datum for elevations; it was determined by leveling networks across the United States and sea-level measurements at 26 coastal tide stations; this reference is now superseded by the North American vertical datum of 1988 (NAVD88)

national tidal datum epoch (NTDE)

the latest 19-year time period over which NOAA has computed and published official tidal datums and local mean sea-level elevations from tide station records; currently, the latest NTDE is 1983-2001

nearshore zone

refers to the zone extending from the shoreline seaward to a short, but indefinite distance offshore, typically confined to depths less than 5 meters (16.5 feet)

nontidal wetlands

wetlands that are not exposed to the periodic change in water level that occurs due to astronomical tides

nor'easter (northeaster)

name given to the strong northeasterly winds associated with extra-tropical cyclones that occur along East Coast of the United States and Canada; these storms often cause beach erosion and structural damage; wind gusts associated with these storms can approach and sometimes exceed hurricane force in intensity

North American vertical datum of 1988 (NAVD88)

a fixed reference for elevations determined by geodetic leveling, derived from a general adjustment of the first-order terrestrial leveling networks of the United States, Canada, and Mexico; NAVD88 supersedes NGVD29

100-year flood

the standard used by the National Flood Insurance Program (NFIP) for floodplain management purposes and to determine the need for flood insurance; a structure located within a special flood hazard area shown on an NFIP map has a 26 percent chance of suffering flood damage during the term of a 30-year mortgage

ordinary high water mark

a demarcation between the publicly owned land along the water and privately owned land which has legal implications regarding public access to the shore; generally based on mean high water, the definition varies by state; along beaches with significant waves, it may be based on the line of vegetation, the water mark caused by wave runup, surveys of the elevation of mean high water, or other procedures

overwash

sediment that is transported from the beach across a barrier, and is deposited in an apron-like accumulation along the backside of the barrier; overwash usually occurs during storms when waves break through the frontal dune ridge and flow landward toward the marsh or lagoon

outwash plain

braided stream deposit beyond the margin of a glacier; it is formed from meltwater flowing away from the glacier, depositing mostly sand and fine gravel in a broad plain

passive margin

type of continental margin occurring in the middle of a tectonic plate, consequently tectonic activity is minimal; these margins are typified along the margins of the Atlantic Ocean and often so it is often termed an Atlantic margin

pocket beach

a typically small, narrow beach formed between two littoral obstacles, such as between rocky headlands or promontories that occur at the shore

Public Trust Doctrine

a legal principle derived from English Common Law; the essence of the doctrine is that the waters of the state are a public resource owned by and available to all citizens equally for the purposes of navigation, hunting, fowling, and fishing, and that this trust is not invalidated by private ownership of the underlying land

relative sea-level rise

the rise in sea level measured with respect to a specified vertical datum relative to the land, which may also be changing elevation over time; typically measured using a tide gauge

retreat

the act of moving inland

revetment

a sloped facing of stone, concrete, etc., built to protect a scarp, embankment, or shore structure against erosion by wave action or currents

river diversion

engineering approaches used to redirect the flow of water from its natural course for a range of purposes; commonly used to by-pass water during dam construction, for flood control, for navigation, or for wetland and floodplain restoration

rip-rap

loose boulders placed on or along the shoreline as a form of *armoring*

riverine flooding

flooding of lands caused by the elevation of nontidal or tidal waters resulting from the drainage of upstream areas, usually after periods of exceptional rainfall

roll-over

see *barrier island roll-over*

rolling easement

an interest in land (by title or interpretation of the *Public Trust Doctrine*) in which a property owner's interest in preventing real estate from eroding or being submerged yields to the public or environmental interest in allowing wetlands or beaches to migrate inland

root mean square error

a measure of statistical error calculated as the square root of the sum of squared errors, where error is the difference between an estimate and the actual value; if the mean error is zero, it also equals the standard deviation of the error

salt marsh

a grassland containing salt tolerant vegetation established on sediments bordering saline water bodies where water level fluctuates either tidally or nontidally (see also *marsh*)

saltwater intrusion

displacement of fresh or ground water by the advance of salt water due to its greater density, usually in coastal and estuarine areas

sand bypassing

hydraulic or mechanical movement of sand from the accreting updrift side to the eroding downdrift side of an inlet or harbor entrance; the hydraulic movement may include natural movement as well as movement caused by man

seawall

a structure, often concrete or stone, built along a portion of a coast to prevent erosion and other damage by wave action, often it retains earth against its shoreward face; a seawall is typically more massive and capable of resisting greater wave forces than a *bulkhead*

sediment(s)

solid materials or fragments that originates from the break up of rock and is transported by air, water or ice, or that accumulates by other natural agents such as chemical precipitation or biological secretions; solid material that has settled from being suspended as in moving water or air

sediment broadcasting

a technique in which sediment from an external source is spread onto salt marshes to supply mineral material to enhance their growth

sediment supply

refers to the abundance or lack of sediment in a coastal system that is available to contribute to the maintenance or evolution of coastal landforms including both exposed features such as beaches and barrier islands, and underwater features such as the seabed

setback

the requirement that construction be located a minimum distance inland from tidal wetlands, tidal water, the primary dune line, or some other definition of the shore

shoal

a relatively shallow place in a stream, lake, sea, or other body of water; a submerged ridge, bank, or bar consisting of or covered by sand

shore

the narrow strip of land immediately bordering any body of water, especially a sea or large lake; the zone over which the ground is alternately exposed and covered by the tides or waves, or the zone between high and low water

shoreface

the narrow relatively steep surface that extends seaward from the beach, often to a depth of 30 to 60 feet, at which point the slope flattens and merges with the continental shelf

shoreline

the intersection of a specified plane of water with the shore or beach; on National Ocean Service nautical charts and surveys, the line representing the shoreline approximates the mean high water line

shoreline armoring (see *armoring*)

a method of shore protection that prevents shore erosion through the use of hardened structures such as seawalls, bulkheads, and revetments

shore protection

refers to a range of activities that focus on protecting land from inundation, erosion, or storm-induced flooding through the construction of various structures such as jetties, groins, or seawalls, or the addition of sediments to the shore (*e.g.*, beach nourishment)

significant wave height
the average height of the highest one-third of waves in a given area

sill
semicontinuous structures placed along the edge of a marsh in order to diminish wave erosion of the marsh, usually made of stone; similar to breakwaters, except that breakwaters are generally farther from the shore and have larger open spaces between them

soft shore protection
a method of shore protection that prevents shore erosion through the use of materials similar to those already found in a given location, such as adding sand to an eroding beach or planting vegetation whose roots will retain soils along the shore

spit
a fingerlike extension of the beach that was formed by longshore sediment transport; typically, it is a curved or hook-like sandbar extending into an inlet

spring high water
the average height of the high waters during the semi-monthly times of spring tides (occurs at the full and new moons)

storm surge
an abnormal rise in sea level accompanying a hurricane or other intense storm, whose height is the difference between the observed level of the sea surface and the level that would have occurred in the absence of the cyclone

subsidence
the downward settling of material with little horizontal movement; the downwarping of the earth's crust relative to the surroundings

submergence
a rise of the water level relative to the land, so that areas that were formerly dry land become inundated; it is the result either of the sinking of the land or a net rise in sea level

supratidal zone
the shore area just above the high-tide level

surf zone
the zone of the nearshore region where bore-like waves occur following breaking waves, extending from the point where waves break to the wet beach

taxon (plural, taxa)
a general term applied to any taxonomic element, population, or group irrespective of its classification level

threshold

in climate change studies, a threshold generally refers to the point at which the climate system begins to change in a marked way because of increased forcing; crossing a climate threshold triggers a transition to a new state of the system at a generally faster rate

tidal datum

a base elevation used as a vertical from which to reckon heights or depths; called a tidal datum when defined in terms of a certain phase of the tide

tidal freshwater marsh

marsh along rivers and estuaries close enough to the coastline to experience significant tides by nonsaline water, vegetation is often similar to nontidal freshwater marshes

tidal inlet

an opening in the shoreline through which water penetrates the land, thereby providing a connection between the ocean and bays, lagoons, and marsh and tidal creek systems; the main channel of a tidal inlet is maintained by tidal currents

tidal range

the vertical difference between normal high and low tides often computed as the elevation difference between mean high water and mean low water; spring tide range is the elevation difference between spring high water and spring low water

tidal wetlands

wetlands that are exposed to the periodic rise and fall of the tides (see *wetlands*)

tide-dominated coast

coast where the morphology is primarily a product of tidal processes

tide gauge

the geographic location where tidal observations are conducted and consisting of a water level sensor, data collection and transmission equipment, and local bench marks that are routinely surveyed into the sensors

tidelands

lands that are flooded during ordinary high water, and hence available to the public under the *Public Trust Doctrine*

tipping point

a critical point in the evolution of a system that leads to new and potentially irreversible effects at a rate that can either be much faster or much slower than forcing

transgression

the spread or extension of the sea over land areas, and the consequent evidence of such advance; also, any change such as a rise in sea level that brings offshore deep-water environments to areas formerly occupied by nearshore, shallow-water environments or that shifts the boundary between marine and nonmarine deposition away from deep water regions

updrift

refers to the location of one section or feature along the coast in relation to another; often used to refer to the direction of net longshore sediment transport between two or more locations (*i.e.*, upstream)

wave-dominated coast

coast where the morphology is primarily a product of wave processes

wave refraction

the process by which a water wave, moving in shallow water as it approaches the shore at an angle, tends to be turned from its original direction so that the wave crest is more parallel to shore; also can refer to the bending of wave crests by currents

wave run-up

the upper levels reached by a wave on a beach or coastal structure, relative to still-water level

wet floodproofing

wetlands

specifies those areas that are inundated or saturated by surface or ground water at a frequency and duration sufficient to support, and that under normal circumstances do support, a prevalence of vegetation typically adapted for life in saturated soils; wetlands generally include swamps, marshes, bogs, and similar areas

wetland accretion

a process by which the surface of wetlands increases in elevation; see also *accretion, vertical*

wetland migration

a process by which tidal wetlands adjust to rising sea level by advancing inland into areas previously above the ebb and flow of the tides

SCIENTIFIC NAMES— CHAPTER 5 SPECIES

American black duck	*Anas rubripes*
American oystercatcher	*Haematopus palliatus*
Atlantic menhaden	*Brevoortia tyrannus*
Atlantic silverside	*Menidia spp.*
bald eagle	*Haliaeetus leucocephalus*

(Continued)

bay anchovy	*Anchoa mitchilli*
belted kingfisher	*Ceryle alcyon*
black rail	*Laterallus jamaicensis*
black skimmer	*Rynchops niger*
bladderwort	*Utricularia spp.*
blue crab	*Callinectes sapidus*
bluefish	*Pomatomus saltatrix*
brant	*Branta bernicla*
canvasback duck	*Aythya valisineria*
carp	*Family Cyprinidae*
catfish	*Order Siluriformes*
clapper rail	*Rallus longirostris*
common tern	*Sterna hirundo*
crappie	*Pomoxis spp.*
diamondback terrapin	*Malaclemys terrapin*
eastern mud turtle	*Kinosternum subrubrum*
elfin skimmer (dragonfly)	*Nannothemis bella*
fiddler crab	*Uca spp.*
Forster's tern	*Sterna forsteri*
fourspine stickleback	*Apeltes quadracus*
grass shrimp	*Hippolyte pleuracanthus*
great blue heron	*Ardea herodias*
gull-billed tern	*Sterna nilotica*
herring	*Clupea harengus*
horseshoe crab	*Limulus polyp hem us*
Kemp's ridley sea turtle	*Lepidochelys kempii*
laughing gull	*Larus atricilla*
least bittern	*Ixobrychus exilis*
meadow vole	*Microtus pennsylvanicus*
minnows	*Family Cyprinidae*
mummichog	*Fundulus herteroclitus*
naked goby	*Gobiosoma bosci*
northern pipefish	*Syngnathus fuscus*
piping plover	*Charadrius melodus*
red drum	*Sciaenops ocellatus*
red knot	*Calidris canutus*
red-winged blackbird	*Agelaius phoeniceus*
ribbed mussel	*Geukensia demissa*

(Continued)

sand digger	*Neohaustorius schmitzi*
sand flea	*Talorchestia spp.*
sandpiper	*Family Scolopacidae*
sea lettuce	*Ulva lactuca*
sea trout	*Salvelinus fontinalis*
shad	*Alosa sapidissima*
sheepshead minnow	*Cyprinodon variegatus*
shiners	*Family Cyprinidae*
spot	*Leiostomus xanthurus*
striped anchovy	*Anchoa hepsetus*
striped bass	*Morone saxatilis*
striped killifish	*Fundulus majalis*
sundew	*Drosera spp.*
sunfish	*Family Centrarchidae*
threespine stickleback	*Gasterosteus aculeatus*
tiger beetle	*Cicindela spp.*
weakfish	*Cynoscion regalis*
white croaker	*Genyonemus lineatus*
white perch	*Morone americana*
widgeon grass	*Ruppia maritima*
willet	*Catoptrophorus*

Acronyms and Abbreviations

A–P Albemarle–Pamlico
ABFE Advisory Base Flood Elevations
AEC Areas of Environmental Concern (
ASFPM Association of State Floodplain Managers
BFE base flood elevation
CAFRA Coastal Facility Review Act
CAMA Coastal Area Management Act
CBRA Coastal Barrier Resources Act
CCMP Comprehensive Coastal Management Plan
CCSP Climate Change Science Program
CORS continuously operating reference stations
CRC Coastal Resources Commission
CTP Cooperative Technical Partnership
CVI Coastal Vulnerability Index
CZM Coastal Zone Management
CZMA Coastal Zone Management Act

DDFW	Delaware Division of Fish and Wildlife
DEC	Department of Environmental Conservation
DEM	Digital elevation Model
DFIRM	digital flood insurance rate maps
FEMA	Federal Emergency Management Agency
FGDC	Federal Geographic Data Committee
FIRM	Flood Insurance Rate Maps
FIS	Flood Insurance Studies
GAO	General Accounting Office (1982)
GAO	General Accountability Office (2007)
GEOSS	Global Earth Observation System of Systems
GIS	geographic information system
GCN	greatest conservation need
GPS	Global Positioning System
HOWL	highest observed water levels
IDA	intensely developed area
IOOS	Integrated Ocean Observing System
IPCC	Intergovernmental Panel on Climate Change
IPCC CZMS	Intergovernmental Panel on Climate Change Coastal Zone Management Subgroup
LDA	limited development area
LMSL	local mean sea level
MHHW	Mean Higher High Water
MHW	Mean High Water
MLW	Mean Low Water
MLLW	Mean Lower Low Water
MSL	mean sea level
NAI	No Adverse Impact
NAS	National Academy of Sciences
NAVD	North American Vertical Datum
NCDC	National Climatic Data Center
NERRS	National Estuarine Research Reserve System
NDEP	National Digital Elevation Program
NED	National Elevation Dataset
NFIP	National Flood Insurance Program
NGVD	National Geodetic Vertical Datum
NHP	National Heritage Program
NHS	National Highway System
NLCD	National Land Cover Data
NMAS	National Map Accuracy Standards
NOAA	National Oceanic and Atmospheric Administration
NPS	National Park Service
NRC	National Research Council
NSSDA	National Standard for Spatial Data Accuracy
NTDE	National Tidal Datum Epoch
NWR	National Wildlife Refuge

NWS	National Weather Service
PORTS	Physical Oceanographic Real-Time System
RCA	resource conservation area
RMSE	root mean square error
RPA	resource protection area
SAV	submerged aquatic vegetation
SFHA	Special Flood Hazard Area
SRTM	Shuttle Radar Topography Mission
SWFL	still water flood level
TNC	The Nature Conservancy
USACE	United States Army Corps of Engineers
U.S. EPA	United States Environmental Protection Agency
U.S. FWS	United States Fish and Wildlife Service
U.S. DOT	United States Department of Transportation
USGS	United States Geological Survey
VA PBB	Virginia Public Beach Board
WRCRA	Waterfront Revitalization and Coastal Resources Act

INDEX

A

B

C

D

E

F

G

H

I

N

O

P

Q

R

S

T

U

V

W

Y

Z